FOSSILS OF THE UPPER ORDOVICIAN PLATTEVILLE FORMATION IN THE UPPER MIDWEST USA: AN OVERVIEW

FOSSILS OF THE UPPER ORDOVICIAN PLATTEVILLE FORMATION IN THE UPPER MIDWEST USA: AN OVERVIEW

Dennis R. Kolata

Bulletin 108 2021

ILLINOIS STATE GEOLOGICAL SURVEY
Prairie Research Institute
University of Illinois at Urbana-Champaign
615 East Peabody Drive
Champaign, IL 61820
217-333-4747
www.isgs.illinois.edu

ILLINOIS
Illinois State Geological Survey
PRAIRIE RESEARCH INSTITUTE

Cover image: Reconstruction of an underwater seascape during deposition of the Platteville Formation showing a subtidal marine environment populated by an abundant and diverse fauna. Created by Dennis R. Kolata.

Printed in the United States of America

ISBN: 978-0-578-91175-5

Library of Congress Control Number: 2021938856

DEDICATION

This report is dedicated to Milton W. Mahlburg (1907–2000), director of the Burpee Museum of Natural History, Rockford, Illinois, from 1942 to 1987. Milt was a man of varied knowledge, interests, and skills who built from scratch a small but popular public museum that became an educational and cultural asset benefiting the citizens of northern Illinois and southern Wisconsin. He was a mentor, inspiration, and friend to many nature-loving enthusiasts of the region. Thank you, Milt, for introducing me to the fascinating fossils of the Platteville Formation and for your support and encouragement over the years.

Milt Mahlburg (circa 1960)

CONTENTS

ACKNOWLEDGMENTS

An overview covering a diverse assemblage of macrofossils like this is impossible without the cooperation of dedicated, well-informed, and experienced reviewers willing to spend their valuable time helping to ensure accuracy. I am grateful to the following people for the many comments and suggestions that helped improve this overview: Lisa Amati, New York State Museum; Jack B. Bailey, Western Illinois University; Richard C. Berg, Illinois State Geological Survey; Stig M. Bergström, The Ohio State University; Daniel B. Blake, University of Illinois at Urbana–Champaign; Marcelo Carrera, Universidad Nacional de Cordoba; John A. Catalani, Woodridge, Illinois; L. Robin M. Cocks, Natural History Museum, London; John C. W. Cope, National Museum Cardiff; Roger J. Cuffey, Pennsylvania State University; Jan Ove R. Ebbestad, Uppsala University; Douglas Eernisse, California State University Fullerton; Robert J. Elias, University of Manitoba; Robert C. Frey, Ohio Department of Health; Jiří Frýda, Czech Geological Survey; Daniel Goldman, University of Dayton; Thomas E. Guensburg, Field Museum of Natural History; Dirk Knaust, Statoil ASA; Steven T. LoDuca, Eastern Michigan University; William C. Miller, Humboldt State University; W. John Nelson, Illinois State Geological Survey; Rodney D. Norby, Illinois State Geological Survey; Mark E. Patzkowsky, Pennsylvania State University; Allan W. Scheer, Stoddard, Wisconsin; David J. Siveter, University of Leicester; Michael J. Vendrasco, Pasadena City College; Olev Vinn, University of Tartu; Barry D. Webby, Macquarie University; and Graham Young, University of Manitoba.

I am indebted to Daniel B. Blake for his scholarly insights and encouragement over our 50 years of association. I am grateful to John A. Catalani and Allan W. Scheer for freely sharing their knowledge of the Platteville Formation and allowing me to photograph specimens from their collections and to Steven T. LoDuca for providing macroalgae photographs. I also thank Brian J. Witzke for donating his collection of Platteville specimens from northeastern Iowa to be used in this overview and for his support and many stimulating discussions about the Upper Ordovician succession of the upper Midwest.

I am grateful to the staff at St Marys Cement Inc., Dixon, Illinois, for help provided during the collecting phase of this project. In particular, a very special thanks to Paul Biggerstaff for his help preparing quarry sites for safe collecting, to Mike Kresanek, Larry Lally, and Dan Stewart for their help excavating fossil sites, and to Ronaldo Rocha dos Santos for his help and encouragement during the early phase of this research.

I thank Sam W. Heads, Curator of Paleontology, and M. Jared Thomas, Paleontology Collections Manager at the Prairie Research Institute, University of Illinois at Urbana–Champaign for their help in establishing a permanent home for the university's Platteville fossil collection and for introducing me to focus stacking and image processing to avoid the dreaded depth of field problems in photographing fossil specimens.

I thank Daniel L. Byers for introducing me to the tools of digital image processing and construction, consulting on the book design and layout, and answering myriad questions about computers and software.

For technical and copy editing, I thank Marguerite Torrey and Susan Krusemark.

Last, but not least, I thank Jacquelyn L. Hannah for her assistance and encouragement throughout this entire effort.

ABSTRACT

The Upper Ordovician Platteville Formation of the Midwest USA consists largely of nodular and tabulate beds of fossiliferous, argillaceous, bioturbated, skeletal lime mudstones and wackestones that are regionally dolomitized. Within the outcrop belt of northern Illinois, southern Wisconsin, northeastern Iowa, and southeastern Minnesota, the Platteville Formation ranges in thickness from 6 to 42 m and is divided into five members, including the Pecatonica, Mifflin, Grand Detour, Nachusa, and Quimbys Mill Members, in ascending order. The latter two members are confined primarily to north-central Illinois and south-central Wisconsin. Fossils occur throughout the Platteville Formation but are particularly well preserved, abundant, and diverse in the Mifflin and Grand Detour Members (combined and recognized as the McGregor Member in southeastern Wisconsin, northeastern Iowa, and southeastern Minnesota). The fossil assemblage consists of sponges, corals, bryozoans, brachiopods, polyplacophorans, gastropods, bivalves, rostroconchs, cephalopods, trilobites, ostracodes, echinoderms, graptolites, hyolithids, cornulitids, and macroalgae as well as a microfauna of conodonts, foraminifera, and scolecodont-bearing polychaetes. A wide variety of trace fossils are also present. Approximately 300 species of macroinvertebrates and a small number of macroalgae are here described and illustrated with high-resolution photographs. At least two distinct faunal associations occur in the Platteville Formation. These include (1) the abundant and diverse macrofossil assemblage in the Pecatonica, Mifflin, and Grand Detour Members occurring throughout the outcrop belt, and (2) the sparse, geographically restricted, low-diversity faunal assemblage in the Nachusa and Quimbys Mill in north-central Illinois and south-central Wisconsin. The overall fossil assemblage is typical of that occurring in the Late Ordovician (Turinian Stage) carbonate-dominated stratigraphic succession in the midcontinent of eastern North America.

The Platteville Formation was deposited largely in open-marine conditions with normal salinity and well-oxygenated waters, largely in subtidal depths below fair-weather wave base but above storm/tsunami wave base and within the photic zone. For the most part, facies belts were broad and transitions very gradual. Exceptions include local, restricted intertidal and supratidal shoal environments along the axis of the Wisconsin Arch during deposition of the Nachusa and Quimbys Mill Members.

CHAPTER 1
INTRODUCTION

The Upper Ordovician Platteville Formation of the upper Midwest USA consists primarily of pure to shaley limestone and dolostone ranging in thickness from 6 to 42 m. The formation is exposed in numerous roadcuts, natural exposures, and quarries throughout the region. It is a primary source of aggregate and for many years was mined for use in the production of cement at Dixon, Illinois. The Platteville Formation contains a diverse, abundant, and locally well-preserved assemblage of macrofossils (Figure 1-1), including sponges, corals, bryozoans, brachiopods, polyplacophorans, gastropods, bivalves, rostroconchs, cephalopods, trilobites, ostracodes, echinoderms, graptolites, and macroalgae as well as a microfauna of conodonts, foraminifera, and scolecodont-bearing polychaetes. A wide variety of trace fossils are also present.

In many respects, the Platteville Formation is an archive of information bearing on the taxonomic, ecologic, and morphologic diversity that peaked between an earlier Ordovician faunal radiation and the end-Ordovician extinction event. This remarkable fossil record offers an excellent perspective on the so-called Great Ordovician Biodiversity Event recognized as one of the most rapid, long-sustained bursts of biotic diversification in the history of marine life on Earth (Webby et al. 2004b).

Many Platteville fossils were described in pre-20th century reconnaissance reports commissioned by the state geological surveys of Illinois, Iowa, Minnesota, and Wisconsin. A historical summary of paleontological investigations of the Upper Ordovician stratigraphic succession can be found in Sloan (1987). A drawback to understanding the documented faunal composition of the Platteville Formation is that many of the fossils are described and illustrated in diverse monographs and shorter papers spanning more than 150 years and published by a variety of organizations. Many of these publications are difficult to find in libraries and are not available online. Furthermore, they tend to focus on specific phyla rather than the overall faunal associations. An added drawback is that many early publications lack high-resolution photographs, making it difficult to compare and contrast fossil specimens.

The purpose of this report is to provide a comprehensive overview of the Platteville macrofauna and trace fossils. The overview is directed to students, collectors, and professional paleontologists. Each chapter begins with a brief introduction to the morphological terms (in boldface) used to describe the various fossil groups. This is followed by systematic descriptions, remarks, and occurrence information. Revealed for the first time in this overview are new and undescribed species representing at least eight phyla. This overview can serve as a starting place and a source of inspiration for future paleontological research on the fossils of the Platteville Formation, offer insights regarding past climate and the environments in which life thrived, and ultimately shed light on the history of life on our planet.

The specimens described and illustrated in this overview are primarily from the Platteville Formation reference collection housed at the University of Illinois at Urbana–Champaign, augmented with specimens accessed from museums and private collections. Specimens are identified in the figure captions by an abbreviation of the repository or collection and a unique number. These are (1) the American Museum of Natural History, AMNH; (2) the Burpee Museum of Natural History, BMNH; (3) the Field Museum of Natural History, FMNH; (4) the University of Iowa Paleontological Repository, SUI; (5) the Illinois State Geological Survey Paleontological Collection, University of Illinois at Urbana–Champaign, ISGS; (6) the Prairie Research Institute Center for Paleontology, University of Illinois at Urbana–Champaign, Illinois, PRIP; (7) the U.S. National Museum of Natural History, USNM; and (8) the University of Wisconsin Geology Museum, UWGM. Private collections include Phil Burgess, PB; John A. Catalani, JAC; and Allan W. Scheer, AWS.

A locality register is provided in Appendix 1 and a complete faunal list with an abundance classification is in Appendix 2. The abundance of each species within the member subdivisions of the Platteville Formation is expressed as rare when <10 specimens are known, common if >10 to <100, and abundant if >100.

The Platteville fossils are illustrated in more than 1,200 high-resolution photographs. Most specimens were photographed by the author with a Canon D60 digital camera. Small specimens (<2 mm) were photographed using a Canon CA6528MP MP-E 65mm f/2.8 1-5× macro photo lens and larger specimens using a Canon EF-S 60mm f/2.8 macro USM lens. Focus stacking of images was obtained with an automated macro rail by StackShot, and images were processed with Helicon software. To reveal small skeletal

details, many specimens were stained with a wash of diluted nonwaterproof black India ink, dried, and coated with ammonium chloride prior to being photographed.

Figure 1-1 (Opposite page) Bedding plane surface of limestone from the Mifflin Member at Dixon, Illinois (locality 4), showing a diverse assemblage of well-preserved fossils. Magnification ×5.

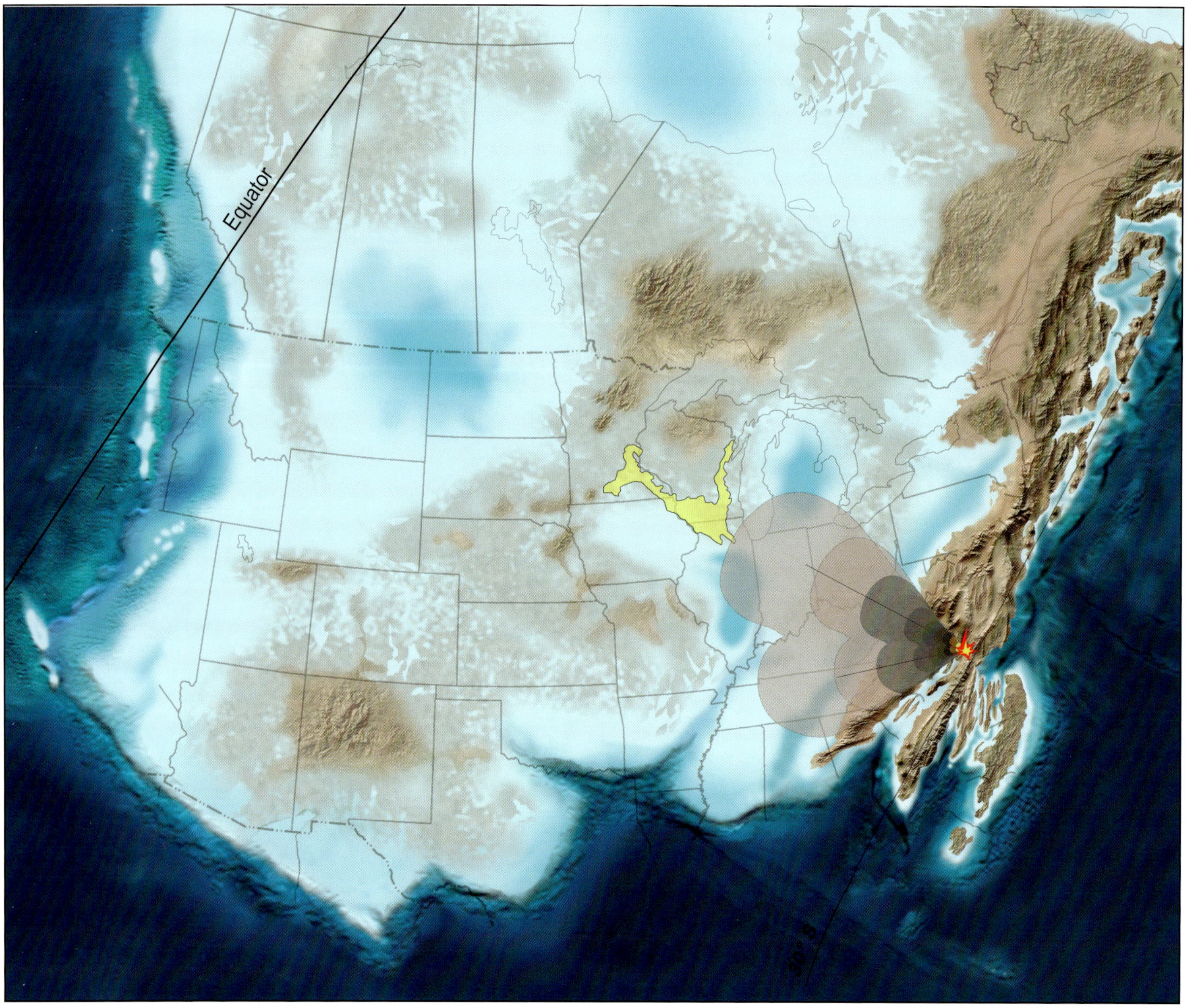

Figure 2-1 Southern part of the Late Ordovician (450 Ma) continent Laurentia, with present state and provincial boundaries for reference. At this time, Laurentia was positioned between the equator and approximately 30° south latitude. Map shows Laurentia covered by vast shallow seas surrounded by deep ocean waters, explosive volcanic eruptions along southeastern margin with ash carried by southeast trade winds across the midcontinent, and present-day outcrop belt of the Platteville Formation in the upper Midwest USA. Copyright © 2013 Colorado Plateau Geosystems Inc. Used by permission of Ron Blakey.

CHAPTER 2
GEOLOGIC SETTING

PALEOGEOGRAPHY

Paleomagnetic evidence and the distribution of paleoclimatically sensitive rock types suggest that during Cambrian and Ordovician time the paleocontinent of Laurentia was situated within 30° latitude of the equator (Golonka 2002; Scotese 2002; Figure 2-1). The combination of equatorial position, greenhouse conditions ($p\mathrm{CO}_2$ 14–16 times preindustrial levels; Crowley and Baum 1995; Berner and Kothavala 2001), and high sea level created favorable conditions for extensive deposition of carbonate rocks across Laurentia (Morgan 2012).

During the Ordovician, the southeastern margin of Laurentia underwent a long and complex episode of both extensional and compressional tectonics. Beginning in the Late Cambrian and Early Ordovician, a narrow strip of continental crust rifted from the southeastern margin. By the Middle Ordovician, tectonic forces reversed and a north-facing arc and subduction zone developed between the rifted microcontinent and Laurentia. Collision progressed from the northeast to the southwest (McKerrow et al. 1991). This was the Taconic orogeny, the first of three major mountain-building episodes in what is now the Appalachian Mountains. By the Late Ordovician (Turinian Stage) explosive volcanism developed in the arc/continent collision zone (Figure 2-1). Prevailing winds carried large quantities of volcanic ash in a northwesterly direction, where it fell into the shallow seas of Laurentia and subsequently was diagenetically altered to widespread beds of K-bentonite (Kolata et al. 1986, 1996, 1998; Huff et al. 1992).

DEPOSITION OF THE SAUK AND TIPPECANOE MEGASEQUENCES

During the Cambrian through the earliest Middle Ordovician time, carbonates of the Sauk megasequence (Figure 2-2) were deposited on and around the Laurentian paleocontinent. This resulted in formation of one of the largest carbonate-dominated platforms in Phanerozoic history, extending across almost the entire North American craton.

By the earliest Middle Ordovician time, sea level dropped, exposing the extensive carbonate platform to an episode of erosion lasting approximately 10 million years. Deep valleys were eroded into the exposed and denuded surface of the Sauk megasequence. Subsurface data from closely spaced drill holes in east-central Indiana reveal a surface of locally occurring steep-sided, ovate, buttelike erosional remnants with vertical relief of 46 m or more (Keller and Abdulkareem 1980). Karst sinkholes and valleys up to 200 m deep are present along the Sauk/Tippecanoe unconformity in Kansas (Merriam and Atkinson 1956) and northern Illinois (Buschbach 1964).

Gradual but punctuated marine transgression of Laurentia was underway by mid-Middle Ordovician time. Initial deposition of the Tippecanoe megasequence was restricted to the continental margin. With rising sea level, the onlapping stratigraphic packages reflected a continuation of depositional styles like that in the underlying Sauk megasequence. Initially, a diachronous sheet of pure quartz sands of the St. Peter Sandstone was deposited in an advancing shoreline in the central midcontinent. An extensive marine bar complex (Starved Rock Bar Complex) developed in southwestern Michigan, prograding through northern Illinois, southeastern Iowa, and northern Missouri into northeastern Kansas (Nunn 1986; Figure 2-3). Shale, dolomitic mudstone, and argillaceous sandstone of the Glenwood Formation were deposited in a restricted lagoon north of the bar complex, whereas marine carbonates of the Dutchtown Limestone and Joachim Dolomite were deposited to the south (Kolata 2010).

The bar complex gradually became less of a barrier with rising sea level. Subsequently, open-marine waters spread northward and deposition of the Platteville Formation carbonates commenced in the upper Midwest. The advancing seaway brought with it a very diverse and locally abundant fauna that is evident even in the basal Pecatonica Member. The "Great Ordovician Biodiversification Event" (Webby et al. 2004b) was in full bloom (Appendix 2).

By late Turinian time, the combination of rising sea level, optimal positioning of the Subtropical Convergence Zone, and preexisting bathymetric depression over the ancient Reelfoot Rift (Figure 2-4) allowed cool, phosphate-rich waters to be funneled into midcontinent Laurentia (Kolata et al. 2001). Accordingly, the midcontinent Laurentian seaway transitioned to an epicontinental estuarine circulation pattern with associated upwelling in a density-stratified water column and concomitant net inflow of deep cool waters. The upper Midwest experienced minor structural deformation during Ordovician time. The sedimentary record indicates long periods of submergence and deposition with only minor

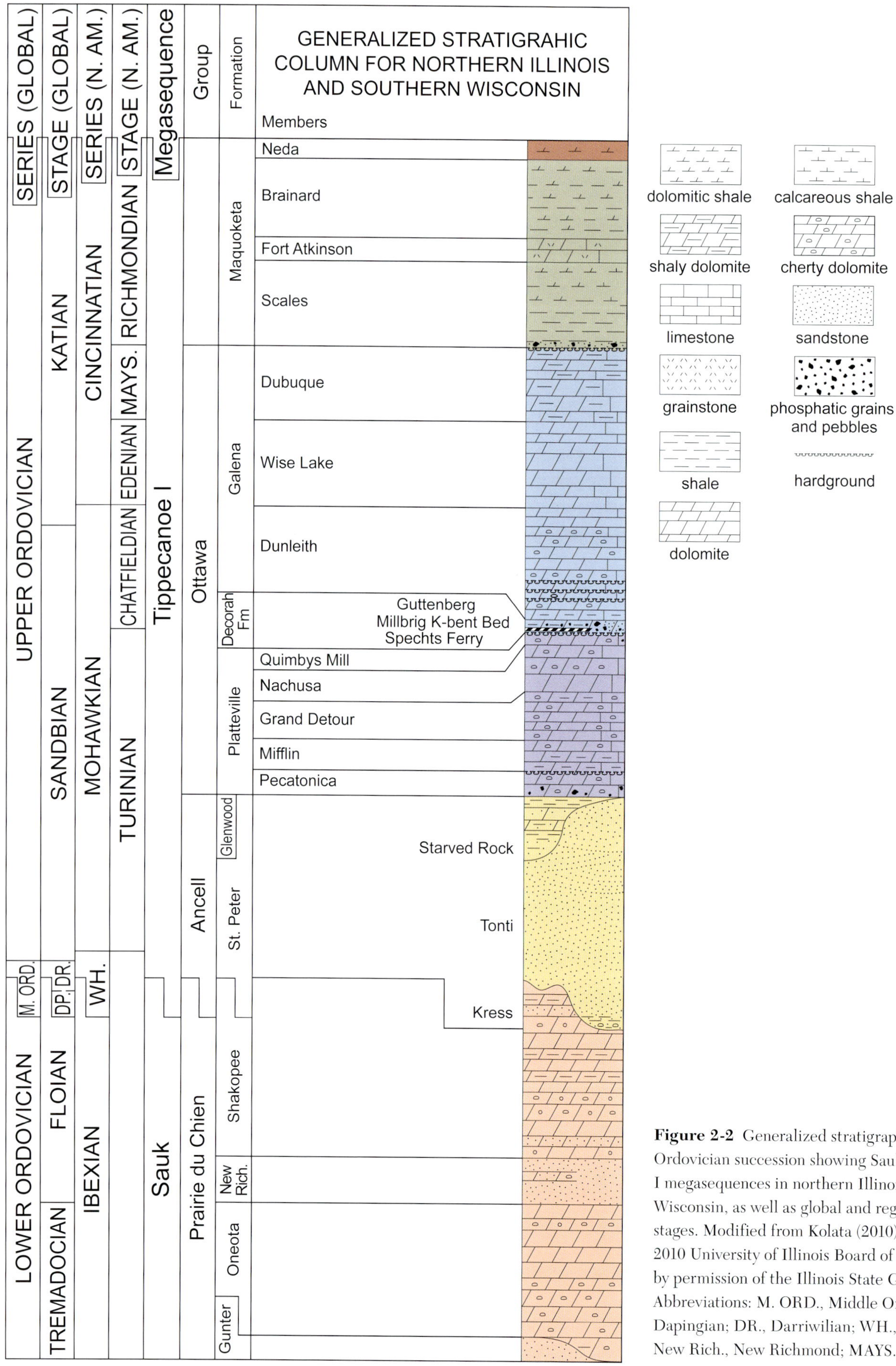

Figure 2-2 Generalized stratigraphic column of the Ordovician succession showing Sauk and Tippecanoe I megasequences in northern Illinois and southern Wisconsin, as well as global and regional series and stages. Modified from Kolata (2010). Copyright © 2010 University of Illinois Board of Trustees. Used by permission of the Illinois State Geological Survey. Abbreviations: M. ORD., Middle Ordovician; DP., Dapingian; DR., Darriwilian; WH., Whiterockian; New Rich., New Richmond; MAYS., Maysvillian.

episodes of uplift on the Wisconsin and Kankakee Arches (Kolata and Nelson 1991; Figure 2-4).

BIOSTRATIGRAPHIC AND K-BENTONITE CHEMOSTRATIGRAPHIC CORRELATIONS

On the basis of conodont biostratigraphy, the Platteville Formation lies entirely within the *Phragmodus undatus* zone (Sweet 1984, 1987; Leslie 2000; Leslie and Bergström 2005). The graptolite *Amplexograptus maxwelli* that occurs in the Mifflin Member of the Platteville Formation (Goldman et al. 2002) is also present in the Balclatchie Group, Glenkiln, and Lower Hartfell Shales of southern Scotland, a stratigraphic interval assigned to the *Climacograptus peltifer* and *C. wilsoni* zones (= *Diplograptus foliaceous* zone; Elles and Wood 1907; Williams 1994). These faunal relations support assignment of the Platteville Formation to the Turinian Stage of the Mohawkian Series.

Regional correlations based on chemical fingerprinting of K-bentonite beds show that throughout the upper Midwest, the Platteville Formation lies below the Millbrig K-bentonite bed (Kolata et al. 1986, 1996, 1998; Huff and Kolata 1990; Figure 2-2). The Millbrig K-bentonite is a confirmed Ordovician Global Boundary Stratotype Section and Point (GSSP) marking the boundary between the Turinian and Chatfieldian Stages of North America (Bergström et al. 2009). The widespread Deicke K-bentonite bed occurs beneath the Millbrig K-bentonite at the top of the Platteville or within the overlying Carimona Member of the Decorah Formation (Kolata et al. 1998). In the nearby Illinois Basin of central Illinois, the Deicke occurs at approximately 6 m below the top of the Platteville Formation ("Black River Limestone"; Kolata et al. 1998). These stratigraphic relations support assignment of the Platteville Formation to the North American Turinian Stage of the Mohawkian Series and the global Sandbian Stage of the Upper Ordovician Series (Figure 2-2).

RADIOMETRIC AGE DATING

Uranium–lead dating of the Deicke K-bentonite bed yields an age of 454.5 ± 0.5 Ma (Tucker and McKerrow 1995). The Millbrig K-bentonite bed has a uranium–lead age of 453.1 ± 1.3 Ma (Tucker and McKerrow 1995) and an argon–argon age of 454.1 ± 2.1 Ma (Kunk et al. 1985).

GAMMA-RAY LOG CORRELATIONS

One of the more effective tools used in the regional correlation of the Paleozoic stratigraphic succession are gamma-ray logs. This is a wireline logging method used in petroleum exploratory or water-well drilling that measures the total natural gamma radiation emanating from deeply buried rock or sediment. The tool is useful in distinguishing units that are relatively high in primary radioactive isotopes (uranium, thorium, potassium-40), such as shale, argillaceous limestone, or K-bentonite beds (altered volcanic ash), whereas pure limestone produces low readings (Figure 2-5). Logs of the Platteville Formation have a characteristic gamma-ray signature that corresponds to the relative amount of disseminated clay in each of the stratigraphic members. For example, computed gamma-ray counts are relatively high in the shale-rich beds of the Mifflin Member and upper part of the Grand Detour Member and relatively low in the pure carbonate rocks of the lower Grand Detour and Nachusa Members (Figure 2-5). These variations are stratigraphically consistent throughout wide areas of northern Illinois, southern Wisconsin, and northeastern Iowa.

LITHOSTRATIGRAPHIC NOMENCLATURE AND CLASSIFICATION

Development of Upper Ordovician rock stratigraphic nomenclature and classification within the U.S. upper Midwest has evolved primarily through the efforts of the state geological surveys of Illinois, Iowa, Minnesota, and Wisconsin. Much of this work is summarized in reports of the (1) Illinois State Geological Survey (Templeton and Willman 1963; Willman et al. 1975; Willman and Kolata 1978; Kolata and Graese 1983; Kolata 2013), (2) Iowa Geological Survey (Bunker et al. 1985), (3) Minnesota Geological Survey (Mossler 1983, 1985, 1987, 2008), and (4) Wisconsin Geological and Natural History Survey (Ostrom 1967, 1969, 1970, 1987). Although there are some differences between these states in applied stratigraphic nomenclature and ranks, most observers agree that the Upper Ordovician carbonate succession, including the Platteville Formation, consists of distinctive rock units that can be traced over wide areas of the U.S. Midcontinent. Primary physical characteristics include (1) the relative amount of disseminated clay, (2) chert content, (3) bedding type, (4) widely traceable K-bentonite beds, (5) hardground omission surfaces, (6) phosphatic grains and

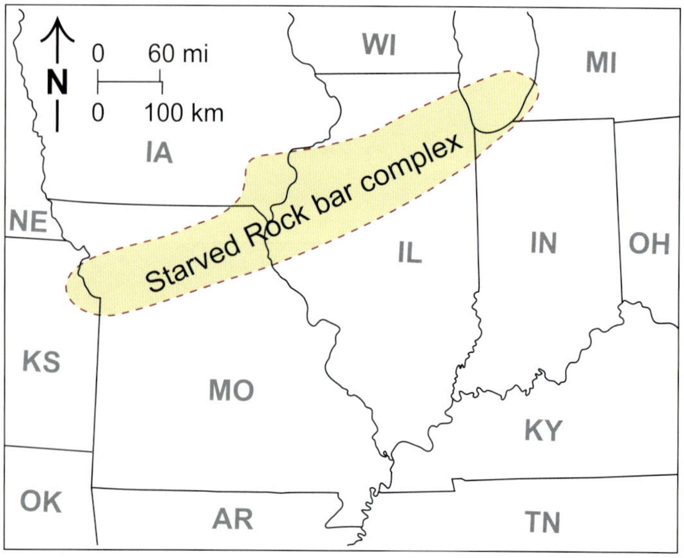

Figure 2-3 Map showing distribution of the Late Ordovician Starved Rock bar complex. Modified from Nunn (1986). Used with permission.

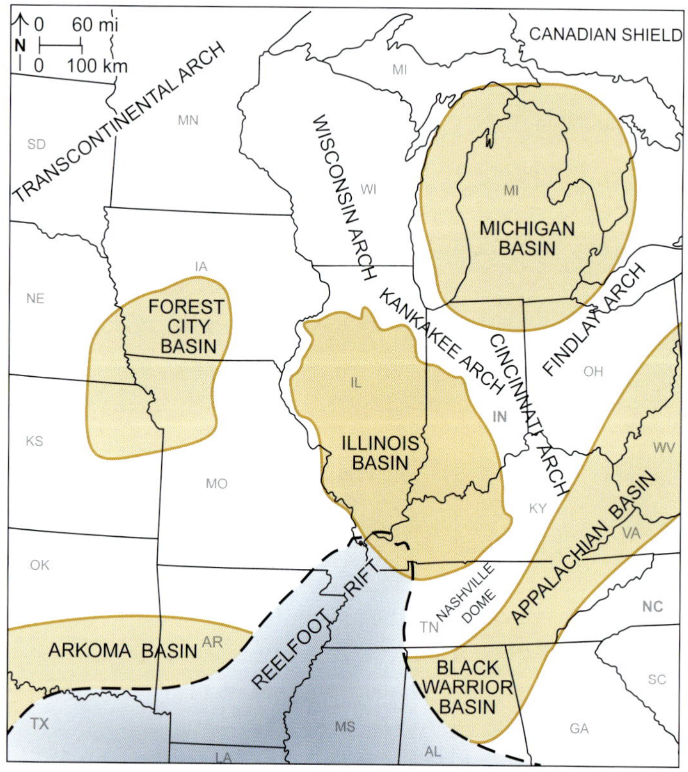

Figure 2-4 Map showing selected regional structural features in the upper Midwest USA. Modified from Nelson (2010). Copyright © 2010 University of Illinois Board of Trustees. Used by permission of the Illinois State Geological Survey.

Figure 2-5 (Opposite page) Stratigraphic column based on a drill core and gamma-ray log from the Illinois State Geological Survey No. 1 Jerry Stuff exploratory boring (API 121412600200) drilled in a small quarry on the west side of Lowell Road, 9 km north of Dixon, Illinois, 41°56′30.76″N, 89°29′39.36″ W. The core and gamma-ray log serve as a reference for the Platteville Formation in the area of Dixon and more generally for north-central Illinois and south-central Wisconsin. Layers containing shale and disseminated clay have a higher computed gamma-ray count, expressed by a deflection of the vertical trace line to the right. Conversely, a deflection to the left indicates relatively pure rocks such as limestone or sandstone. The gamma-ray trace can serve as a profile of a typical weathered outcrop of the Platteville Formation, with the deepest weathering (curve to the right) occurring in the less resistant shale and argillaceous limestone/dolomite. Listed on the right side of the column are the most common and widespread macrofossils typically found in the Platteville Formation of north-central Illinois and south-central Wisconsin. Also shown are the stratigraphic positions of unique faunas, sedimentary features, and key beds, such as K-bentonites (altered volcanic ash), that occur in the region.

pebbles, and (7) fossil content. Imprinted on these features are the effects of local and regional diagenetic alteration.

The Illinois State Geological Survey has long followed the Upper Ordovician stratigraphic classification system proposed by Templeton and Willman (1963), with updates by Willman and Kolata (1978). This system has worked well in geological investigations, particularly within northern Illinois, and some of the subdivisions have been recognized and applied to other parts of the upper Midwest (e.g., Ludvigson et al. 2004). Many of the formations, however, are too thin to map at the current scale of 1:24,000 and therefore do not meet the current "test of mappability" recommended in the code of the North American Commission on Stratigraphic Nomenclature (2005). Furthermore, the ranks of "megagroup" and "subgroup" are not recognized in the stratigraphic code (2005). Adjustments in the ranks of the Illinois Upper Ordovician lithostratigraphic units were made by Kolata (2013) to bring the Templeton and Willman (1963) and Willman and Kolata (1978) classification schemes up to code and to maintain the usefulness of the numerous recognized units. Kolata (2013) proposed the following changes in stratigraphic rank: (1) eliminate megagroups and subgroups, (2) reassign the Ottawa megagroup to the rank of group, (3) reassign the Platteville, Decorah, and Galena Groups to formations, (4) reassign the Pecatonica, Mifflin, Grand Detour, Nachusa, and Quimbys Mill Formations to members of the Platteville Formation, (5) reassign the Spechts Ferry, Kings Lake, and Guttenberg Formations to members of the Decorah Formation, (6) reassign the Dunleith, Wise Lake, and Dubuque Formations to members of the Galena Formation, (7) reassign all 32 corresponding members to beds, and (8) keep the nine named K-bentonite beds (Willman and Kolata 1978) at bed rank (Figure 2-6). Likewise, the Maquoketa Group is reassigned to the rank of formation and the Cape, Scales, Fort Atkinson, Brainard, and Neda Formations to the rank of members. The series and stage names as well

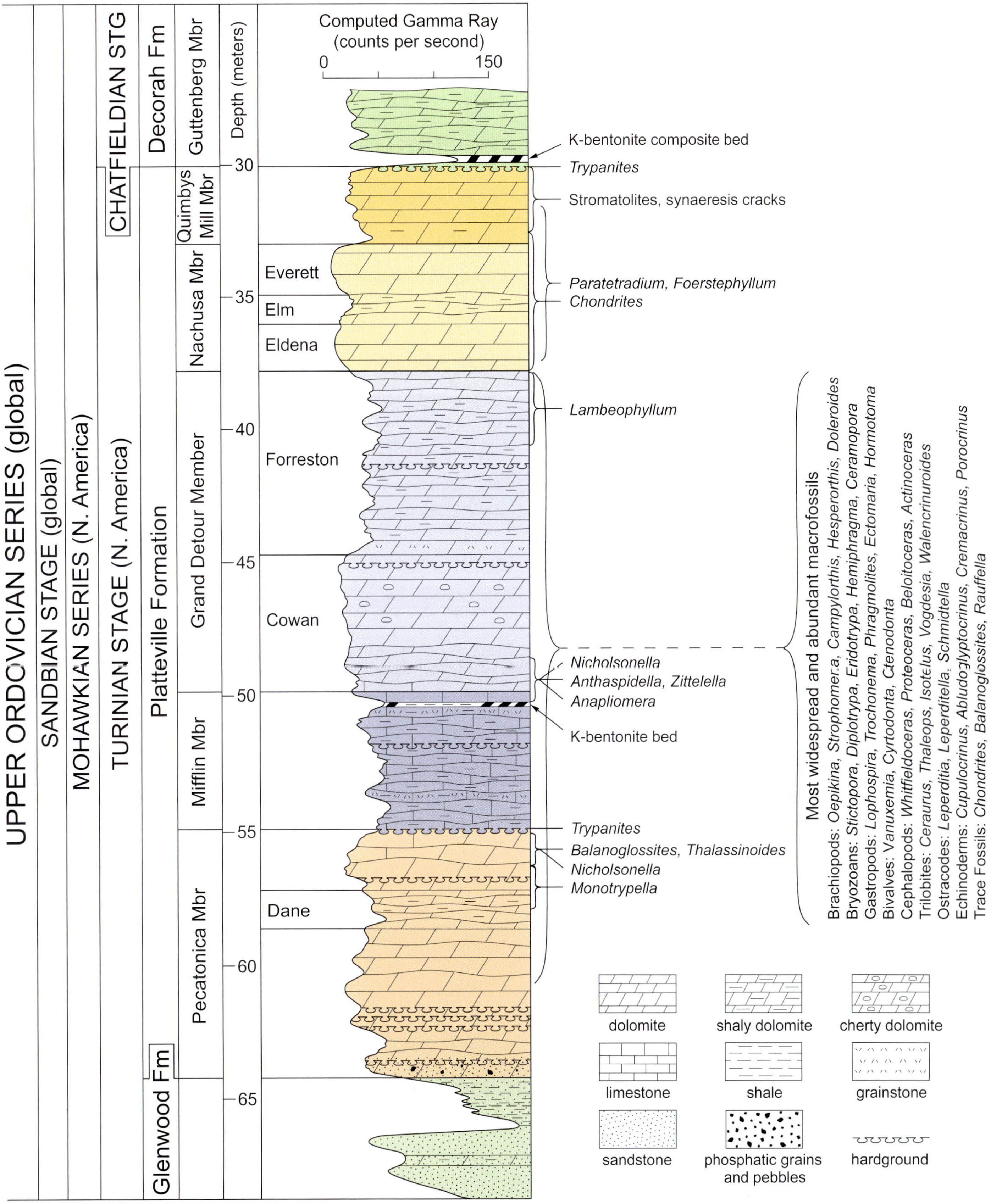
UPPER ORDOVICIAN SERIES (global)
SANDBIAN STAGE (global)
MOHAWKIAN SERIES (N. America)
TURINIAN STAGE (N. America)
CHATFIELDIAN STG
Decorah Fm
Platteville Formation
Glenwood Fm
Guttenberg Mbr
Quimbys Mill Mbr
Nachusa Mbr
Grand Detour Member
Mifflin Mbr
Pecatonica Mbr
Depth (meters)
30
35
40
45
50
55
60
65
Computed Gamma Ray (counts per second)
0
150
Everett
Elm
Eldena
Forreston
Cowan
Dane
K-bentonite composite bed
Trypanites
Stromatolites, synaeresis cracks
Paratetradium, Foerstephyllum
Chondrites
Lambeophyllum
Nicholsonella
Anthaspidella, Zittelella
Anapliomera
K-bentonite bed
Trypanites
Balanoglossites, Thalassinoides
Nicholsonella
Monotrypella
Most widespread and abundant macrofossils
Brachiopods: Oepikina, Strophomena, Camplyorthis, Hesperorthis, Doleroides
Bryozoans: Stictopora, Diplotrypa, Eridotrypa, Hemiphragma, Ceramopora
Gastropods: Lophospira, Trochonema, Phragmolites, Ectomaria, Hormotoma
Bivalves: Vanuxemia, Cyrtodonta, Ctenodonta
Cephalopods: Whitfieldoceras, Proteoceras, Beloitoceras, Actinoceras
Trilobites: Ceraurus, Thaleops, Isotelus, Vogdesia, Walencrinuroides
Ostracodes: Leperditia, Leperditella, Schmidtella
Echinoderms: Cupulocrinus, Abludoglyptocrinus, Cremacrinus, Porocrinus
Trace Fossils: Chondrites, Balanoglossites, Rauffella
dolomite
shaly dolomite
cherty dolomite
limestone
shale
grainstone
sandstone
phosphatic grains and pebbles
hardground

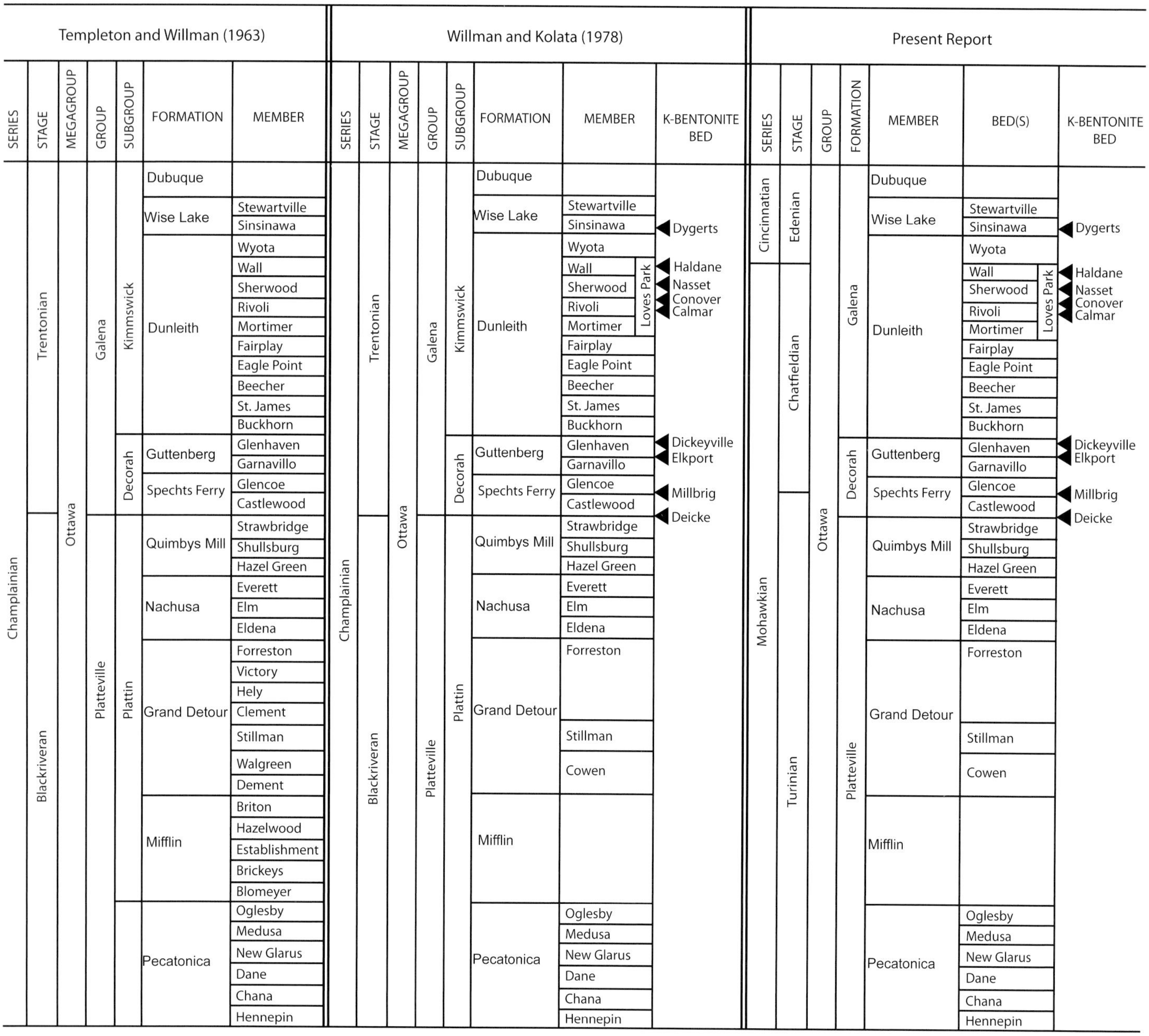

Figure 2-6 A comparison of stratigraphic nomenclature and rank used in past classifications of the Platteville and Galena Formations in Illinois (Templeton and Willman 1963; Willman and Kolata 1978). Modified from Kolata (2013). Copyright © 2013, University of Illinois Board of Trustees. Used by permission of the Illinois State Geological Survey.

as boundaries are also updated in recognition of the recent advancements in Ordovician chronostratigraphy. The historical development of Platteville stratigraphic nomenclature is illustrated in Templeton and Willman (1963, figures 16-A through 16-D) and Mossler (2008, figure 10).

Platteville Formation

The Platteville Formation was named by Bain (1905) after the city of Platteville, Grant County, Wisconsin, located approximately 6.5 km east of exposures along the Little Platte River. Because Bain did not designate a precise outcrop locality, Kay (1931) proposed the principal reference section as a ravine exposure situated 3.6 km northwest of the intersection of township Route D and U.S. Highway 151 (42°43′24.44″N, 90°31′33.71″W). The formation is well exposed and easily accessible nearby in a series of roadcuts on the west side of U.S. Highway 151, 9 km south of Platteville (42°40′3.26″N, 90°31′3.15″W; Figure 2-7). The entire formation is here exposed, including the underlying Glenwood Formation and St. Peter Sandstone and overlying Decorah and Galena Formations.

Distribution of the Platteville Formation at the bedrock surface is related principally to the southward projection of the Wisconsin Arch (Figure 2-4). The arch extends from northwestern Wisconsin southeastward into north-central Illinois, terminating in LaSalle County, Illinois. Throughout most of its extent, the arch has gentle slopes, averaging 5–10 m/km. Erosional truncation of the arch has produced a V-shaped outcrop belt of Platteville strata extending from north-central Illinois northeastward through Green Bay, Wisconsin, into the upper peninsula of Michigan and on eastward into southern Ontario, Canada. The northwest-trending belt extends from north-central Illinois through southwestern Wisconsin and northeastern Iowa, and into southeastern Minnesota up to the area of St. Paul and Minneapolis (Figure 2-8).

The Platteville consists largely of fossiliferous, argillaceous, heavily bioturbated, skeletal calcareous mudstone and wackestone that are regionally dolomitized. Within the outcrop belt, the formation ranges in thickness from 6 m in southeastern Minnesota to 42 m in northern Illinois. The Platteville Formation thickens southward through the subsurface of the Illinois Basin to approximately 200 m in southernmost Illinois (Willman et al. 1975).

Stratigraphic nomenclature of member-ranked units in the Platteville Formation varies between states (Figure 2-9). In Illinois, the Platteville is subdivided into the Pecatonica, Mifflin, Grand Detour, Nachusa, and Quimbys Mill Members in ascending order (Figure 2-5). The Pecatonica Member is recognized throughout most of the outcrop belt. In Wisconsin (Ostrom 1967, 1969, 1970), Iowa (Bunker et al. 1985), and southeasternmost Minnesota (Mossler 2008), the Mifflin and Grand Detour stratigraphic interval is assigned to the McGregor Member. In the area of Minneapolis, Minnesota, the McGregor equivalent strata are divided into the Mifflin, Hidden Falls, and Magnolia Members in ascending order (Mossler 2008). The Nachusa and Quimbys Mill Members, best developed in northern Illinois, thin northwestward and pinch out in southwestern Wisconsin.

The Platteville is underlain by the siliciclastic-rich Glenwood Formation of the Ancell Group, and the hardground omission surface at the top of the Platteville is overlain by a northwest to southeast downlapping succession of siliciclastic and carbonate rocks of the Decorah Formation (Ludvigson et al. 1996).

Pecatonica Member. The Pecatonica consists largely of relatively thick, tabular beds of brown, fine- to medium-grained, relatively pure dolostone that is locally cherty. Throughout most of northern Illinois, southern Wisconsin, and northeastern Iowa, the Pecatonica consists of dolomitized mudstone and wackestone locally containing a diverse and

Figure 2-7 Outcrop of the Pecatonica, Mifflin, and Grand Detour Member of the Platteville Formation in a roadcut on the west side of U.S. Highway 151, 9 km south of Platteville (42°40′7.46″N, 90°33′3.96″W). The Quimbys Mill Member is exposed in the hillside to the right of this photograph.

well-preserved fauna particularly rich in mollusks. The Pecatonica ranges in thickness from 8–10 m in northern Illinois to a meter or less in southeastern Minnesota.

The lower third of the Pecatonica typically consists of nonfossiliferous, sandy, argillaceous dolostone, one or more hardground surfaces, and phosphatic grains and pebbles. A relatively argillaceous interval called the Dane Beds is recognized in the upper part of the Pecatonica Member. The top of the Pecatonica is capped by a prominent pyrite-rich hardground surface that is present throughout most of the outcrop belt. This surface typically is covered with vertical borings of *Trypanites* (Byers and Stasko 1978). Locally, as near Dixon, Illinois (localities 1 through 4), the upper 2 m of the Pecatonica Member has a well-developed assemblage of *Balanoglossites* and *Thalassinoides* burrows as well as

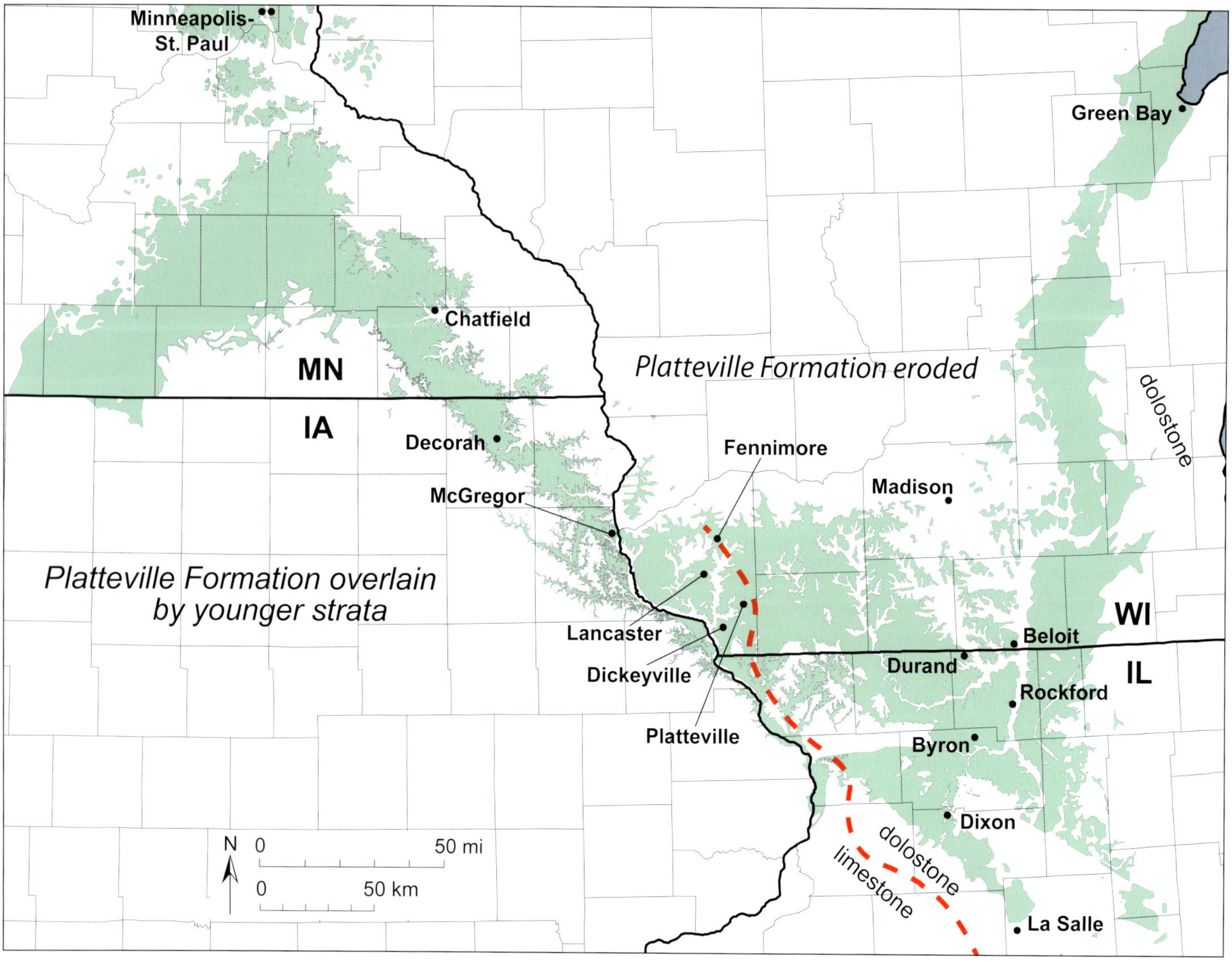

Figure 2-8 Map showing the distribution of the Upper Ordovician Platteville and Galena Formations at the bedrock surface (light green) in the upper Midwest USA. The Platteville and Galena Formations have been pervasively dolomitized in the region east (right) of the dashed red line. The Platteville and Galena Formations continue into the subsurface to the southwest, south, and east of the outcrop belt and are eroded north of the outcrop belt. Outcrop map data are from the state geological surveys of Illinois, Iowa, Minnesota, and Wisconsin.

Trypanites borings. These same strata contain a unique association of the bryozoans *Nicholsonella* and *Monotrypella*.

Mifflin Member. The Mifflin consists mainly of thin, argillaceous, wavy- to nodular-bedded lime mudstone and wackestone with relatively thick greenish-gray shale partings and a few thin interbeds of packstone and grainstone (Figures 2-10 and 2-11). The style of bedding readily distinguishes the Mifflin from members above and below. Throughout northern Illinois, southern Wisconsin, and northeastern Iowa, the Mifflin has a consistent thickness of 5–6 m. The member contains a very diverse, abundant, and locally well-preserved fauna represented by corals, bryozoans, brachiopods, polyplacophorans, gastropods, bivalves, rostroconchs, nautiloid cephalopods, trilobites, ostracodes, echinoderms, graptolites, hyolithids, cornulitids, macroalgae, various microfossils, and a rich trace fossil record (Appendix 2). The Mifflin is regionally dolomitized in northern Illinois and southern Wisconsin.

One of the more notable outcrops of the Mifflin Member is in the numerous open-pit cement mines exposed on the north side of Dixon, Illinois (includes localities 1-4, Appendix 1). Here the Mifflin Member consists of 5.5 m of thin, wavy, nodular beds containing a diverse, abundant, and well-preserved invertebrate fauna. Local structural deformation has folded the strata into small anticlines and domes, some with associated fractures that have exposed the calcium-rich Mifflin limestone to a long episode of weathering and dissolution (Figure 2-12). Particularly notable are localized pockets of unconsolidated "marl" formed by diagenetic alteration of the limestone. Fossils in these deposits tend to weather free of their matrix and can be extraordinarily well preserved.

Grand Detour Member. The Grand Detour is largely dolomitized in its type area of northern Illinois and in adjacent parts of southern Wisconsin. The lower part is pure, fine to medium crystalline dolomite with 10–25 cm thick tabular beds (Cowen and Stillman Beds) grading upward to argillaceous, finely crystalline, wavy, nodular beds (Forreston Beds) mostly 5–15 cm thick. Relict bioclastic textures suggest that prior to dolomitization, the Grand Detour was a skeletal lime mudstone and wackestone with rare interbeds of packstone and grainstone. White to medium gray chert nodules are locally present at all levels of the member. The Grand Detour Member can be distinguished in the outcrop belt of southwestern Wisconsin and northeastern Iowa, where it is common practice to combine it with the Mifflin Member to form the McGregor Member (Figure 2-9). The Grand Detour varies in thickness from 14 m in northern Illinois to approximately 4 m in southwestern Wisconsin and northeastern Iowa.

The Grand Detour contains the same dominant fossil groups as the Mifflin, and in total number of identified species it has the most diverse fauna in the Platteville Formation. The echinoderms alone have at least 10 species, including crinoids, cystoids, and a paracrinoid, that are known only from the Grand Detour Member. The lower 1 m of the Grand Detour contains a locally abundant lithistid fossil sponge fauna, including *Anthaspidella*, *Zittelella*, and *Dystactospongia*, that has been observed near Dixon (localities 1–4), Byron, (locality 9), and Durand (locality 11), Illinois. This same interval has yielded specimens of the relatively rare trilobite *Anapliomera*. The Grand Detour Member in north-central Illinois and south-central Wisconsin is particularly diverse in mollusks, including bivalves, gastropods, and nautiloid cephalopods.

Nachusa Member. The Nachusa consists primarily of dolomite that is pure to slightly argillaceous, fine to medium grained, locally cherty, and has 10–20 cm thick tabular beds. Relict bioclastic textures in the chert indicate that the initial carbonate deposit was lime mudstone. The base is commonly marked by an abrupt change from the underlying clay and silt-rich dolomite of the Grand Detour Member to relatively pure dolomite of the Nachusa. The Nachusa ranges in thickness from 4.5 to 7.5 m in the outcrop belt of north-central Illinois, gradually diminishing in thickness toward the northwest and pinching out in Grant County, Wisconsin (Figure 2-9). Three subdivisions of the Nachusa Member are recognized in northern Illinois and southern Wisconsin: the Eldena, Elm, and Everett Beds (Figure 2-5).

The Nachusa fauna is relatively sparse and low in diversity. The most notable fossils are large colonies of the tabulate coral *Foerstephyllum* and tetradiid *Paratetradium*. Bedding planes are commonly crowded with conspicuous horizontal burrows of the trace fossil *Chondrites*. Overall, the Nachusa was deposited in a more restricted marine environment than the Grand Detour and Mifflin Members.

Quimbys Mill Member. At least two distinct depositional/faunal facies are present in the Quimbys Mill. The first is positioned along the axis and east of the Wisconsin Arch in north-central Illinois and south-central Wisconsin. The Quimbys Mill is here characterized by yellowish buff, slightly argillaceous to relatively pure, dense, fine to very fine grained dolomite in 5–10 cm thick tabular beds separated by light greenish gray shale partings mostly less than 3 mm thick. Typical examples can be seen in outcrops between LaSalle, Illinois, and Madison, Wisconsin, where the Quimbys Mill is 2–3.5 m thick. The member locally contains white chert nodules with relict bioclastic textures, indicating an initial carbonate sediment of lime mudstone. Abundant vertically oriented lenticular cavities 5–20 mm across and up to 3 mm wide are also present locally (Figure 2-13-3). These closely resemble "intrastratal shrinkage cracks," an indicator of microbial mat formation in salinity-stressed marginal-marine depositional environments (Harazim et al. 2013). Also present are low-profile, dome-shaped stromatolites (Figure 2-13-2) with internal laminations (Figure 2-13-1). In this region, the Quimbys Mill contains a sparse, low-diversity invertebrate fauna. Together, the lithic and faunal evidence suggests deposition in relatively shallow water with restricted marine conditions.

Northwestward in southwestern Wisconsin, the Quimbys Mill gradually transitions to a relatively deep-water carbonate facies characterized by medium to dark brown, argillaceous, shaley, very dense, "glass rock," carbonaceous, fossiliferous lime mudstone. Here the fauna is dominated by asaphid trilobites and, to a lesser extent, strophomenid brachiopods *Oepikina* and *Strophomena*. At roadcuts on U.S. Highway 151 south of Platteville, Wisconsin, the Quimbys Mill is about 1 m thick, pinching out a short distance northwest of Dickeyville, Wisconsin. The emerging picture of deposition of the Quimbys Mill and Nachusa Members in the upper Midwest is one of shoal environments positioned along the Wisconsin Arch, possibly extending southeastward along the Kankakee Arch. Carbonate mud was deposited here in restricted intertidal and supratidal settings while

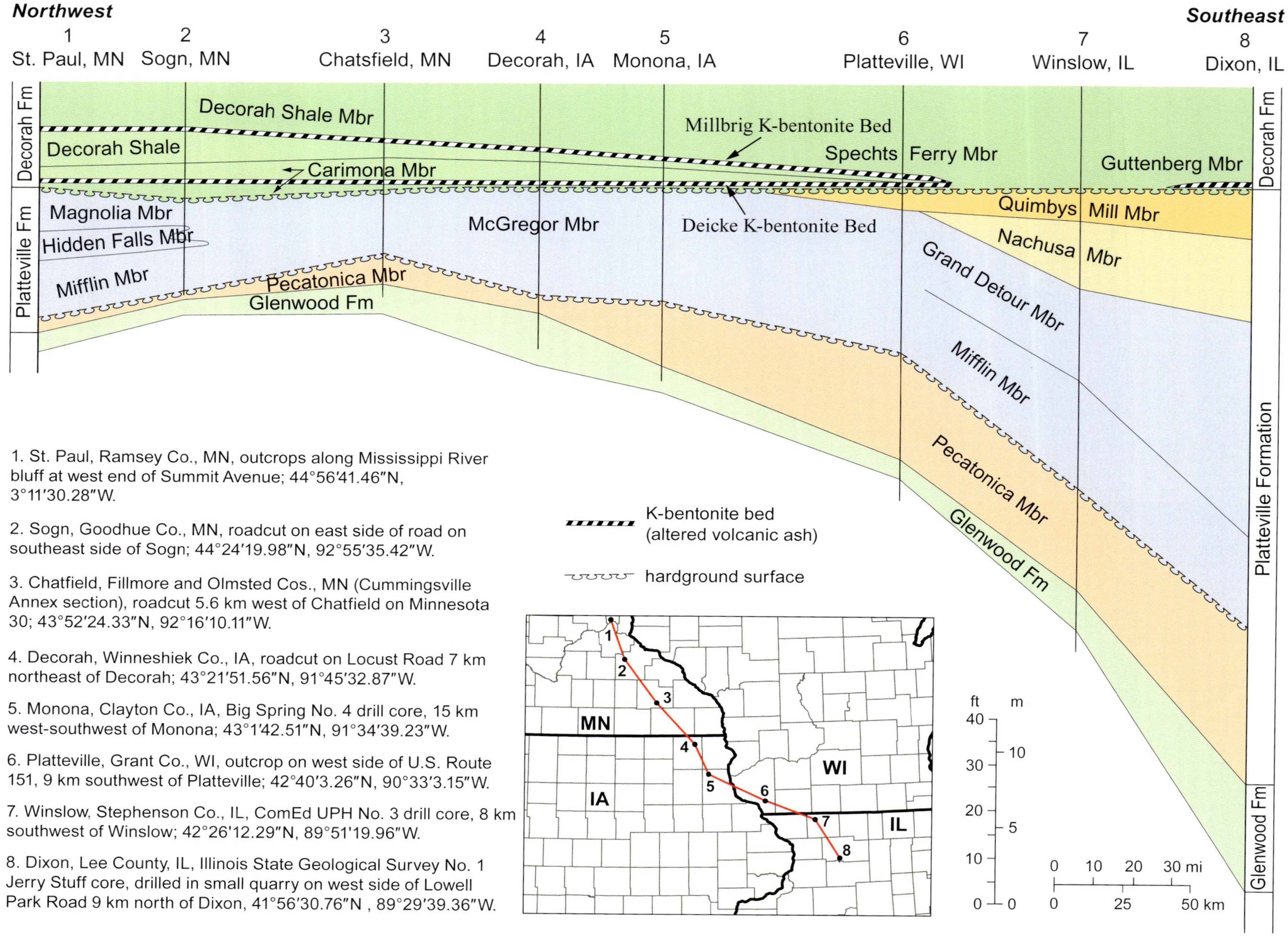

Figure 2-9 Cross section of the Platteville Formation and adjacent parts of the underlying Glenwood and overlying Decorah Formations. Transect based on measured sections in outcrops and drill cores. Baseline of cross section is the base of the Deicke K-bentonite or top of the Platteville Formation where the Deicke is absent. Data for the sections in Minnesota are from Mossler (1985) and Sloan et al. (1987), and in Iowa from Witzke et al. (2005) and Ludvigson and Witzke (2005).

skeletal-rich carbonates were being deposited in normal open-marine subtidal conditions in southwestern Wisconsin and adjacent parts of northeastern Iowa.

The top of the Quimbys Mill Member is commonly marked by a prominent pyrite-rich hardground omission surface with vertical borings of *Trypanites* and is in places overlain by phosphatic grains and pebbles. Locally, as in the ISGS #1 Jerry Stuff boring (API 121412600200) in Ogle County, Illinois, the Quimbys Mill Member is overlain by a K-bentonite bed. It is uncertain whether this represents a single ash fall or the stacking of multiple ash beds over a relatively long period of sediment starvation that occurred prior to deposition of the overlying Decorah Formation (Kolata et al. 1998).

Dolostone/Limestone Facies

The Upper Ordovician stratigraphic succession, including the Platteville Formation, has been pervasively dolomitized in parts of northern Illinois as well as southern and eastern Wisconsin (Figure 2-8). The dolostone facies generally coincides with the crest and flanks of the Wisconsin Arch. The dolostone grades to low-magnesium limestone eastward in the Michigan Basin, southward in the Illinois Basin, and west-southwest in the Forest City Basin (Pratt and Wandrey 1996). Small, isolated pockets of limestone are present locally in the dolostone facies. One such occurrence is present on the north side of Dixon in Lee County, Illinois; others are shown on Pratt and Wandrey's (1996) regional map.

One of the more often cited dolomitization models involves mixing of marine water with fresh meteoric water in a phreatic zone, a mechanism Badiozamani (1973) referred to as the "Dorag model." While the Dorag model predicts that mixing-zone fluids are capable of producing dolomite, there is some doubt as to whether it can explain the pervasive dolomite that is observed. Recent petrographic and fluid inclusion evidence from investigations in eastern Wisconsin suggests that the Platteville Formation lime mudstone and wackestone were dolomitized by hydrothermal fluids (60–120 °C) that migrated westward from the Michigan Basin during Devonian and Mississippian time (Luczaj 2006). Similarly, the northward flow of hot brines from the Illinois Basin during the late Paleozoic (Bethke 1986; Rowan et al. 2002) likely was responsible for emplacement of the Mississippi Valley-type ore deposits and dolomitization of limestone in northwestern Illinois and southwestern Wisconsin.

Fossil Preservation

The shells or skeletons of Platteville invertebrate fossils originally consisted of a variety of minerals, including

Figure 2-10 Outcrop of the Mifflin Member near Fennimore, Grant County, Wisconsin (locality 26), showing typical wavy, nodular bedding. Shown with 1 m (3.3 ft) scale divided in decimeter increments.

aragonite (gastropods, bivalves, nautiloid cephalopods), calcite (sponges, corals, brachiopods, bryozoans, echinoderms), calcium phosphate (trilobites, conularids, conodonts), and silica (some sponges). To varying degrees, these compositions have been altered through time by diagenetic processes involving heat, pressure, and the infiltration of mineral-rich waters. This is particularly evident in the limestone facies of the Platteville Formation, where the original aragonite shells of mollusks have been replaced by coarsely crystalline calcite (**sparite**). More extreme effects of diagenesis are seen in the dolostone facies, where the original skeletal material has dissolved and is lost entirely, leaving only an impression or mold of the fossil. These are preserved as **external** and **internal molds**, the latter referred to as **steinkerns**.

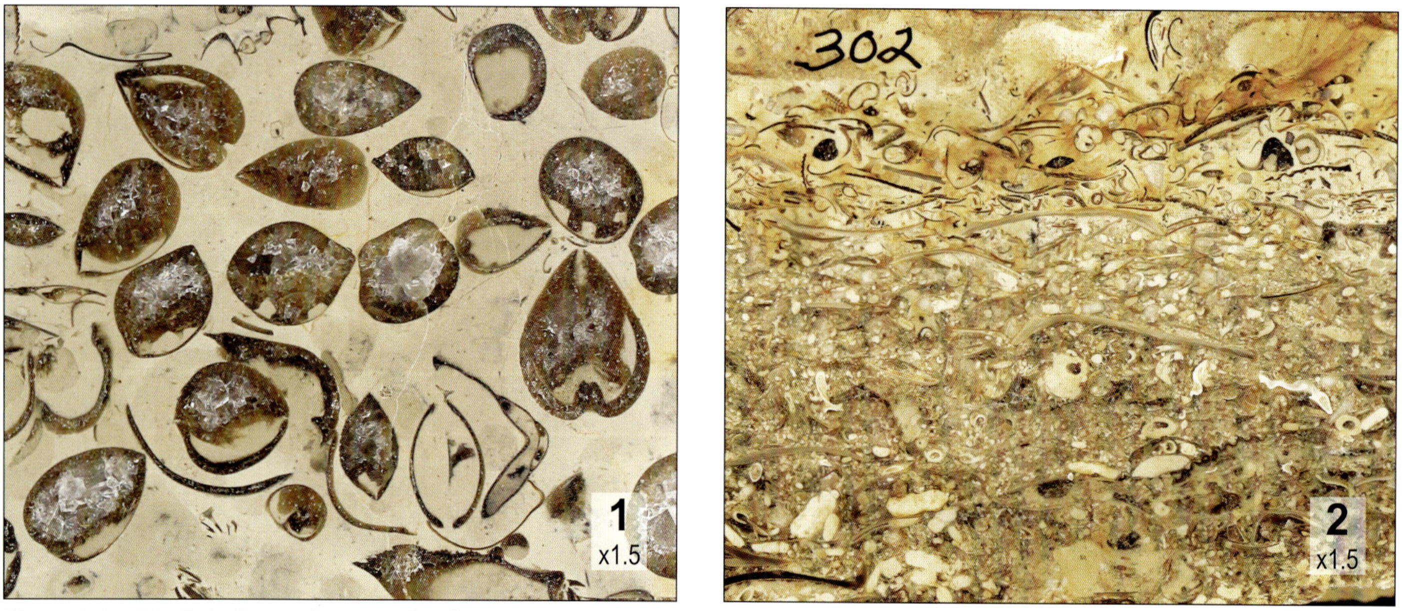

Figure 2-11 (**1**) Polished vertical section of wackestone with numerous articulated shells of bivalve *Vanuxemia dixonensis*; (**2**) polished vertical section of shell fragments forming a grainstone at base grading upward to packstone and wackestone. Mifflin Member, Dixon, Illinois (locality 2).

FAUNAL AND FLORAL ASSEMBLAGES

Approximately 300 species of macroinvertebrates and a small number of macroalgae have been identified in this overview. At least two distinct faunal associations occur in the Platteville Formation of the upper Midwest USA (Appendix 2). These include (1) the abundant and diverse macrofossil assemblage in the Pecatonica, Mifflin, and Grand Detour Members that occurs throughout the outcrop belt, and (2) the sparse, low-diversity faunal assemblage in the Nachusa and Quimbys Mill in north-central Illinois and south-central Wisconsin.

Faunal associations in the Pecatonica, Mifflin, and Grand Detour Members consist of a ubiquitous assemblage of the following dominant faunal groups:

Brachiopods: *Oepikina, Strophomena, Campylorthis, Hesperorthis, Doleroides*

Bryozoans: *Stictopora, Diplotrypa, Eridotrypa, Hemiphragma, Ceramopora*

Gastropods: *Lophospira, Trochonema, Phragmolites, Ectomaria, Hormotoma*

Bivalves: *Vanuxemia, Crytodonta, Ctenodonta*

Cephalopods: *Whitfieldoceras, Proteoceras, Beloitoceras, Actinoceras*

Trilobites: *Ceraurus, Thaleops, Isotelus, Vogdesia, Walencrinuroides*

Ostracodes: *Leperditia, Leperditella, Schmidtella*

Echinoderms: *Cupulocrinus, Abludoglyptocrinus, Cremacrinus, Porocrinus*

Trace Fossils: *Chondrites, Balanoglossites, Rauffella*

These are the most likely genera to be encountered in any given Platteville outcrop, but there are local, numerically large occurrences of much less representative faunal elements. For example, the gastropod *Sinuites* is very abundant in the Mifflin Member in the area around Lancaster and Fennimore, Grant County, Wisconsin, but is relatively rare in other areas of the outcrop belt.

Faunal associations in the Nachusa and Quimbys Mill Members are characterized by a relatively sparse, low-diversity assemblage consisting primarily of the following groups:

Corals and Tetradiids: *Foerstephyllum, Paratetradium*

Brachiopods: *Oepikina*

Ostracodes: *Leperditia*

Trace Fossils: *Chondrites*

In a regional examination of the molluscan- and brachiopod-dominated biofacies of the Mifflin Member, Bretsky et al. (1977) concluded, "The overlapping of the various biofacies seems to have taken place at the scale of individual 3–5 cm thick carbonate layers, and there was no consistent stratigraphic pattern of biofacies that encompass the entire Mifflin" (p. 115). This is consistent with observations made in this overview. They also suggested that there was a positive association between the predominance of mollusks and dolostone facies of the Mifflin Member. This association appears to be fortuitous and more likely due to postdepositional

Figure 2-12 A small anticline exposed in outcrop of Mifflin Member limestone at Dixon, Illinois. Top of Pecatonica Member is exposed in foreground. Limestone along axis of anticline is fractured and deeply weathered, forming a residuum of unconsolidated marl consisting of calcareous clay and silt as well as a well-preserved fauna of mollusks, bryozoans, and brachiopods. Photograph by John A. Catalani. Used with permission.

dolomitization. For example, the molluscan faunas in the Mifflin Member at Dixon, Illinois (locality 2; limestone facies) and Lee Center, Illinois (locality 31; dolostone facies), situated 20 km to the southeast, are very similar in overall diversity and abundance as well as sedimentologic and stratigraphic parameters. This suggests that the depositional environments were essentially identical and that dolomitization had little to do with the faunal content.

PALEOENVIRONMENTS

The Late Ordovician cratonic interior seaway of Laurentia was characterized by low wave energy and low tidal ranges. For the most part, facies belts were broad and transitions very gradual. The Platteville Formation was deposited largely in open-marine conditions with normal salinity and well-oxygenated waters. Exceptions include local, restricted intertidal, and supratidal environments along the axis of the Wisconsin Arch during deposition of the Nachusa and Quimbys Mill Members. By analogy to the physical, chemical, and biological conditions in modern marine environments, the Pecatonica, Mifflin, and Grand Detour were likely deposited in warm subtidal seas below the fair-weather base but above the wave base of powerful hurricanes and tsunamis (Figure 2-14). This is evidenced by (1) the abundance of lime mud, (2) the presence of numerous hardground surfaces, (3) locally abundant fecal pellets, (4) abundant and diverse fauna consisting largely of whole, unabraded, commonly articulated macroinvertebrates composed of thin, delicate shells or skeletons, and (5) infrequent interbeds of packstone and grainstone formed by relatively large storm- or tsunami-generated waves that occasionally scoured the seafloor, concentrating shell fragments and locally forming graded shell beds.

By far the bulk of sediment in the Platteville Formation originated from marine lime mud and lesser amounts of invertebrate shelly debris and terrigenous clay and silt. What was the source of the lime mud? The Platteville Formation lime mudstone is relatively high in strontium and has microfabrics, suggesting a probable biogenic aragonite precursor dissolved in shallow-water calcite seas (Lasemi 1983; Palmer and Wilson 2004). In modern carbonate muds, codiacean green algae produce major amounts of aragonite, and it is likely that the source of the lime mud in the Platteville Formation was from postmortem disintegration of aragonite-producing algae. Modern aragonite-secreting algae are common in warm-water equatorial latitudes, which is consistent with paleogeographic reconstructions of the Late Ordovician midcontinent Laurentia. The warm-water conditions suggest a lagoonal-type circulation pattern characterized by a net outflow of deep platform waters from the midcontinent Laurentia (Kolata et al. 2001). Somewhat restricted circulation between epicontinental and oceanic waters in Late Ordovician Laurentia is supported by neodymium and carbon isotope compositions (Holmden et al. 1998).

Figure 2-13 (**1**) Polished vertical section of algal-bound dolostone composed of uniform, partly discontinuous, millimeter-sized laminae; (**2**) in situ flat-topped algal dome with hammer for scale; (**3**) bedding surface of dolostone showing numerous intrastratal shrinkage cracks, an indicator of microbial mat formation in salinity-stressed marginal-marine environments. Quimbys Mill Member, Dixon, Illinois (locality 48).

CHAPTER 3

PORIFERANS (SPONGES, STROMATOPOROIDS, CHAETETIDS)

The Platteville poriferans are relatively diverse and locally abundant (Appendix 2). At least three classes are present: the Demospongea, Heteractinida, and Stromatoporoidea. The lithistid demosponges are particularly diverse, many of which were described from the Platteville limestone at Dixon, Illinois, by Ulrich and Everett (1890). The heteractinida, represented by *Astraeospongium*, are rare in the Platteville. Locally common are stromatoporoids, including branched aulaceratids and domal stromatocerids as well as the chaetetid *Solenopora*. The overall composition of the Platteville poriferan fauna is most like that from the coeval Upper Ordovician Stones River Group of Tennessee.

Poriferans are the simplest and most primitive of the multicellular animals. Feeding and respiration are accomplished by drawing water into and out of their bodies by the vibrating action of flagella in cells termed **choanocytes**. Inhalant pores called **ostia** (**sg. ostium**) are located on the exterior walls leading inward to a network of **canals** and on through cavities lined with choanocytes that extract food and oxygen. Waste materials and carbon dioxide exit the body by way of exhalant canals that empty through terminal pores called **oscules**. The **spongocoel** is a large central exhalant cavity. **Dermal** refers to the external or inhalant surface, whereas **gastral** refers to the internal or exhalent surface. Most poriferans have an internal skeleton consisting of calcareous or siliceous **spicules**. Morphologically, spicules are very diverse and sponge specialists have developed an extensive nomenclature to describe them. These details are beyond the scope of this overview, and the reader is referred to the *Treatise on Invertebrate Paleontology* (Finks and Rigby 2004) for a more extensive discussion of spicules and sponge morphology in general. **Lithistid** sponges, which are one of the most common sponges in the Platteville Formation, possess relatively large, irregularly branched, articulated spicules called **desmas** that form a rigid skeletal framework. Rodlike vertical elements called **trabs**, produced by fusion of desma tips, are cross-connected by horizontal **dendroclones** that, in total, form a ladderlike framework.

In addition to having spicules, some Platteville sponges secreted massive skeletons of calcium carbonate and are referred to as hypercalcified sponges. Platteville forms include chaetetids and stromatoporoids. These groups have unique skeletal features that are discussed in the systematic descriptions.

The earliest reported occurrence of fossil sponges from the Platteville Formation was by Worthen and Meek (1875), who described the anthaspidellid *Cnemidium*? *trentonensis* (=*Zittelella typicalis*) from near Dixon, Illinois. Subsequently, the Dixon fossil site yielded an unusually diverse and well-preserved sponge fauna that was documented by Ulrich and Everett (1890). These authors described the sponges as occurring 3 miles northeast of Dixon, Illinois, in a clay-rich bed up to 13 cm (5 in.) thick in the Platteville Formation ("Trenton limestone") 7.6 m (25 ft) above the St. Peter Sandstone. Taking into account the changes in stratigraphic nomenclature and classification that have been made to this succession of rocks since the late 1800s, it can be assumed that Ulrich and Everett believed the Platteville Formation rested directly on the St. Peter Sandstone (the upper Glenwood sandstone, shale, and dolomite included). Accordingly, an elevation of 7.6 m (25 ft) above the base of the Platteville would place Ulrich and Everett's sponge bed in the Pecatonica Member 3.1 m (10 ft) below the Mifflin Member. Regarding where Ulrich and Everett collected the sponges, it most likely was in the southern half of sec. 27, T 22 N, R 9 E, the site of mining operation in the Platteville Formation during the late 1800s. Currently, the Pecatonica Member is sparsely exposed in sec. 27 and Ulrich and Everett's (1890) original sponge-bearing rocks have long since been mined out. During the past 100 years, the Pecatonica Member has been mined at many locations in the sprawling quarries north of Dixon, Illinois. Over this period, a small number of isolated lithistid sponges have been discovered in the Pecatonica Member, but Ulrich and Everett's clay-rich bed with abundant lithistid sponges has not been observed. Possible evidence of Ulrich and Everett's sponge bed is suggested by the presence of *Dystactospongia* at an elevation of 5.5 m (18 ft) above the base of the Pecatonica Member in a 2-in. drill core (SM-04) obtained by St Marys Cement Company (SW¼ sec. 16, T 22 N, R 9 E, Grand Detour Quadrangle).

A clay-rich bed containing abundant specimens of the lithistids *Anthaspidella* and *Zittelella* was discovered at the top of the Mifflin Member near Dixon at an elevation of 18.3 m (60 ft) above the base of the Platteville (Willman and Kolata 1978). Sponges also have been observed at the same stratigraphic position in other parts of northern Illinois (Willman and Kolata 1978). This brings to question whether there are two distinct clay-rich sponge beds in the Platteville Formation near Dixon, Illinois, one occurring in the Pecatonica Member 3.1 m (10 ft) below the Mifflin Member and another at the top of the Mifflin.

SYSTEMATIC DESCRIPTIONS

Phylum PORIFERA

Class DEMOSPONGEA

Subclass "LITHISTIDA"

Remarks. The lithistids are a polyphyletic group of demosponges characterized by a solid, rigid, siliceous skeleton consisting of structural spicules or desmas that are articulate. The distinctiveness of the lithistid skeleton and rarity of intermediary forms have resulted in this artificial taxon being almost universally retained.

Order ORCHOCLADINA

Family ANTHASPIDELLIDAE

Description. Saucer-, funnel-, or frond-shaped sponge supported by short subcylindrical stem, ladderlike skeleton composed of vertical trabs produced by union of ray tips of horizontal runglike structures, small canals parallel the trabs.

Remarks. The Platteville Formation anthaspidellid sponge fauna includes *Anthaspidella*, *Zittelella*, *Strotospongia*, and *Patellispongia*. Not included in the following descriptions are Ulrich and Everett's (1890) rare and poorly known *Calathium*? (*Zittelella*) *infelix* and *Strotospongia maculosa* as well as the streptosolenid *Edriospongia basalis*. Middle to Late Ordovician anthaspidellid sponge faunas also are known from Tennessee, Virginia, Quebec, Newfoundland, and the Canadian Arctic (Raymond and Okulitch 1940; Rigby 1967, 1977; Klappa and James 1980; Rigby and Desrochers 1995; Carrera and Sumrall 2020). The overall biogeography of Ordovician sponges is discussed in Carrera and Rigby (1999).

Genus ANTHASPIDELLA

Description. Funnel- or saucer-shaped, attached to substrate by a short stalk, more than one oscule present on upper surface, radially positioned exhalent canals converge on oscules.

Remarks. Nine Platteville species of *Anthaspidella*, including the type species *A. mammulata*, were described by Ulrich and Everett (1890) from Dixon, Illinois. They stated that the internal structures in all eight species were practically identical and that their species identifications were based entirely upon external features. They further suggested, "As these characters are often unreliable among the sponges, we will not be surprised if future investigations prove that we have made more species and, perhaps, even genera than was necessary" (p. 257). Since 1890, numerous additional specimens have been collected from the Platteville Formation at Dixon and other localities in northern Illinois and southern Wisconsin. Examination of the new and old material suggests that there is a significant amount of intraspecific variation in the external features of Ulrich and Everett's (1890) *Anthaspidella* species and that overall size is not a reliable criterion for species recognition. Six of the nine *Anthaspidella* species defined by Ulrich and Everett (1890) are included in the following descriptions and are photographically illustrated for the first time. These six species form natural groupings into which most known Platteville specimens of *Anthaspidella* can be placed. Excluded from this report are *A. scutula*, *A. grandis*, and *A. magnifica* (Appendix 2).

Anthaspidella firma Ulrich and Everett, 1890

Figures 3-1-1, 3-1-3, and 3-1-4

Description. Conical-shaped, shallow depression on upper surface, short stem, oscules on upper surface deeply impressed, irregularly stellate, and relatively numerous. Holotype 92 mm in diameter, depth of upper surface 13 mm, stem diameter 25 mm.

Remarks. Based on a single specimen, differs from other *Anthaspidella* species in having larger and more widely separated canals (Ulrich and Everett 1890).

Occurrence. Present but rare in the lower part of the Grand Detour Member near Dixon, Illinois. Holotype ISGS (ISM) 2650 described by Ulrich and Everett (1890) as occurring in the Platteville Formation 7.6 m (25 ft) above the St. Peter Sandstone near Dixon, Illinois. More recently collected specimens have been reported from a 15 cm K-bentonite-rich (volcanic ash) layer at the top of the Mifflin Member continuing up into the lower 1 m (3.3 ft) of the Grand Detour Member (Willman and Kolata 1978).

Anthaspidella obliqua Ulrich and Everett, 1890

Figure 3-1-2

Description. Funnel-shaped, upper surface depressed and sloping gradually toward center where well-preserved specimens have a group of three or four oscules, each surrounded by 12–16 radiating canals. The figured syntype is 40 mm in diameter, 53 mm high, stem diameter 15 mm, depth of cup 9 mm.

Remarks. Based on two imperfect specimens. Distinguished from *A. fenestrata* by its smaller size, more expanded cup, and different arrangement of the canals and oscules (Ulrich and Everett 1890).

Occurrence. Present but rare in the lower part of the Grand Detour Member near Dixon, Illinois. Syntype ISGS (ISM) 2653 described by Ulrich and Everett (1890) as occurring in the Platteville Formation 7.6 m (25 ft) above the St. Peter Sandstone near Dixon, Illinois. More recently collected specimens have been reported from a 15 cm K-bentonite-rich (volcanic ash) layer at the top of the Mifflin Member continuing up into the lower 1 m (3.3 ft) of the Grand Detour Member (Willman and Kolata 1978).

Anthaspidella parvistellata Ulrich and Everett, 1890

Figure 3-1-5

Description. Dish-shaped, very shallow depression on upper surface, abruptly expanded above short stem. Holotype measures 88 mm in diameter, height of expanded portion of sponge 13 mm.

Remarks. The small size and great number of oscules characterize this species and serve to distinguish it from *A. scutula* and *A. florifera*, "both of which have a somewhat similar form" (Ulrich and Everett 1890, p. 261).

Occurrence. Present but rare in the lower part of the Grand Detour Member near Dixon, Illinois. Holotype ISGS (ISM) 2651 described by Ulrich and Everett (1890) as occurring in the Platteville Formation 7.6 m (25 ft) above the St. Peter Sandstone near Dixon, Illinois. More recently collected specimens have been reported from a 15 cm K-bentonite-rich (volcanic ash) layer at the top of the Mifflin Member continuing up into the lower 1 m (3.3 ft) of the Grand Detour Member (Willman and Kolata 1978).

Anthaspidella florifera Ulrich and Everett, 1890

Figures 3-2-1 and 3-2-2

Description. Saucer-shaped, expanding abruptly above short central stalk, upper surface gently concave with a central oscule surrounded by similar but slightly smaller oscules approximately 12 mm apart. Holotype 37 mm in height, stem 12 mm long, cup diameter 88 mm, depth of cup 10 mm, thickness of wall 12 mm (Ulrich and Everett 1890).

Remarks. *A. florifera* differs from other Platteville species of *Anthaspidella* in having comparatively few oscules and greater regularity of radiating canals around the central oscule (Ulrich and Everett 1890).

Occurrence. Present but rare in the lower part of the Grand Detour Member near Dixon, Illinois. Holotype ISGS (ISM) 2648 described by Ulrich and Everett (1890) as occurring in the Platteville Formation 7.6 m (25 ft) above the St. Peter Sandstone near Dixon, Illinois. More recently collected specimens have been reported from a 15 cm K-bentonite-rich (volcanic ash) layer at the top of the Mifflin Member continuing up into the lower 1 m (3.3 ft) of the Grand Detour Member (Willman and Kolata 1978).

Anthaspidella fenestrata Ulrich and Everett, 1890

Figures 3-2-3 and 3-2-4

Description. Funnel-shaped, relatively deep cup with central oscule from which 10–12 channels radiate, more than 50 smaller oscules distributed over the inner cup, each with 5–10 radiating canals, large subcylindrical stem, sides expand upward with gentle outward curve to rounded and slightly oblique margins of cup. Holotype height about 90 mm, greatest diameter at cup margin 80 mm, stem diameter 20 mm, depth of cup 23 mm.

Remarks. *A. fenestrata* resembles *A. obliqua* but differs from the latter in its larger size, less expanded cup, and much more numerous oscules. *A. fenestrata* also resembles *A. magnifica* but differs from the latter in its smaller size, furrowed outer surface, and generally larger canals and oscules (Ulrich and Everett 1890).

Occurrence. Present but rare in the lower part of the Grand Detour Member near Dixon, Illinois. Holotype ISGS (ISM) 2649 described by Ulrich and Everett (1890) as occurring in the Platteville Formation 7.6 m (25 ft) above the St. Peter Sandstone near Dixon, Illinois. More recently collected specimens have been reported from a 15 cm K-bentonite-rich (volcanic ash) layer at the top of the Mifflin Member

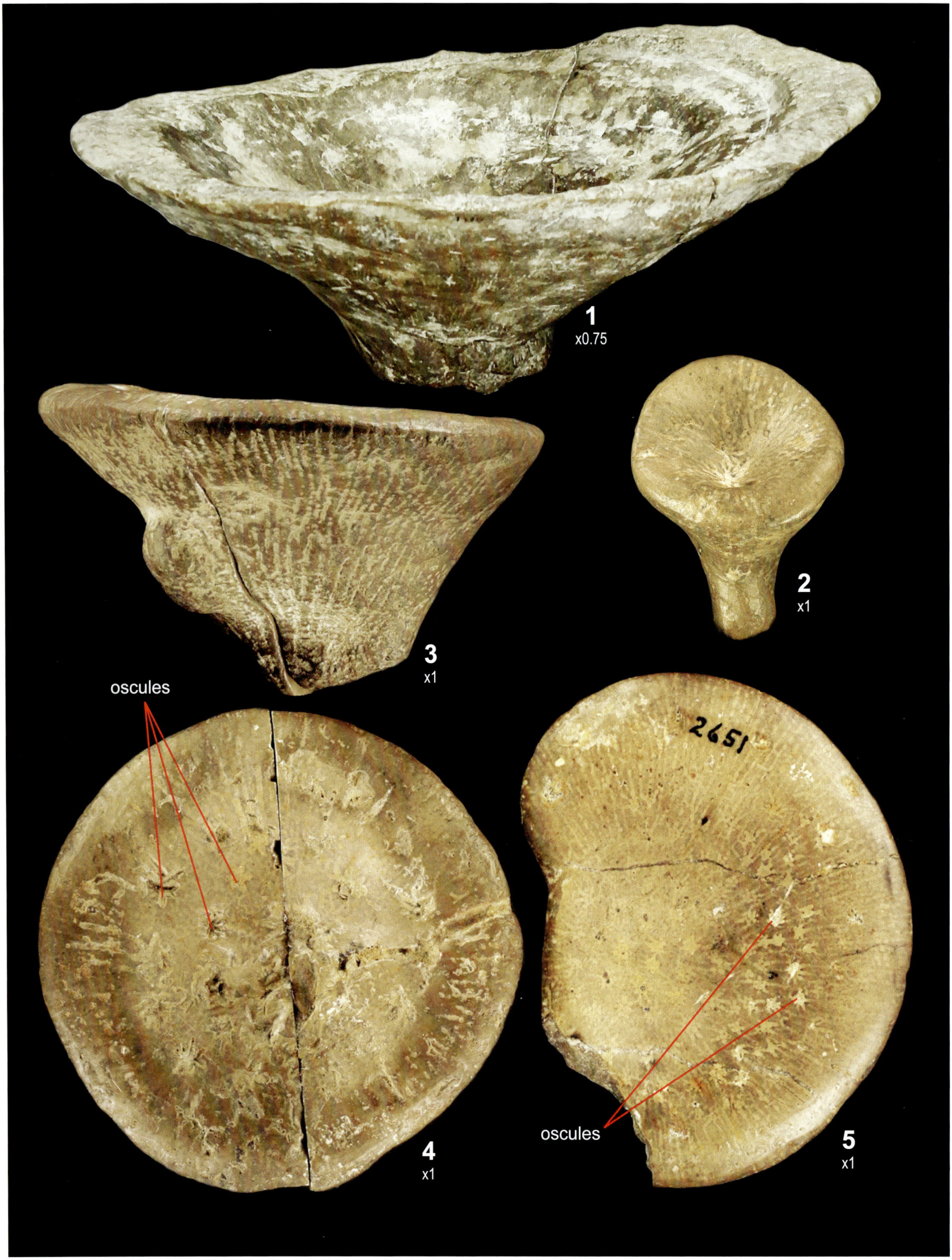
1
x0.75
2
x1
3
x1
oscules
2651
oscules
4
x1
5
x1

continuing up into the lower 1 m (3.3 ft) of the Grand Detour Member (Willman and Kolata 1978).

Anthaspidella mammulata Ulrich and Everett, 1890

Figure 3-2-5 and 3-2-6

Description. Saucer-shaped, attached to substrate by short central stalk, concave upper surface covered by numerous protuberant oscules some with radially arranged exhalent canals, oscules have apertures that are approximately 2.5 mm in diameter and 5–20 mm apart, trabs arranged in rows that radiate from center of sponge. Holotype 100 mm in diameter, 40 mm high, 25 mm deep with walls 7–10 mm thick.

Remarks. The prominent mounded oscules differentiate *A. mammulata* from other Platteville species of *Anthaspidella* (Ulrich and Everett 1890).

Occurrence. Present but rare in the lower part of the Grand Detour Member near Dixon, Illinois. Holotype ISGS (ISM) 2642 was not available during this study. However, the species characteristics are well displayed in topotype ISGS K523. The holotype was described by Ulrich and Everett (1890) as occurring in the Platteville Formation 7.6 m (25 ft) above the St. Peter Sandstone near Dixon, Illinois. More recently collected specimens have been reported from a 15 cm K-bentonite-rich (volcanic ash) layer at the top of the Mifflin Member continuing up into the lower 1 m (3.3 ft) of the Grand Detour Member (Willman and Kolata 1978).

Genus ZITTELELLA

Description. Conical or flattened shape with a prominent stem, upper surface slightly convex to deeply concave, numerous exhalent canals converge on a single centrally located oscule, thus differing from *Anthaspidella*.

Figure 3-1 (Opposite page)
1, 3, 4. ***Anthaspidella firma*** Ulrich and Everett. (**1**) Side view of large, conical specimen; (**3**) side view of conical specimen showing large, widely separated inhalant canals; (**4**) upper surface showing deeply depressed oscules. 7.6 m (25 ft) above base of Platteville Formation, Dixon, Illinois, **1**, ISGS 11P-1; **3**, **4**, holotype ISGS (ISM) 2650.

2. ***Anthaspidella obliqua*** Ulrich and Everett. (**2**) Side view showing characteristic oblique angle of gastral (exhalant) surface. 7.6 m (25 ft) above base of Platteville Formation, Dixon, Illinois, **2**, Syntype ISGS (ISM) 2653.

5. ***Anthaspidella parvistellata*** Ulrich and Everett. (**5**) Upper surface (gastral/exhalant) showing numerous small stellate oscules. 7.6 m (25 ft) above base of Platteville Formation, Dixon, Illinois, **5**, Holotype ISGS (ISM) 2651.

Zittelella typicalis Ulrich and Everett, 1890

Figures 3-3-1 through 3-3-4

Description. Funnel-shaped, with relatively long stem, upper surface slightly depressed, with radiating canals that are laterally joined. The figured syntype height 44 mm and 60 mm in diameter; stem 18 mm long and 20 mm in diameter.

Remarks. Three varieties of *Z. typicalis* were described and illustrated by Ulrich and Everett (1890): *pistilliformis*, *turbinata*, and *subrotunda*.

Occurrence. Syntypes ISGS (ISM) 2640-1/2 described by Ulrich and Everett (1890) as occurring in the Platteville Formation 7.6 m (25 ft) above the St. Peter Sandstone near Dixon, Illinois. More recently collected specimens have been reported from a 15 cm K-bentonite-rich (volcanic ash) layer at the top of the Mifflin Member continuing up into the lower 1 m (3.3 ft) of the Grand Detour Member (Willman and Kolata 1978).

Zittelella inosculata Ulrich and Everett, 1890

Figure 3-3-5

Description. Pestle-shaped, relatively large stem, margin between stem and convex upper surface narrowly rounded, sides and top of sponge marked by irregular inosculate channels, center of upper surface abruptly depressed and stellate. The figured syntype height 33 mm, upper surface 27 mm in diameter, stem 14–18 mm in diameter.

Remarks. *Z. inosculata* is distinguished from other species of *Zittelella* in having wavy inosculating surface channels separated by wide interspaces (Ulrich and Everett 1890).

Occurrence. Present but rare in the lower part of the Grand Detour Member near Dixon, Illinois. Syntype ISGS (ISM) 2656 was described by Ulrich and Everett (1890) as occurring in the Platteville Formation 7.6 m (25 ft) above the St. Peter Sandstone near Dixon, Illinois. More recently collected specimens have been reported from a 15 cm K-bentonite-rich (volcanic ash) layer at the top of the Mifflin Member continuing up into the lower 1 m (3.3 ft) of the Grand Detour Member (Willman and Kolata 1978).

Zittelella lobata Ulrich and Everett, 1890

Figure 3-3-6

Description. Low-profile conical form, upper surface flat to slightly convex, expands abruptly above the stem, radiating channels on upper surface with an average of seven within 10 mm, outer surface irregular and marked by prominent protrusions. Figured syntype height above stem 30 mm, greatest diameter 90 mm, stem diameter 15–20 mm.

Remarks. *Z. lobata* is distinguished from other species of *Zittelella* by the presence of lobes and protrusions along the margins and on the lower surface (Ulrich and Everett 1890).

Occurrence. Present but rare in the lower part of the Grand Detour Member near Dixon, Illinois. Syntype ISGS (ISM) 2647 was described by Ulrich and Everett (1890) as occurring in the Platteville Formation 7.6 m (25 ft) above the St. Peter Sandstone near Dixon, Illinois. More recently collected specimens have been reported from a 15 cm K-bentonite-rich (volcanic ash) layer at the top of the Mifflin Member continuing up into the lower 1 m (3.3 ft) of the Grand Detour Member (Willman and Kolata 1978).

Genus PATELLISPONGIA

Description. Explanate fronds, upper surface bearing numerous radial rows of small, closely spaced exhalant pores without conspicuous exhalant canals convergent on them, thus differing from *Anthaspidella* (Finks and Rigby 2004).

Patellispongia cf. *oculata* Bassler, 1927

Figures 3-3-7 through 3-3-9

Description. Broad, flat, frond- or disc-shaped sponge, skeletal structure typical of the anthaspidellids, trabs are mostly 2 mm apart and vertical with typical ladderlike structure, numerous round openings mostly 1 mm in diameter situated between trabs on both surfaces.

Figured specimen height 120 mm, 60 mm wide, 10 mm thick.

Remarks. The Platteville specimens are comparable in overall shape and skeletal structure to *Patellispongia oculata* of Bassler (1927).

Occurrence. Present but rare in the lower 1 m (3.3 ft) of the Grand Detour Member at Dixon, Illinois (locality 4).

Family STREPTOSOLENIDAE

Description. Discoidal, palmate, funnel, branched or irregular, some with surficial nodes and ridges; canal systems complex and intertwining.

Remarks. The Platteville Formation streptosolenid sponges include *Streptosolen* and *Edriospongia*. The latter is a poorly known Platteville sponge and is not included in the following descriptions.

Genus STREPTOSOLEN

Remarks. *Streptosolen* differs from *Anthaspidella*, *Zittelella*, and *Patellispongia* in having an irregular array of canals that pass through the sponge in every direction.

Streptosolen obconicus Ulrich and Everett, 1890

Figures 3-4-1 and 3-4-2

Description. Broad, asymmetrical, conical-shaped sponge with shallow concave upper surface; short, irregular discontinuous trabs; axial cluster of vertical exhalant canals connected to sinuous branching tributary canals, some of which are subradial, branched and visible on the upper surface; outer surface marked by similar sinuous, branching canals without definite orientation (Finks and Rigby 2004). Holotype 75 mm in height, 60 mm wide, stem diameter 17 mm by 24 mm (Ulrich and Everett 1890).

Remarks. *Streptosolen obconicus* is the type species of the genus.

Occurrence. Present but rare in the lower part of the Grand Detour Member near Dixon, Illinois. Syntype ISGS (ISM) 2660 described by Ulrich and Everett (1890) as occurring in the Platteville Formation 7.6 m (25 ft) above the St. Peter Sandstone near Dixon, Illinois. More recently collected specimens have been reported from a 15 cm K-bentonite-rich (volcanic ash) layer at the top of the Mifflin Member continuing up into the lower 1 m (3.3 ft) of the Grand Detour Member (Willman and Kolata 1978).

Figure 3-2 (Opposite page)
1, 2. ***Anthaspidella florifera*** Ulrich and Everett. (**1**) Side view showing broad conical shape and inhalant canals; (**2**) upper or exhalant surface showing large central oscule surrounded by slightly smaller oscules. 7.6 m (25 ft) above base of Platteville Formation, Dixon, Illinois, **1**, **2**, Holotype ISGS (ISM) 2648.

3, 4. ***Anthaspidella fenestrata*** Ulrich and Everett. (**3**) Side view showing part of upper or exhalant surface; (**4**) upper or exhalant surface showing bell shape of margins and stellate oscules. 7.6 m (25 ft) above base of Platteville Formation, Dixon, Illinois, **3**, **4**, Holotype ISGS (ISM) 2649.

5, 6. ***Anthaspidella mammulata*** Ulrich and Everett. (**5**) Upper or exhalant surface showing numerous small oscules surrounded by slightly raised rims; (**6**) cross section showing typical anthaspidellid skeletal structure with interconnected desmas and current canals. Sponge bed at top of Mifflin Member, Dixon, Illinois (locality 4), **5**, **6**, ISGS K523.

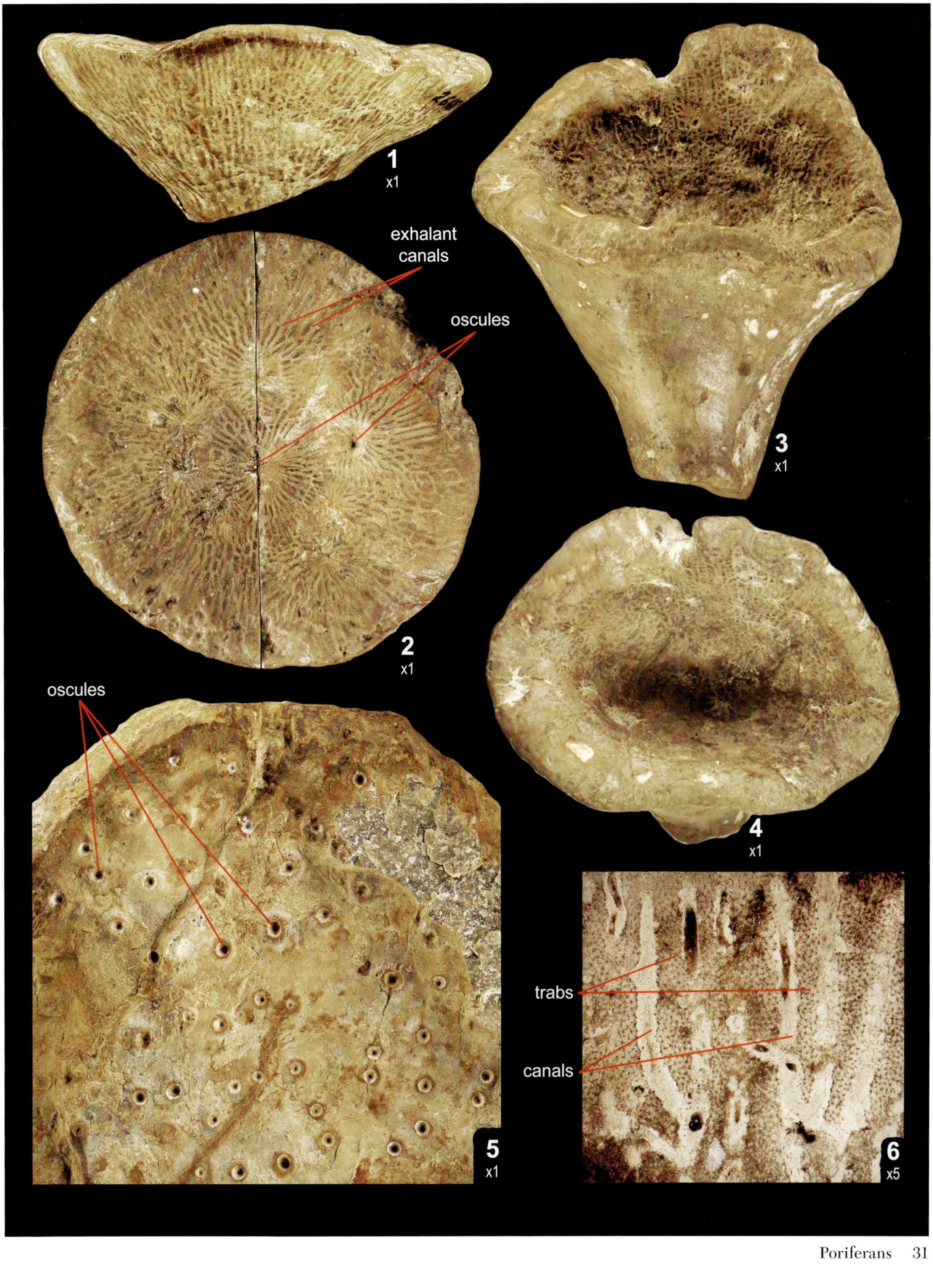
1
x1
exhalant
canals
oscules
2
x1
3
x1
4
x1
oscules
5
x1
trabs
canals
6
x5

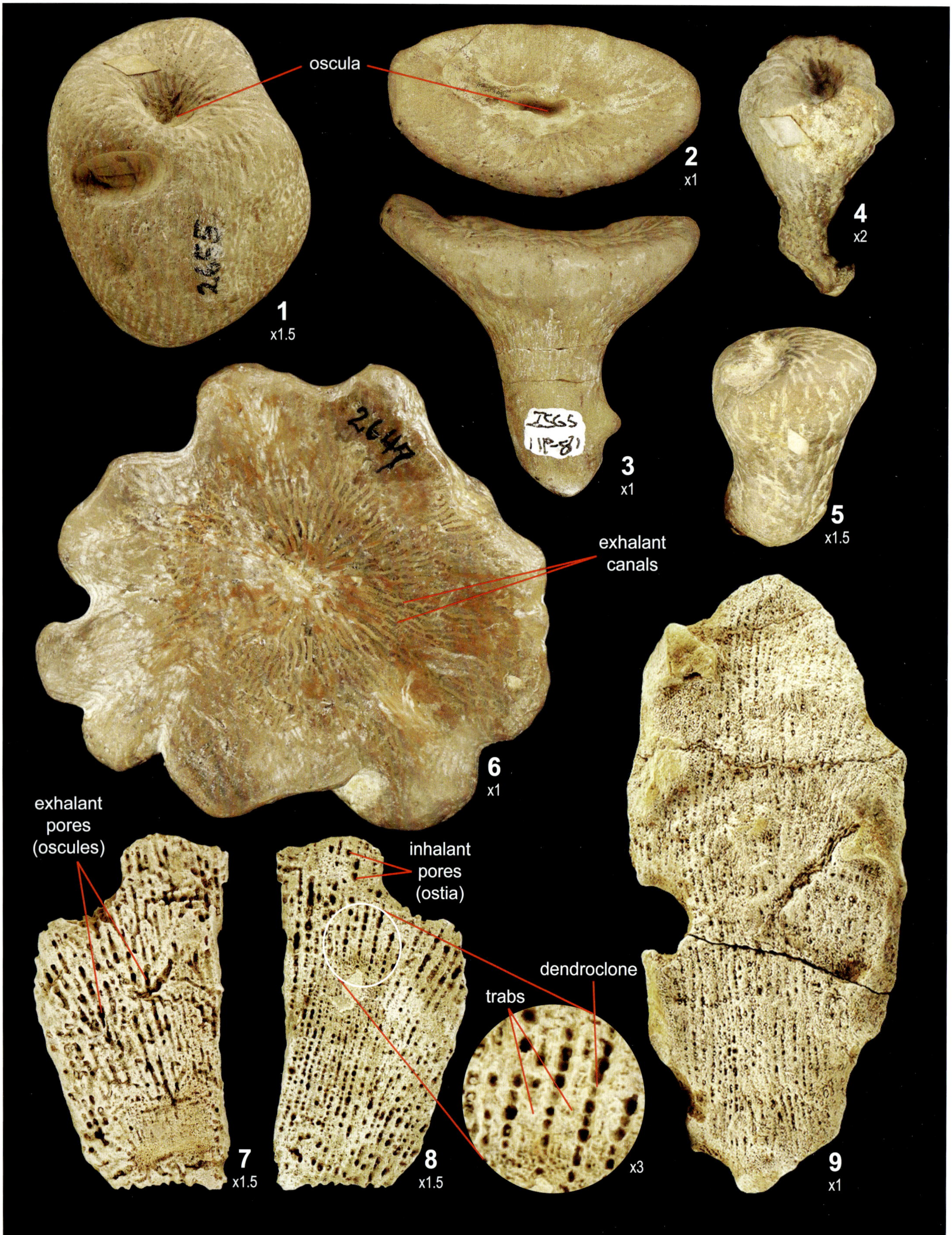
oscula
1
x1.5
2
x1
3
x1
4
x2
5
x1.5
exhalant
canals
6
x1
exhalant
pores
(oscules)
inhalant
pores
(ostia)
dendroclone
trabs
7
x1.5
8
x1.5
x3
9
x1

Order STREPTOSCLEROPHORIDA

Family HINDIIDAE

Genus HINDIA

Hindia inaequalis Ulrich and Everett, 1890

Figures 3-4-3 and 3-4-4

Description. Spherical free-living sponge, skeleton consists of three-armed tricranoclones forming stacked series that radiate from the center to outer surface, small (inhalant) and somewhat larger (exhalant) canals positioned between radiating skeletal framework. Holotype ISGS (ISM) 2661 is 13 mm in diameter. Other Platteville specimens range from 5 to 20 mm in diameter.

Occurrence. Common in the Mifflin and Grand Detour Members in northern Illinois and southern Wisconsin and the McGregor Member of southwestern Wisconsin, northeastern Iowa, and southeastern Minnesota. Holotype ISGS (ISM) 2661 described by Ulrich and Everett (1890) as occurring in the Platteville Formation 7.6 m (25 ft) above the St. Peter Sandstone near Dixon, Illinois.

Figure 3-3 (Opposite page)

1. ***Zittelella typicalis subrotunda*** Ulrich and Everett. (**1**) Side view showing single depressed oscule at top. 7.6 m (25 ft) above base of Platteville Formation, Dixon, Illinois, **1**, holotype ISGS (ISM) 2655.

2–4. ***Zittelella typicalis pistilliformis*** Ulrich and Everett. (**2**) Upper or exhalant surface showing single central oscula; (**3**) side view of the same specimen showing flared out upper margins; (**4**) small specimen with long, slender stem. 7.6 m (25 ft) above base of Platteville Formation, Dixon, Illinois, **2**, **3**, ISGS 11P-8; **4**, syntype ISGS (ISM) 2657.

5. ***Zittelella inosculata*** Ulrich and Everett. (**5**) Side view showing inosculated canals. 7.6 m (25 ft) above base of Platteville Formation, Dixon, Illinois, **5**, Syntype ISGS (ISM) 2656.

6. ***Zittelella lobata*** Ulrich and Everett. (**6**) Top or exhalant surface showing large central oscule, radiating canals and lobate margins. 7.6 m (25 ft) above base of Platteville Formation, Dixon, Illinois, **6**, Syntype ISGS (ISM) 2647.

7–9. ***Patellispongia* cf. *oculata*** Bassler. (**7**) Exhalant surface showing numerous oscules; (**8**) inhalant surface showing round ostia, vertical trabs, and horizontal dendroclones; (**9**) inhalant surface of broad, flat, frond-shaped skeleton. Sponge bed at top of Mifflin Member, Dixon, Illinois (locality 4), **7**, **8**, ISGS K 548; **9**, ISGS K 521.

Subclass CERACTINOMORPHA

Order SIGMATOSCLEROPHORIDA

Family DYSTACTOSPONGIIDAE

Genus DYSTACTOSPONGIA

Description. Massive, tuberous, fibrous skeleton; surface covered with circular, polygonal, or submeandriform pores of varying sizes.

Dystactospongia minor Ulrich and Everett, 1890

Figures 3-4-7 through 3-4-10

Description. Massive lobate sponge, skeleton replaced entirely with clear sparite, forming a network of tubes 0.5–0.8 mm in diameter that radiate from the base to upper surface and are laterally interconnected, microskeletal structures are poorly preserved, inhalant and exhalant canals filled with very fine limestone matrix. The calcite-filled tubes and associated canals form a distinctive labyrinth-like network.

Remarks. Differs from *D. rudis* in having smaller canals and from *D. insolens* in lacking large radiate oscules.

Occurrence. Present but rare in the lower part of the Grand Detour Member near Dixon, Illinois. Syntype ISGS (ISM) 2667 described by Ulrich and Everett (1890) as occurring in the Platteville Formation 7.6 m (25 ft) above the St. Peter Sandstone near Dixon, Illinois. More recently collected specimens have been reported from a 15 cm K-bentonite-rich (volcanic ash) layer at the top of the Mifflin Member continuing up into the lower 1 m (3.3 ft) of the Grand Detour Member (Willman and Kolata 1978).

Dystactospongia rudis Ulrich and Everett, 1890

(not figured)

Description. Massive, oval-shaped, skeleton replaced with sparite, forming a network of ill-defined latticelike structures and pores. Holotype 89 mm long, 64 mm wide, and 40 mm high.

Remarks. The single poorly preserved specimen differs from *D. minor* in having larger canals. The holotype lacks sufficient structural detail to warrant a photographic illustration.

Occurrence. Present but rare in the lower part of the Grand Detour Member near Dixon, Illinois. Holotype ISGS (ISM) 2646 described by Ulrich and Everett (1890) as occurring in the Platteville Formation 7.6 m (25 ft) above the St. Peter Sandstone near Dixon, Illinois.

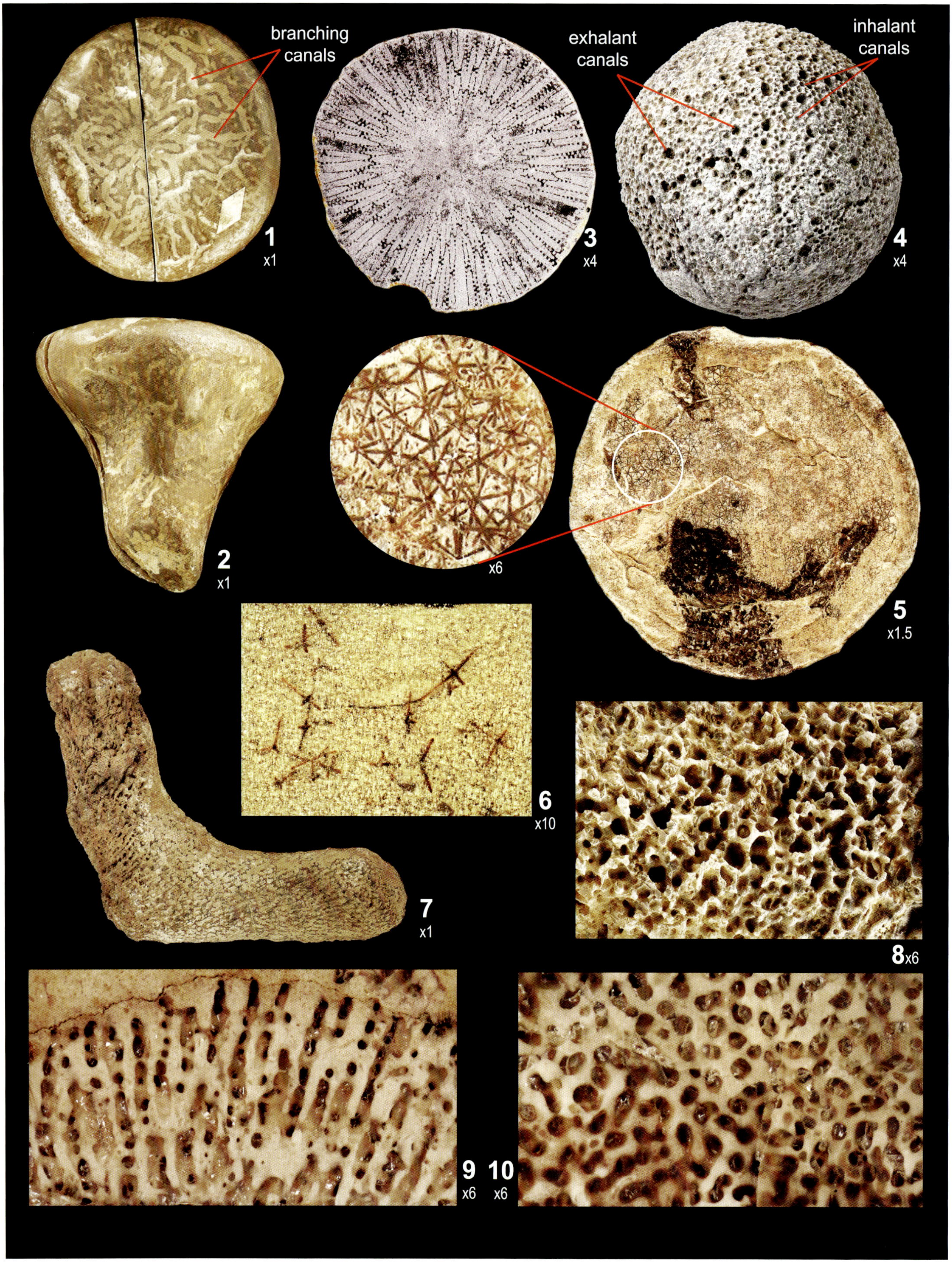
branching canals
exhalant canals
inhalant canals
1
x1
3
x4
4
x4
2
x1
x6
5
x1.5
6
x10
7
x1
8x6
9
x6
10
x6

Class HETERACTINIDA

Order OCTACTINELLIDA

Family ASTRAEOSPONGIIDAE

Genus ASTRAEOSPONGIA

Astraeospongia dixonensis Gutschick, 1954a

Figure 3-4-6

Description. Gutschick (1954a) based the species on 11 detached octactinellid spicules found in acetic acid residues. The form of the sponge is unknown. The spicules are mostly less than 1 mm across and have six horizontal rays that may or may not have a vertical ray oriented normal to the plane of the disc and the other six rays.

Remarks. On the basis of size and shape of the spicules, Gutschick (1954a) suggested that *Astraeospongia dixonensis* is very closely comparable, if not conspecific with, *A. patina* Roemer, 1861. *A. dixonensis* differs from other species in having slender spicular rays.

Occurrence. *A. dixonensis* type specimens were collected from the upper part of the Mifflin Member near the center of the N½ sec. 27, T22N, R9E, in the rock quarry of the Medusa Portland Cement Company about 3 miles northeast of the bridge (U.S. Highway 52) over the Rock River, Dixon, Lee County, Illinois. Subsequently, a small cluster of suspected spicules of *A. dixonensis* was found in the Mifflin Member at the Dixon cement works nearby at locality 4.

Figure 3-4 (Opposite page)

1, 2. ***Streptosolen obconicus*** Ulrich and Everett. (**1**) Top or exhalant surface showing the characteristic branching tributary canals; (**2**) side view showing the conical-shaped skeleton with expanding upper margins. 7.6 m (25 ft) above base of Platteville Formation, Dixon, Illinois, **1**, **2**, Syntype ISGS (ISM) 2660.

3, 4. ***Hindia inaequalis*** Ulrich and Everett. (**3**) Acetate peel of cross section through center of skeleton showing stacked tricranoclones that radiate from the center; (**4**) complete specimen showing small (inhalant) and somewhat larger (exhalant) canals positioned between radiating skeletal framework. 7.6 m (25 ft) above base of Platteville Formation, Dixon, Illinois, **3**, Holotype ISGS (ISM) 2661. Mifflin Member, Dixon, Illinois (locality 2), **4**, ISGS 1546.

5. ***Astraeospongia* cf. *dixonensis*** Gutschick. (**5**) Shallow saucer-shaped skeleton; photographic enlargement showing the characteristic spicules with six equally spaced rays. Quimbys Mill Member, found near the Platteville Formation type section southwest of Platteville, Wisconsin (locality 30), **5**, UWGM 1463.56.

6. ***Astraeospongia dixonensis*** Gutschick. (**6**) Cluster of spicules. Mifflin Member, Dixon, Illinois (locality 4). **6**, ISGS K597.

7–10. ***Dystactospongia minor*** Ulrich and Everett. (**7**) Specimen showing typical tuberous lobate skeleton; (**8**) photographic enlargement of surface showing polygonal or submeandriform pores of varying sizes; (**9**) cross section showing sparite-filled skeletal framework; (**10**) tangential section showing skeletal framework surrounded by lighter colored matrix. Lower 1 m (3.3 ft) of Grand Detour Member at Dixon, Illinois (locality 4), **7**, **8**, ISGS 368; **9**, ISGS 372; **10**, ISGS 371.

Astraeospongia* cf. *dixonensis Gutschick, 1954a

Figure 3-4-5

Description. Shallow saucer-shaped sponge, body wall thin and packed with irregularly arranged spicules that have six equally spaced horizontal rays, spicules are mostly 1.2 mm across. It is uncertain whether vertical rays are present.

Remarks. This Platteville sponge is represented by a single nearly complete specimen. The specimen label states that identification and description are by J. Keith Rigby. The overall size and shape of the spicules are nearly identical to those of *A. dixonensis*.

Occurrence. As indicated on the museum label, the single known specimen (UWGM 1463.56) was found in a "stream bank five miles south-southwest of Platteville, Grant County, Wisconsin." In this area, the Platteville Formation occurs at the bedrock surface. The matrix surrounding the sponge consists of lithographic lime mudstone that is medium to dark brown in color. Lithologically, the unique matrix is the same as that occurring in the Quimbys Mill Member at the nearby type section of the Platteville Formation (locality 30).

HYPERCALCIFIED SPONGES

Remarks. Hypercalcified sponges are a polyphyletic group having a rigid calcified skeleton that may or may not have a complementary spicular skeleton. This informal group includes living hypercalcified members of the classes Demospongiae and Calcarea as well as a rich Paleozoic fossil record of morphological types or grades represented by the archaeocyathids, stromatoporoids, chaetetids, sphinctozoans, and iozoans. Hypercalcified sponges in the Platteville Formation include stromatoporoids and chaetetids.

Class STROMATOPOROIDEA

Description. Stromatoporoids grew by secreting thin, calcareous layers called **laminae** parallel to the substrate with rodlike **pillars** perpendicular to the laminae. Living tissue occurred primarily at the upper surface of the skeleton.

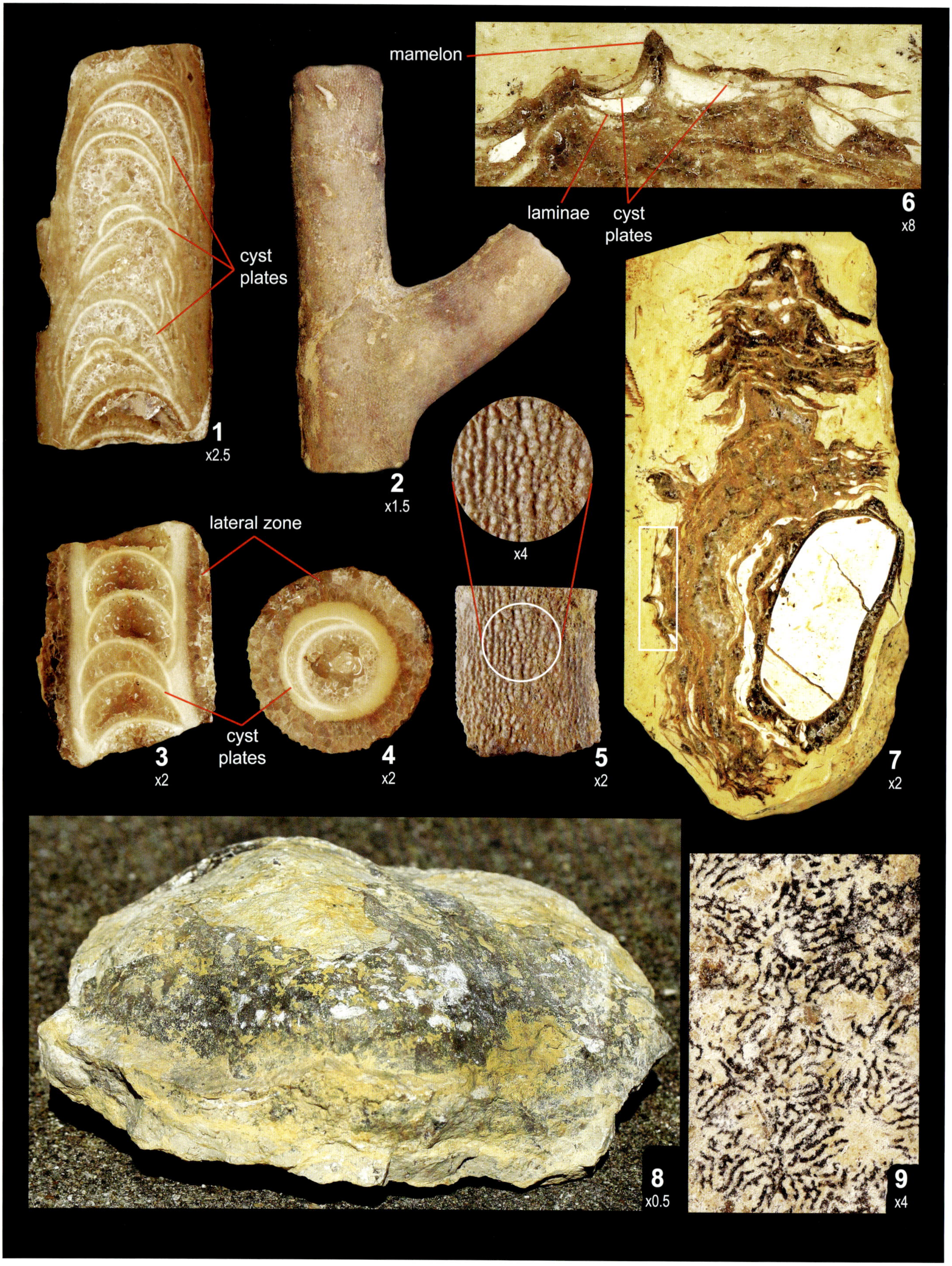
mamelon
laminae
cyst
plates
6
x8
cyst
plates
1
x2.5
2
x1.5
x4
lateral zone
cyst
plates
3
x2
4
x2
5
x2
7
x2
8
x0.5
9
x4

Round or irregular elevations on the upper surface called mamelons apparently were the sites of excurrent openings. Associated with the **mamelons** are a network of canals called **astrorhizae**.

Order LABECHIIDA

Family AULACERATIDAE

Description. Branched or unbranched columnar skeletons with well-differentiated axial and lateral zones; axial zone consists of large stacked or overlapping cyst plates; lateral zone with small compressed cysts plates (Webby 2015).

Genus THAMNOBEATRICEA

Thamnobeatricea sp.

Figures 3-5-1 through 3-5-5

Description. Branched or unbranched cylindrical skeleton with axial column composed of large, stacked horseshoe-shaped cyst plates (in longitudinal section) that occupy more than half of the overall diameter of the column, lateral zone composed of an inner layer of small cyst plates and faint remains of pillars; outer surface smooth or marked by longitudinal rows of small tubercles; largest specimen is 15 mm in diameter and more than 300 mm in length with only two bifurcations.

Remarks. This hypercalcified sponge is most like the aulaceratid *Thamnobeatricea* (Raymond 1931) in having a branching skeleton with wide axial column, large horseshoe-shaped cyst plates, and a narrow lateral zone. It differs from the Ordovician genus *Cryptophragmus* in its branching form and relatively wide axial zone.

Occurrence. Common in the Pecatonica and Grand Detour Members at Dixon, Illinois (locality 2).

Family STYLOSTROMATIDAE

Genus PACHYSTYLOSTROMA

Pachystylostroma cf. *ungerni* Rosen, 1867

Figures 3-5-6 and 3-5-7

Description. Skeleton strongly mamelonate, cysts of variable sizes, cyst plates alternating between thickened wavy laminae, small denticles present on laminae.

Remarks. The Platteville species is most similar to *Pachystylostroma ungerni* Rosen, 1867, from the Lower Silurian Hilliste Formation, Hiiumaa Island, Estonia.

Occurrence. Rare in the lower 5 m of the Grand Detour Member at Dixon, Illinois (locality 4).

Family STROMATOCERIIDAE

Genus STROMATOCERIUM

Stromatocerium rugosum Hall, 1847

Figures 3-5-8 and 3-5-9

Description. Large dome-shaped stromatoporoid, astrorhizae consist of radial branching ridges that join to form a stellate pattern on the terminal growth surface; centers spaced 5–10 mm apart; lacks denticles and mamelon columns. Specimen ISGS 478 is 30 cm wide and 12 cm high.

Remarks. Most interior skeletal structures have been destroyed by recrystallization in the single available specimen. The domal shape of this Platteville stromatoporoid, the vermicular pattern of the pillars on the outer surface, and the lack of denticles or mamelon columns are nearly similar to *S. rugosum* from the Upper Ordovician Black River Group of Watertown, New York (Hall 1847).

Occurrence. Rare in the lower 2 m (6.6 ft) of the Grand Detour Member at Dixon, Illinois (locality 4).

HYPERCALCIFIED DEMOSPONGES WITH A CHAETETID SKELETON

Description. Chaetetids are laminar or dome-shaped sponges having a honeycomb structure consisting of microscopic hexagonal tubules partitioned by thin, planar, horizontal layers of skeletal material called tabulae. The rigid

Figure 3-5 (Opposite page)
1–5. ***Thamnobeatricea* sp.** (**1**) Longitudinal section showing typical upward arched cyst plates at center and relatively narrow lateral zone consisting of very small overlapping cyst plates; (**2**) bifurcated specimen with smooth outer surface; (**3**) longitudinal section showing arched cyst plates and relatively large lateral zone; (**4**) cross section showing central cyst plates and surrounding lateral zone; (**5**) outer surface showing longitudinal rows of small tubercles in photographic enlargement. Pecatonica Member, Dixon, Illinois (locality 2), **1**, ISGS 542; **2**, ISGS 505; **3**, ISGS 504; **4**, **5**, ISGS 506.

6, 7. ***Pachystylostroma* cf. *ungerni*** Rosen. (**6**) Photographic enlargement showing characteristic prominent mamelon, laminae, and cyst plates; (**7**) encrustation on gastropod *Trochonema* shell; white rectangular box shows position of Figure 6 enlargement. Grand Detour Member, Dixon, Illinois (locality 4), **6**, **7**, ISGS 469.

8, 9. ***Stromatocerium rugosum*** Hall. (**8**) Mound-shaped mass consisting of alternating layers of skeletal laminae and limestone matrix; (**9**) photographic enlargement showing typical vermicular shape of pillars on outer surface (dark colored). Lower part of Grand Detour Member, Dixon, Illinois (locality 4), **8**, **9**, ISGS 478.

skeleton is calcareous. The living tissue forms a thin layer at the surface and in the outer parts of the tubules.

Remarks. Chaetetids possess a relatively simple skeleton. The lack of unique skeletal features has resulted in a long and varied history of attempts to classify them. They have been assigned to the sponges, corals, bryozoans, foraminifera, and algae. The discovery of tabulae in the tubules of the living chaetetid *Acanthochaetetes wellsi* (Hartman and Goreau, 1975) has strengthened the idea that fossil chaetetids have a close affinity with the sponges. Likewise, a sponge relationship has been suggested recently by the discovery of spicules in some Carboniferous chaetetids (Gray 1980) and astrorhizae in Carboniferous (West and Clark 1983) and Mesozoic (Cuif et al. 1973) chaetetids.

Order CHAETETIDA

Genus SOLENOPORA

Remarks. The type species of *Solenopora* was collected from a glacial erratic in northern Estonia and subsequently interpreted as originating from the Upper Ordovician Saku Member, Vasalemma Formation. It was originally described as a chaetetid (Dybowski 1877) but was reinterpreted by Brown (1894a, 1894b) as red algae, an idea that persisted for many years. As pointed out recently by Riding (2004), the internal micromorphology of *Solenopora* is not characteristic of calcified red algae but is consistent with that of a chaetetid. Accordingly, being a chaetetid, *Solenopora* is herein assigned to the sponges.

Solenopora* cf. *spongioides Dybowski, 1877

Figure 3-6-1 through 3-6-4

Description. Skeleton consists of lobate and somewhat flattened nodules mostly less than 25 mm in diameter; free-living or attached to molluscan shells; internal structure consists of narrow polygonal (some hexagonal) to slightly rounded tubules in transverse section ranging from 0.10 to 0.16 mm in diameter; shared walls between adjacent tubules; tubules straight or gently curved longitudinally and are arranged perpendicular to the surface of the skeleton, resulting in a characteristic honeycomb pattern; tubules may be partitioned horizontally at different levels by poorly developed tabulae that may occur at the same or different levels as those in adjacent tubules.

Remarks. The Platteville species has a skeleton very much like the type species *Solenopora spongioides* Dybowski, 1877.

Occurrence. Common in the Mifflin and Grand Detour Members in northern Illinois and southern Wisconsin and equivalent strata of the McGregor Member of southwestern Wisconsin, northeastern Iowa, and southeastern Minnesota.

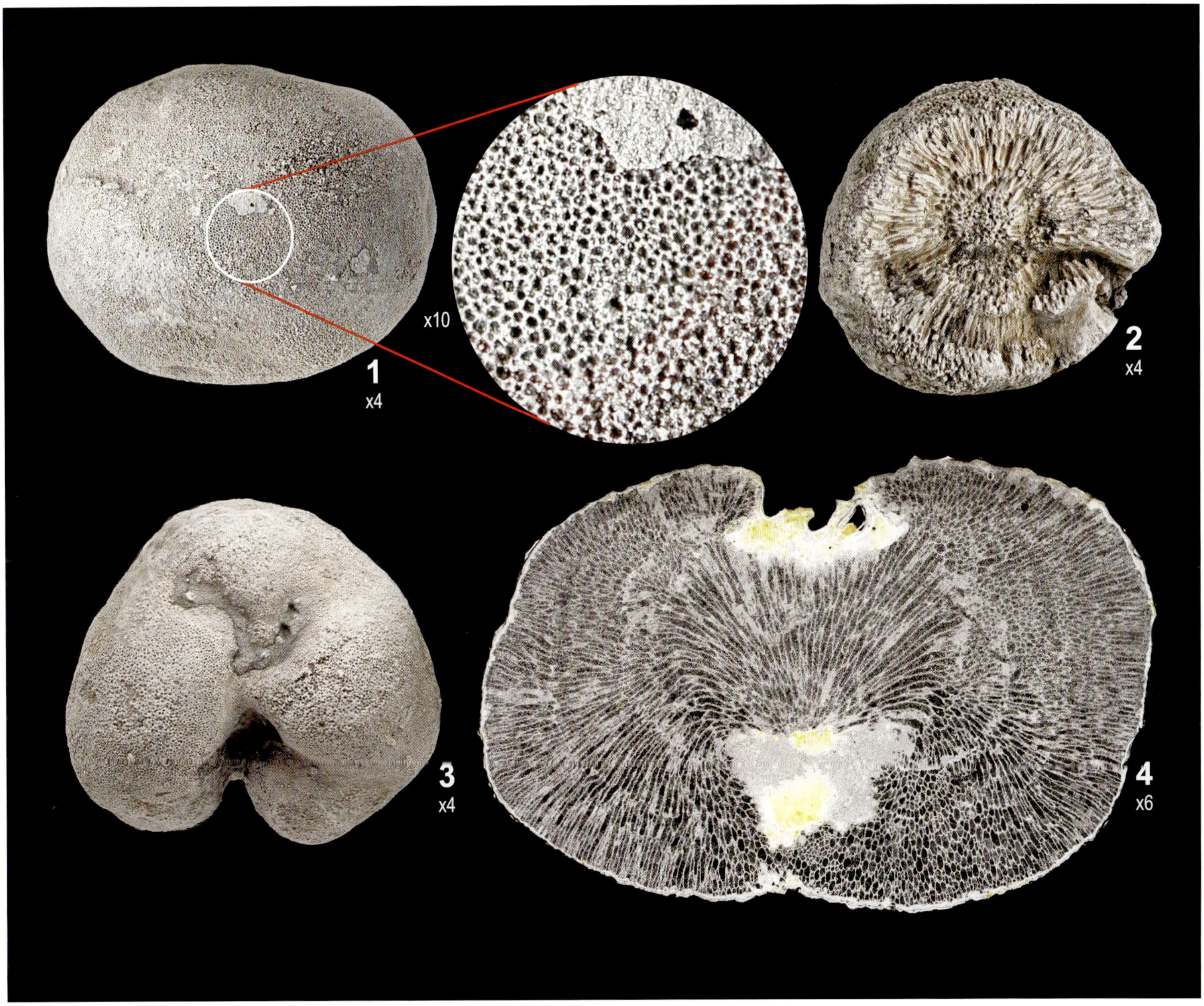

Figure 3-6

1–4. ***Solenopora* cf. *spongioides*** Dybowski. (**1**) Small nodular colony, photographic enlargement showing small surface apertures; (**2**) lower surface of nodular colony showing radiating tubules; (**3**) typical invaginated colony; (**4**) acetate peel of a nodular colony showing radiating tubules. Mifflin Member, Dixon, Illinois (locality 2), **1**, ISGS 1550; **2**, ISGS 1549; **3**, ISGS 1548; **4**, ISGS 683.

CHAPTER 4
CNIDARIANS (CORALS, TETRADIIDS, CONULARIIDS)

The Platteville Formation contains a moderately diverse cnidarian fauna, including rugose and tabulate corals, tetradiids, and conulariids (Appendix 2). Preservation ranges from relatively poor in the dolostone facies to excellent in the limestone. The overall cnidarian fauna is typical of that occurring in the Late Ordovician (Turinian Stage) carbonate-dominated succession of eastern North America.

The cnidarians are simple animals that have existed from Late Precambrian to the present and include corals, hydrozoans, anemones, and jellyfish. They are characterized by (1) radial symmetry, (2) a digestive cavity with the mouth being the only orifice, (3) a lack of highly developed organ systems, (4) the presence of tentacles with cnidocytes (stinging cells), and (5) in the case of corals, a calcareous skeleton originally consisting of calcite or aragonite. All are aquatic and most are marine.

Tabulate and rugose corals range in age from the Ordovician to Permian and are believed to have had life cycles that began with microscopic larvae settling on the seafloor and developing into a soft body **polyp** that began to secrete a calcareous exoskeleton called a **corallum** (pl. **coralla**). The polyp gradually formed an outer wall as well as vertically oriented **septa** (sing. **septum**) that radiated from the inner wall toward the center. Symmetrically positioned spaces between the septa are called **cardinal** and **alar fossulae**. In some rugose corals, the septa met in the center of the corallum, forming a dome-shaped structure called the **columella**. The polyp rested in a depression called the **calice** at the top of the corallum. A **solitary coral** consisted of a single polyp and its corallum, whereas the more common **colonial corals** consisted of multiple polyps that formed the corallum. A **corallite** is the skeleton formed from a single polyp. Colonial corals formed by the budding of polyps within or at the margins of the colony. As the skeleton grew upward, the polyp periodically secreted, on its lower surface, a calcareous platform called a **tabula** (pl. **tabulae**). In some rugose corals, curved plates called **dissepiments** connected to the septa and tabulae. The outer surface of the exoskeleton is called the **epitheca**. A colonial coral is described as **phaceloid** if the walls are tall, tubular, and subparallel and **cerioid** if the corallites share a common wall. Some cnidarians, such as living hydrozoans, are polymorphic, having either a hollow, branching, sessile polyp or a tentacled, free-swimming **medusa** form.

Papers by Hill (1981) and Scrutton (1997, 1998) are excellent sources of information on the structure and classification of Paleozoic corals. Likewise, helpful overviews of Ordovician corals are presented in Webby et al. (2004a) and in descriptive papers by Winchell and Schuchert (1895a), Bassler (1950), Browne (1965), and Elias (1983).

SYSTEMATIC DESCRIPTIONS

Phylum CNIDARIA

Class ANTHOZOA

Subclass ZOANTHARIA

Order RUGOSA

Remarks. The Rugosa are an extinct order of solitary and colonial corals that were abundant in Middle Ordovician to Late Permian seas. They possess septa, tabulae, dissepiments, and in some cases a columella. Although fundamentally radially symmetrical, some rugose corals display a degree of bilateral symmetry as determined by orientation of the septa in a transverse section of a corallite. Solitary rugosans are commonly referred to as horn corals because of their horn-shaped form with a wrinkled, or rugose, outer wall.

Suborder STREPTELASMATINA

Family LAMBELASMATIDAE

Genus LAMBEOPHYLLUM

Lambeophyllum profundum (Conrad, 1843)

Figures 4-1-1 through 4-1-5

Description. Solitary rugose coral, conical except for tendency to be curved in early ontogenetic stages, maximum observed length and diameter 40 mm and 28 mm, respectively. Average depth of calice in mature specimens 50%–70% of corallum length. Major septa are straight, typically 30–36 in number; commonly separated by low-profile minor septa, major and minor septa may have denticulate edges; cardinal and alar fossulae narrow but distinct; tabulae absent.

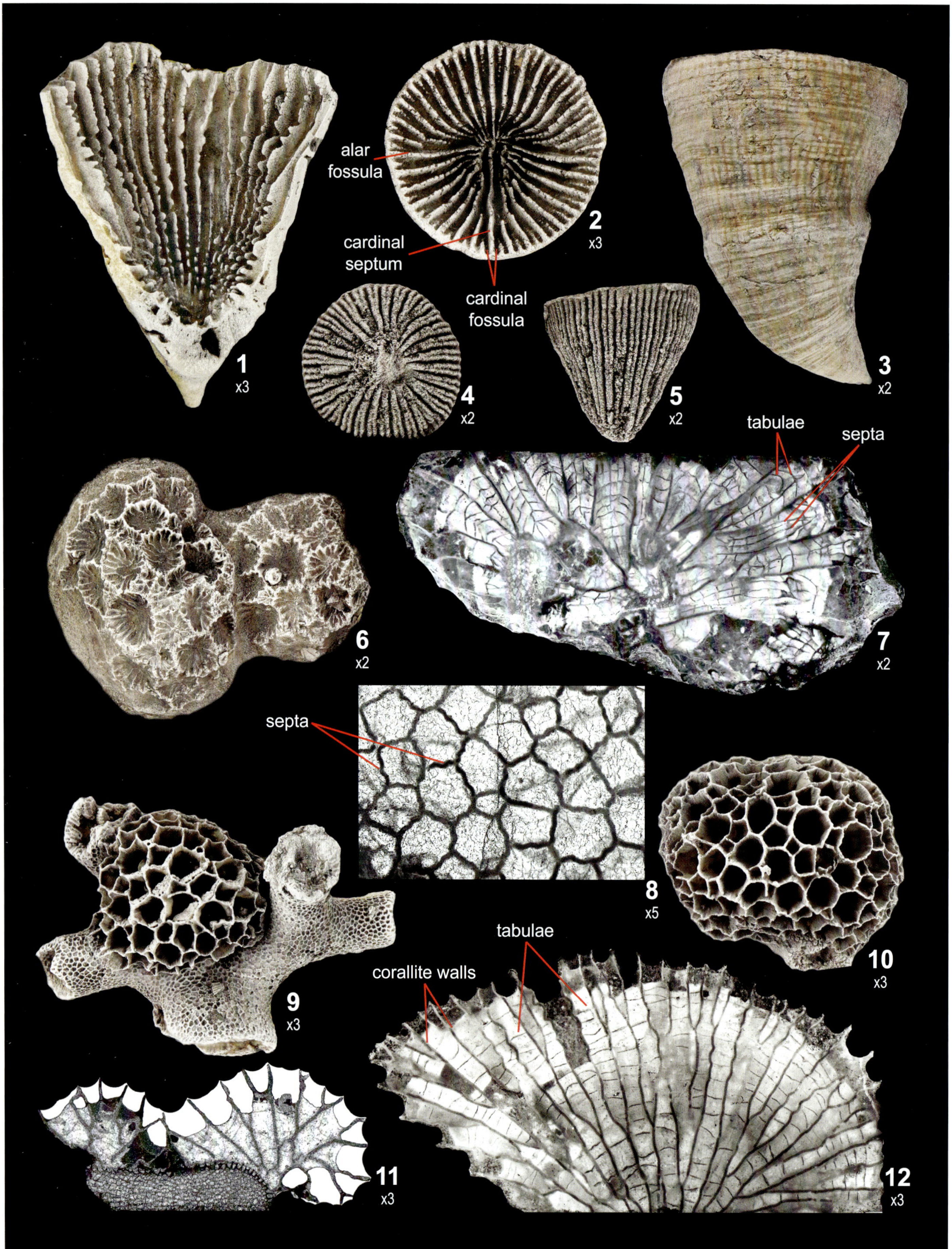
alar
fossula
cardinal
septum
cardinal
fossula
1
x3
2
x3
3
x2
4
x2
5
x2
tabulae
septa
6
x2
7
x2
septa
8
x5
9
x3
10
x3
tabulae
corallite walls
11
x3
12
x3

Exterior relatively smooth with faint, closely set growth lines visible in some areas.

Remarks. Conrad's (1843) original designation of *Cyathophyllum profundum* is based on specimens from the Platteville Formation at Mineral Point, Wisconsin. Okulitch (1938) reassigned this species to the genus *Lambeophyllum*, an assignment that has been challenged by Weyer (1973). *Lambeophyllum* appears to be the most acceptable name until a thorough investigation of the North American rugose corals is conducted (Elias 1983).

Occurrence. Common in the Mifflin and Grand Detour Members in northern Illinois and southern Wisconsin and equivalent strata of the McGregor Member of southwestern Wisconsin, northeastern Iowa, and southeastern Minnesota. Particularly abundant in the upper part of the Forreston Beds, Grand Detour Member in north-central Illinois and south-central Wisconsin.

Figure 4-1 (Opposite page)

1–5. ***Lambeophyllum profundum*** (Conrad). (**1**) Solitary coral showing longitudinally sectioned corallum and deep calice with denticulate septa; (**2**) top view of calice showing radial pattern of septa and bilateral symmetry, cardinal fossula (space) in which the cardinal septum is located, and alar fossula; (**3**) side view of corallum showing growth lines, apical end typically curved in the plane of the cardinal septum; (**4**) apical end of a dolomitized steinkern with septal impressions; the steinkern actually preserves the shape of the polyp that once occupied the calice at top of the corallum; (**5**) same specimen as 4 showing side view of the steinkern. Mifflin Member, Dixon, Illinois (locality 2), **1**, ISGS 1158; **2**, ISGS 1148. Grand Detour Member, Beloit, Wisconsin (locality 21), **3**, ISGS 1154. Mifflin Member, Lee Center, Illinois (locality 31), **4**, **5**, ISGS 1156.

6, 7. ***Cyathophylloides* cf. *stellata*** (Hall). (**6**) Corallum showing numerous calices with prominent septa extending to center of each corallite; (**7**) negative photograph of longitudinal section of corallum showing tabulae and septa. Grand Detour Member, Dixon, Illinois (locality 4), **6**, ISGS 1145; **7**, ISGS 1144.

8. ***Saffordophyllum undulatum*** Bassler. (**8**) Transverse thin section showing the very subtle septa (undulations) in corallum walls. "Sponge bed," Platteville Formation, Dixon, Illinois, **8**, holotype USNM 90994.

9–12. ***Lichenaria* cf. *typa*** Winchell and Schuchert. (**9**) Corallum encrusted on a trepostome bryozoan colony; (**10**) well-preserved corallum; (**11**) acetate peel of small corallum encrusted on a trepostome bryozoan colony; (**12**) negative photograph of longitudinal section of a relatively large corallum showing shared corallite walls and thin tabulae. Mifflin Member, Dixon, Illinois (locality 2), **9**, ISGS 1138; **10**, ISGS 1137; **11**, ISGS 1135. Mifflin Member, Fennimore, Wisconsin (locality 39), **12**, ISGS 1142.

Suborder STAURIINA

Family STAURIIDAE

Genus CYATHOPHYLLOIDES

Cyathophylloides* cf. *stellata (Hall, 1847)

Figures 4-1-6 and 4-1-7

Description. Colonial rugose coral with broad, hemispherical, ceroid corallum; most commonly less than 80 mm in diameter. Corallite diameter 4–5 mm, some as much as 8 mm; thin-walled, shallow calice. Number of major septa 14 or 15; in mature corallites, most major septa are close to or meet at corallite centers and are undulate; minor septa very small or absent; approximately 5 tabulae in 5 mm longitudinal distance.

Remarks. One of the most detailed comparisons of eastern North American *Cyathophylloides* species was published by Browne (1965, table 2). On the basis of corallite diameter, number and length of major and minor septa, number and zonation of tabulae, and wall characteristics, the Platteville species is most similar to *C. stellata* from the Whitewater Formation (Richmondian Stage) of north-central Kentucky.

Occurrence. Rare in the Grand Detour Member of northern Illinois and southern Wisconsin.

Order TABULATA

Remarks. Tabulata are all colonial and are characterized by the absence or weak development of septa, the presence of tabulae, and radial symmetry. They lack a columella.

Suborder SARCINULINA

Family BILLINGSARIIDAE

Genus FOERSTEPHYLLUM

Foerstephyllum magnificum (Okulitch, 1938)

Figures 4-2-4 through 4-2-6

Description. Relatively large cerioid tabulate coral with numerous horizontal tabulae and short, straight septa commonly with denticulate free edges; some coralla reach 30 cm or more in diameter and 25 cm in height. Mature corallite diameters mainly 4–6 mm, but some may reach diameters of 10–12 mm; approximately 30 short sharp-edged septa extend very slightly into the corallite cavity. Tabulae horizontal and complete; 3 or 4 occur in longitudinal distance equal to tube diameter in immature stage and 6 to 8 in the mature zone.

Remarks. Bassler (1950) noted the presence of *F. magnificum* in the Platteville Formation at Dixon, Illinois, and commented that the unusually large corallites distinguish this species of *Foerstephyllum*.

Occurrence. Particularly common in the Grand Detour and Nachusa Members of north-central Illinois and south-central Wisconsin.

Suborder LICHENARIIDA

Family CRYPTOLICHENARIDAE

Genus LICHENARIA

Lichenaria cf. *typa* Winchell and Schuchert, 1895a

Figures 4-1-9 through 4-1-12

Description. Encrusting colonial coral; cerioid corallum small, irregular, hemispherical, typically not exceeding 35 mm in diameter, commonly attached to trepostome bryozoans. Addition of new corallites is either interstitial or along the periphery. Corallite walls approximately 0.15 mm thick, imperforate; septa very subtle to absent, rarely two or three very faint rounded septa can be observed just inside the calice. Corallites polygonal, small but variable, the average adult size about 2 mm or a little less in diameter, but as much as 2.5 mm in the largest specimens. Tabulae thin, horizontal, and complete.

Remarks. Except for having slightly larger corallites, this Platteville species is very similar to *L. typa* from the overlying Decorah Formation near Minneapolis, Minnesota. Note that *Lichenaria* cf. *typa* is very similar to the tabulate coral *Saffordophyllum undulatum* (Figure 4-1-8) described by Bassler (1950) from the "sponge bed" in the Platteville Formation at Dixon, Illinois. Both genera are similar in corallite size, tabulae, and overall growth form. *S. undulatum* has very subtle septa compared with other species of *Saffordophyllum*, making it difficult to distinguish this Platteville tabulate from *Lichenaria*. In part, the problem may be due to cyclomorphic variation during growth of the coral. Accordingly, these variations could result in the development of thinner, straighter walls with reduced formation of septa followed by growth of thicker walls and more prominent septa (Robert J. Elias, personal communication, 2018).

Occurrence. Rare in the Mifflin and Grand Detour Members in northern Illinois and southern Wisconsin and equivalent strata of the McGregor Member of southwestern Wisconsin, northeastern Iowa, and southeastern Minnesota.

Order TETRADIIDA

Remarks. The tetradiids are a small, monophyletic group of cnidarians with modular skeletons and a distinctive quadripartite longitudinal mode of fission (Webby et al. 2004a). There is some suggestion that their skeleton may have been composed of aragonite, thus differing from other tabulate corals, which have a calcitic skeleton (Semeniuk 1971). The traditional interpretation of tetradiids as cnidarians is followed here, although it has been proposed recently that they are florideophycean rhodophyte algae (Steele-Petrovich 2009, 2011).

Family TETRADIIDAE

Genus RHABDOTETRADIUM

Rhabdotetradium sp.

Figures 4-3-1 through 4-3-5

Description. Tetradiid with corallum consisting of predominantly tubular corallites 0.4–0.8 mm in diameter and subcircular in cross section; the single observed corallum is approximately 10 × 50 × 70 mm in dimensions. Corallites are subparallel and have phaceloid-type branching. In a peripheral area of the corallum, corallites are closely packed and may have a crescentic side. Corallites are covered by an epitheca that has a granular surface texture. Corallite walls uniformly 0.10 mm thick. Calices open obliquely to the corallum surface. Immature corallites are 0.4 mm in diameter and appear to lack septa. Thin tabulae are present, typically in 0.5 mm intervals. During growth, four slender septa grew adaxially from the interior wall toward the center. Upon reaching a mature diameter of approximately 0.8 mm, the four septa joined at the center of the corallite in a quadripartite manner typical of the tetradiids. At this point, two, three, or four new corallites formed and bent sinuously

Figure 4-2 (Opposite page)

1–3. ***Conularia* cf. *trentonensis*** Hall. (**1**) Flattened specimen showing two faces separated by a marginal corner and furrow; (**2**) side view of uncrushed three-dimensional specimen; (**3**) photographic enlargement showing a midline, ribs, and nodes. Mifflin Member, Dixon, Illinois (locality 2), **1, 3**, JAC 146. Grand Detour Member, Lightsville, Illinois (locality 7), **2**, JAC 147.

4–6. ***Foerstephyllum magnificum*** (Okulitch). (**4**) Surface of colony showing mature and immature corallites; (**5**) enlargement of corallum showing the horizontal tabulae and vertical septa within the corallites; (**6**) transverse section through a dolomitized corallum. Grand Detour Member, Lightsville, Illinois (locality 7), **4–6**, ISGS 1146.

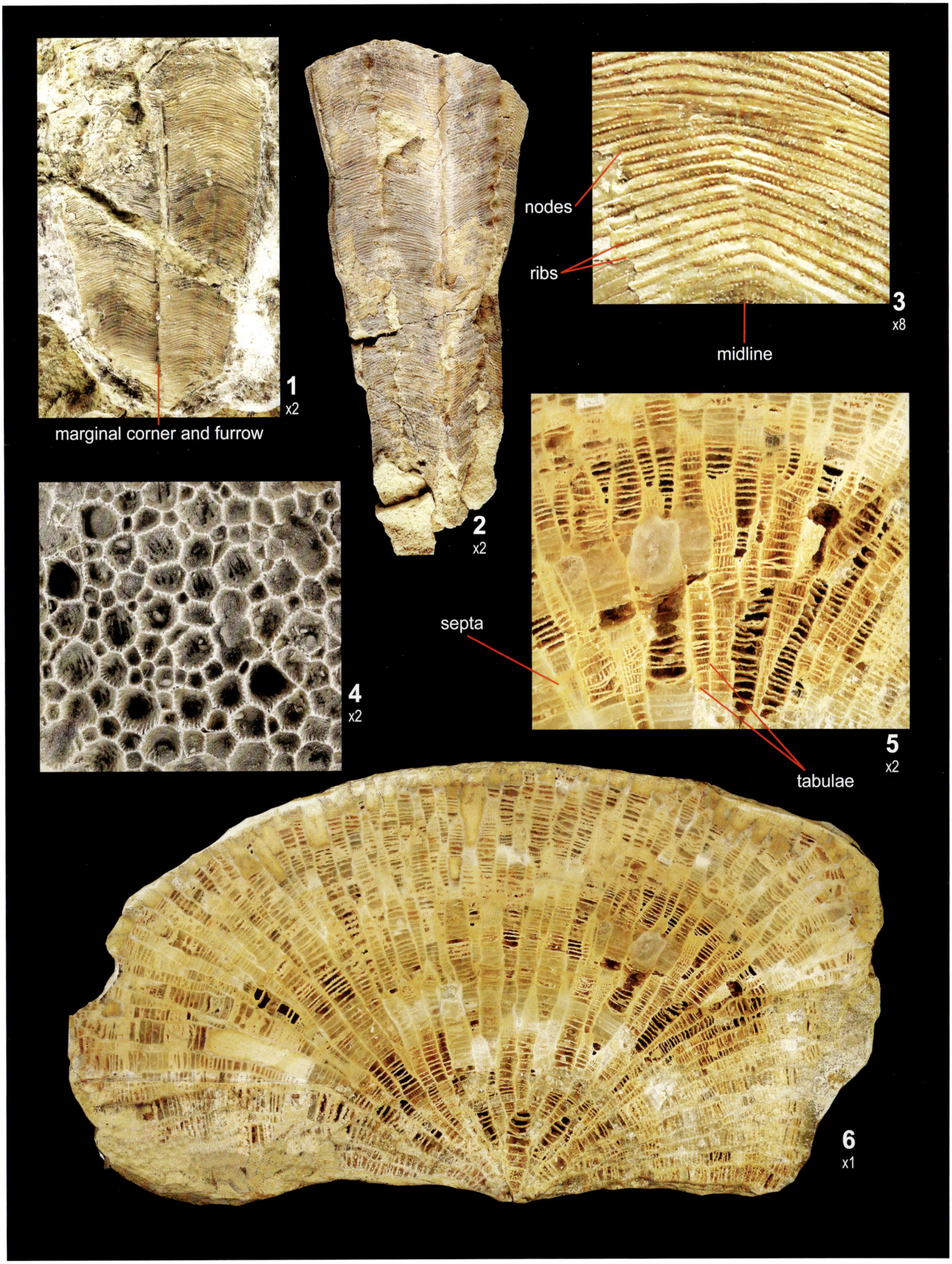
1
x2
marginal corner and furrow
2
x2
nodes
ribs
midline
3
x8
4
x2
septa
tabulae
5
x2
6
x1

away from each other with their calices facing away from the corallum.

Remarks. *Rhabdotetradium* is unique among the tetradiids in having a phaceloid corallum with long, meandering free corallites and outwardly flexed calices. Closely packed corallites in a peripheral area of the described specimen resemble those of *Paleoalveolites*, but the latter genus differs from *Rhabdotetradium* in having closely packed polygonal corallites in the central area of the corallum and its fingerlike branches (Bassler 1950; Hill 1981).

Occurrence. Currently known from a single well-preserved calcitic corallum from the Mifflin Member near Dixon, Illinois (locality 4).

Genus PARATETRADIUM

Paratetradium sp.

Figures 4-3-6 through 4-3-11

Description. Tetradiid with corallum composed of single, double, or multiple rows of corallites branching from a central axis and enclosing empty spaces (sediment-filled lacunae). Coralla are somewhat flattened and hemispherical, up to 25 cm across and 10 cm high. Corallites mostly four-sided with diameters of 1.0–1.5 mm. Septa clearly developed, thin, and relatively straight. Tabulae very thin, delicate, and spaced at 1–1.5 mm.

Remarks. *Paratetradium* differs from other Platteville tetradiids in having coralla with fencelike chains of corallites and sediment-filled lacunae.

Occurrence. Locally common in the Mifflin, Grand Detour, Nachusa, and Quimbys Mill Members of north-central Illinois and south-central Wisconsin.

Class SCYPHOZOA

Order CONULARIIDA

Remarks. Conulariids are an extinct group of marine invertebrates characterized by a mineralized exoskeleton with quadrate transverse cross section of the body. They inhabited benthic marine environments from Late Cambrian to near the end of the Triassic (Lucas 2012). According to recent cladistic analyses, there is growing consensus that conulariids are most closely related to cnidarian coronate scyphozoans (Marques and Collins 2004; Van Iten et al. 2006). Robson and Young (2013) have suggested that conulariids represent the exoskeletons of a cnidarian polypoid stage that likely attached to, or was recumbent upon, the seafloor.

The exoskeleton (**periderm**) tapers from the main opening, or **aperture**, down to the **apical end**, which is presumably attached to the substrate. The four sides, or **faces**, are crossed by narrow transverse ridges, or **ribs**, that are arched away from the aperture, meeting at a distinct longitudinal **midline**. Small, closely placed **nodes** commonly occur on the crests of the ribs. A **marginal furrow** is present at the four **marginal corners** of the periderm.

Family CONULARIIDAE

Subfamily CONULARIINAE

Genus CONULARIA

Conularia cf. *trentonensis* Hall, 1847

Figures 4-2-1 through 4-2-3

Description. Steeply pyramidal, bilaterally symmetrical, four-sided with thin chitinophosphatic periderm, corners diverge at 5°–10°; corner grooves and midlines lack internal thickening; periderm 65–70 mm in length, 35 mm wide at aperture; indented along marginal corners. Faces bearing transverse ribs (lirae), arched adaperturally at midline and continuous along midline of faces, forming angle of approximately 145°, lacking a longitudinal ridge; transverse ribs reduced in height and deflected toward aperture in corner; 20–30 ribs in longitudinal distance of 1 cm. Rib nodes rounded, positioned on crest of transverse ribs, seven or eight in a distance of 1 mm, longitudinal ridges between nodes and adjacent ribs very subtle or absent.

Figure 4-3 (Opposite page)

1–5. ***Rhabdotetradium* sp.** (**1**) Outer surface of an exceptionally well-preserved corallum showing the long, tubular, subparallel corallites with calices open obliquely to the corallum surface; (**2**) enlargement of corallum showing quadripartite budding of a single corallite; (**3**) enlargement of corallum showing separation of corallites; (**4**) enlargement of corallum showing the septa (vertical) and tabulae (horizontal) and granular texture of the epitheca; (**5**) peripheral view of corallum. Mifflin Member, Dixon, Illinois (locality 4), **1–5**, ISGS 1136.

6–11. ***Paratetradium* sp.** (**6**) Calice view of sheetlike calcitic corallum showing quadripartite corallites; (**7**) transverse section showing quadripartite corallites; (**8**) lateral surface of well-preserved corallum; (**9**) enlargement of lateral surface showing several corallites that are missing their outer walls and exposing the septa and tabulae; (**10**) section cut perpendicular to bedding showing chains of corallites (dark) surrounded by lime mud sediment; (**11**) section cut parallel to bedding showing chains of calcitic corallites (dark) surrounded by lime mud sediment. Mifflin Member, Dixon, Illinois (locality 4), **6**, ISGS 1152; **8**, **9**, ISGS 1151. Grand Detour Member (locality 4), **7**, ISGS 1143; **10**, ISGS 1157; **11**, ISGS 1141.

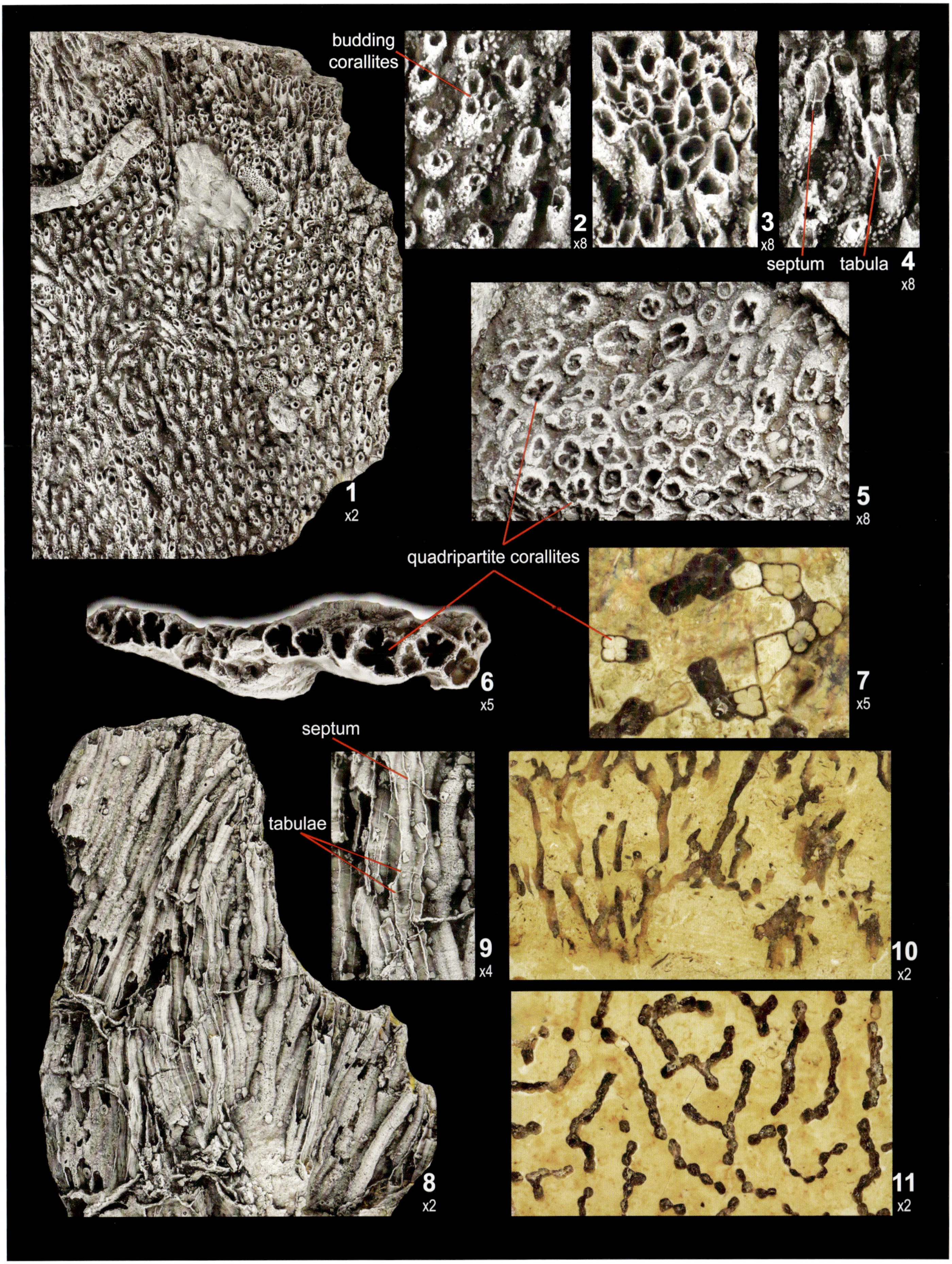
budding corallites
1
x2
2
x8
3
x8
septum
tabula
4
x8
5
x8
quadripartite corallites
6
x5
7
x5
septum
tabulae
8
x2
9
x4
10
x2
11
x2

Remarks. This Platteville conulariid is most similar to *Conularia trentonensis* Hall, 1847, a widespread species in the Middle and Upper Ordovician of North America (Van Iten et al. 2006).

Occurrence. Rare in the Mifflin and Grand Detour Members of northern Illinois and southern Wisconsin.

CHAPTER 5
BRYOZOANS

Bryozoans are one of the most common fossil groups in the Platteville Formation (Figure 5-1). The phylum is represented by the class Stenolaemata, including the orders Cyclostomata, Trepostomatida, Fenestrida, Cryptostomata (suborders Ptilodictyina and Rhabdomesina), and Cystoporata (Appendix 2). The Platteville bryozoan fauna is typical of that found in the North American Late Ordovician (Turinian Stage) cratonic stratigraphic succession. Preservation ranges from exceptionally good in the limestone facies to poor in the dolomite. The highest concentrations of bryozoan fossils are in the Mifflin Member, followed by the Grand Detour, Pecatonica, Quimbys Mill, and Nachusa Members, in that order.

Bryozoans are microscopic, colonial, filter-feeding animals. The main body of an individual bryozoan animal is called a **zooid**. It consists of a protractible tentacle-bearing feeding organ called the **lophophore**. The tentacles have numerous current-producing cilia that direct food particles toward the mouth, which is positioned at the base of the lophophore and opens into a complete U-shaped digestive tract. Zooids are further identified by their structure and function. Those primarily involved in feeding are referred to as **autozooids**. Others, called **polymorphs**, differ morphologically from autozooids and are specialized for other purposes, including sexual, supportive, connective, cleaning, and defensive functions (Boardman et al. 1983). The skeleton of an individual zooid is a **zooecium** (pl. **zooecia**) and the entire colonial skeleton is a **zoarium** (pl. **zoaria**). The skeleton of an individual autozooid is an **autozooecium** (pl. **autozooecia**, and the polymorphs include **mesozooecia** and **exilazooecia** (Figures 5-2-1, 5-2-2, and 5-2-5). The **zooecial wall** is the calcified wall surrounding a zooid, and the **zooecial aperture** is the outer pore or opening of a zooecium. Apertures can vary in shape from circular to five- or six-sided polygons. In the ceramoporid bryozoans, the apertures have a crescent-shaped hood called a **lunarium** on one side. **Endozone** and **exozone** refer to the inner and outer parts of a zoarium, respectively (Figures 5-2-3 and 5-2-4). A **monticule** is a cluster of polymorphs (typically mesozooids and megazooecia), which commonly form a prominence on the surface of a colony. **Styles**, also referred to as stylets, acanthopores, acanthorods, or acanthostyles, are spinelike structures adjoining the zooecial wall and commonly projecting slightly above the zoarial surface. A **diaphragm** is a flat skeletal partition that extends transversely across a zooecial chamber, and a **cystiphragm** is a curved partition that extends partway across a zooecial chamber. **Hemiphragms** are relatively short, thick, shelflike projections in a zooecial living chamber, particularly well developed in the Platteville genus *Hemiphragma* (Figure 5-2-6). The most common bryozoan colony growth forms include (1) ramose or branching, (2) frond- or leaflike, (3) encrusting, either sheetlike or threadlike, (4) hemispherical or domal, and (5) massive or irregular. A **bifoliate** is a frond- or leaflike form that grew erect and budded on two sides of a central plane.

Key features used in the identification of bryozoans include (1) colony size and form, (2) branch thickness, (3) exozonal wall thickness, (4) wall microstructure, (5) zooecial diameter, (6) mesozooecia abundance and appearance, (7) abundance and size of styles, (8) cystiphragms, (9) curved diaphragms, and (10) hemiphragms. Many of these features are pointed out on the photographs.

Compared with other Platteville phyla, bryozoans are one of the most challenging invertebrate groups to identify at the genus and species levels. The primary distinguishing characteristics are based largely on internal microscopic skeletal wall structures. Accurate identification requires the preparation of thin sections or acetate peels, a good-quality microscope, hours of specimen preparation, access to paleontological literature, which commonly is somewhat esoteric, and for comparative purposes, access to the institutional collections where the type specimens are held. Identification is further hindered by the wide range of structural variations that commonly are present within a single colony. Despite these difficulties, preliminary identifications can be made by examining the microscopic skeletal structures with a good-quality hand lens. Key morphologic features are discussed and illustrated in this chapter. More conclusive identifications require the preparation of thin sections or acetate peels. Bryozoan morphologic terms and biology can be found in Bassler (1953), Boardman et al. (1983), and Boardman and Cheetham (1983), as well as in invertebrate paleontology textbooks and internet sources. Key references in identifying the Platteville bryozoans include Ulrich (1886, 1893), Coryell (1921), Perry (1962), Brown (1965), Karklins (1969, 1984, 1985), Utgaard (1983), and Bork and Perry (1967, 1968a, 1968b).

In this report, family and higher level assignments follow Bassler (1953) except for the orders Cystoporida and

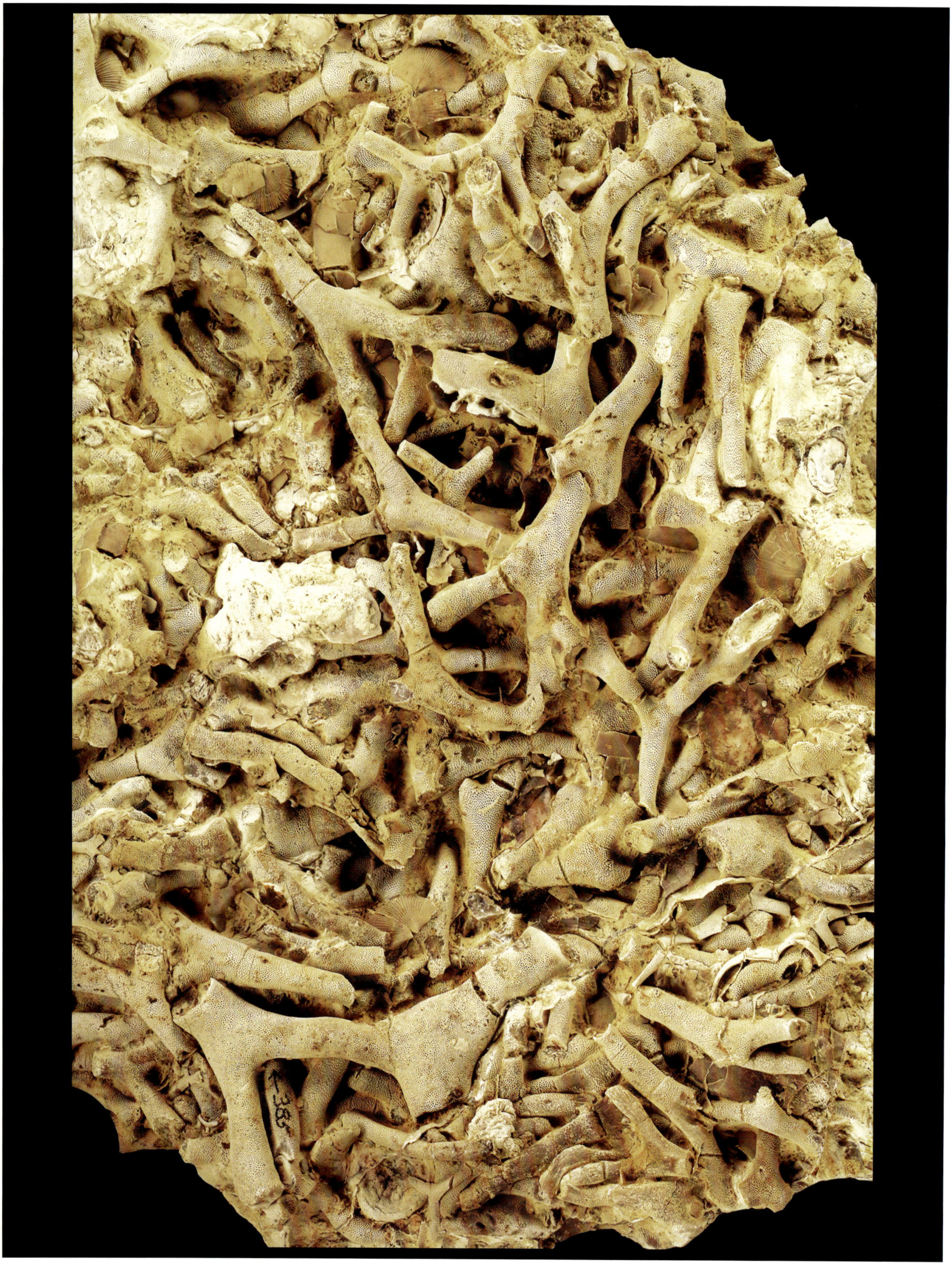

Cryptostomida, which follow the revised classification scheme of Boardman et al. (1983), who followed but did not credit Blake (1975). Taxonomy of the Paleozoic Stenolaemata, particularly the early Trepostomata, has been unstable in the literature, with authors employing widely varied familial concepts. The classification of the 1953 *Treatise on Invertebrate Paleontology* volume by R. S. Bassler is followed here because it is widely available and provides an objective reference for the reader. Usage here does not argue that any, or all, of the cited families are accurate phylogenetic clades. Guidance on familial concepts of the Trepostomata is anticipated in the forthcoming revised *Treatise on Invertebrate Paleontology*.

Most bryozoans described herein are identified to genus. Species assignments are avoided or restricted to comparisons (cf.) only unless identified in previous peer-reviewed literature as being from the Platteville Formation in the Upper Mississippi Valley region. Previously published Platteville species include nine trepostomes described by Bork and Perry (1967, 1968a, 1968b) and two trepostomes by Ulrich (1890d). Described and figured herein are approximately 20 additional species that were previously not known to occur in the Platteville Formation. Overall, the Platteville bryozoan fauna is most like that described from the Upper Ordovician Stones River Group of Tennessee (Karklins 1985) and younger rocks (Mohawkian and Cincinnatian Series) of Minnesota and central Tennessee (Ulrich 1886; Karklins 1969), as well as Kentucky (Karklins 1984).

SYSTEMATIC DESCRIPTIONS

Phylum BRYOZOA

Class STENOLAEMATA

Description. According to McKinney and Jackson (1989), Zooids are elongate cylindrical and continue to lengthen through ontogeny, with long axis at angle to local colony growth direction; basal and vertical walls rigidly calcified; interzooidal communication through space between outer ends of vertical walls in most, but through pores in vertical walls in some. Lower Ordovician – Recent. (p. 21)

Figure 5-1 (Opposite page) Slab of fossiliferous limestone showing a concentration of well-preserved trepostome bryozoans. Mifflin Member of the Platteville Formation, Dixon, Illinois (locality 4).

Order TREPOSTOMATIDA

Description. McKinney and Jackson (1989) describe trepostomes thus: "Colonies encrusting or erect; autozooids generally containing basal diaphragms and commonly other lateral structures; extrazooidal skeleton in some; skeletons without communication pores, typically laminated. Lower Ordovician to Upper Triassic (or Recent?)" (p. 21).

Remarks. Trepostomes are the most diverse and abundant of the Platteville bryozoans. A wide range of forms are present, including robust, dendroid, ramose, encrusting, and massive colonies. Because of their subtle morphologic features, trepostomes are difficult to identify. This is particularly evident at family-level classification. Recognizing these difficulties, Bork and Perry (1967, 1968a, 1968b), in their analysis of Upper Ordovician bryozoans of the upper Mississippi Valley, avoided assigning their trepostomes to either families or suborders. Taxonomic inadequacies of trepostome families were also recognized by McKinney (1971) and Karklins (1984). Bassler's (1953) widely cited classification scheme in the *Treatise on Invertebrate Paleontology* has significant problems and is long overdue for revision but is still a useful starting point. For reference convenience in this overview, Platteville trepostome bryozoan genera are assigned informally to families in the classification system of Bassler (1953), although significant future revision is likely.

Several Platteville trepostomes from northern Illinois and southern Wisconsin were described by Bork and Perry (1967, 1968a, 1968b). Thin sections of their type specimens reside in the collections of the Illinois State Geological Survey and were examined in this investigation.

Family MONTICULIPORIDAE

Genus MONTICULIPORA

Monticulipora sp.

Figures 5-3-7 and 5-3-8

Description. Zoarium frondose, commonly 20 mm or longer and 6 mm wide. Autozooecial apertures angular to subangular, typically eight or nine autozooecia in a distance of 2 mm in areas between macula; erect in endozone bending gently upward with new autozooecia added distally; autozooecial walls thin in endozone, thickening slightly in exozone, walls granular with poorly defined median line. Diaphragms abundant throughout and generally straight. Cystiphragms well-rounded and abundant throughout endozone and exozone in long overlapping series. Mesozooecia common and less than half the size of the autozooecia. Monticules

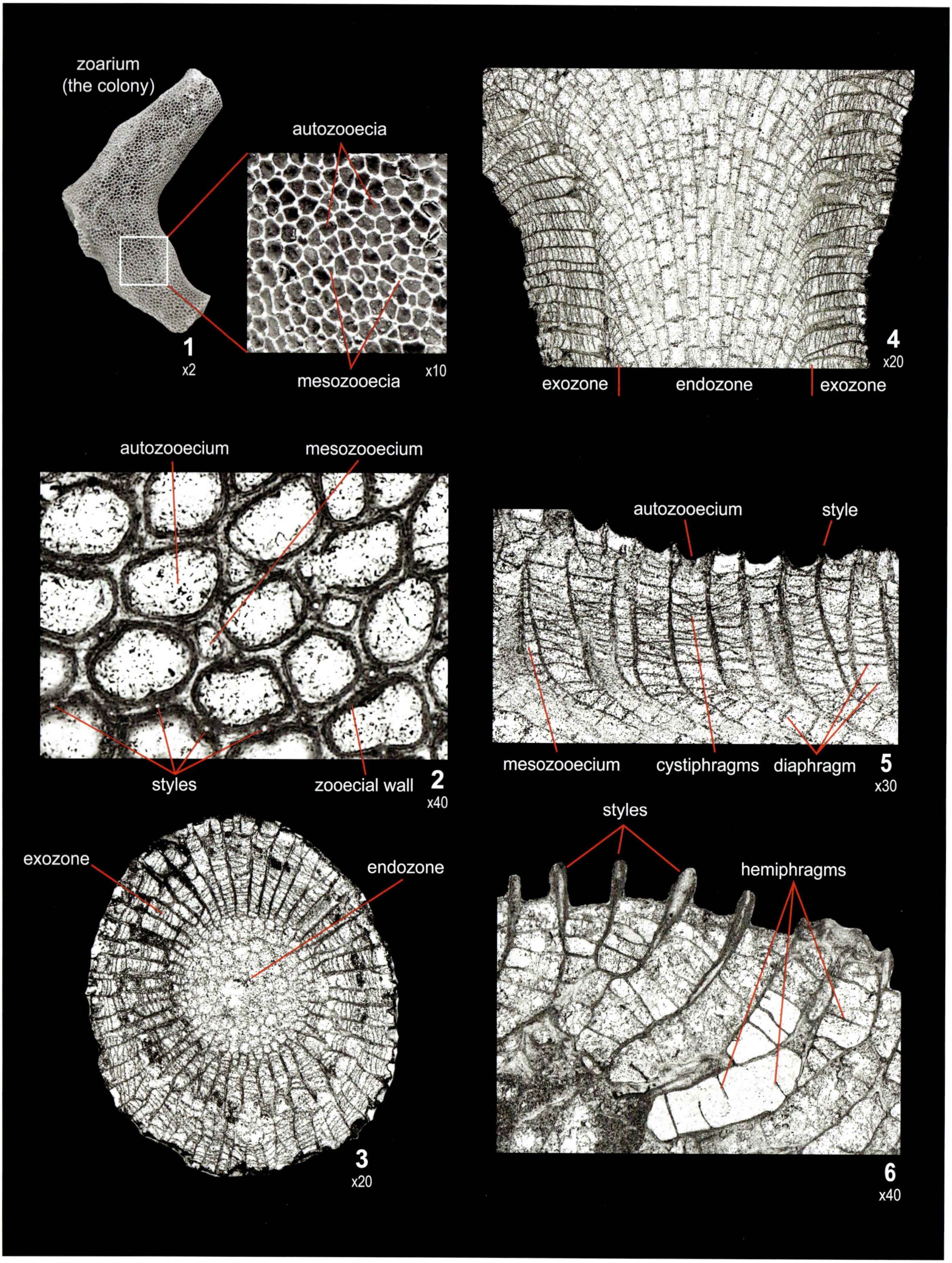
zoarium
(the colony)
autozooecia
1
x2
mesozooecia
x10
4
x20
exozone
endozone
exozone
autozooecium
mesozooecium
styles
zooecial wall
2
x40
autozooecium
style
mesozooecium
cystiphragms
diaphragm
5
x30
styles
hemiphragms
exozone
endozone
3
x20
6
x40

unevenly distributed, elevated slightly, and consisting of normal-sized autozooecia and mesozooecia with relatively thick walls. Styles small and uncommon.

Remarks. The abundant cystiphragms in the endozone and exozone are distinguishing features of this Platteville trepostome.

Occurrence. Rare in the Mifflin Member in northern Illinois and southern Wisconsin.

Genus ATACTOPORELLA

Atactoporella sp.

Figures 5-3-5 and 5-3-6

Description. Zoarium encrusting, approximately 1.0 mm thick. Autozooecial apertures angular to subangular, typically eight autozooecia in a distance of 2 mm, walls approximately 0.05 mm thick. Diaphragms present. Cystiphragms form overlapping series. Mesozooecia common, angular, three- to five-sided, average diameter 0.10 mm. Monticules slightly elevated, gently rounded, approximately 1.0 mm in diameter, consisting of mesozooecia and relatively large autozooecia up to 0.30 mm in diameter. Styles prominent and abundant, ranging from 0.05 to 0.08 mm in diameter.

Remarks. The encrusting form of the colony and presence of overlapping cystiphragms distinguish *Atactoporella* sp. from other trepostomes in the Platteville Formation.

Occurrence. A rare bryozoan in the Mifflin Member in northern Illinois and southern Wisconsin.

Genus ORBIGNYELLA

Orbignyella sublamellosa Ulrich and Bassler, 1904

Figures 5-3-3 and 5-3-4

Description. Zoarium encrusting, 1–2 mm thick, commonly a laminate expansion on cephalopods and gastropods. Autozooecial apertures angular, wide variety of polygonal shapes, typically eight autozooecia in a distance of 2 mm; walls thin, average thickness 0.01 mm. Diaphragms straight to curved, many inclined rather than perpendicular to zooecial axis. Mesozooecia few or lacking. Monticules characterized by evenly spaced flat areas consisting of relatively large autozooecia. Styles abundant, small, slightly protuberant.

Remarks. Comparable in most features to *Orbignyella sublamellosa* Ulrich and Bassler, 1904, from the Pierce and Ridley Formations of Kentucky and Tennessee.

Occurrence. Abundant in the Mifflin and Grand Detour Members in northern Illinois and southern Wisconsin, and equivalent strata of the McGregor Member of southwestern Wisconsin, northeastern Iowa, and southeastern Minnesota. Common locally in the Pecatonica Member of northern Illinois and southern Wisconsin.

Genus PERONOPORA

Peronopora sp. 1

Figures 5-4-1 through 5-4-3

Description. Zoarium bifoliate, fan-shaped, flattened to slightly wavy. Mesotheca (wall between bifoliate halves) straight to slightly sinuous. Autozooecial apertures round to oval at the colony surface, typically six in a 2 mm distance; subrectangular in transverse section; autozooecia in exozone approximately at right angles to mesotheca; thick walled in exozone, relatively thin walled in endozone. Diaphragms straight to slightly curved, three or four occurring in most autozooecia. Cystiphragms commonly form overlapping series. Mesozooecia common, partly surrounding autozooecia; diaphragms abundant, straight to slightly concave upward. Monticules common, flat to slightly elevated, consisting of central cluster of mesozooecia, larger autozooecia, or both, approximately 1.0 mm in diameter and 3.0 mm apart. Styles small and sparse.

Remarks. *Peronopora* sp. 1 differs from *Peronopora* sp. 2 in having an overall thinner, more delicate zoarium.

Occurrence. Common in the Mifflin Member, northern Illinois, and southern Wisconsin.

Peronopora sp. 2

Figures 5-4-4 and 5-4-5

Description. Zoarium bifoliate, fan-shaped, convoluted, relatively massive. Mesotheca (wall between bifoliate halves) straight to slightly sinuous, consisting of abundant, tightly packed circular tubes. Autozooecial apertures round to oval at the colony surface, typically five in a 2 mm distance; autozooecia in exozone approximately at right angles to mesotheca; thick walled in exozone. Diaphragms straight to

Figure 5-2 (Opposite page)
1–6. Bryozoan terminology. (**1**) Zoarium of *Heterotrypa* showing typical colony form with photographic enlargement showing autozooecia and mesozooecia; (**2**) acetate peel of shallow tangential section through *Heterotrypa* showing autozooecia, mesozooecia, styles, and zooecial walls; (**3**) transverse section of *Heterotrypa* zoarium showing the endozone and exozone; (**4**) longitudinal section of *Heterotrypa* showing the endozone and exozone; (**5**) exozone of *Heterotrypa* showing autozooecia, mesozooecia, styles, cystiphragms, and diaphragms; (**6**) exozone of *Hemiphragma* showing styles and hemiphragms.

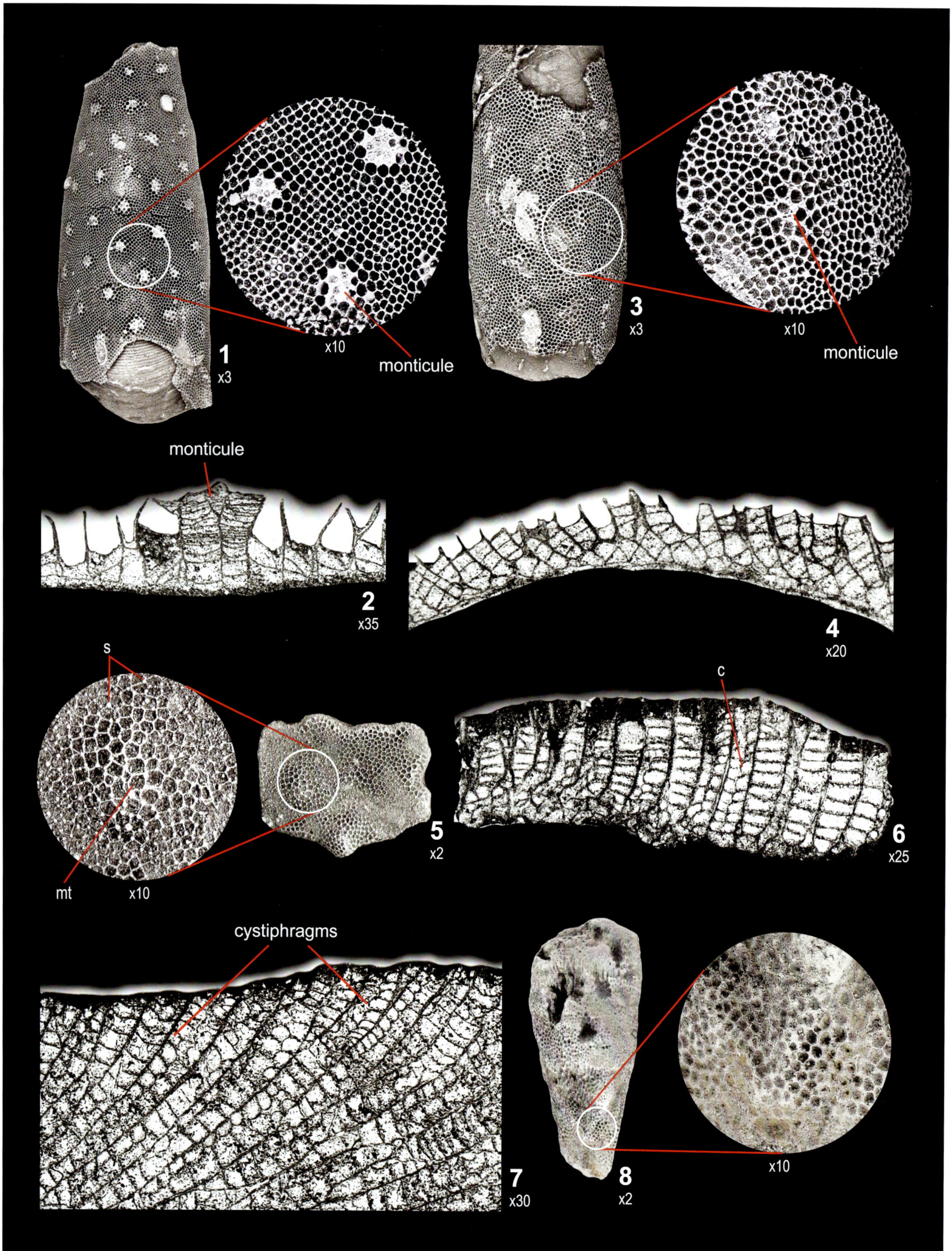
monticule
1
x3
x10
monticule
3
x3
x10
monticule
monticule
2
x35
4
x20
s
mt
x10
5
x2
c
6
x25
cystiphragms
7
x30
8
x2
x10

slightly curved, numerous. Cystiphragms common, forming overlapping series in some autozooecia. Mesozooecia common, partly surrounding autozooecia; diaphragms abundant, straight to slightly concave upward. Monticules common, flat to slightly elevated, consisting of central cluster of mesozooecia, larger autozooecia, or both, approximately 1.0 mm in diameter and 3.0 mm apart. Styles small and sparse.

Remarks. *Peronopora* sp. 2 differs from *Peronopora* sp. 1 in having a more massive and convoluted zoarium, a mesotheca characterized by tightly packed circular tubes, and slightly larger autozooecial apertures. The two species are alike in having small and sparse styles.

Occurrence. Rare in the Grand Detour Member, Dixon, Illinois.

Genus HOMOTRYPA

Homotrypa exilis Bork and Perry, 1968a

Figures 5-4-6 through 5-4-8

Description. Zoarium ramose, round to broadly oval in transverse section, branches 4–7 mm in diameter; larger specimens can be in excess of 40 mm in length. Autozooecial apertures subround to subangular, typically seven or eight autozooecia in a distance of 2 mm; autozooecia erect in endozone, bending somewhat abruptly in the exozone to a 90° angle at the zoarial surface; walls thin, amalgamate, and flexuous in the endozone, becoming thicker and integrated in the exozone. Diaphragms lacking in the endozone and numerous in the exozone. Cystiphragms present in the outer exozone. Mesozooecia very few, apertures mostly less than half the diameter of autozooecia. Monticules consisting of small clusters of mesozooecia and autozooecia that are noticeably larger than those in the intermacular areas; low, rounded profile. Styles lacking.

Remarks. Specimens described and illustrated here are most like Bork and Perry's (1968a) *H. exilis* from the Quimbys Mill Member in southwestern Wisconsin and northwestern Illinois.

Occurrence. Common in the Mifflin, Grand Detour, and Quimbys Mill Members in northern Illinois and southern Wisconsin.

Figure 5-3 (Opposite page)
1, 2. ***Leptotrypa hexagonalis*** Ulrich. (**1**) A colony encrusted on a hyolithid shell, the most common occurrence of *Leptotrypa* in the Platteville Formation; photographic enlargement showing the predominantly hexagonal shape of the autozooecial apertures; (**2**) acetate peel showing transverse section through a monticule consisting of several occluded mesozooecia with numerous diaphragms. Mifflin Member, Dixon, Illinois (locality 2), **1**, ISGS 423. Mifflin Member, Fennimore, Wisconsin (locality 25), **2**, ISGS 379.

3, 4. ***Orbignyella sublamellosa*** Ulrich and Bassler. (**3**) A colony encrusted on the shell of *Whitfieldoceras*; photographic enlargement showing the wide variations in shapes and sizes of the autozooecial apertures; also shown are several low-profile monticules distinguished by their large apertures; (**4**) acetate peel showing transverse section of the colony. Mifflin Member, Dixon, Illinois (locality 2), **3**, ISGS 1110; **4**, ISGS 172.

5, 6. ***Atactoporella* sp.** (**5**) Encrusting colony; photographic enlargement showing angular autozooecial apertures, prominent styles (s) and small low-profile monticule (mt); (**6**) acetate peel of colony transverse section showing characteristic overlapping series of cystiphragms (c). Mifflin Member, Fennimore, Wisconsin (locality 24), **5**, **6**, ISGS 410.

7, 8. ***Monticulipora* sp.** (**7**) Acetate peel of a longitudinal section through a colony showing the characteristic overlapping series of cystiphragms; (**8**) a small, poorly preserved frond-shaped colony; photographic enlargement showing the variations in size of the mostly subangular autozooecial apertures. Mifflin Member, Fennimore, Wisconsin (locality 26), **7**, **8**, ISGS 160.

Family HETEROTRYPIDAE

Genus HETEROTRYPA

Heterotrypa sp.

Figures 5-5-5 through 5-5-8

Description. Zoarium ramose to fan-shaped, round to broadly ovate in transverse section, branches typically 4–7 mm in diameter. Autozooecial apertures mostly subangular, typically six or seven autozooecia in a distance of 2 mm; erect and 0.01 mm thick in the endozone, bending rather abruptly in the exozone at a 90° angle to the zoarial surface; moderately thick walls in the exozone; the relatively wide exozone is clearly differentiated from the endozone. Diaphragms present in the endozone and exozone, straight and spaced approximately one autozooecial diameter apart in the endozone, straight to concave outward in the exozone, five in a distance of one autozooecial diameter, a few cystiphragm-like diaphragms in the exozone. Mesozooecia numerous, apertures subangular and mostly less than half the diameter of autozooecial apertures. Monticules consist of small clusters of mesozooecia and autozooecia that are noticeably larger than the average zooid and have a low, rounded profile. Styles abundant, protuberant above zoarial surface, commonly inflecting autozooecial wall, some show concentric laminae and central lumen.

Remarks. The relatively wide exozone with abundant diaphragms is one of the most distinguishing characteristics of this species of *Heterotrypa*. The specimens described and

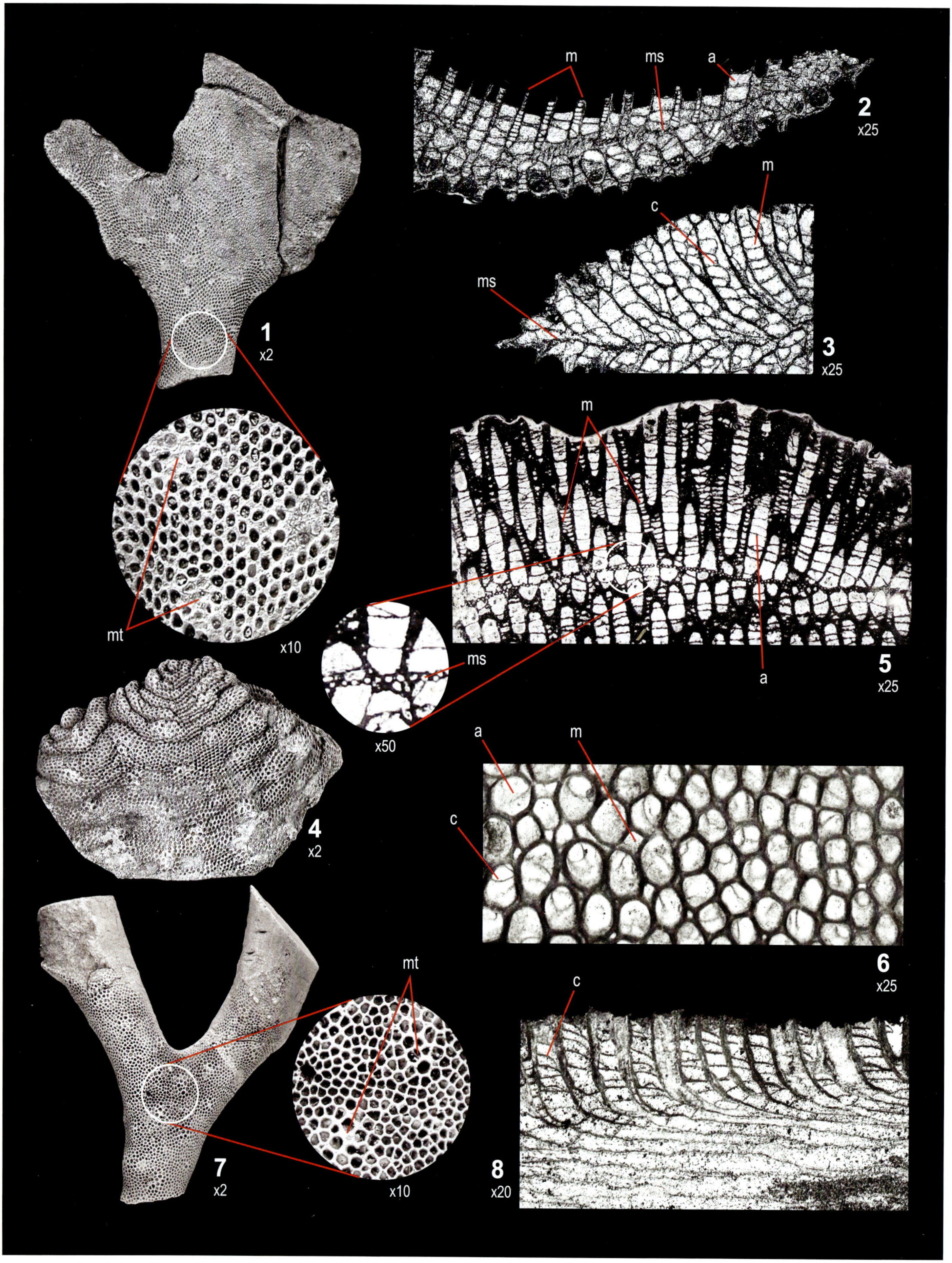
m
ms
a
2
x25
m
c
ms
3
x25
1
x2
mt
x10
m
ms
x50
5
x25
a
4
x2
a
m
c
6
x25
mt
c
7
x2
x10
8
x20

illustrated here are essentially identical to Bork and Perry's (1968a) *Heterotrypa* sp. from the Mifflin Member of northern Illinois.

Occurrence. Rare in the Mifflin and Grand Detour Members in northern Illinois and southern Wisconsin and equivalent strata of the McGregor Member of southwestern Wisconsin, northeastern Iowa, and southeastern Minnesota.

Genus LEPTOTRYPA

Leptotrypa hexagonalis Ulrich, 1890d

Figures 5-3-1 and 5-3-2

Description. Zoarium encrusting, typically less than 1 mm thick and commonly spread over the surface of hyolithid shells; some ramose. Autozooecial apertures angular, hexagonal, ranging in size from 0.2 to 0.35 mm in diameter; typically eight to nine autozooecia in a distance of 2 mm; walls thin throughout. Diaphragms very sparse and straight. Mesozooecia less than half the size of the autozooecia, angular and commonly four sided, with numerous diaphragms. Monticules prominently elevated, arranged in diagonally intersecting rows approximately 3 mm apart, and consisting of relatively large autozooecia and small mesozooecia. Styles sparsely distributed throughout zoarium.

Remarks. Specimens from the Upper Mississippi Valley region are indistinguishable from the type specimen of *L. hexagonalis* Ulrich, 1890d, from the Platteville Formation, Calhoun County, Illinois. In more than two dozen specimens collected during the preparation of this overview, all are encrusted on the surfaces of the hyolithid *Solenotheca baconi* (Whitfield, 1878).

Occurrence. Common in the Mifflin and Grand Detour Members in northern Illinois and southern Wisconsin and equivalent strata of the McGregor Member of southwestern Wisconsin, northeastern Iowa, and southeastern Minnesota. Particularly abundant in the Mifflin Member in the area of Fennimore, Wisconsin.

Family TREMATOPORIDAE

Genus HEMIPHRAGMA

***Hemiphragma* sp.**

Figures 5-5-1 through 5-5-4

Description. Zoarium ramose, round to broadly oval in transverse section, branches 3–5 mm in diameter; largest observed complete specimen is 45 mm in length with four bifurcations. Autozooecia apertures round to ovate and 0.15–0.20 mm in diameter; autozooecial walls are erect and 0.01 mm thick in the endozone, bending in a gradual arc in the exozone, where they thicken to about 0.10 mm. Diaphragms and hemiphragms present in the endozone; four to six hemiphragms per autozooecium in the exozone. Mesozooecia abundant, round to ovate, less than half the size of autozooecia. Monticules low profile, consisting of slightly larger than average-sized autozooecia surrounded by several mesozooecia; some flat areas (maculae) lack autozooecia and mesozooecia and have clusters of styles. Styles abundant, somewhat protuberant above zoarial surface, approximately four or five per autozooecia, commonly inflecting the autozooecial wall.

Remarks. Two variants have been observed in the Platteville Formation. One has a slender zoarium with very thick autozooecial walls and large styles that project relatively high above the zoarial surface. The other has more robust branches with relatively thin autozooecial walls and low-profile styles.

Occurrence. Common in the Pecatonica, Mifflin, and Grand Detour Members throughout the outcrop belt.

Figure 5-4 (Opposite page)

1–3. ***Peronopora* sp. 1.** (**1**) Partially weathered colony; photographic enlargement showing oval autozooecial apertures and low-profile monticules (mt); (**2**) acetate peel showing mesozooecia with numerous diaphragms (m), autozooecia (a), and mesotheca (ms); (**3**) acetate peels showing overlapping cystiphragms (c). Mifflin Member, Dixon, Illinois (locality 3), **1**, **2**, ISGS 427; **3**, ISGS 175.

4, 5. ***Peronopora* sp. 2.** (**4**) Silicified specimen showing characteristic convoluted surface of colony; (**5**) acetate peel of tangential section showing autozooecia (a) and mesozooecia (m) containing numerous closely spaced diaphragms; photographic enlargement of mesotheca (ms, wall between bifoliate halves of the colony) showing transverse section of the small, circular tubes. Grand Detour Member, Dixon, Illinois (locality 4), **4**, **5**, ISGS 554.

6–8. ***Homotrypa exilis*** Bork and Perry. (**6**) Shallow tangential thin section showing cystiphragm (c), autozooid (a), and mesozooid (m); (**7**) typical ramose colony; photographic enlargement showing the characteristic oval autozooecial apertures, thick autozooecial walls, and low-profile monticules (mt); (**8**) acetate peel of longitudinal section showing lack of diaphragms in the endozone, thick zooecial walls, and cystiphragms (c). Mifflin Member, Dixon, Illinois (locality 2), **6**, Bork and Perry (1968a) paratype ISGS 12P 352; **7**, **8**, ISGS 380.

Genus DIPLOTRYPA

Diplotrypa catenulata Coryell, 1921

Figures 5-6-1 through 5-6-4

Description. Zoarium hemispherical, slightly concave at base; colony bases were recognized encrusting mollusks and brachiopods, radiating outward to form massive colonies; some colonies are as much as 8–10 cm wide and 4–5 cm high. Autozooecial apertures sharply angular, mostly hexagonal or pentagonal, typically four or five autozooecia in a distance of 2 mm; walls have consistent thickness of approximately 0.01 mm throughout the endozone and exozone; walls are mostly straight but in places are flexuous to crenulated, forming a beaded appearance in longitudinal section that commonly can be seen on broken edges with a hand lens. Diaphragms straight and slightly less thick than the autozooecial walls; present in the endozone at regular but widely spaced intervals; approximately four or five diaphragms present in indistinct exozone. Mesozooecia common, angular, conspicuously smaller than autozooecia, three- to four-sided. Monticules slightly elevated, gently rounded, approximately 1.0 mm in diameter, consisting of small mesozooecia and relatively large autozooecia. Styles absent.

Remarks. Like *D. catenulata* Coryell, 1921, from the Stones River Group of Tennessee, this Platteville species has a massive hemispherical zoarium, relatively large, sharply angular apertures, and relatively small mesozooecia. *D. catenulata* is sometimes misidentified in the field as *Prasopora* but differs from the latter in lacking overlapping series of cystiphragms.

Occurrence. Abundant in the Mifflin and Grand Detour Members in northern Illinois and southern Wisconsin and equivalent strata of the McGregor Member of southwestern Wisconsin, northeastern Iowa, and southeastern Minnesota.

Family BATOSTOMELLIDAE

Genus ERIDOTRYPA

Eridotrypa subtilis Bork and Perry, 1968a

Figures 5-6-5 through 5-6-7

Description. Zoarium ramose, round to broadly oval in transverse section, branches 3–5 mm in diameter. Autozooecial apertures teardrop-shaped, angular to subangular, typically six autozooecia in a distance of 2 mm; walls less than 0.02 mm in the endozone, thickening significantly in the exozone, where they are club-shaped in transverse section; erect in axis, curving very gradually to intersect the zoarial surface at an angle of 20°; exozone narrow. In the axial parts of the endozone, the autozooecia are larger than at the periphery of the zoarium; this feature commonly can be seen on the broken end (transverse fracture) of a branch and is a fairly good identifier of *Eridotrypa* in the field. Diaphragms lacking in the endozone and commonly limited to less than five per autozooecia in the exozone. Mesozooecia subangular, less than half the size of autozooecium, and not abundant. Monticules flat, consisting of small clusters of mesozooecia and autozooecia that are noticeably larger than those in the intermacular areas. Styles not recognized.

Remarks. *Eridotrypa subtilis* differs from most other Platteville trepostomes by having a wide endozone with large autozooecia that lack diaphragms.

Occurrence. The type species *E. subtilis* was described by Bork and Perry (1968a) from the Mifflin Member, at Dixon, Illinois. Very abundant in the Mifflin and Grand Detour Members in north central Illinois.

Family AMPLEXOPORIDAE

Genus AMPLEXOPORA

Amplexopora winchelli spinulosum (Ulrich, 1886)

Figures 5-7-1 through 5-7-3

Description. Zoarium ramose, branching unevenly, round to slightly compressed, observed length 45 mm and branch diameters of 5–7 mm. Autozooecial apertures angular to subangular, typically six autozooecia in a distance of 2 mm; autozooecia erect in the endozone, curving gently in the exozone at a 45°–60° angle; walls integrated with a dark line of demarcation between adjacent autozooecia;

Figure 5-5 (Opposite page)
1–4. ***Hemiphragma* sp.** (**1**) Typical slender, ramose colony; photographic enlargement showing thick autozooecial walls and styles protruding above colony surface; (**2**) colony surface showing prominent styles (s) and macula; (**3**) complete colony of a species with thick branches, low-profile styles, and basal attachment encrusted on bivalve shell; (**4**) acetate peel of longitudinal section showing thick, blunt hemiphragms (h). Mifflin Member, Dixon, Illinois (locality 2), **1**, **2**, ISGS 452. Mifflin Member, Fennimore, Wisconsin (locality 24), **3**, ISGS 158; **4**, ISGS 1131.

5–8. ***Heterotrypa* sp.** (**5**) Typical knotty, ramose colony; photographic enlargement showing polygonal autozooecial apertures; (**6**) photographic enlargement of colony surface showing autozooecial apertures and small styles (s); (**7**) acetate peel of longitudinal section showing autozooecia (a), cystiphragms (c), diaphragms (d), styles (s), and mesozooid (m) with numerous diaphragms; (**8**) acetate peel of transverse section showing the characteristic wide exozone with thick autozooecial walls and numerous diaphragms. Mifflin Member, Dixon, Illinois (locality 2), **5**, ISGS 412; **6**, ISGS 454; **7**, ISGS 423; **8**, ISGS 106.

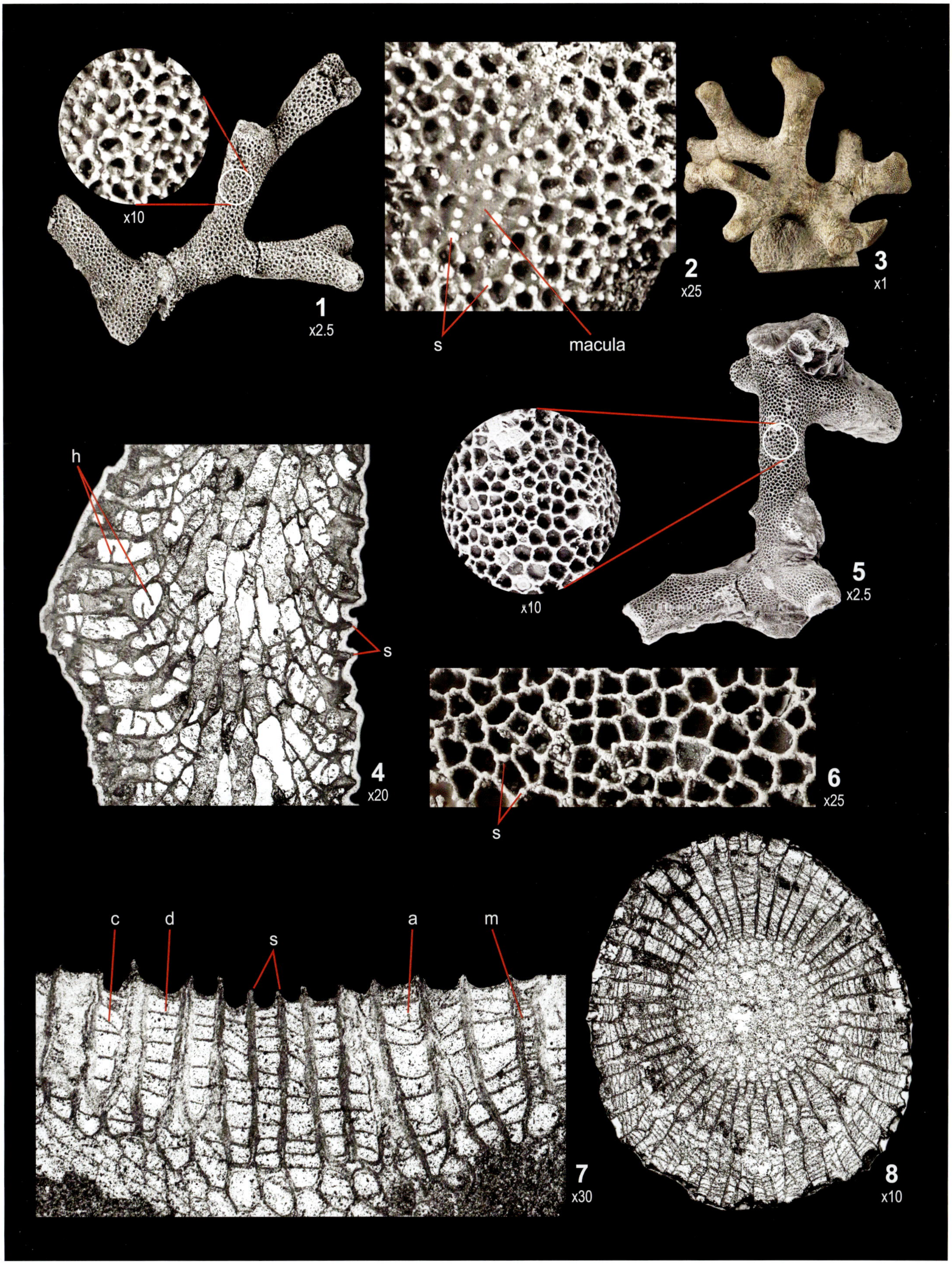
x10
1
x2.5
2
x25
s
macula
3
x1
h
s
4
x20
x10
5
x2.5
6
x25
s
c
d
s
a
m
7
x30
8
x10

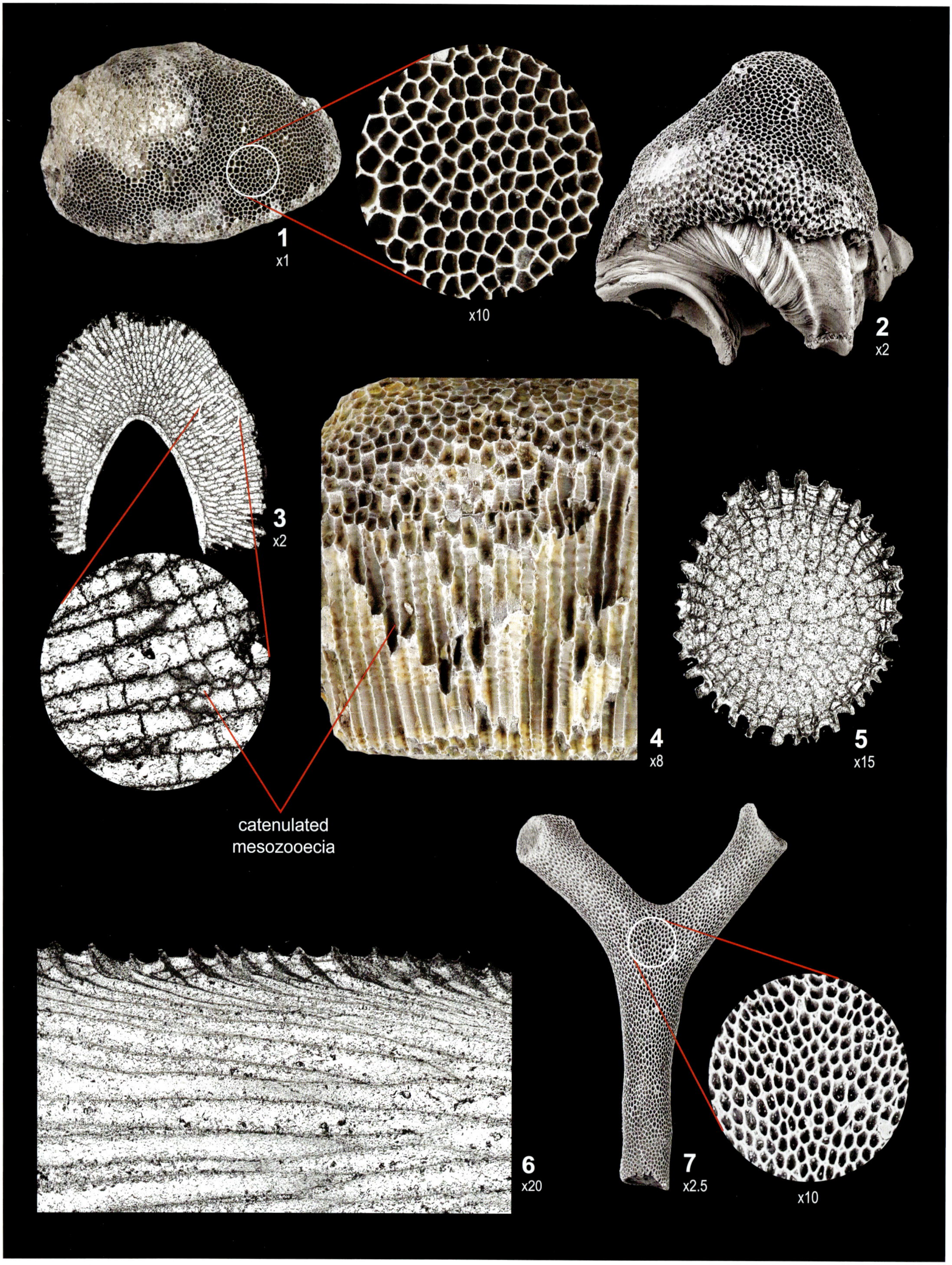
1
x1
x10
2
x2
3
x2
4
x8
5
x15
catenulated
mesozooecia
6
x20
7
x2.5
x10

walls thin and straight in the endozone, curving outward in the exozone, where they are thicker. Diaphragms variable in number; in the endozone, spaced approximately three to five zooecial diameters apart; in the exozone, three or four diaphragms in a distance of one zooecial diameter. Mesozooecia sparse, angular, average aperture diameter 0.15 mm. Monticules sparse to abundant, elevated slightly, gently rounded, approximately 1.5 mm in diameter, consisting of six to eight mesozooecia. Styles prominent and abundant, ranging from 0.03 to 0.08 mm in diameter, many extend above the zoarial surface.

Remarks. The specimens described and illustrated here are similar in most respects to *A. winchelli spinulosum*, as described by Bork and Perry (1967) from the Mifflin Member, Platteville Formation at Dixon, Lee County, Illinois.

Occurrence. Abundant in the Mifflin Member at Dixon, Illinois, and Fennimore, Wisconsin.

Genus MONOTRYPELLA

Monotrypella aequalis grandis Bork and Perry, 1967

Figures 5-7-4 through 5-7-7

Description. Zoarium ramose, round to suboval in transverse section, branches 9–12 mm in diameter. Autozooecial apertures polygonal, mostly sharply angular, a few subrounded; approximately four or five autozooecia in a distance of 2 mm; autozooecia erect and straight to mildly flexuous in the endozone, bending gradually outward distally to the exozone, where they bend abruptly, meeting the zoarial surface at a 70°–90° angle; walls thin and granular with poorly defined median line in the endozone and exozone; locally small crenulations in the walls near junction of the endozone and exozone. Diaphragms lacking in the endozone and numerous in the exozone, where they are mostly straight and closely spaced. Newly budded autozooecia resembling mesozooecia in being sparse, angular, about half the size of full-grown autozooecia. Monticules not obvious because of the overall uniform size of autozooecia. Styles absent.

Remarks. This is one of the largest ramose trepostomes in the Platteville Formation. It is distinguished from other trepostomes in the formation by having relatively large, angular autozooecial apertures and by lacking mesozooecia and styles. *M. aequalis grandis* was described by Bork and Perry (1967) as occurring in the Quimbys Mill Member near Rewey, Iowa County, Wisconsin.

Occurrence. Common in the Quimbys Mill of northwestern Illinois and southwestern Wisconsin; rare in the Mifflin and Grand Detour Members throughout the outcrop belt; particularly abundant in the upper 2 m of the Pecatonica Member at Dixon, Illinois (localities 1–4).

Figure 5-6 (Opposite page)
1–4. ***Diplotrypa catenulata*** Coryell. (**1**) Small hemispherical colony; photographic enlargement showing the relatively large polygonal zooecial apertures; (**2**) small, massive colony encrusted on a *Lophospira* shell; (**3**) acetate peel showing massive encrustation on monoplacorphoran shell; photographic enlargement showing catenulated (beaded) mesozooecia typical of this species; (**4**) weathered edge of specimen showing catenulated mesozooecia. Mifflin Member, Dixon, Illinois (locality 2), **1**, ISGS 414; **2**, ISGS 415; **3**, ISGS 398; **4**, ISGS 420.

5–7. ***Eridotrypa subtilis*** Bork and Perry. (**5**) Transverse section showing relatively large diameter of autozooecia in the endozone; (**6**) longitudinal section showing the low-angle intersection of the autozooecia at the colony surface; specimen shows the typical lack of diaphragms in the endozone and the greater diameter of autozooecia in the center of the endozone; (**7**) well-preserved ramose colony; photographic enlargement showing the angular zooecial apertures at surface of colony. Mifflin Member, Dixon, Illinois (locality 2), **5**, ISGS 1117; **6**, ISGS 402; **7**, ISGS 655.

Family HALLOPORIDAE

Genus TARPHOPHRAGMA

Tarphophragma incontroversa (*Hallopora incontroversa*) (Ulrich, 1886)

Figures 5-8-1 and 5-8-2

Description. Zoarium ramose, ovate in transverse section, branches typically 3–9 mm in diameter; largest observed incomplete specimen is 32 mm in length and bifurcates twice; encrusting growth mode seen in some specimens. Autozooecial apertures subpolygonal to subcircular, typically seven to eight autozooecia in a 2 mm distance; budding in the endozone from junctions of the preceding autozooecia, gradually expanding distally, curving outward in the exozone; autozooecial walls thicken in the outer exozone, distinct boundaries are present. Diaphragms commonly one to three zooecial diameters apart in the endozone; in the exozone, the diaphragms increase in number and are closely spaced; in the outer parts of the exozone, diaphragms commonly curve downward, uniting with previously formed diaphragms, forming structures that resemble cystiphragms. Mesozooecia less than half the size of the autozooecia, subangular to ovate, abundant, and contain numerous diaphragms. Monticules low-profile mounds approximately 1.5 mm in diameter regularly spaced at 4 mm apart, consisting of 5–20 small mesozooecia surrounded by polymorphs slightly larger than autozooecia. Styles absent.

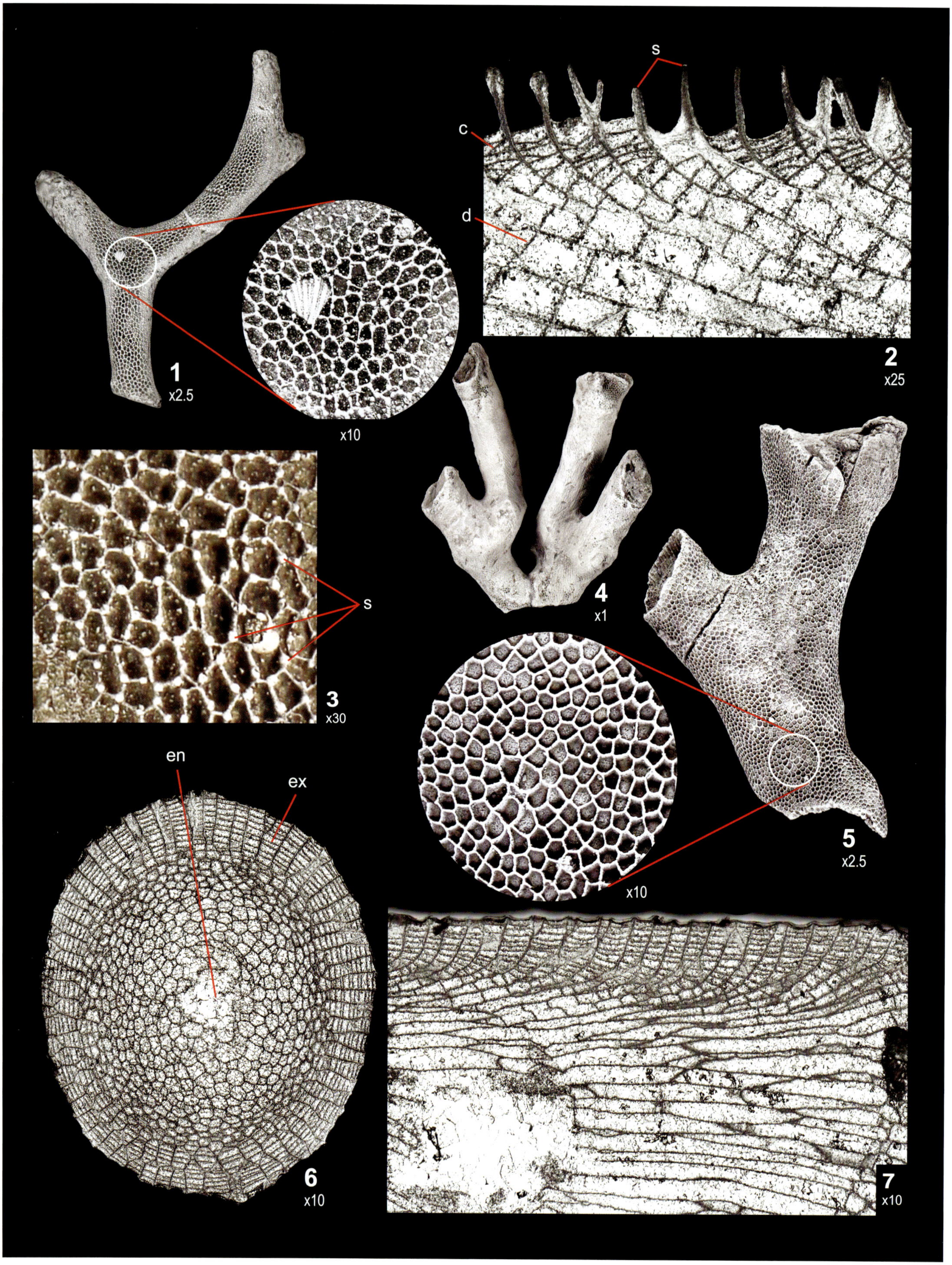
1
x2.5
x10
s
c
d
2
x25
4
x1
3
s
x30
5
x2.5
x10
en
ex
6
x10
7
x10

Remarks. Differs from *T.* cf. *multitabulata* in having a larger zoarium, more robust and multiple branches as well as numerous small mesozooecia.

Occurrence. Common in the Mifflin and Grand Detour Members in northern Illinois and southern Wisconsin and equivalent strata of the McGregor Member of southwestern Wisconsin, northeastern Iowa, and southeastern Minnesota.

Tarphophragma cf. *multitabulata* (*Hallopora multitabulata*) (Ulrich, 1886)

Figures 5-8-5 through 5-8-7

Description. Zoarium ramose and encrusting, consisting of small (6–25 mm in longest dimension) globular colonies most commonly found as encrustations on mollusks eventually forming short, round, typically nonbifurcating branches; some individuals lack an attachment scar and appear to have been free living. Mature colonies commonly consist of a mass of anastomosing branches that surround small internal cavities. Autozooecial apertures subpolygonal to subcircular, typically eight autozooecia in a 2 mm distance; budding regularly in the endozone from junctions of the preceding autozooecia, gradually expanding distally, curving outward in the exozone; autozooecial walls thicken in the outer exozone, distinct boundaries are present. Diaphragms planar, abundant in endozone, increasing in number in exozone, slightly thinner than zooecial walls. Mesozooecia common in the exozone and typically less than half the diameter of the autozooecia. Monticules slightly raised and consisting of several mesozooecia and one or two relatively large autozooecia. Styles absent.

Figure 5-7 (Opposite page)
1–3. ***Amplexopora winchelli spinulosum*** (Ulrich). (**1**) Well-preserved ramose colony; photographic enlargement showing zooecial apertures; (**2**) acetate peel of longitudinal section showing curvature of autozooecia, widely spaced diaphragms (d) in the endozone and closely spaced in the exozone, styles (s) and several cystiphragms (c) in the exozone; (**3**) enlargement of colony surface showing prominent styles (s) at corner intersections between zooids. Mifflin Member, Dixon, Illinois (locality 2), **1**, **3**, ISGS 165; **2**, ISGS 461.

4–7. ***Monotrypella aequalis grandis*** Bork and Perry. (**4**) Typical colony form; (**5**) well-preserved ramose colony with photographic enlargement showing the relatively large polygonal zooecial apertures; (**6**) acetate peel of zoarium transverse section showing the inner endozone (en) surrounded by the outer exozone (ex) with closely spaced diaphragms; (**7**) acetate peel of longitudinal section showing curvature of autozooecia, lack of diaphragms in the endozone, and closely spaced diaphragms in the exozone. Upper 3 m of Pecatonica Member, Dixon, Illinois (locality 2), **4**, ISGS 416; **5**, ISGS 389; **6**, ISGS 411; **7**, ISGS 191.

Remarks. Differs from *T. incontroversa* in having smaller zoaria with anastomosed branches.

Occurrence. Common in the Mifflin and Grand Detour Members and equivalent strata throughout the outcrop belt. Particularly abundant in the lower part of the Grand Detour Member at Dixon, Illinois.

Genus PARVOHALLOPORA

Parvohallopora cf. *spissata* (Coryell, 1921)

Figures 5-8-3 and 5-8-4

Description. Zoarium ramose, round branches approximately 4 mm in diameter. Autozooecial apertures relatively large, round to oval at the colony surface, typically five in a 2 mm distance; transverse section polygonal in the endozone and round in the exozone; autozooecial wall laminae V-shaped; autozooecial lining absent; budding in the endozone from junctions of preceding autozooecia. Diaphragms present in the proximal endozone but commonly lacking in the outer endozone and exozone. Mesozooecia numerous, occurring mainly in the inner exozone between corners of autozooecia and merging with autozooecial walls in the outer exozone; numerous closely spaced mesozooecial diaphragms present in the exozone. Monticules evenly spaced and low in profile, consisting of slightly larger autozooecia and numerous mesozooecia. Styles absent.

Remarks. *Parvohallopora* is distinguished from other ramose trepostomes in the Platteville by its relatively large autozooecial apertures, lack of styles, lack of autozooecial diaphragms in the exozone (which is unusual for this genus) and numerous closely spaced diaphragms in the exozonal mesozooecia. Like *P. spissata* (Coryell) from the approximately coeval Pierce Limestone in Murfreesboro, Tennessee, this Platteville species has mesozooecia with numerous diaphragms and autozooids lacking diaphragms in the exozone.

Occurrence. Common in the Mifflin and Grand Detour Members of northern Illinois and southern Wisconsin.

Family NICHOLSONELLIDAE

Nicholsonella cf. *ponderosa* Ulrich, 1890d

Figures 5-9-4 through 5-9-6

Description. Zoarium ramose, round to elliptical in the transverse section, branches 15–30 mm in diameter; largest observed fragmentary specimen 100+ mm long, 22 mm wide, and bifurcated twice in 90 mm. Autozooecial apertures rounded to subrounded, typically seven autozooecia

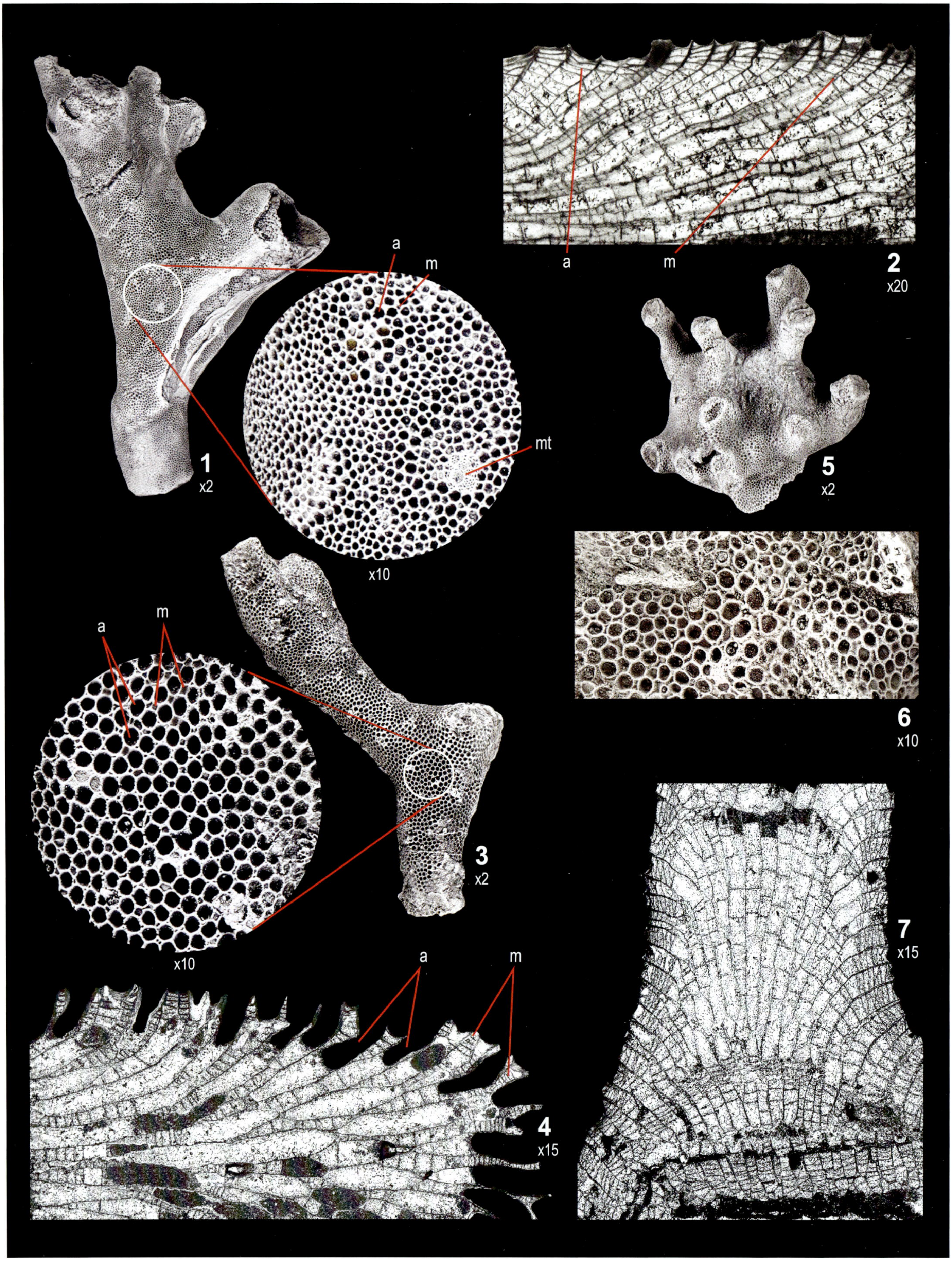
a
m
a
m
2
x20
1
x2
mt
x10
5
x2
a
m
6
x10
3
x2
x10
7
x15
a
m
4
x15

in a distance of 2 mm in intermacular areas, erect in the endozone; walls 0.30 mm thick in the exozone; wall structure indefinite due to recrystallization. Diaphragms straight, present in the endozone and exozone. Mesozooecia present but not common, smaller than autozooecia with somewhat angular apertures. Monticules not obvious. Styles present, 0.10 mm in diameter, 4–7 shared with adjacent autozooecia.

Remarks. *Nicholsonella* is one of the largest ramose trepostome bryozoan in the Platteville Formation. *N. ponderosa* was originally described from the Platteville Formation at Dixon, Illinois, by Ulrich (1890d).

Occurrence. Particularly abundant in the upper 2 m (6.6 ft) of the Pecatonica Member and lower 5 m (16 ft) of the Grand Detour Member at Dixon, Illinois.

Family PHYLLOPORINIDAE (reassigned to Order CRYPTOSTOMATA)

Remarks. *Phylloporina* (Figures 5-9-1 through 5-9-3), a common bryozoan in the Platteville Formation, has long been considered intermediate between the cryptostomatids, which it resembles in zoarial form, and trepostomatids, which it resembles in internal structure (Bassler 1953). Bassler argued for assignment to the trepostomatids, stating that *Phylloporina* "lacks proven diagnostic cryptostomatid structures" (p. G116). Blake (1975) resurrected the order Cryptostomata and assigned to it the suborders Ptilodictyina, Rhabdomesina, and Fenestellina (which includes the phylloporinids), a classification system followed in the *Treatise on Invertebrate Paleontology, Revised: Bryozoa* (Boardman et al. 1983).

Figure 5-8 (Opposite page)
1, 2. ***Tarphophragma incontroversa*** (Ulrich). (**1**) Well-preserved colony showing typical robust branches; photographic enlargement showing the relatively small autozooid (a) and mesozooid (m) apertures as well as a low-profile monticule (mt); (**2**) thin section showing the unique budding of autozooecia (a) from junctions of preceding autozooecia as well as mesozooecia (m) with numerous diaphragms. Mifflin Member, Dixon, Illinois (locality 2), **1**, ISGS 382; **2**, Bork and Perry (1967) paratype ISGS 12P 301A.

3, 4. ***Parvohallopora* cf. *spissata*** (Coryell). (**3**) Typical ramose colony; photographic enlargement showing the oval autozooecial apertures; numerous occluded mesozooecia (m) distributed between autozooecia (a); (**4**) acetate peel of longitudinal section showing typical budding of zooids in the endozone from junctions of preceding zooids; autozooecial apertures (a) are filled with matrix and mesozooecia (m) are filled with sparry calcite. Mifflin Member, Dixon, Illinois (locality 2), **3**, ISGS 1113; **4**, ISGS 144.

5–7. ***Tarphophragma* cf. *multitabulata*** (Ulrich). (**5**) Typical mature colony showing globular form with numerous short branches; (**6**) close-up showing subcircular autozooecia; (**7**), acetate peel of longitudinal section showing the endozone and exozone. Mifflin Member, Dixon, Illinois (locality 2), **5**, ISGS 400; **6**, ISGS 1111; **7**, ISGS 241.

Order CYSTOPORATA

Description. McKinney and Jackson (1989) provided the following description:

> Colonies encrusting or erect, autozooids short and without, or long and with, basal diaphragms; most autozooids with thickened strip (lunarium) of different microstructure along one side in thick-walled outer zone; skeletal structure laminated, granular or granular-prismatic, some with communication pores; vesicular extrazooidal skeleton common and abundant. Lower Ordovician–Upper Permian (or Cretaceous?). (p. 22)

Family CERAMOPORIDAE

Description. The ceramoporids are an abundant but inconspicuous group of bryozoans in the Platteville Formation. They are mostly microscopic and exhibit a wide variety of zoarial growth forms. Important characteristics of the Ceramoporidae include tubelike autozooecia, few to absent diaphragms, relatively thick walls in the exozone, and the presence of communication pores and lunaria (hoodlike structures extending above the zooecial apertures). Specific identification is difficult because of the wide range of morphologic variations within a colony. In addition, much of the definitive ceramoporid literature lacks high-resolution photographs of the external and internal skeletal features that are critical to accurate identification. An excellent starting point in understanding the ceramoporid bryozoans can be found in Utgaard (1968a, 1968b, 1969, 1983). Platteville ceramoporids include *Ceramopora, Ceramoporella*, and a variety of small, thin, hollow, ramose ceramoporids.

Genus CERAMOPORA

Ceramopora sp.

Figures 5-10-1 through 5-10-3

Description. Zoarium thin, discoidal, encrusting expansion with a prominent central monticule. Autozooecia tubelike, rhombically arranged, apertures elliptical, typically six autozooecia in a 2 mm distance in area adjacent to monticule, gradually becoming smaller in size distal to the monticule; communication pores abundant in autozooecial walls. Lunaria small in inner exozone, large at zoarial surface;

partial to complete radial arrangement around central monticule, radius of lunarium conspicuously smaller than radius of autozooecium. Exilazooecia smaller than autozooecia, subcircular in transverse section, occurring mainly in monticules, lacking diaphragms. Basal diaphragms generally absent. Monticule single or multiple. Styles absent.

Remarks. Some specimens possess a single centrally positioned monticule and overall colony form very similar to *Ceramopora imbricata* Hall, 1852, from the Middle Silurian (Wenlockian) Rochester Shale of New York.

Occurrence. Abundant in the Mifflin and Grand Detour Members in northern Illinois and southern Wisconsin and equivalent strata of the McGregor Member of southwestern Wisconsin, northeastern Iowa, and southeastern Minnesota.

Genus CERAMOPORELLA

Ceramoporella sp.

Figures 5-10-4 through 5-10-7

Description. Zoarium thin, discoidal, encrusting expansions consisting of several superimposed layers of autozooecia; individual layers approximately 0.5 mm thick. Autozooecial apertures mainly elliptical, typically eight autozooecia in a 2 mm distance, relatively thick walls, communication pores rare to absent in autozooecial walls. Some autozooecia have a prominent outward-projecting tubular extension or rim of the zooidal body that extends beyond the surface of the colony ("peristomes" of Boardman and Cheetham 1983). Lunaria erect, small in inner exozone, large at zoarial surface; not forming a prominent hood over the autozooecial aperture, partial radial arrangement around larger monticules. Exilazooecia abundant, subpolygonal to subcircular in transverse section, variable in size, locally aligned, generally surrounding the autozooecia. Diaphragms sparse to absent. Monticules irregularly spaced, consisting of small exilazooecia surrounded by approximately 10 relatively large autozooecia; depressed in center. Styles absent.

Remarks. The Platteville species is most like *Ceramoporella distincta* Ulrich (1890d) from the Upper Ordovician Decorah Formation of Minnesota, in having superimposed layers of autozooecia and abundant exilazooecia.

Occurrence. Abundant in the Mifflin and Grand Detour Members in northern Illinois and southern Wisconsin and equivalent strata of the McGregor Member of southwestern Wisconsin, northeastern Iowa, and southeastern Minnesota.

Ceramoporid spp.

Figures 5-11-1 through 5-11-14

Description. Zoarium thin, hollow, ramose, main stalk 1.0–1.5 mm in diameter, typically no more than one bifurcation; some zooaria are cap-shaped or bulbous; very thin basal layer. Autozooecia small, typically 10–12 in a 2 mm distance, angular apertures, forming indistinct ranges separated by exilazooecia, communication pores indistinct; well-laminated walls. Lunaria erect, oriented proximal to monticule, small in inner exozone, large at zoarial surface; irregular alignment of lunaria in exozone; central core of lunarium projects outward, forming a blunt rod- or bladelike spine at zoarial surface. Zoarial surfaces in some specimens show gradation from prominent to inconspicuous or absent lunaria. Exilazooecia sparse to common, subpolygonal to subcircular in section, lacking lunaria; irregular distribution of exilazooecia in exozone. Diaphragms sparse or absent. Monticules inconspicuous, consisting of six to eight radiating autozooecia. Styles possibly represented by blunt rod- or spinelike structures present at the corners of the autozooecial apertures in some specimens.

Remarks. This group of Platteville ceramoporids is distinguished by having a hollow ramose zoarium that appears to be overgrowths on a nonskeletal organism, such as a small branching alga. The wide variations in lunaria, exilazooecia, and stylelike structures suggest that more than one species may be present in the Platteville.

Occurrence. Particularly abundant in the Mifflin Member throughout the outcrop belt.

Figure 5-9 (Opposite page)
1–3. ***Phylloporina* sp.** (**1**) Typical fenestrate colony with autozooid apertures facing upward (front); photographic enlargement showing median ridge flanked on either side by one or two rows of autozooecia; (**2**) longitudinally striated back side of a colony with a few autozooecia (a) visible along the edge; (**3**) back side of a well-preserved colony. Pecatonica Member, Forreston, Illinois (locality 6), **1**, ISGS 431. Mifflin Member, Dixon, Illinois (locality 2), **2**, ISGS 455; **3**, ISGS 364.

4–6. ***Nicholsonella* cf. *ponderosa*** Ulrich. (**4**) Acetate peel of shallow tangential section showing the autozooecial apertures (a) surrounded by styles (s); (**5**) a typical robust colony; (**6**) photographic enlargement shows the relatively small, round autozooecial apertures and thick walls. Pecatonica Member, Dixon, Illinois (locality 2), **4**, ISGS 1133; **5**, ISGS 459. Grand Detour Member, Dixon, Illinois (locality 4), **6**, ISGS 1116.

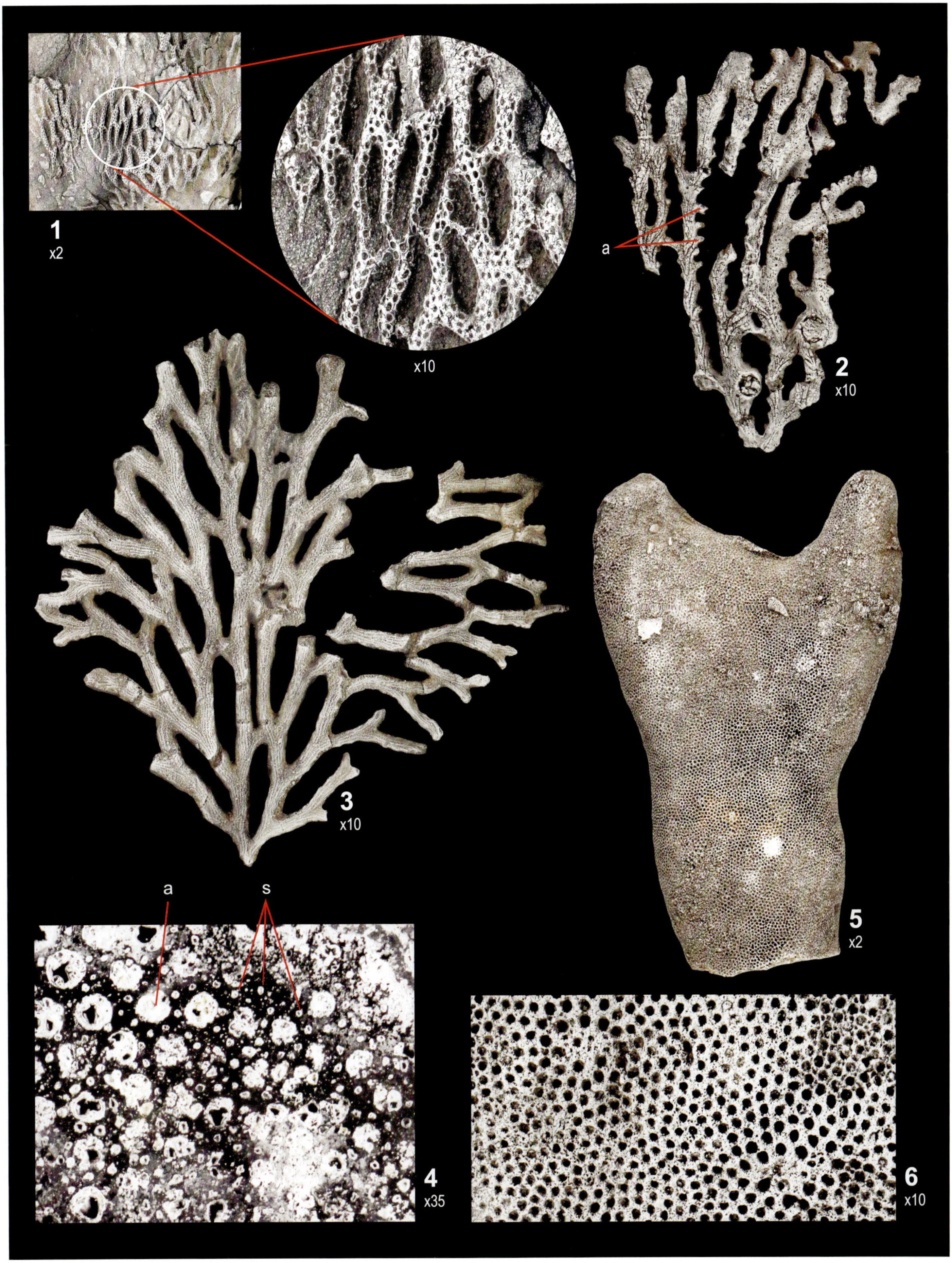
1
x2
x10
a
2
x10
3
x10
a
s
5
x2
4
x35
6
x10

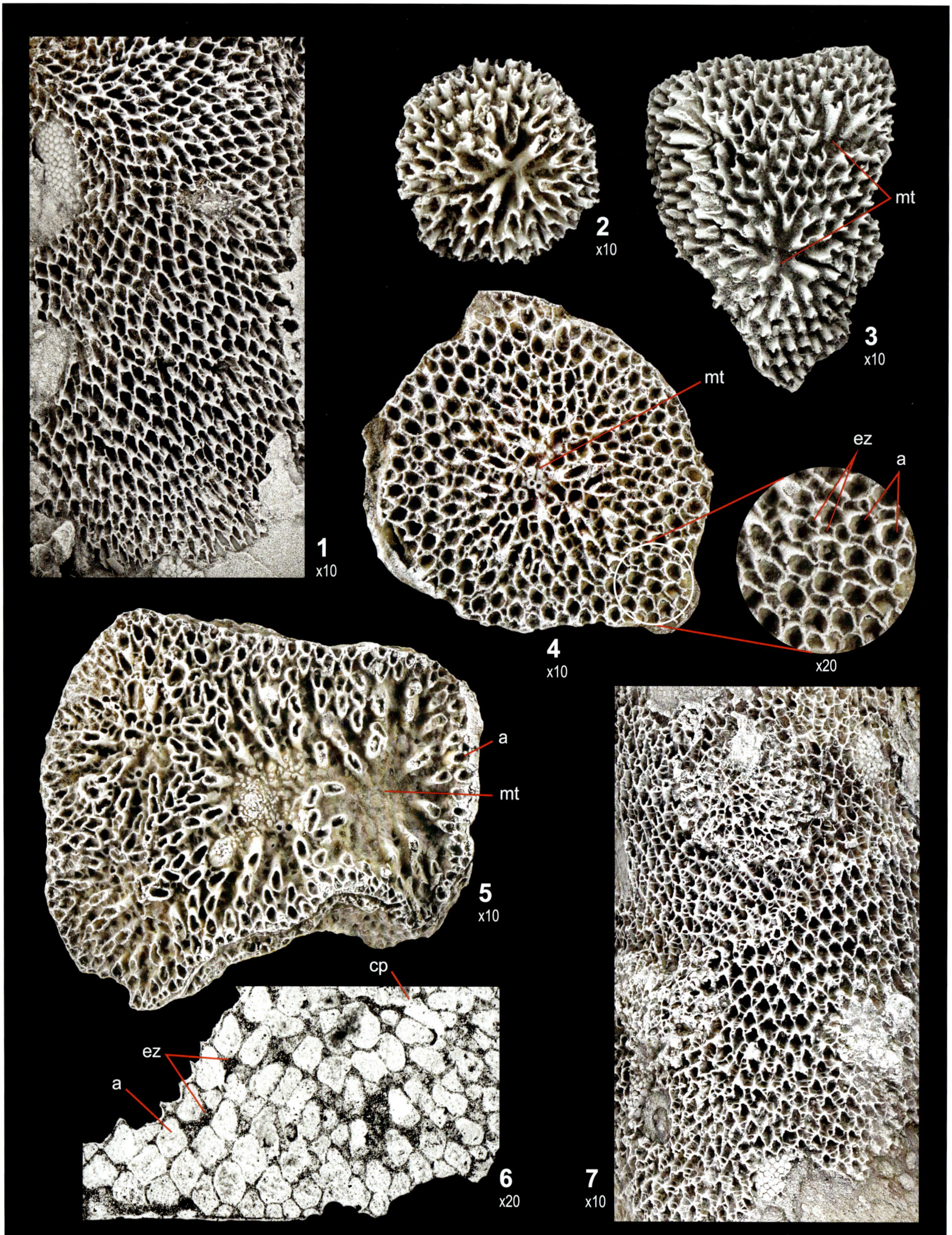
1
x10
2
x10
mt
3
x10
mt
ez
a
4
x10
x20
a
mt
5
x10
cp
ez
a
6
x20
7
x10

Family ANOLOTICHIIDAE

Genus BYTHOTRYPA

Bythotrypa* cf. *laxata (Ulrich, 1889a)

Figure 5-12-3

Description. Zoarium hemispherical, plane or concave base and convex upper surface. Autozooecial apertures mainly elliptical or pear-shaped, typically four or five granular zooecial walls in a 2 mm distance. Diaphragms very thin, slightly concave, generally 1.5–3 autozooecial diameters apart, more closely spaced in the exilazooecia (mesozooecia). Exilazooecia (mesozooecia) abundant, angular to subangular, lacking lunaria, ranging in size from less than one-tenth of a typical autozooecium to slightly larger than an autozooecium. Lunaria erect, small in inner exozone, large at zoarial surface; forming a small hood over part of the autozooecial apertures. Monticules irregularly distributed, approximately 1.7 mm in diameter, consisting of exilazooecia of irregular size and shape; flanked by autozooecia having lunaria directed toward center of monticule. Styles absent.

Remarks. The overall colony form, erect lunaria, and abundant exilazooecia of the Platteville specimens are most like *Bythotrypa laxata* (Ulrich, 1889a) from the Guttenberg Formation of southwestern Wisconsin (Bork and Perry 1968a).

Occurrence. Present but rare in the lower 2 m of the Grand Detour Member near Dixon, Illinois (locality 2).

Order CRYPTOSTOMATA

Description. McKinney and Jackson (1989) provided the following description: "Colonies are erect, occurring in arborescent or bilaminar sheets; generally short autozooids, some with basal diaphragms or hemisepta (incomplete lateral partitions). Skeletons without communication pores, typically laminated; extrazooidal skeleton may be present. Lower Ordovician–Upper Permian (or Cretaceous?)" (p. 22). Included are the suborders Ptilodictyina, Rhabdomesina, and Fenestellina.

Suborder PTILODICTYINA

Family ESCHAROPORIDAE

Description. The escharoporids are largely characterized by zoaria that are bifoliate, unbranched, lance-shaped, slender sheets that taper to a point.

Genus ESCHAROPORA

Escharopora angularis Ulrich, 1893

Figure 5-12-1

Description. Zoarium unbranched and lanceolate, slightly curved, tapered proximal tip, typically 1.2 mm wide and 10 mm long, some specimens greater than 80 mm long and 25 mm wide. Autozooecial apertures hexagonal, typically 0.25 mm long and 0.18 mm wide, aligned in distinct diagonal rows that intersect the zoarial margins at a 25° angle; zooids form a linear-rhombic pattern. Diaphragms rare to absent. Mesozooecia sparse to absent. Monticules indistinct. Styles absent.

Remarks. The overall size and shape of the zoarium and distinct hexagonal outline of the autozooecial apertures are like that of *E. angularis* Ulrich, 1893, from the Platteville Formation of Minnesota.

Occurrence. Common to locally abundant Mifflin and Grand Detour Members in northern Illinois and southern Wisconsin and equivalent strata of the McGregor Member of southwestern Wisconsin, northeastern Iowa, and southeastern Minnesota.

Family RHINIDICTYIDAE

Description. The rhinidictyids possess bifoliate zoaria that are branched or unbranched and flat. The autozooecia are organized in ranges that alternate on opposite sides of the mesotheca (median plane) and are separated by continuous longitudinal walls.

Figure 5-10 (Opposite page)
1–3. ***Ceramopora* sp.** (**1**) Encrusting colony showing radial arrangement of autozooecia; (**2**) small colony showing radial arrangement of autozooecia around central monticule; (**3**) thin encrusting colony with multiple concave monticules (mt). Mifflin Member, Dixon, Illinois (locality 2), **1**, ISGS 1120; **2**, ISGS 284; **3**, ISGS 267.

4–7. ***Ceramoporella* sp.** (**4**) Thin discoid encrusting colony showing a centrally positioned concave monticule (mt) and radiating autozooecia (a) separated by smaller exilazooecia (ez); (**5**) large colony with several generations of overgrowth and multiple concave monticules (mt); many autozooecia (a) have a prominent tubular extension that projects upward from the colony surface; (**6**) acetate peel of deep tangential section showing autozooecia (a) surrounded by exilazooecia (ez) and communication pores (cp); (**7**) thin encrusting colony. Mifflin Member, Fennimore, Wisconsin (locality 26), **4**, ISGS 349. Mifflin Member, Dixon, Illinois (locality 4), **5**, ISGS 276; **6**, ISGS 560. Mifflin Member, Dixon, Illinois (locality 2), **7**, ISGS 660.

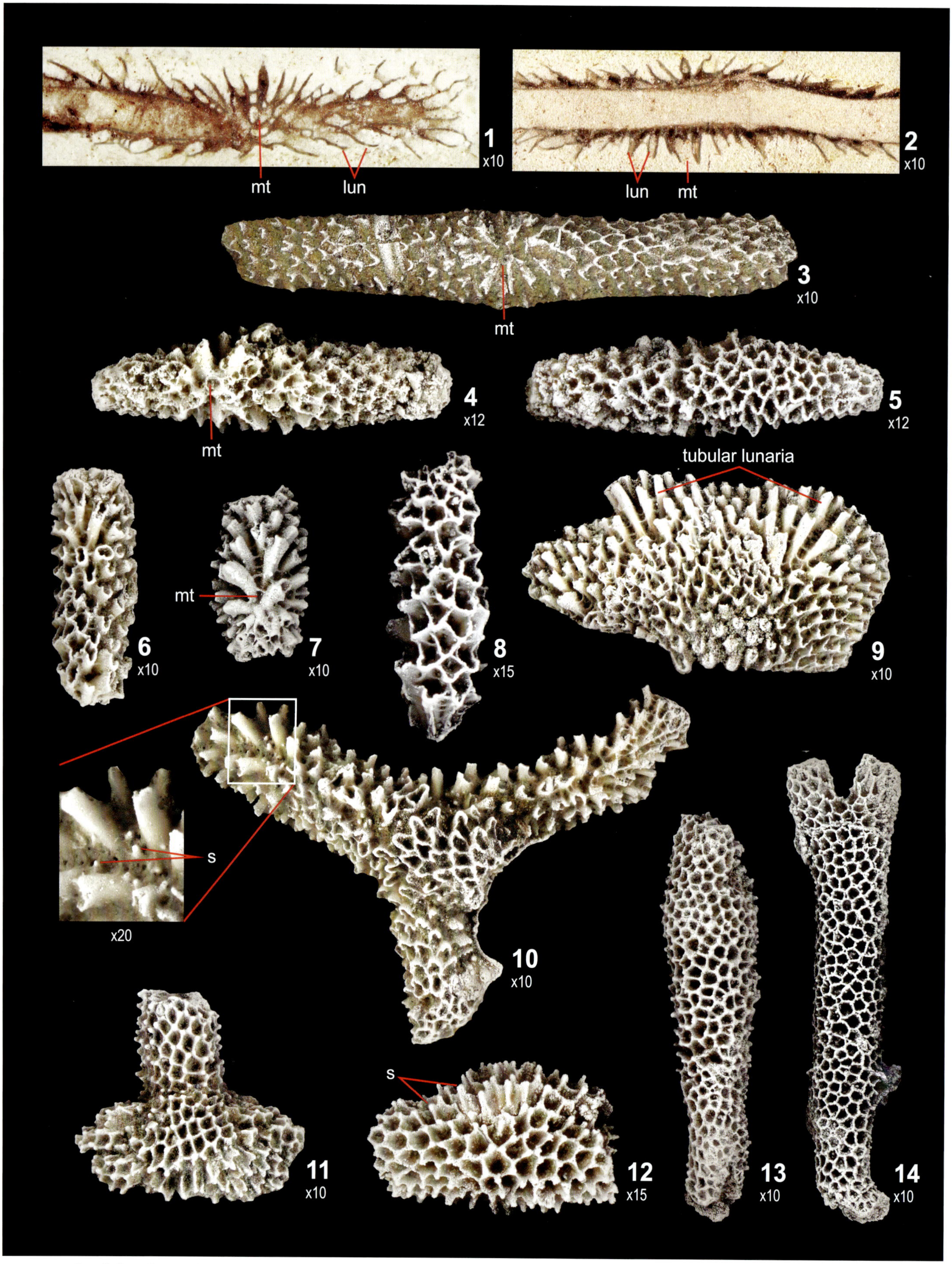
1
x10
mt
lun
2
x10
lun
mt
3
x10
mt
4
x12
mt
5
x12
tubular lunaria
6
x10
mt
7
x10
8
x15
9
x10
s
x20
10
x10
11
x10
s
12
x15
13
x10
14
x10

Genus PACHYDICTYA

Pachydictya sp.

Figures 5-12-4 through 5-12-8

Description. Zoarium relatively small and branching with wide margins, typically about 5 mm long and 1.3 mm wide. Zoaria may possess a holdfast that is shaped like a hollow claw grip. Autozooecial apertures oval, typically 0.2 mm long and 0.1 mm wide, separated by relatively thick walls; longitudinally sinuous striae between autozooecia and along zoarial margins. Diaphragms absent. Mesozooecia sparse. Monticules absent. Styles surround the zooecial apertures and are arranged in longitudinal rows between the zooecia.

Remarks. Differs from *Stictopora* in lacking hemisepta and in lacking thick, continuous longitudinal walls with zooecia arranged in distinct ranges. Well-preserved microskeletal structural details are needed to differentiate this form from other members of the Rhinidictyidae.

Occurrence. Abundant to common on bedding planes in the Mifflin and Grand Detour Members in northern Illinois and southern Wisconsin and equivalent strata of the McGregor Member of southwestern Wisconsin, northeastern Iowa, and southeastern Minnesota.

Figure 5-11 (Opposite page)
1–14. **Ceramoporid spp.** (**1**) Polished limestone slab showing longitudinal section through hollow cylindrical colony with the lunaria (lun) dipping toward a central monticule (mt); (**2**), longitudinal section through hollow cylindrical colony showing the lunaria (lun) dipping toward a central monticule (mt); (**3**) elongate colony with a single centrally positioned monticule (mt); (**4**) cylindrical colony with a single monticule (mt) facing upward; (**5**) reverse side of same colony; (**6**) small, hollow, cylindrical colony; (**7**) small, hollow, cylindrical colony with a single monticule (mt); (**8**) small, hollow, cylindrical colony with irregularly shaped autozooecial apertures; (**9**) encrusted colony showing elongation of tubular lunaria; (**10**) hollow, branching colony with variably shaped lunaria; photographic enlargement showing apparent styles (s); (**11**) encrusted colony (at base) projecting upward into hollow cylindrical colony; (**12**) small encrusted colony with protruding styles (s); (**13**,**14**), hollow cylindrical colonies with angular autozooecial apertures and poorly developed lunaria. Mifflin Member, Dixon, Illinois (locality 1), **1**, ISGS 1134; **2**, ISGS 205; **4**, **5**, ISGS 278; **6**, ISGS 238; **7**, ISGS 274; **8**, ISGS 1118; **9**, ISGS 271; **10**, ISGS 265; **11**, ISGS 289; **12**, ISGS 254; **13**, ISGS 270; **14**, ISGS 286. Mifflin Member, Fennimore, Wisconsin (locality 25), **3**, ISGS 1119.

Genus ATHROPHRAGMA

Athrophragma cf. *foliata* Ulrich, 1886

Figure 5-12-2

Description. Zoarium flattened, slightly lobate, and undulating; expands distally into fanlike shape; 24 mm long and 22 mm wide. Autozooecial apertures variable in shape, commonly round or elliptical, less commonly subangular, typically 0.35 mm long and 0.30 mm wide; autozooecia in indistinct ranges without continuous longitudinal walls. Diaphragms straight to slightly curved. Mesozooecia sparse. Monticules flat, marked by three or four small mesozooids ("vesicles" of Karklins 1983). Styles absent.

Remarks. This Platteville species is comparable to *A. foliata* (Ulrich, 1886) from the Decorah Formation of Minnesota.

Occurrence. Common in the Mifflin Member in northern Illinois and southern Wisconsin.

Genus STICTOPORA

Stictopora spp.

Figures 5-13-1 through 5-13-3

Description. Zoarium branched or unbranched and flattened, typical zoarium is less than 12 mm long, 1.5 mm wide, and 0.7 mm thick; larger specimens are in excess of 40 mm long by 2 mm wide and have five or more bifurcations; holdfasts are variable in shape and possibly species specific, common forms include flat encrusting expansions, bulbous inflated bases, and hollow clawlike grips. Autozooecial apertures round to elliptical, typically 0.30 mm long and 0.20 mm wide, autozooecia separated by relatively thick continuous longitudinal walls that form ranges and short straight to slightly curved transverse walls. Diaphragms unknown. Mesozooecia sparse. Monticules absent. Styles absent.

Remarks. *Stictopora* is one of the most common bryozoans in the Platteville Formation. Given the wide range in morphologic variations, it is possible that several species might be recognized in the Platteville Formation.

Occurrence. Abundant in the Mifflin and Grand Detour Members in northern Illinois and southern Wisconsin and equivalent strata of the McGregor Member of southwestern Wisconsin, northeastern Iowa, and southeastern Minnesota.

Family STICTOPORELLIDAE

Remarks. The stictoporellids differ from other Platteville bifoliates in microstructure, transverse sectional shape of autozooecia, and distribution of mesozooids ("exilazooecia" of Karklins 1983) in the exozone. The basal attachments are continuous skeletally with the erect parts of the zoaria.

Genus STICTOPORELLA

***Stictoporella* spp.**

Figures 5-13-4 through 5-13-6

Description. Zoarium small, slender, branched or unbranched, typically 0.5 mm wide by 5–10 mm long. Autozooecial apertures approximately hexagonal as demarcated by the zooecial boundary; each boundary segment is relatively straight and is shared by two adjacent zooecia or a zooecium and one or two mesozooecia. Diaphragms absent. Mesozooecia common. Monticules absent. Styles absent.

Remarks. The presence of one or more mesozooecia situated between zooecia is a primary distinguishing characteristic of *Stictoporella*. The smaller specimens tend to have thin zooecial walls, whereas the large specimens are thick walled.

Occurrence. Common in the Mifflin and Grand Detour Members in northern Illinois and southern Wisconsin and equivalent strata of the McGregor Member of southwestern Wisconsin, northeastern Iowa, and southeastern Minnesota throughout the outcrop belt.

Suborder FENESTELLINA

Family PHYLLOPORINIDAE

Genus PHYLLOPORINA

***Phylloporina* sp.**

Figures 5-9-1 through 5-9-3

Description. Zoarium slender, anastomosing branches, 0.8–1.0 mm wide; some zoaria wider than 20 mm wide. Autozooecial apertures round to elliptical, typically 10 in a distance of 2 mm, fenestrules long, narrow, and irregularly shaped; autozooecia in three or four rows on front side and none on the back, which is striated. Diaphragms undetermined. Mesozooecia less than half the diameter of autozooecia. Monticules absent. Styles absent.

Remarks. Because of poor preservation in the available specimens, the microskeletal structures are not fully known.

Occurrence. Common in the Pecatonica, Mifflin, and Grand Detour Members, northern Illinois and southern Wisconsin.

Suborder RHABDOMESINA

Family ARTHROSTYLIDAE

Description. Arthrostylids have erect, branching zooaria consisting of small, discrete sticklike branch segments. Mostly branching, some are articulated, many very slender.

Genus ARTHROCLEMA

Arthroclema* cf. *armatum Ulrich, 1893

Figures 5-13-7 and 5-13-8

Description. Zoarium branched, jointed, typical segments are 0.3–0.5 mm in diameter, 4 mm long and straight to slightly curved; autozooecia arranged in two to four longitudinal rows separated by longitudinal ridges, obverse side lacks autozooecia. Autozooecial apertures rounded to subpolygonal, 0.1 mm wide and 0.2 mm long; apertural arrangement in longitudinal rows separated by a prominent ridge; apertures positioned side by side or slightly offset; curved ridge (peristome) developed on proximal edge of apertures. Diaphragms undetermined. Mesozooecia absent. Monticules absent. Styles undetermined.

Remarks. This Platteville arthrostylid most resembles *A. armatum* Ulrich, 1893, from the Decorah Formation of southeastern Minnesota, in having a well-defined axial stem

Figure 5-12 (Opposite page)

1. ***Escharopora angularis*** Ulrich. (**1**) Well-preserved colony showing rhombic arrangement of autozooecia; photographic enlargement showing the predominantly hexagonal autozooecial apertures. Mifflin Member, Dixon, Illinois (locality 2), **1**, ISGS 341.

2. ***Athrophragma* cf. *foliata*** Ulrich. (**2**) Well-preserved colony; photographic enlargement shows subangular autozooecial apertures with thick walls and small localized mesozooecia (m). Mifflin Member, Fennimore, Wisconsin (locality 25), **2**, ISGS K 311.

3. ***Bythotrypa* cf. *laxata*** (Ulrich). (**3**) Well-preserved specimen showing the massive hemispherical form of the colony; photographic enlargement shows the angular apertures of the autozooid (a) and mesozooid (m) apertures. Grand Detour Member, Dixon, Illinois, (locality 4), **3**, ISGS 1127.

4–8. ***Pachydictya* sp.** (**4**) Dolomitized colony showing anastomosing branches; photographic enlargement shows oval autozooecial apertures, thick walls, and longitudinally sinuous striae between apertures; (**5**) small bifurcated colony with holdfast intact; (**6**) small colony with wide margins; photographic enlargement showing linear arrangement of mural styles (mst) between autozooecial apertures; (**7**) small colony with wide margins and basal holdfast; (**8**) typical colony with wide margins. Grand Detour Member, South Beloit, Illinois (locality 20), **4**, ISGS 440. Mifflin Member, Dixon, Illinois (locality 2), **5**, ISGS 1123; **6**, ISGS 325; **7**, ISGS 335; **8**, ISGS 314.

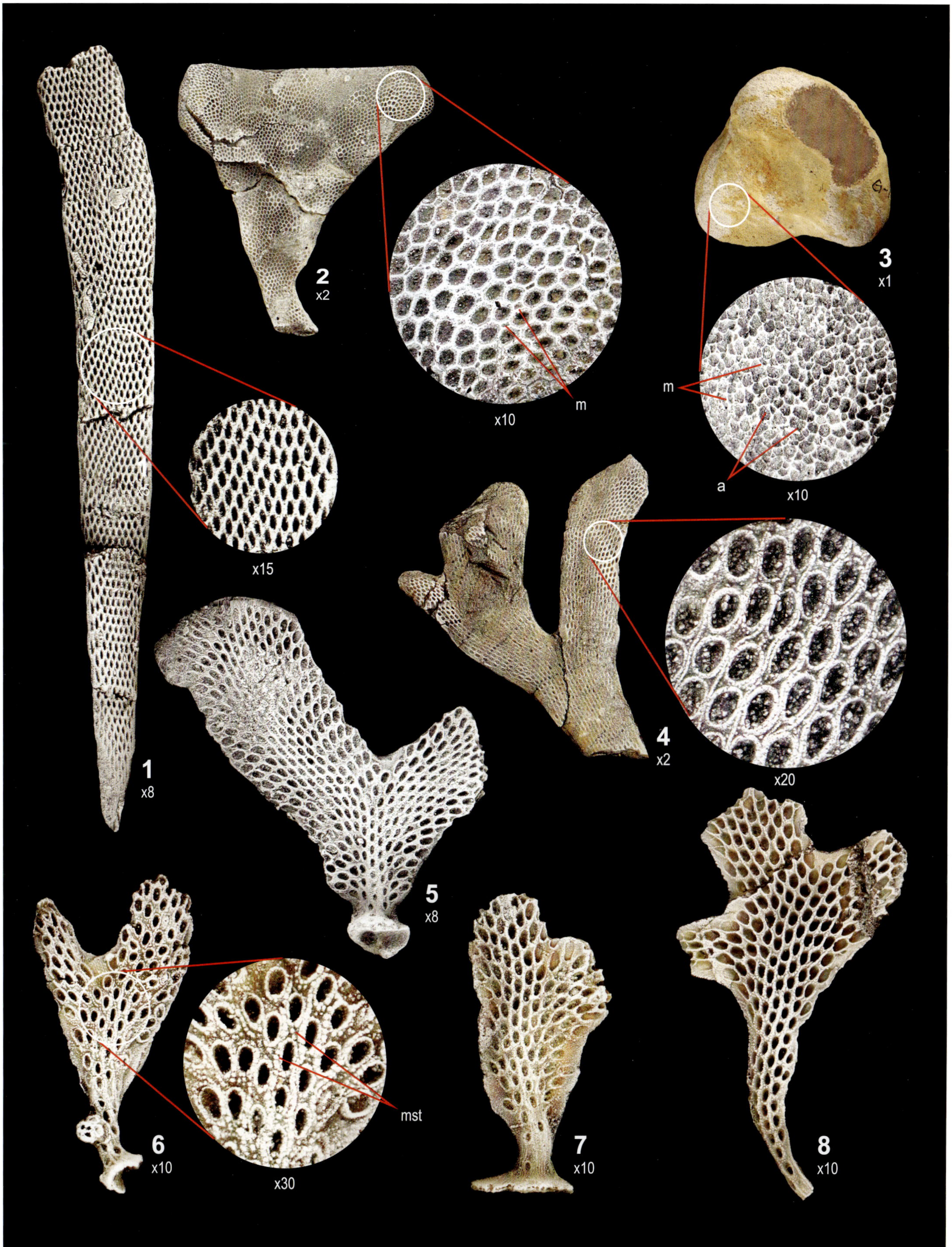
1
x8
x15
2
x2
m
x10
3
x1
m
a
x10
4
x2
x20
5
x8
6
x10
mst
x30
7
x10
8
x10

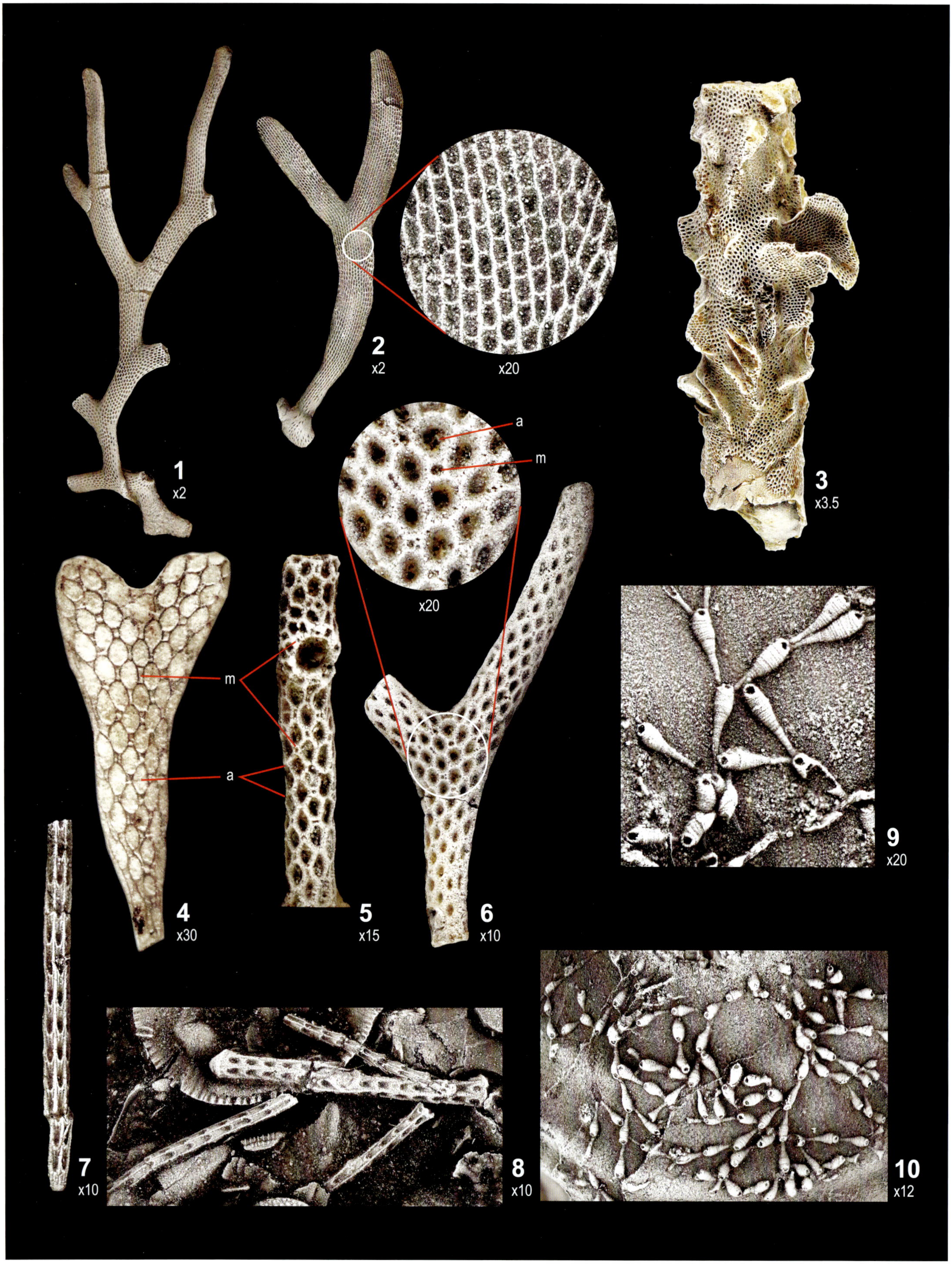
1
x2
2
x2
x20
3
x3.5
a
m
x20
m
a
4
x30
5
x15
6
x10
9
x20
7
x10
8
x10
10
x12

and alternate secondary and tertiary branches. Precise identification will require preparation of thin sections.

Occurrence. Common in the Mifflin and Grand Detour Members in northern Illinois and southern Wisconsin and equivalent strata of the McGregor Member of southwestern Wisconsin, northeastern Iowa, and southeastern Minnesota.

Order TUBULIPORATA

Description. Zoaria consists of simple thin-walled calcareous tubes generally without diaphragms.

Family CORYNOTRYPIDAE

Description. The corynotrypids have encrusting threadlike colonies of uniserial, club-shaped, entirely exterior-walled autozooecia arranged in a branching pattern.

Genus CORYNOTRYPA

Corynotrypa* cf. *delicatula (James, 1878)

Figures 5-13-9 and 5-13-10

Description. Zoarium encrusting, runner-like, uniserial; branches bifurcate at regular intervals; branching angles mostly 70°–80°. Autozooecia consist of a linear series connected by a narrow canal; elongate and club-shaped, 0.5 mm long and 0.2 mm wide, proximal end 0.08 mm wide. Diaphragms, mesozooids, monticules, and styles absent.

Remarks. The size, shape and branching pattern are most like those of *Corynotrypa delicatula* from the Eden Shale of Ohio (James 1878) and Platteville Formation at St. Paul and Minneapolis, Minnesota (Ulrich 1886). Specimens are commonly found encrusting the interiors of gastropods and cephalopods.

Occurrence. Common in the Mifflin and Grand Detour Members in northern Illinois and southern Wisconsin and equivalent strata of the McGregor Member of southwestern Wisconsin, northeastern Iowa, and southeastern Minnesota.

Figure 5-13 (Opposite page)
1–3. ***Stictopora* spp.** (**1**) Typical ramose colony; (**2**) complete colony with clawlike holdfast; photographic enlargement showing continuous longitudinal walls with short slightly curved transverse walls; (**3**) encrustation of budding colonies on small cephalopod. Mifflin Member, Dixon, Illinois (locality 2), **1**, ISGS 370; **3**, ISGS 1102. Mifflin Member, Fennimore, Wisconsin (locality 25), **2**, ISGS 438.

4–6. ***Stictoporella* spp.** (**4**) Small colony photographed under glycerin showing mesozooecia (m) between adjacent autozooecia (a); (**5**) colony showing well-developed autozooecia (a) and mesozooecia (m); (**6**) mature colony; photographic enlargement showing the thick zooid walls and distinct autozooecia (a) and mesozooecia (m). Mifflin Member, Dixon, Illinois (locality 2), **4**, ISGS 336; **5**, ISGS 443; **6**, ISGS 315.

7, 8. ***Arthroclema* cf. *armatum*** Ulrich. (**7**) Single branch showing the autozooecia; (**8**) partly disarticulated colony with primary and secondary branch segments showing autozooecia arranged in two or three longitudinal rows separated by longitudinal ridges. Mifflin Member, Dixon, Illinois (locality 2), **7**, ISGS 365. Mifflin Member, Fennimore, Wisconsin (locality 25), **8**, ISGS 356.

9, 10. ***Corynotrypa* cf. *delicatula*** (James). (**9**) Threadlike colonies of uniserial, entirely exterior-walled autozooecia encrusted on a *Trochonema* shell; (**10**) encrustation of autozooecia on *Tetranota* shell. Mifflin Member, Dixon, Illinois (locality 2), **9**, ISGS K442; **10**, ISGS 1121.

CHAPTER 6
BRACHIOPODS

Brachiopods are among the most common fossils in the Platteville Formation. Approximately 20 species are present, including members of the orders Lingulida, Craniida, Strophomenida, Orthotetida, Protorthida, Orthida, Rhynchonellida, Pentamerida, and Atrypida (Appendix 2). On the basis of overall distribution, abundance, and diversity, the brachiopod fauna is a strophomenid–orthid-dominated association typical of that in the Midcontinent North American Late Ordovician carbonate succession. In the limestone facies, brachiopod shell material can be very well preserved and only slightly altered by diagenetic processes. In contrast, brachiopod fossils in the Platteville dolostone facies typically are preserved as external or internal molds (steinkerns) and casts.

Brachiopods are attached or free-living marine animals with a fossil record beginning in Early Cambrian time and continuing to the present. All soft tissues are enclosed in a pair of valves (shells) that are bilaterally symmetrical. Unlike bivalve mollusks, the plane of symmetry extends through the two valves rather than between them. The two valves are commonly dissimilar in size, shape, and ornament and are anatomically designated as **dorsal** and **ventral** (Figures 6-1 and 6-2). Most brachiopods are attached to the substrate by a fleshy stalk called a **pedicle**, which extends from near the **posterior end** of the ventral valve often through an opening named the **pedicle foramen**. Hence, the ventral shell is also referred to as the **pedicle valve**. Likewise, the dorsal valve is referred to as the **brachial valve** because it supports the two brachial arms of the lophophore, a ciliated organ that extracts food particles from the surrounding seawater and passes them into the stomach near the posterior end (Figure 6-1). "Dorsal" and "ventral" are the preferred descriptive terms used in this report. In life, the two valves gape apart at the anterior when feeding, allowing nutrient-rich seawater to circulate through the body cavity; their adjacent edges where the valves close is called the **commissure**. The ventral and dorsal valves each have unique features that are important for identification and classification. These include shell composition, overall valve shape or profile, exterior shell ornamentation, hinge morphology, and muscle scars.

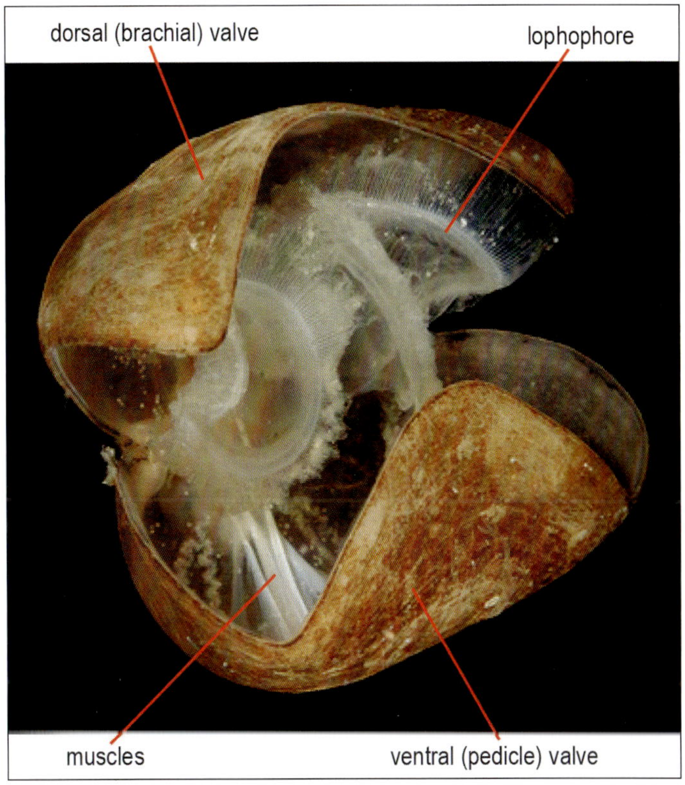

Figure 6-1 A living brachiopod with gaping dorsal and ventral valves exposing internal organs, including ciliated lophophore and muscles. Photograph by Arthur Anker. Used with pemission.

Brachiopod specialists traditionally have classified the phylum into two main subdivisions, the classes Articulata and Inarticulata. The "articulates" have a well-developed hinge with interlocking ventral valve **teeth** and dorsal valve **sockets**, whereas the "inarticulates" lack a hinge. Advances in molecular biology and a phylogenetic approach have put less emphasis on the presence or absence of a hinge (Williams et al. 1996). The revision of the brachiopods in the *Treatise on Invertebrate Paleontology* (Williams et al. 2000) supports a classification scheme based on the recognition of three subphyla: the Linguliformea (lingulides, acrotretides, siphonotretides, and paterinides), Craniiformea (craniides), and Rhynchonelliformea (all other "articulate" brachiopods). However, most authors recognize "articulate" and "inarticulate" as useful descriptive terms but avoid their use in formal classification. The classification scheme in this report follows Williams et al. (2000). All three subphyla are represented in the Platteville Formation.

The Platteville Linguliformea consist of a small number of lingulides characterized by organophosphatic shells lacking a tooth and socket hinge. The Platteville craniides are largely of dorsal shells that are cap-shaped, encrusting, and calcareous. The Rhynchonelliformea, which possess

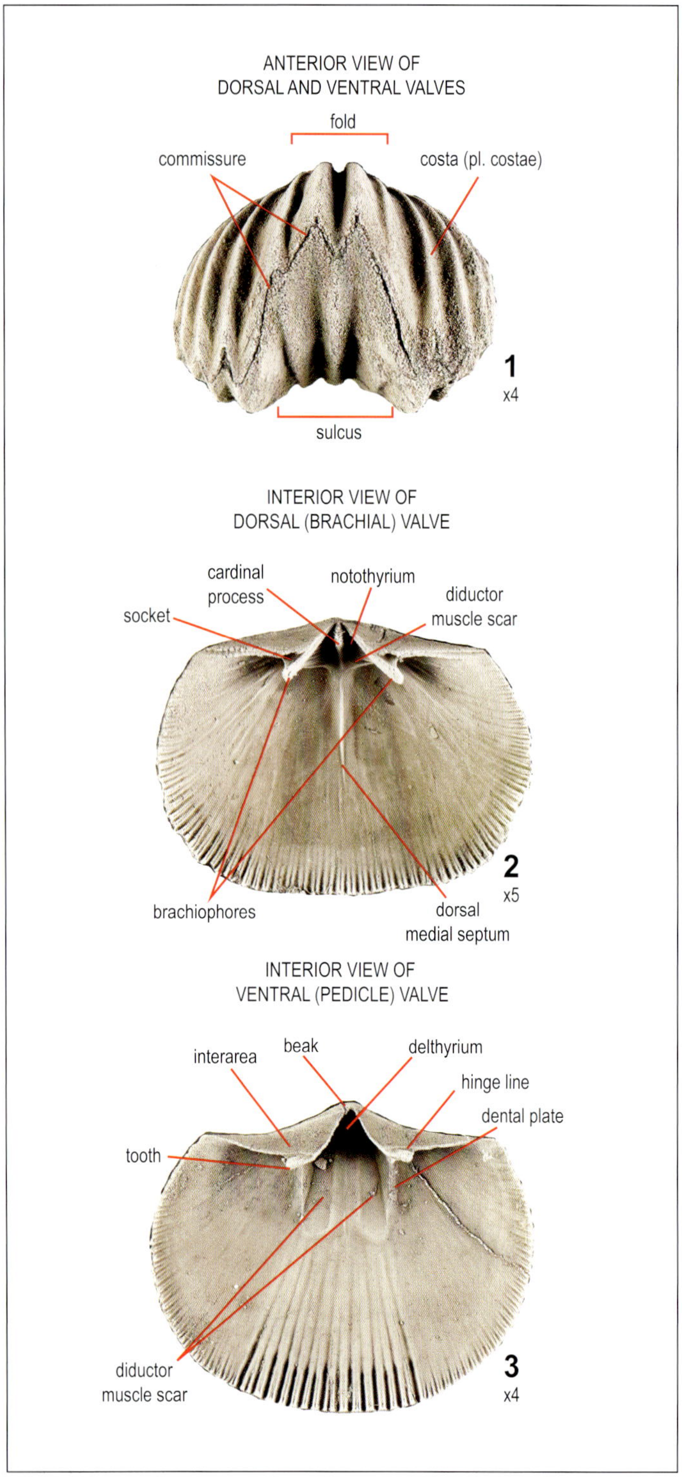

Figure 6-2

1–3. Key morphologic features used in the identification of Platteville brachiopods. All specimens are from the Mifflin Member, Dixon, Illinois (locality 2). (**1**) Anterior view of *Rostricellula minnesotensis* showing closed dorsal (above) and ventral (below) valves (ISGS 646); (**2**) dorsal valve interior of *Doleroides pervetus* (ISGS 669); (**3**) ventral valve interior of *Doleroides pervetus* (ISGS 600).

calcareous shells and a **tooth** and **socket** hinge, are by far the most abundant and diverse brachiopods in the formation. Key features in their identification are shown in Figure 6-2 and include (1) the presence of deflections of the anterior valve edge forming a median **fold** or **sulcus**; (2) radial ridges or ribs called **costae** (a condition called **costate**) on the external surface of both valves extending outward from the medial posterior end; (3) structures of the shell in the medial posterior of the dorsal valve associated with articulation, support of the lophophore, and muscle attachment, including the **cardinal process**, **socket ridges**, **crural bases**, and **brachiophores** (support for the lophophore) as well as an opening in the posterior end called the **notothyrium**—together, referred to as the **cardinalia**; (4) structures in the medial posterior pedicle valve associated with the pedicle opening, including the **delthyrium**, a triangular aperture often serving as the pedicle opening, and **dental ridges** or plates underlying the hinge teeth and extending to the floor of the ventral valve; (5) size and shape of the surface lying between the beak and hinge line in both valves, called the **interarea**; (6) shape and position of **muscle scars**; (7) raised structures named **septa** in the interior of either valve; and (8) microscopic perforations in the shell generally referred to as **punctae**, which are variably present. More in-depth morphology can be found in Williams et al. (1997).

One of the primary literature sources for the identification of brachiopods in the Platteville Formation, and the Ordovician in general, is the monumental work of G. Arthur Cooper (1956) titled *Chazyan and Related Brachiopods*. That volume is particularly rich in high-quality photographs, and the descriptions are clear and concise. Other important taxonomic works on Platteville and related brachiopods include Fenton (1928), Salmon (1942), Wilson (1945), Weiss (1955), Bretsky and Bretsky (1975), and Congreve et al. (2015, 2019).

SYSTEMATIC DESCRIPTIONS

Phylum BRACHIOPODA

Subphylum LINGULIFORMEA

Class LINGULATA

Order LINGULIDA

Remarks. The Platteville Formation contains a small number of lingulide brachiopod species characterized by organophosphatic shells that are subrectangular or oval in outline. Most specimens are difficult to assign to a particular genus or species because key morphologic features, such as muscle scars, septa, pedicle openings, and posterior shell margins, are poorly preserved and very subtly expressed. On the basis of size, shape, and shell ornamentation, observed Platteville lingulides are most similar to *Pseudolingula* and *Plectoglossa*.

Family PSEUDOLINGULIDAE

Genus PSEUDOLINGULA

***Pseudolingula* sp.**

Figures 6-3-1 and 6-3-2

Description. Shell outline subrectangular to subelliptic, organophosphatic shell, typically 25–30 mm long, ornamentation consists of concentric ridges.

Remarks. This lingulide appears to be most like the widespread genus *Pseudolingula* found in Late Ordovician rocks of North America and elsewhere. Most Platteville specimens are preserved perpendicular to bedding in their original life position.

Occurrence. Common in the Mifflin and Grand Detour Members in northern Illinois and southern Wisconsin and equivalent strata of the McGregor Member of southwestern Wisconsin, northeastern Iowa, and southeastern Minnesota.

Family OBOLIDAE

Genus PLECTOGLOSSA

Plectoglossa* cf. *oklahomensis Cooper, 1956

Figures 6-3-3 and 6-3-4

Description. Shell outline elongate oval, biconvex in profile, organophosphatic shell, typically 10–15 mm long, with fine concentric ridges.

Remarks. This lingulide is morphologically most like *Plectoglossa* cf. *oklahomensis* from the coeval Pooleville Member of the Bromide Formation in the Criner Hills, Carter County, Oklahoma.

Occurrence. Common in the Mifflin Member of northern Illinois and southern Wisconsin. Typically found in clusters on shaley bedding planes.

Subphylum CRANIIFORMEA

Class CRANIATA

Order CRANIIDA

Family CRANIIDAE

Genus ACANTHOCRANIA

Acanthocrania granulosa (Hall and Clarke, 1892)

Figures 6-3-5 through 6-3-7

Description. Dorsal valve convex to subconical, typically 5–10 mm in diameter; encrusting; ornament of concentrically arranged hollow spines, in some cases simulating ornament of the host shell; beak off-center; ventral valve unknown.

Remarks. Assignment to *Acanthocrania* is based on the convex dorsal valve, with numerous hollow spines and lack of a ventral valve. Like *Petrocrania* and *Philhedra* described below, the ventral valves are almost never found as fossils but are known in other members of the family. The size and density of spines are most like *A. granulosa* (Hall and Clarke, 1892) from the McGregor Member in the upper Mississippi Valley. The spines in the Platteville species are also similar to *A. grandis* Cooper, 1956, from the approximately coeval Late Ordovician Benbolt Formation of Virginia.

Occurrence. Common in the Mifflin and Grand Detour Members in north-central Illinois and south-central Wisconsin and the McGregor Member of southwestern Wisconsin, northeastern Iowa, and southeastern Minnesota. Locally common in the Pecatonica Member in north-central Illinois.

Genus PETROCRANIA

Petrocrania dixonensis Cooper, 1956

Figures 6-3-8 through 6-3-12

Description. Dorsal valve convex to subconical, calcified and punctate; encrusting; 5–10 mm in diameter; exterior surface can be smooth or marked by numerous irregular indentations; in some specimens, the shell conforms to the ornament of the host shell, whereas others have concentrically arranged growth lines; beak slightly off-center; ventral valve unknown.

1
x2
2
x2
3
x2
x10
5
x4
4
x4
x5
9
x1.5
punctae
x20
8
x2
6
x3.5
x10
beak
10
x3
12
x1.5
7
x3.5
beak
11
x3
13
x12

Remarks. *Petrocrania* is similar to *Acanthocrania* but lacks the hollow spines. The species is named for Dixon, Illinois, where it occurs in the Platteville Formation (Cooper 1956).

Occurrence. Common in the Mifflin and Grand Detour Members in north-central Illinois and south-central Wisconsin and the McGregor Member of southwestern Wisconsin, northeastern Iowa, and southeastern Minnesota. Locally common in the Pecatonica Member in north-central Illinois.

Genus PHILHEDRA

Philhedra cf. *minor* Cooper, 1956

Figure 6-3-13

Description. Dorsal valve small, slightly longer than wide, forming an eccentric cone; encrusting. Apex smooth, positioned about one-fifth the length from the posterior margin and slightly overhanging the posterior slope. Ventral valve unknown. Costae nearly reaching the beak, increasing in number by intercalation and in width from center to margin.

Remarks. The Platteville species is most similar to *Philhedra minor* Cooper, 1956, from the slightly older rocks of the Upper Ordovician Pierce Formation in central Tennessee. It differs from *Acanthocrania* and *Petrocrania* in having costae that radiate from the beak.

Occurrence. Rare in the Platteville Formation.

Figure 6-3 (Opposite page)

1, 2. ***Pseudolingula* sp.** (**1**) Specimen in apparent living position perpendicular to bedding, with anterior end oriented upward; (**2**) lateral view of specimen oriented with anterior end at top. Grand Detour Member, Dixon, Illinois (locality 2), **1**, ISGS 639; **2**, ISGS 620.

3, 4. ***Plectoglossa* cf. *oklahomensis*** Cooper. (**3**) Bedding plane surface with small cluster of specimens; (**4**) well-preserved specimen showing organophosphatic shell and concentric growth lines. Mifflin Member, Fennimore, Wisconsin, **3**, JAC 104. Mifflin Member, Dixon, Illinois (locality 2), **4**, ISGS 622.

5–7. ***Acanthocrania granulosa*** (Hall and Clark). (**5**) Dorsal valve exterior; photographic enlargement showing small, hollow spines; (**6**) dorsal valve exterior; (**7**) lateral view of dorsal valve exterior showing small, hollow spines in photographic enlargement. Mifflin Member, Dixon, Illinois (locality 2), **5**, ISGS 609; **6**, **7**, ISGS 1574.

8–12. ***Petrocrania dixonensis*** Cooper. (**8**) Outer surface of dorsal valve; photographic enlargement showing numerous punctae (small pores); (**9**) dorsal valve encrusted on gastropod *Clathrospira conica*; photographic enlargement showing concentric growth rings; outer surface of dorsal valve showing beak and prominent irregular indentations; (**10**) angled profile view and (**11**) plan view showing beak and prominent irregular indentations; (**12**) profile of dorsal surface encrusted on bivalve *Cyrtodonta billingsi*. Mifflin Member, Dixon, Illinois (locality 2), **8**, ISGS 616; **9**, ISGS 610; **10**, **11**, ISGS 611; **12**, ISGS 606.

13. ***Philhedra* cf. *minor*** Cooper. (**13**) Dorsal valve showing prominent radiating costae. Mifflin Member, Dixon, Illinois (locality 2), **13**, ISGS 676.

Subphylum RHYNCHONELLIFORMEA

Class STROPHOMENATA

Order STROPHOMENIDA

Superfamily STROPHOMENOIDEA

Family STROPHOMENIDAE

Remarks. In *Strophomena* and *Trigrammaria* the dorsal and ventral valves have a resupinate shell profile characterized by a reversal in the relative convexity in which the convex juvenile stage ventral valve becomes concave and the normally concave dorsal valve becomes convex during growth. The opposite condition, called "concavo-convex," characterizes some other strophomenides, such as *Oepikina*. A distinguishing characteristic of the Platteville strophomenide brachiopods is that the pedicle opening or delthyrium and corresponding notothyrium on the dorsal valve are covered by small plates, the pseudodeltidium and chilidium, respectively. Although juveniles had a small apical pedicle, in adults that pedicle opening became sealed over with shell material, so their increased weight at their posterior margins kept them correctly oriented for feeding, with their anterior margins extending above the sea floor.

Genus OEPIKINA

Oepikina minnesotensis (Winchell, 1881)

Figures 6-4-1 through 6-4-8

Description. Outline semicircular and truncated by a straight hinge line; profile concavo-convex; ornament consists of fine radiating costae arising entirely by intercalation between coarse costae; pseudodeltidium and small chilidium present; teeth crenulated; dental plates short and weakly developed; ventral muscle field strongly impressed; dorsal valve interior has a large bilobed cardinal process near the hinge line, well-developed muscle scars separated by a prominent median septum, and a distinctive subperipheral rim.

Remarks. The genus *Oepikina* was introduced by Salmon (1942) in her review of *Rafinesquina* and related species of the North American Middle Ordovician. Using the point of greatest curvature and degree of inflation of the shell, she recognized three species in the Platteville Formation: *O. minnesotensis* (Winchell, 1881), *O. transitionalis* (Okulitch,

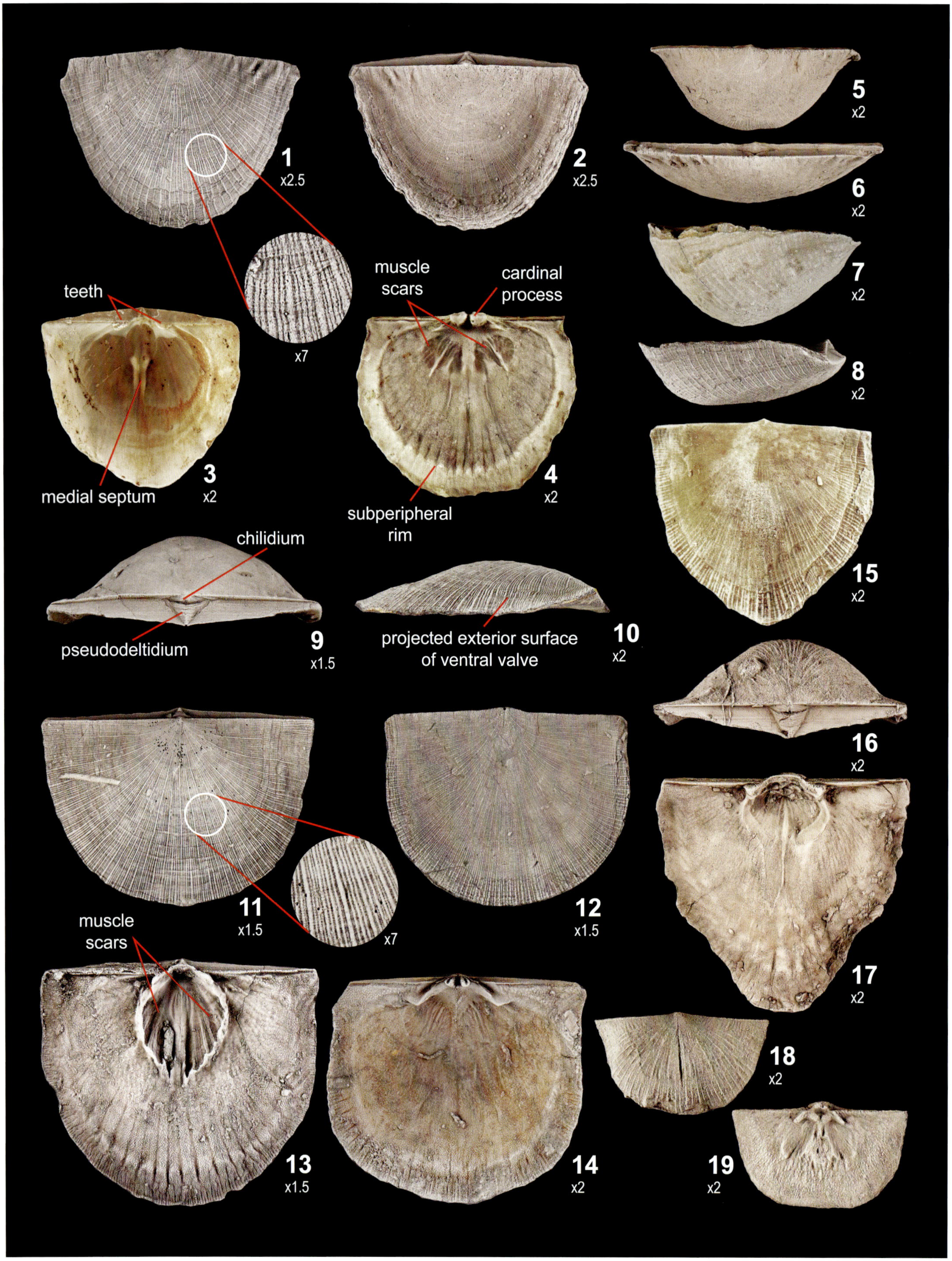
1
x2.5
2
x2.5
5
x2
6
x2
7
x2
8
x2
x7
teeth
muscle
scars
cardinal
process
3
x2
medial septum
4
x2
subperipheral
rim
15
x2
chilidium
pseudodeltidium
9
x1.5
projected exterior surface
of ventral valve
10
x2
16
x2
11
x1.5
x7
12
x1.5
17
x2
muscle
scars
18
x2
13
x1.5
14
x2
19
x2

1935), and *O. inquassa* (Sardeson, 1892). Employing a multivariate analysis of 437 *Oepikina* specimens from the Platteville Formation, Bretsky and Bretsky (1975) found that Salmon's (1942) species concepts were impossible to apply in practice and chose to follow Cooper (1956) in recognizing only *O. minnesotensis*.

Occurrence. Very abundant in the Mifflin and Grand Detour Members in north-central Illinois and south-central Wisconsin and the McGregor Member of southwestern Wisconsin, northeastern Iowa, and southeastern Minnesota. Locally common in the Pecatonica and Quimbys Mill Members in north-central Illinois.

Figure 6-4 (Opposite page)
1–8. ***Oepikina minnesotensis*** (Winchell). (**1**) Ventral valve exterior; photographic enlargement showing costae; (**2**) dorsal valve exterior with small part of ventral valve interarea showing at top; (**3**) ventral valve interior showing medial septum and teeth; (**4**) dorsal valve interior showing cardinal process, muscle scars, and subperipheral rim; (**5**) posterior end of a highly inflated ventral valve; (**6**) posterior end of a moderately inflated ventral valve; (**7**) profile of a highly inflated ventral valve with posterior end on the right; (**8**) profile of a moderately inflated ventral valve with posterior end on the right. Mifflin Member, Fennimore, Wisconsin (locality 24), **1**, **2**, ISGS 582. Mifflin Member, Dixon, Illinois (locality 2), **3**, **8**, ISGS 590; **4**, ISGS 570; **5**, ISGS 648; **6**, ISGS 649; **7**, ISGS 589.

9–14. ***Strophomena plattinensis*** Fenton. (**9**) Posterior end (dorsal valve facing up) showing the chilidium and pseudodeltidium; (**10**) profile view (posterior end to the right) showing the reversal in relative convexity in which the ventral valve becomes concave during growth and the dorsal valve becomes convex; (**11**) dorsal valve exterior; photographic enlargement showing two or three thin costae alternating with a relatively wide costa; (**12**) ventral valve exterior; (**13**) ventral valve interior showing prominent muscle scars; (**14**) dorsal valve interior. Mifflin Member, Dixon, Illinois (locality 2), **9**, ISGS 583; **13**, ISGS 559; **14**, ISGS 644. Mifflin Member, Fennimore, Wisconsin (locality 24), **10**, **11**, ISGS 560; **12**, ISGS 561.

15–17. ***Trigrammaria winchelli*** (Hall and Clarke). (**15**) Dorsal valve exterior showing typical triangular outline of *Trigrammaria*; (**16**) posterior end (dorsal valve facing up) showing interarea and moderately inflated dorsal valve; (**17**) ventral valve interior showing prominent muscle scars. Mifflin Member, Dixon, Illinois (locality 2), **15**, ISGS 1346; **16**, ISGS 647; **17**, ISGS 562.

18, 19. ***Sowerbyella* sp.** (**18**) Ventral valve exterior; (**19**) dorsal valve interior. Upper Mifflin Member, Lancaster, Wisconsin (locality 27), **18**, ISGS 1564; **19**, ISGS 1563.

Genus STROPHOMENA

Strophomena plattinensis Fenton, 1928

Figures 6-4-9 through 6-4-14

Description. Relatively large shell; profile gently resupinate; ornament typically consists of fine radiating costae arising entirely by intercalation between coarse costae; prominent pseudodeltidium (cover over the delthyrium) and small chilidium (cover over the notothyrium); dental plates curve anteriorly from teeth into elevated bounding ridges surrounding a prominent subcircular ventral muscle field; dorsal muscle field is gently impressed, fading anteriorly.

Remarks. Previous brachiopod specialists have identified several species and subspecies of *Strophomena* in the Platteville Formation of the Upper Mississippi Valley region, including *S. plattinensis* Fenton, 1928, *S. plattinensis crassa* (Shrock and Raasch, 1937), *S. auburnensis* (Fenton, 1928), and *S. auburnensis impressa* (Shrock and Raasch, 1937). In a study of phenetic variation in *S. plattinensis*, Bretsky and Bretsky (1975) determined that key diagnostic features, such as a valve profile, intergrade between described species of *Strophomena*. They concluded that morphology was largely environmentally controlled and only recognized the single species *S. plattinensis* in their analysis.

Occurrence. Abundant in the Mifflin and Grand Detour Members in north-central Illinois and south-central Wisconsin and the McGregor Member of southwestern Wisconsin, northeastern Iowa, and southeastern Minnesota. Locally common in the Pecatonica Member in north-central Illinois.

Genus TRIGRAMMARIA

Trigrammaria winchelli (Hall and Clarke, 1892)

Figures 6-4-15 through 6-4-17

Description. Outline triangular; profile gently resupinate anteriorly, with variably developed dorsal fold and corresponding ventral sulcus at the anterior end; ornament typically consists of fine radiating costae arising entirely by intercalation between coarse costae; pseudodeltidium and small chilidium present; dental plates curve anteriorly from teeth into elevated bounding ridges surrounding a subcircular ventral muscle field; dorsal muscle field gently impressed.

Remarks. *Trigrammaria* differs from *Strophomena* in having (1) dorsal and ventral valves that are triangular in outline, (2) a more substantial dorsal fold and corresponding ventral sulcus, and (3) fine short septa within the dorsal muscle field.

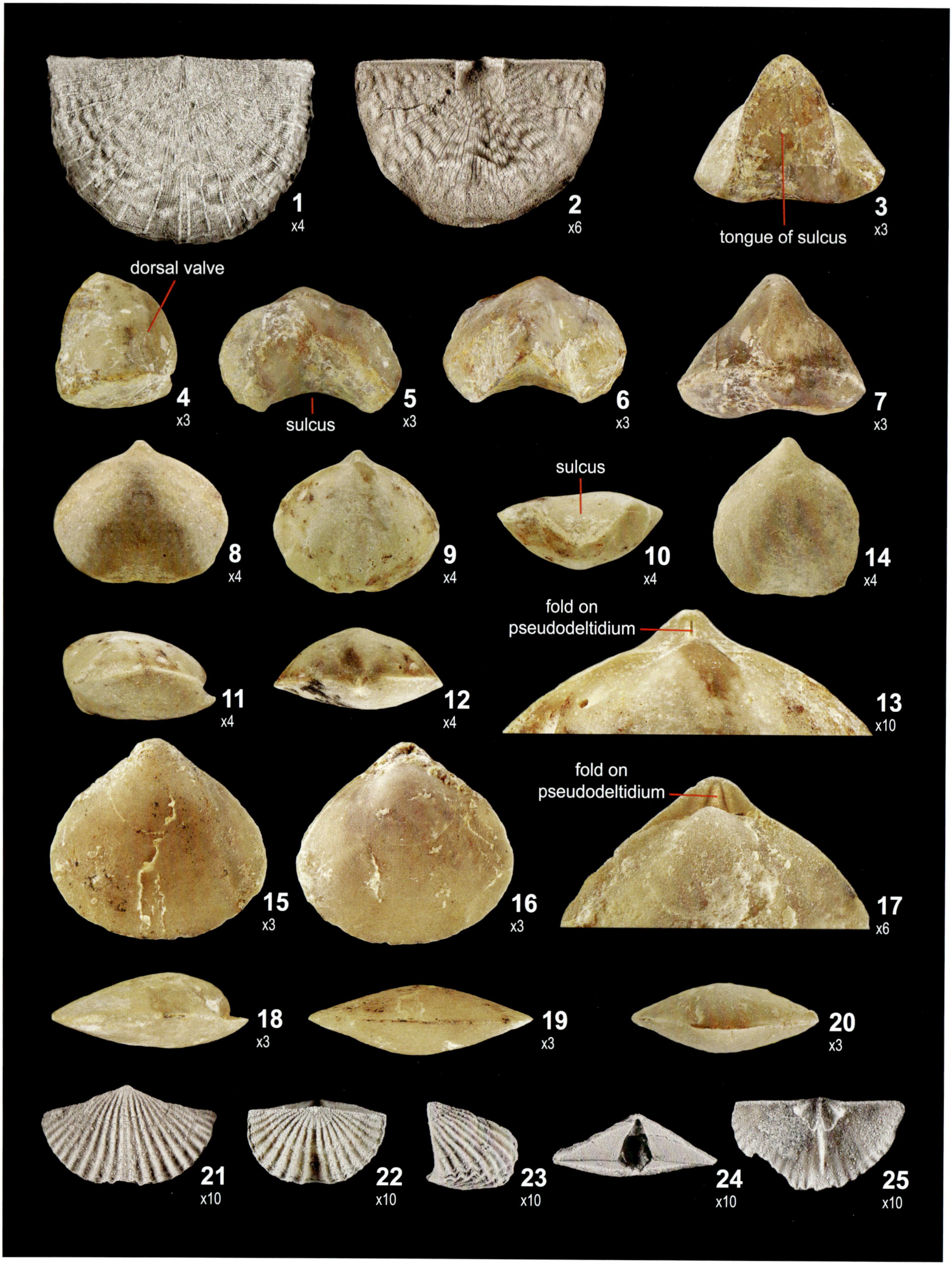
1
x4
2
x6
3
x3
tongue of sulcus
dorsal valve
4
x3
5
x3
sulcus
6
x3
7
x3
8
x4
9
x4
sulcus
10
x4
14
x4
fold on pseudodeltidium
11
x4
12
x4
13
x10
15
x3
16
x3
fold on pseudodeltidium
17
x6
18
x3
19
x3
20
x3
21
x10
22
x10
23
x10
24
x10
25
x10

Occurrence. Common to locally abundant in the Mifflin and Grand Detour Members in north-central Illinois and south-central Wisconsin and the McGregor Member of southwestern Wisconsin, northeastern Iowa, and southeastern Minnesota. Locally common in the Pecatonica Member in north-central Illinois. The type locality is from the McGregor Member of the Platteville Formation at Clifton, Wisconsin.

Genus BELLIMURINA

Bellimurina subquadrata Cooper, 1956

Figures 6-5-1 and 6-5-2

Description. Outline is semicircular and truncated by a straight hinge line; profile concavo-convex to planoconvex; shell thin and delicate, approximately 10–15 mm wide at the hinge line; ornament consists of relatively coarse costae separated by 5–10 fine radiating costae that are interrupted by small irregular wrinkles; teeth small, dental plates short and weakly developed; poorly impressed muscle fields; some specimens show pronounced geniculation of both of the ventral and dorsal valve at the anterior end.

Remarks. This Platteville strophomenide brachiopod is a typical representative of the genus *Bellimurina*, which is known from Middle Ordovician to Early Silurian rocks in many places. Cooper (1956) noted that the genus *Bellimurina* occurs in the Platteville Formation of the Upper Mississippi Valley, but the specimens available to him were not suitable for description. The Platteville species can be placed within the species *B. subquadrata* from the approximately coeval Mountain Lake Member of the Bromide Formation of Oklahoma in its size, outline, and distinctive irregular wrinkles (small costellae interrupted by the radial ribs).

Occurrence. Locally common but inconspicuous Platteville brachiopod presently known from the Pecatonica and Mifflin Members throughout the outcrop belt.

Figure 6-5 (Opposite page)

1, 2. ***Bellimurina subquadrata*** Cooper. (**1**) Ventral valve exterior showing fine costae and irregular wrinkles; (**2**) ventral valve interior. Pecatonica Member, Forreston, Illinois (locality 6), **1**, ISGS 565. Mifflin Member, Dixon, Illinois (locality 3), **2**, ISGS 566.

3–7. ***Triplesia* cf. *carinata*** Cooper. (**3**) Anterior view of a complete specimen showing the unusually long sulcus tongue extending upward from the ventral valve; (**4**) profile showing the highly inflated dorsal valve; (**5**) ventral valve exterior showing prominent sulcus; (**6**) dorsal valve exterior; (**7**) posterior view showing inflated dorsal valve. Grand Detour Member, Dixon, Illinois (locality 1), **3–7**, ISGS 638.

8–14. ***Triplesia* cf. *subcarinata*** Cooper. (**8**) Ventral valve exterior; (**9**) dorsal valve exterior; (**10**) anterior view showing sulcus; (**11**) profile view of the complete specimen; (**12**) posterior end of the complete specimen; (**13**) enlargement of the beak area showing characteristic small fold on the pseudodeltidium; (**14**) ventral valve exterior. Identified only as Platteville Formation, Little Grant Township, Grant County, Wisconsin, **8–13**, University of Iowa Paleontological Repository 87999. Grand Detour Member, Durand, Illinois (locality 12), **14**, ISGS 661.

15–20. ***Onychoplecia* sp.** (**15**) Ventral valve exterior; (**16**) dorsal valve exterior; (**17**) photographic enlargement of hinge area showing a prominent fold on the pseudodeltidium; (**18**) profile view showing posteriorly projecting beak on the ventral valve; (**19**) anterior view of the complete specimen showing the rectilinear (straight) commissure; (**20**) posterior view of the complete specimen. Grand Detour Member, Dixon, Illinois (locality 2), **15–20**, JAC 106.

21–25. ***Skenidioides anthonensis*** (Sardeson). (**21**) Ventral valve exterior; (**22**) dorsal valve exterior; (**23**) lateral view of dorsal and ventral valves; (**24**) posterior end of dorsal and ventral valves (interior) showing large, triangular-shaped pedicle opening; (**25**) interior surface of dorsal valve. Mifflin Member, Dixon, Illinois (locality 2), **21**, ISGS 1570; **22**, **23**, ISGS 1554; **24**, ISGS 1569; **25**, ISGS 1572.

Superfamily PLECTAMBONITOIDEA

Family SOWERBYELLIDAE

Genus SOWERBYELLA

***Sowerbyella* sp.**

Figures 6-4-18 and 6-4-19

Description. Shell length two-thirds the width, typically 15–18 mm wide, 10 mm long; resupinate profile; greatest width at hinge line; anterior margin broadly rounded to slightly truncated; ornament typically consists of one to five fine radiating costae arising entirely by intercalation between coarse costae. Ventral valve with moderate curvature in lateral profile, greatest curvature positioned just anterior of the hinge line. Dorsal valve exterior moderately concave; interior with short, widely divergent dental plates; medium septa short and stout, separating large muscle scars; inner surface with abundant small papillae.

Remarks. Additional specimens are needed to make a useful comparison to other species of *Sowerbyella*.

Occurrence. This Platteville species is relatively uncommon and confined largely to the middle and upper parts of the McGregor Member in southwestern Wisconsin and southeastern Minnesota.

Order ORTHOTETIDA

Description. Biconvex strophomenate brachiopods with hinge line parallel to hinge axis, ventral interarea well developed with pseudodeltidium; dorsal interarea short or absent; cardinal process bilobed to forked; variably punctate shell.

Superfamily TRIPLESIOIDEA

Family TRIPLESIIDAE

Genus TRIPLESIA

Triplesia* cf. *carinata Cooper, 1956

Figures 6-5-3 through 6-5-7

Description. Shell strongly trilobate with pronounced dorsal fold and ventral sulcus; wider than long, 10–15 mm wide; oval in outline; sides narrowly rounded; hinge narrow; smooth surface with only faint growth lines; anterior commissure deeply uniplicate with prominent dorsal fold and deep ventral sulcus. Ventral valve has an elongate beak; sulcus originating halfway between the posterior and anterior ends, deep central groove with extremely long and angular tongue. Dorsal valve widest in middle; lateral profile greatly convex with greatest depth in the center; narrow swollen beak; prominent crested fold originating on the beak; marked on both sides by obliquely oriented grooves.

Remarks. Observed specimens show external features only. The size and shape of the ventral and dorsal valves are very similar to *Triplesia carinata* Cooper, 1956, from the older Pratt Ferry Formation of Alabama.

Occurrence. A rare brachiopod in the Grand Detour Member in north-central Illinois.

Triplesia* cf. *subcarinata Cooper, 1956

Figures 6-5-8 through 6-5-14

Description. Shell oval in outline, surface smooth, wider than long, sides rounded, hinge narrow, valves unequally convex; anterior commissure uniplicate, with well-developed fold and sulcus; ranging in size from 5 to 15 mm in width. Ventral valve slightly convex in lateral profile, with central tongue sharply bent at the anterior end; sulcus begins at middle of shell and widens rapidly; tongue marked by shallow median groove. Dorsal valve is more convex than the ventral; widest at middle; prominent median fold.

Remarks. Observed specimens show external features only. The size and shape of the ventral and dorsal valves are comparable to *Triplesia subcarinata* Cooper, 1956, from the Turinian-age Upper Ordovician Lebanon Formation in Tennessee. This species differs from the Platteville species *T.* cf. *carinata* in lacking (1) a sharp median crest on the dorsal valve and (2) shorter tongue on the ventral sulcus.

Occurrence. A rare brachiopod in the Grand Detour Member in north-central Illinois and equivalent strata of the McGregor Member in northeastern Iowa.

Genus ONYCHOPLECIA

***Onychoplecia* sp.**

Figures 6-5-15 through 6-5-20

Description. Shell 15 mm at widest point, lenticular in profile, valves unequally convex; sides rounded, with greatest width anterior to middle; anterior commissure rectimarginate (lacking sulcus and fold); no ornament except for concentric growth lines. Ventral valve somewhat tear-shaped in outline with elongate and slender beak; delthyrium covered by flat pseudodeltidium marked along center line by narrow fold; interior of ventral valve unknown. Dorsal valve subcircular in outline, slightly more convex than ventral valve; interior unknown.

Remarks. This species differs from other Platteville triplesiids in having an anterior commissure that is rectimarginate (straight) rather than uniplicate (single fold).

Occurrence. Rare in the Platteville Formation and presently known only from the Grand Detour Member at Dixon, Illinois.

Order PROTORTHIDA

Description. The protorthides are small brachiopods with a hinge line that is coincident with the hinge axis; open delthyrium and notothyrium; wide interarea; simple teeth with dental ridges.

Superfamily SKENIDIOIDEA

Family SKENIDIIDAE

Genus SKENIDIOIDES

Skenidioides anthonensis (Sardeson, 1892)

Figures 6-5-21 through 6-5-25

Description. Shell small for the genus, length one-half to two-thirds the width, typically 3–4 mm wide, sides obliquely rounded and sloping toward the middle; anterior margin sulcate; 10 or 11 costae on each side of fold and sulcus. Ventral valve triangular, fold only slightly elevated; interarea long, flat, triangular, and nearly perpendicular to plane of commissure; pedicle opening relatively large and triangular in outline. Dorsal valve interior with prominent medial septum that extends from the dorsal margin to near the anterior margin of the shell; septum bisects small spoon-like depression beneath socket ridges.

Remarks. A species of *Skenidioides* distinguished by its considerable width, prominent costae, and low fold (Cooper 1956, p. 492).

Occurrence. Very abundant in the Mifflin Member at Dixon, Illinois (locality 2). The type specimens are from the Platteville Formation at Minneapolis, Minnesota (Sardeson 1892).

Order ORTHIDA

Description. The Platteville orthide brachiopods have a shell that is biconvex and roughly rectangular in outline, with maximum-width anterior of a well-developed hinge line, an inward curved interarea on the ventral valve and outward curved interarea on the dorsal valve, well-developed dorsal sulcus, and costellate ornamentation.

Superfamily ORTHOIDEA

Family HESPERORTHIDAE

Genus HESPERORTHIS

Hesperorthis concava Cooper, 1956

Figures 6-6-1 through 6-6-6

Description. Semioval in outline, length and width about equal, sides gently rounded, anterior margin narrowly rounded; planoconvex; anterior commissure rectimarginate; long, flat interareas. Dorsal and ventral surfaces marked by approximately 30 prominent narrowly rounded costae separated by spaces slightly narrower than the costae. Ventral valve strongly convex with maximum convexity slightly posterior to the middle; lateral slopes gently convex and steep; beak somewhat elongate, forming an angle of approximately 130°; interarea long and prominently curved; posterior pair of dorsal adductor scars larger than anterior. Dorsal valve concave in lateral profile, with deepest part slightly posterior of middle, beak gently concave; narrow sulcus expanding anteriorly and containing two costae; interior with long brachiophores and long, low median ridge.

Remarks. The type specimens of *Hesperorthis concava* were originally described by Cooper (1956) from the Platteville Formation (McGregor Member) found in quarries near Platteville and Lancaster, Grant County, Wisconsin.

Occurrence. Abundant in the Mifflin and Grand Detour Members in north-central Illinois and south-central Wisconsin and the McGregor Member of southwestern Wisconsin, northeastern Iowa, and southeastern Minnesota. Locally common in the Pecatonica Member in north-central Illinois.

Family GLYPTORTHIDAE

Genus ERIDORTHIS

Eridorthis costellata (Cooper, 1956)

Figures 6-6-7 through 6-6-12

Description. Shell subquadrate in outline and slightly wider than long; lateral margins slightly rounded to straight; valves nearly equal in convexity; anterior commissure rectimarginate (straight) to slightly sulcate; surface covered with 40–50 costae with strong concentric ornamentation developed as lamellose frills. Ventral valve convexity greatest just posterior of the middle; interarea long and slightly curved inward. Dorsal valve evenly convex with greatest convexity near the middle; sulcus very shallow; cardinal process relatively massive; interarea relatively long and very slightly curved inward; beak slightly incurved.

Remarks. The genera *Eridorthis* and *Glyptorthis* were both erected by A.F. Foerste. Although *Glyptorthis* has been more widely cited, *Eridorthis* predates it by 5 years and thus should be used because the two are now considered synonyms by modern authors. This Platteville form can be identified as "*Glyptorthis*" *costellata*, originally defined from the Turinian-age Pooleville Member of the Bromide Formation in Oklahoma because they are identical in shell outline, convexity, ornamentation, morphology of the cardinal process, and adductor muscle fields.

Occurrence. Common in the Mifflin and Grand Detour Members in north-central Illinois and south-central Wisconsin and the McGregor Member of southwestern Wisconsin, northeastern Iowa, and southeastern Minnesota. Locally common in the Pecatonica Member in north-central Illinois.

Family PLAESIOMYIDAE

Genus CAMPYLORTHIS

Campylorthis deflecta (Conrad, 1843)

Figures 6-6-13 through 6-6-22

Description. Shell subelliptic in outline; large relative to other Platteville orthid brachiopods; hinge generally narrower than shell at the middle; sides and anterior gently rounded; anterior commissure strongly uniplicate; lateral margin

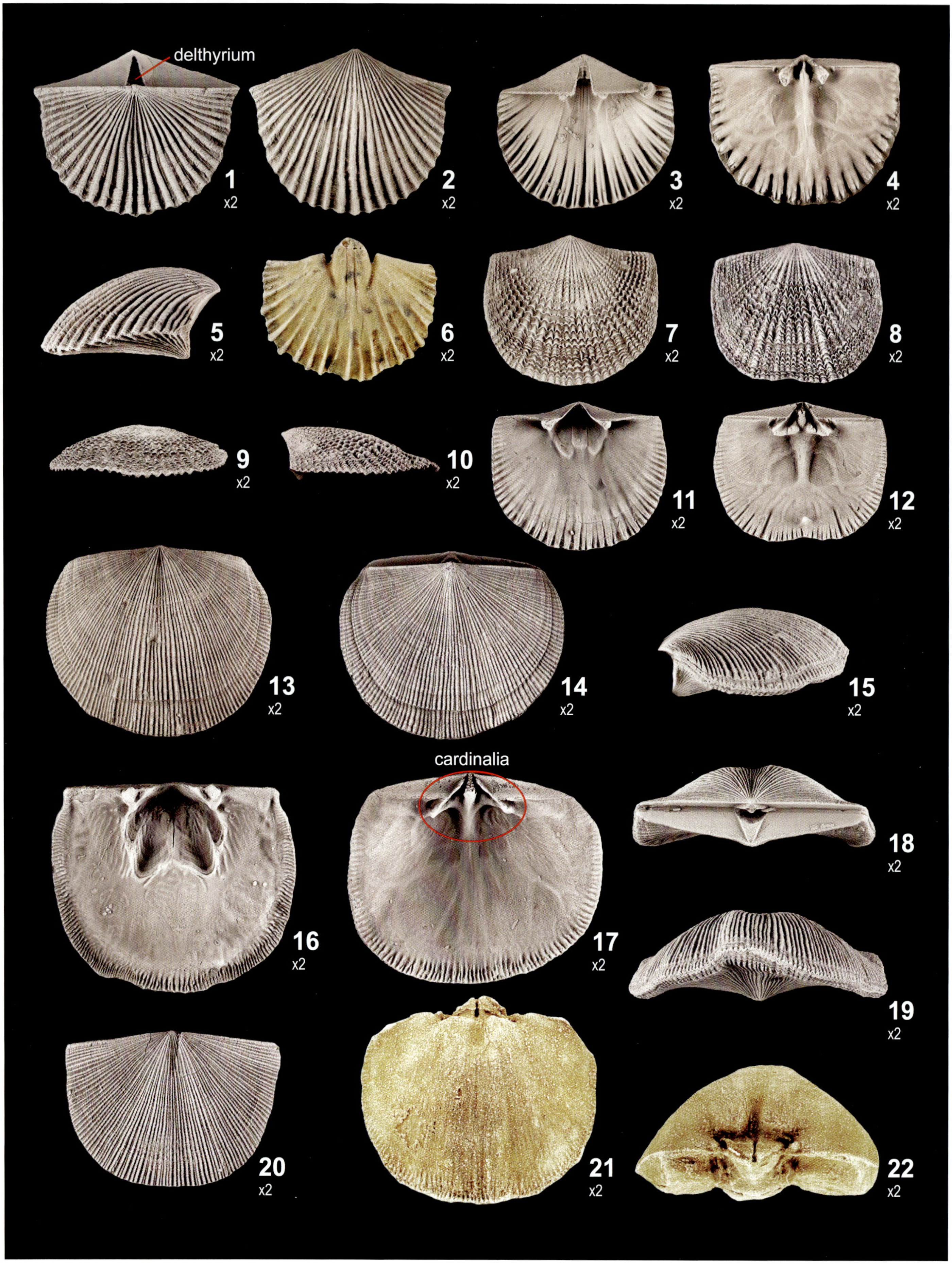

delthyrium
1
x2
2
x2
3
x2
4
x2
5
x2
6
x2
7
x2
8
x2
9
x2
10
x2
11
x2
12
x2
13
x2
14
x2
15
x2
cardinalia
16
x2
17
x2
18
x2
19
x2
20
x2
21
x2
22
x2

strongly deflected toward the dorsal valve; costae broadly rounded; about 10 costae in 5 mm width at the anterior end of a large adult. Ventral valve gently convex at the posterior end and deeply concave in the anterior two-thirds; deeply sulcate from the middle to anterior end; delthyrium covered by a large convex pseudodeltidium. Dorsal valve unevenly convex in lateral profile; median region convex and front flattened; anterior profile strongly convex; interarea long and flat with prominent chilidium; notothyrial region thickened with prominent median ridge; brachiophores stout; cardinal process moderately long and thick.

Remarks. The type specimens of *Campylorthis deflecta* were originally described by Conrad (1843) from the Platteville Formation (McGregor Member) from localities in southwestern Wisconsin. Smaller, immature individuals tend to have an interarea and hinge line that form the widest part of the shell, grading to mature individuals that are widest midway between the posterior and anterior ends. As pointed out by Cooper (1956), this may have led Conrad (1843) to assign some immature specimens incorrectly to *Strophomena recta*.

Occurrence. Abundant in the Mifflin and Grand Detour Members in north-central Illinois and south-central Wisconsin and the McGregor Member of southwestern Wisconsin, northeastern Iowa, and southeastern Minnesota. Locally common in the Pecatonica Member in north-central Illinois.

Figure 6-6 (Opposite page)
1–6. ***Hesperorthis concava*** Cooper. (**1**) Dorsal valve exterior showing wide interarea, posteriorly projecting beak of the ventral valve, and delthyrium; (**2**) ventral valve exterior; (**3**) ventral valve interior; (**4**) dorsal valve interior; (**5**) profile of complete specimen showing the distinctive posteriorly projecting beak; (**6**) dolomitized steinkern of a ventral valve interior. Mifflin Member, Dixon, Illinois (locality 2), **1**, **2**, **5**, ISGS 588; **3**, ISGS 668; **4**, ISGS 586. Mifflin Member, Lee Center, Illinois (locality 31), **6**, JAC 107.

7–12. ***Eridorthis costellata*** Cooper. (**7**) Ventral valve exterior showing distinctive concentric ornamentation developed as lamellose frills; (**8**) dorsal valve exterior; (**9**) anterior end of ventral valve; (**10**) profile of ventral valve; (**11**) ventral valve interior; (**12**) dorsal valve interior. Mifflin Member, Dixon, Illinois (locality 2), **7**, **9–11**, ISGS 664; **8**, **12**, ISGS 625.

13–22. ***Campylorthis deflecta*** (Conrad). (**13**) Dorsal valve exterior; (**14**) ventral valve exterior; (**15**) profile view of the complete specimen; (**16**) ventral valve interior showing prominent muscle scars; (**17**) dorsal valve interior showing the cardinalia; (**18**) posterior end of the complete specimen showing the relatively wide interarea and delthyrium partially covered by the pseudodeltidium; (**19**) anterior of the complete specimen showing strongly uniplicate commissure; (**20**) relatively small and immature dorsal valve showing the hinge line as the widest part of the shell; (**21**) dolomitized steinkern (internal cast) of a dorsal valve interior; (**22**) posterior of internal cast. Mifflin Member, Dixon, Illinois (locality 2), **13**, **14**, ISGS 592; **15**, **19**, ISGS 666; **16**, ISGS 595; **17**, ISGS 594; **18**, ISGS 593; **20**, ISGS 596. Grand Detour Member, Beloit, Wisconsin (locality 20), **21**, **22**, JAC 108.

Superfamily PLECTORTHOIDEA

Family PLECTORTHIDAE

Genus DOLEROIDES

Doleroides pervetus (Conrad, 1843)

Figures 6-7-15 through 6-7-23

Description. Subcircular to subelliptic in outline; biconvex in profile; shell punctate; hinge narrower than the greatest shell width, which is situated midway between the posterior and anterior ends; anterior margin broadly rounded; anterior commissure uniplicate; about nine costae in 5 mm width in the ventral sulcus; costae alternate in size on the flanks. Ventral valve moderately convex in lateral profile, with greatest convexity in the posterior half; sulcus generally shallow, originates just anterior to the beak; interarea relatively long and curved inward. Dorsal valve moderately convex; beak marked by a shallow and narrow sulcus; shallow fold originating posterior to the middle.

Remarks. This species was previously assigned to the genus *Pionodema* based on specimens from the Platteville Formation of Minnesota. It was reassigned to the new genus *Doleroides* when the species was discovered to have a punctate shell as well as cardinalia and ventral musculature that differ greatly from *Pionodema* (Cooper 1930). *D. pervetus* is similar to *D. gibbosus* from the Upper Ordovician Decorah Formation of southeastern Minnesota but differs in its generally smaller size (Cooper 1956).

Occurrence. Abundant in the Mifflin and Grand Detour Members in north-central Illinois and south-central Wisconsin and the McGregor Member of southwestern Wisconsin, northeastern Iowa, and southeastern Minnesota. Locally common in the Pecatonica Member in north-central Illinois. Specimens often occur in clusters on shaley bedding plane surfaces.

Order RHYNCHONELLIDA

Description. Articulate brachiopods with biconvex shells, dorsal fold, and ventral sulcus, and a prominent beak on the ventral valve that projects over the cardinal margin. The shells typically have coarse costae.

Superfamily RHYNCHOTREMATOIDEA

Family TRIGONIRHYNCHIIDAE

Genus ROSTRICELLULA

Rostricellula minnesotensis (Sardeson, 1892)

Figures 6-7-1 through 6-7-9

Description. Shell subtriangular to subpentagonal in outline; biconvex; apical angle slightly greater than 90°; rounded sides; four costae in the fold and two or three costae in the sulcus, flanks commonly have six costae; anterior commissure uniplicate. Ventral valve has a pointed and incurved beak; delthyrium open or partially constricted; interarea very narrow; sulcus tongue long and serrate at the commissure. Ventral valve with thick dental plates separated from sides of the valve by narrow posterior cavities; teeth small and curved. Dorsal valve beak sulcate, with the two median costae lower than lateral costae and extending to about the middle of the valve; fold is low and flattened in profile. Dorsal valve interior with divided hinge plate connected to the median septum; cardinal process absent.

Remarks. Typical preservation is with both valves articulated, thus concealing the internal features. *R. minnesotensis* differs from other species by its rotund and robust form, combined with its swollen flanks and low, wide fold (Cooper 1956). The type specimens were collected from the Platteville Formation in southwestern and south-central Wisconsin as well as northeastern Iowa (Sardeson 1892).

Occurrence. Abundant in the Mifflin and Grand Detour Members in north-central Illinois and south-central Wisconsin and the McGregor Member of southwestern Wisconsin, northeastern Iowa, and southeastern Minnesota. Locally common in the Pecatonica Member in north-central Illinois.

Order PENTAMERIDA

Description. Articulated brachiopods, subquadrate to subpentagonal to elongate-oval in outline; strongly biconvex; anterior commissure usually uniplicate; exterior smooth or ribbed; delthyrium open; dental plates well developed, commonly merging with the median septum, forming a spoon-like structure called a spondylium.

Superfamily CAMERELLOIDEA

Family CAMERELLIDAE

Genus CAMERELLA

Camerella gregeri Cooper, 1956

Figures 6-7-10 through 6-7-14

Description. Shell relatively large for the genus; length and width approximately the same; pentagonal outline; angle of beak approximately 90°; lateral anterior margins narrowly rounded; anterior margin truncated; greatest width anterior to the middle; smooth surface in the posterior half but marked by narrowly rounded costae at the anterior end; five costae on the fold and four costae in the sulcus; three or four costae on the flanks. Ventral valve convex in lateral profile, with the greatest convexity just anterior to the beak; sulcus

Figure 6-7 (Opposite page)

1–9. ***Rostricellula minnesotensis*** (Sardeson). (**1**) Posterior end of the complete specimen; (**2**) ventral valve exterior; (**3**) anterior view of the complete specimen showing prominent sulcus and serrated commissure; (**4**) profile of the complete specimen showing inflated dorsal valve and curved beak on the ventral valve; (**5**) dorsal valve exterior with small portion of ventral valve and ovate foramen exposed at top; (**6**) complete specimen showing low-profile dorsal valve; (**7**) dolomitized steinkern (internal cast) of the posterior end of a complete specimen; (**8**) dolomitized steinkern (internal cast) of a dorsal valve interior; (**9**) small cluster of complete specimens. Mifflin Member, Dixon, Illinois (locality 2), **1**, **2**, ISGS 630; **3**, ISGS 631; **4**, ISGS 632; **5**, **6**, ISGS 633; **9**, ISGS 673. Mifflin Member, Lee Center, Illinois (locality 31), **7**, **8**, JAC 109.

10–14. ***Camerella gregeri*** Cooper. (**10**) Dorsal valve exterior; (**11**) ventral valve exterior; (**12**) profile view of the complete specimen; (**13**) posterior view of the complete specimen; (**14**) anterior view of the complete specimen showing prominent sulcus and serrated commissure. Grand Detour Member, Dixon, Illinois (locality 1), **10–14**, ISGS 637.

15–23. ***Doleroides pervetus*** (Conrad). (**15**) Dorsal valve interior showing cardinalia; (**16**) ventral valve interior; (**17**) profile of complete specimen showing inflated ventral valve with prominent recurved beak; (**18**) anterior view of the complete specimen showing uniplicate commissure; (**19**) ventral valve exterior; (**20**) dorsal valve exterior, photographic enlargement showing the microscopic exopunctae that characterize *Doleroides*; (**21**, **22**) showing the posterior end and dorsal valve exterior of a dolomitized steinkern (internal cast); (**23**) posterior end of a complete specimen showing the interarea. Mifflin Member, Dixon, Illinois (locality 2), **15**, ISGS 669; **16**, ISGS 600; **17**, **18**, ISGS 667; **19**, ISGS 599; **20**, **23**, ISGS 598. Mifflin Member, Lee Center, Illinois (locality 31), **21**, **22**, JAC 110.

24–31. ***Protozyga nicolleti*** (Winchell and Schuchert). (**24**) Profile of the complete specimen showing prominent recurved beak; (**25**) dorsal valve exterior; (**26**) ventral valve exterior; (**27**) posterior end; (**28**) anterior end showing serrated commissure; (**29**) ventral valve exterior; (**30**) profile showing recurved beak; (**31**) dorsal valve exterior. Mifflin Member, Dixon, Illinois (locality 2), **24–28**, ISGS 641; **29–31**, ISGS 672.

1
x2
2
x2
3
x2
4
x2
5
x2
6
x2
7
x2
8
x2
9
x1.5
10
x2
11
x2
12
x2
13
x2
14
x2
cardinalia
15
x3
16
x2
17
x2.5
18
x2.5
21
x2.5
22
x2.5
19
x2.5
20
x2.5
x20
exopunctae
23
x2.5
24
x8
25
x8
26
x8
27
x8
28
x8
29
x8
30
x8
31
x8

originates near the middle and deepens toward the anterior end, where it occupies a little more than half the width of the valve; sulcus tongue moderately long; flanks convex with gentle slopes to the margins. Dorsal valve about twice as deep as ventral valve; strongly arched in anterior profile; beak area strongly convex (Cooper 1956).

Remarks. This Platteville species of *Camerella* is identical to *C. gregeri* from the coeval Plattin Group (Macy Formation) of eastern Missouri.

Occurrence. Present but rare in the Grand Detour Member at Dixon, Illinois.

Order ATRYPIDA

Description. The atrypids are articulate brachiopods with a complex spirally, coiled, calcareous support for the lophophore called the brachidium.

Superfamily PROTOZYGOIDEA

Family PROTOZYGIDAE

Genus PROTOZYGA

Protozyga nicolleti (Winchell and Schuchert, 1895b)

Figures 6-7-24 through 6-7-31

Description. Shell small to minute, typically 3–5 mm in length, subcircular to oval in outline, biconvex with ventral valve generally the deeper. Ventral valve with strongly incurved beak and oval pedicle foramen; moderately elevated fold with narrow sulcus and two or three broadly rounded costae on the flanks. Dorsal valve has a sulcus occupied by a subtle fold corresponding to the median sulcus on the ventral valve.

Remarks. Typical preservation is with both valves articulated, thus concealing the minute internal structures. The type specimens of *P. nicolleti* were collected from the Platteville Formation in southeastern Minnesota.

Occurrence. Abundant in the Mifflin and Grand Detour Members in north-central Illinois and south-central Wisconsin and the McGregor Member of southwestern Wisconsin, northeastern Iowa, and southeastern Minnesota. Locally common in the Pecatonica Member in north-central Illinois.

CHAPTER 7
POLYPLACOPHORANS

The Platteville Formation contains a relatively diverse and well-preserved polyplacophoran fauna representing the orders Paleoloricata and Multiplacophora. Three paleoloricates were described previously from the Grand Detour Member of the Platteville Formation near Beloit, Wisconsin (locality 21). These include *Echinochiton dufoei* (Pojeta et al., 2003), *Helminthochiton* (=*Preacanthochiton*) *baueri* (Hoare and Pojeta, 2006), and *Litochiton* (=*Helminthochiton*) *marginatus* (Hoare and Pojeta, 2006). Recently, the Mifflin Member of the Platteville Formation near Dixon, Illinois (localities 2 and 4), has yielded a diverse assemblage of well-preserved but disarticulated and commingled calcareous plates of *Litochiton* cf. *marginatus* (Figures 7-1 and 7-2) as well as two new and distinct multiplacophoran species (Figures 7-3). All three species have a differentiated series of dorsal plates or valves consisting of a head and tail as well as multiple intermediate valves. Species assignment for the associated spines and scutes is challenging, relying largely on variations in surface sculpture. Multiplacophoran sp. 1 is characterized by shell plates whose dorsal surface contains small granules, radiating tubules, and a prominent mucro. This species also is represented by hollow, asymmetrical spines that show right- and left-handedness and have granules and tubules on the outer surfaces. Multiplacophoran sp. 2 is characterized by both symmetrical and asymmetrical plates consisting of a very fine latticelike framework.

Polyplacophorans, commonly referred to as chitons, inhabit marine environments and have a fossil record extending back approximately 500 million years to Late Cambrian time. Typically, they are flattened dorsoventrally, have a broad muscular foot, and have an expanded part of the body wall called the **girdle** covered by an outer cuticle that can be embedded with small, pointed, calcified **spicules**, also referred to as **scutes** or **spines**. The dorsal surface is typically shielded by relatively thick calcareous plates differentiated into **head**, **intermediate**, and **tail valves**. The posterior edges of the head and intermediate valves fold abruptly toward the ventral surface, forming the **apical area**. The **jugum** is a longitudinal ridge along the axis of the intermediate valves that can extend to the head or tail. Some chitons, like Platteville species *Echinochiton dufoei*, possess prominent **lateral spines** along the periphery of the girdle. In the Platteville specimens, the **mucro** is a small elevation near the posterior part of the tail valve. The valves of modern chitons (order Neoloricata) have an inner layer called the **articulamentum** that bears a pair of anterior projections on the ventral parts of the intermediate valves called **sutural laminae** that extend forward beneath the preceding valve, forming a surface of articulation. They also have laterally positioned **insertion plates** (**articulaments** of Vendrasco et al. 2004) that attach the valves to the mantle. Morphologic terms used herein follow Smith (1960) and classification follows Pojeta et al. (2010).

SYSTEMATIC DESCRIPTIONS

Phylum MOLLUSCA

Class POLYPLACOPHORA

Remarks. Newly discovered specimens and recent studies show that a significant amount of diversification took place among the polyplacophorans during Cambrian and Ordovician time (Pojeta et al. 2003, 2010; Vendrasco and Runnegar 2004; Hoare and Pojeta 2006). These investigations shed light on the poorly understood systematics of multivalved mollusks. Added to the three polyplacophoran species described previously from the Grand Detour Member of the Platteville Formation near Beloit, Wisconsin (Pojeta et al. 2003; Hoare and Pojeta 2006) are two new multiplacophoran species recently collected from the Mifflin Member at Dixon, Illinois.

Order PALEOLORICATA

Remarks. Paleoloricates are chitons that lack an articulamentum as well as the associated insertion plates and sutural laminae. Bergenhayn (1960) and Pojeta et al. (2010) recognized four paleoloricate families, the Mattheviidae, Echinochitonidae, Alastegiidae, and Litochitonidae, as being well conceptualized and based on type genera whose type species are well known. These authors accepted the Paleoloricata as a useful morphologic grouping but suggested that it is almost certainly paraphyletic. The Echinochitonidae and Litochitonidae are present in the Platteville Formation.

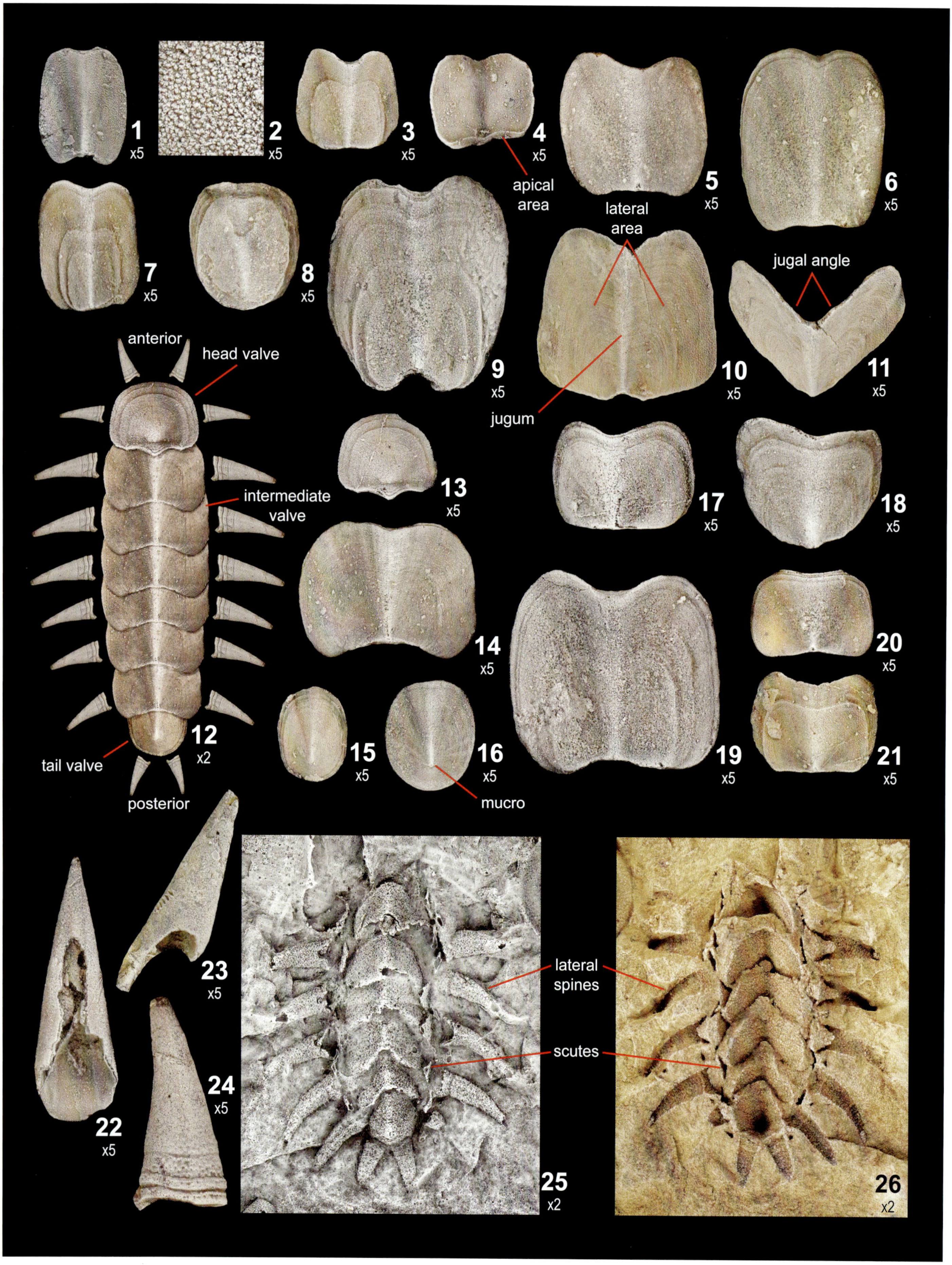

1
x5
2
x5
3
x5
4
x5
apical
area
5
x5
6
x5
7
x5
8
x5
9
x5
lateral
area
10
x5
jugal angle
11
x5
jugum
anterior
head valve
intermediate
valve
tail valve
12
x2
posterior
13
x5
14
x5
15
x5
16
x5
mucro
17
x5
18
x5
19
x5
20
x5
21
x5
22
x5
23
x5
24
x5
lateral
spines
scutes
25
x2
26
x2

Family ECHINOCHITONIDAE

Genus ECHINOCHITON

Echinochiton dufoei Pojeta et al., 2003

Figures 7-1-25 and 7-1-26

Description. A paleoloricate chiton with eight overlapping symmetrical dorsal plates consisting of a differentiated head, six intermediate valves, and a tail valve with a mucro; articulated specimens approximately 35 mm in length. Spines relatively large, hollow; four extend laterally from the head, one on each side of the six intermediate valves, and two from the tail valve. Scutes triangular in profile, dorsally projecting, pointed, compressed, erect; occurring on right and left sides of the intermediate and tail valves; positioned between the valves and hollow lateral spines; 10 or more scutes in each row.

Remarks. Largely on the basis of a single specimen (paratype USNM 517481), Pojeta and DuFoe (2008) and Pojeta et al. (2010) argued that *E. dufoei* has intermediate valves that have a large apical area preserved as molds and casts in sugary dolomite. The specimen is assumed to be fractured along or near the longitudinal axis (sagittally) of a series of articulated intermediate plates. On the basis of the size, shape, and orientation of the molds and casts, it is also possible, and probably more likely, that the fracture occurred through an aligned series of laterally compressed scutes that clearly can be seen in the holotype (USNM 517482) and are noted in the original description (Pojeta et al. 2003). Very similar calcareous scutes that were recently found in the subjacent Mifflin Member of the Platteville Formation in northern Illinois are hollow, compressed cones with slightly curved tips that in longitudinal section are very similar to dolomitic impressions in the paratype (USNM 517481). Furthermore, the relatively flat overlapping profiles of the intermediate valves in the holotype (USNM 517482) and paratype (USNM 517480) suggest a chiton with small apical area. A similar observation was made by Sirenko (2006), who noted:

> The recently described Ordovician chiton *Echinochiton dufoei* (Pojeta et al., 2003) according to the description has more advanced valve shape, a smaller apical area and antemucronal area, and it is herein with *some* doubt placed in order Septemchitonida because the valves of chitons of order Chelodida have a large apical area and no antemucronal area. On the other hand, the description of *E. dufoei* was based mainly on an external *mold* and in my opinion its reconstruction is not successful, because instead of the upper surface of valves with a projected apex we see the internal part of the valves with incisura. (p. 33)

Additional specimens are needed to determine whether *E. dufoei* has intermediate plates with large apical areas or whether they are small or absent like those in the family Litochitonidae, including the Platteville Formation species *Litochiton* cf. *marginatus* (Hoare and Pojeta, 2006).

Occurrence. *E. dufoei* Pojeta et al., 2003, is known from several nearly complete and articulated specimens found in dolostone of the Forreston Beds, Grand Detour Member near Beloit, Wisconsin (locality 21). Isolated intermediate valves are common locally in the Grand Detour and Mifflin Members of northern Illinois and southern Wisconsin.

Figure 7-1 (Opposite page)

1–24. Litochiton **cf.** marginatus (Hoare and Pojeta). (**1**) Dorsal surface of intermediate valve; (**2**) close-up of same valve showing very fine granular surface sculpture; (**3**) dorsal surface of intermediate valve; (**4**) ventral surface of intermediate valve showing small and narrow apical area; (**5**) dorsal surface of intermediate valve; (**6**) dorsal surface of intermediate valve; (**7**) dorsal surface of intermediate valve; (**8**) dorsal surface of tail valve; (**9**) dorsal surface of relatively large intermediate valve; (**10**) dorsal surface of intermediate valve showing jugum and lateral areas; (**11**) anterior end of intermediate valve showing the jugal angle; (**12**) reconstruction of *Litochiton* cf. *marginatus* based on isolated but associated head, intermediate, and tail valves as well as lateral spines; (**13**) dorsal surface of head valve; (**14**) dorsal surface of intermediate valve; (**15**) dorsal surface of tail valve; (**16**) dorsal surface of tail valve showing mucro; (**17**) dorsal surface of intermediate valve; (**18**) dorsal surface of intermediate valve; (**19**) dorsal surface of intermediate valve; (**20**) dorsal surface of intermediate valve; (**21**) dorsal surface of intermediate valve; (**22**) associated spicule of possible lateral spine; (**23**) associated spicule of possible lateral spine; (**24**) associated spicule of possible lateral spine showing comarginal growth lines. Mifflin Member, Dixon, Illinois (locality 4), **1**, **2**, ISGS 1379; **3**, ISGS 1197; **4**, ISGS 1183; **5**, ISGS 1184; **6**, ISGS 1200; **7**, ISGS 1182; **8**, ISGS 1186; **9**, ISGS 1190; **10**, **11**, ISGS 1210; **12**, composite based on ISGS 1196, ISGS 1206, ISGS 1202, and ISGS 1219 showing the head, intermediate, and tail valves as well as lateral spines and anterior/posterior orientation; **13**, ISGS 1196; **14**, ISHS 1206; **15**, ISGS 1209; **16**, ISGS 1202; **17**, ISGS 1189; **18**, 1193; **19**, ISGS 1194; **20**, ISGS 1207; **21**, ISGS 1208; **22**, ISGS 1195; **23**, ISGS 1203; **24**, ISGS 1219. Forreston Beds, Grand Detour Member, Beloit, Wisconsin (locality 21).

25, 26. ***Echinochiton dufoei*** Pojeta et al. (**25**) Latex impression of holotype (BMNH 1996.045.01) showing lateral spines and scutes and relatively flat overlapping profile of the intermediate valves; (**26**) the holotype preserved as molds and casts in relatively coarse-grained dolomite.

Family LITOCHITONIDAE

Genus LITOCHITON

Litochiton* cf. *marginatus (Hoare and Pojeta, 2006)

Figures 7-1-1 through 7-1-24; 7-2-26 through 7-2-29

Description. A paleoloricate chiton represented by differentiated head, intermediate valves, and tail valve as well as hollow lateral spines and probable scutes, all showing comarginal growth lines and surface sculpture of very small granules. Head evenly rounded anteriorly with small medial posterior projection. Intermediate valves variably quadrate, rectangular, or triangular with defined jugum and lateral areas; reduced apical area turned under slightly at posterior end. Tail valve slightly arched, longer than wide with gently convex lateral margins; convex posterior margin; mucro positioned anterior of posterior margin. Lateral spines straight or slightly curved. Scutes small, laterally compressed, triangular, pointed, asymmetrical, exhibiting right- and left-handedness; surface sculpture of relatively small granules and lacking tubules (Figures 7-2-26 through 7-2-29).

Remarks. Observed intermediate valves show considerable variation in length and width. Isolated head, intermediate valves, and tail valve are very similar to and probably conspecific with those of *Litochiton marginatus* (Hoare and Pojeta, 2006) found in the overlying Grand Detour Member of the Platteville Formation. Suspected scutes are found on bedding planes with *Litochiton* and multiplacophoran valves and spines. It remains to be determined whether *Litochiton* and *Echinochiton* have the same or different characteristics in the apical area.

Occurrence. Isolated intermediate valves are locally common on bedding planes in the Mifflin Member at Dixon, Illinois (localities 2 and 4).

Order MULTIPLACOPHORA

Remarks. Multiplacophorans are an early offshoot of chitons having seven transverse rows of valves, including one head and one tail valve separated by five rows of valves, each with large left and right outer valves and smaller median valves, for a total of 17 dorsal plates (Pojeta et al. 2010).

Multiplacophoran sp. 1

Figures 7-2-1 through 7-2-25; 7-3-1 through 7-3-22

Description. A multiplacophoran chiton characterized by isolated, symmetrical, and asymmetrical valves with dorsal surfaces marked by small granules and a network of tubules that radiate from a prominent central mucro-like elevation. Also associated are apparent lateral spines and laterally compressed, pointed, hollow, asymmetrical scutes that show right- and left-handedness and surfaces covered with small granules and tubules. All valves and spicules show comarginal growth lines.

Remarks. The surface features of the bilaterally symmetrical valves and closely associated spines and scutes strongly suggest they are part of the same organism. Many of the scutes are similar in size, shape, and surface sculpture to those of *Dimorphoconus granulatus* Donovan and Paul, 1985, from the Lower Ordovician rocks of Shropshire, England.

Occurrence. Isolated scutes and intermediate valves are common locally on bedding planes in the Mifflin Member at Dixon, Illinois (localities 2 and 4).

Multiplacophoran sp. 2

Figures 7-3-23 through 7-3-26

Description. A multiplacophoran chiton characterized by isolated symmetrical and asymmetrical valves with dorsal surfaces marked by a very fine geometrical latticework of curved ridges and uniform array of pores. The symmetrical valves have a prominent central mucro-like elevation and small denticles 0.05 mm in length and width projecting from the margins. The asymmetrical valves show right- and

Figure 7-2 (Opposite page)

1–25. **Multiplacophoran sp. 1.** (**1**, **2**) Lateral views of spicule showing surface tubules; (**3**) top view showing laterally compressed spicule with tubules; (**4**) lateral view of spicule showing comarginal growth lines; (**5**) end view of laterally compressed spicule; (**6**) lateral view of spicule; (**7**) lateral view of spicule; (**8**) lateral view of spicule; (**9**) lateral view of spicule; (**10**) lateral view of spicule; (**11**) lateral view of spicule; (**12**) lateral view of spicule; (**13**, **14**) lateral and end views showing compressed spicule; (**15**) lateral view of spicule; (**16**) lateral view of spicule; (**17**) lateral view of spicule; (**18**) lateral view of spicule; (**19**) lateral view of spicule; (**20**) lateral view of spicule; (**21**) lateral view of spicule showing comarginal growth lines; (**22**) lateral view of spicule; (**23**) lateral view of spicule; (**24**) top view of spicule; (**25**) lateral view of spicule. Mifflin Member, Dixon, Illinois (localities 2 and 4), **1–3**, ISGS 1402; **4**, **5**, ISGS 1422; **6**, ISGS 1237; **7**, ISGS 1393; **8**, ISGS 1230; **9**, ISGS 1223; **10**, ISGS 1227; **11**, ISGS 1401; **12**, ISGS 1383; **13**, **14**, ISGS 1218; **15**, ISGS 1222; **16**, ISGS 1187; **17**, ISGS 1229; **18**, ISGS 1385; **19**, ISGS 1399; **20**, ISGS 1391; **21**, ISGS 1226; **22**, ISGS 1239; **23–25**, ISGS 1388.

26–29. ***Litochiton* cf. *marginatus*** (Hoare and Pojeta). (**26–29**) Lateral view of spicules with surface sculpture of relatively small granules. Mifflin Member, Dixon, Illinois (localities 2 and 4), **26**, ISGS 1394; **27**, ISGS 1395; **28**, ISGS 1423; **29**, ISGS 1225.

tubules
tubule
1
x8
2
x8
3
x8
4
x8
5
x8
6
x8
7
x8
8
x8
9
x8
10
x8
11
x8
12
x8
13
x8
14
x8
15
x8
16
x8
17
x8
18
x8
19
x8
20
x8
21
x8
comarginal
growth lines
22
x8
27
x8
23
x8
26
x8
28
x8
24
x8
25
x8
29
x8

left-handedness and possess a pair of enlarged articulaments along the two long lateral edges.

Remarks. The articulaments and surface pores are somewhat reminiscent of the dorsal plates of the multiplacophoran *Strobilepis spinigera* from the Middle Devonian Hamilton Group of New York, as shown and described in Hoare and Mapes (1995) and Vendrasco et al. (2004).

Occurrence. Isolated scutes and intermediate valves are common locally on bedding planes in the Mifflin Member at Dixon, Illinois (localities 2 and 4).

Figure 7-3 (Opposite page)
1–22. **Multiplacophoran sp. 1.** (**1**) Dorsal surface of lateral intermediate valve; (**2**) dorsal surface of intermediate valve; (**3**) dorsal surface of intermediate valve; (**4**) dorsal surface of lateral intermediate valve; (**5**) dorsal surface of possible head valve; (**6**) dorsal surface of possible head valve; (**7**) dorsal surface of possible right anterior intermediate valve; (**8**) dorsal surface of intermediate valve; (**9**) enlargement of dorsal surface of lateral intermediate valve showing tubules radiating from mucro-like prominence; (**10**) dorsal surface of lateral intermediate valve; (**11**) dorsal surface of lateral intermediate valve; (**12**) dorsal surface of central intermediate valve; (**13**) lateral view of a spicule showing tubules and granular surface sculpture; (**14**) ventral surface showing spicule to be hollow; (**15**) lateral view of spicule showing granular surface sculpture; (**16**) lateral view of spicule with tubules projecting from edge; (**17**) lateral view of sharply bent spicule with tubules projecting from edge and comarginal growth lines; (**18**) lateral view of spicule showing granular surface sculpture; (**19**) dorsal view showing spicule to be laterally compressed; (**20**) lateral view of spicule showing granular surface sculpture; (**21**) ventral surface showing spicule to be hollow and laterally compressed; (**22**) lateral spine showing granular surface sculpture. Mifflin Member, Dixon, Illinois (localities 2 and 4), **1**, ISGS 1198; **2**, ISGS 1180; **3**, ISGS 1397; **4**, ISGS 1389; **5**, ISGS 1204; **6**, ISGS 1231; **7**, ISGS 1410; **8**, ISGS 1224; **9**, **10**, ISGS 1188; **11**, ISGS 1390; **12**, ISGS 1411; **13–15**, ISGS 1227; **16** ISGS 1228; **17**, ISGS 1220; **18–21**, ISGS 1380; **22**, ISGS 1205.

23–26. **Multiplacophoran sp. 2.** (**23**) Three plates preserved in juxtaposition showing the distinct articulaments along the margins; (**24**) dorsal surface of possible tail valve; (**25**) close-up of dorsal surface showing the distinct latticework of very fine ridges and very small geometrically distributed pores as well as marginal articulaments; (**26**) dorsal surface of possible tail valve. Mifflin Member, Dixon, Illinois (locality 2), **23**, ISGS 1379; **24**, **25**, ISGS 1421; **26**, ISGS 1201.

1
x6
2
x6
3
x6
4
x6
5
x6
6
x6
mucro
tubule
7
x6
8
x6
9
x12
10
x6
11
x6
12
x8
13
x8
14
x8
16
x8
17
x8
15
x8
18
x8
19
x8
20
x8
21
x8
22
x8
articulaments
26
x10
articulaments
23
x6
24
x6
25
x8

Figure 8-1 Depiction of representative Platteville seafloor (450 million years ago) in the upper Midwest USA showing the life mode of various gastropods. Created by Dennis R. Kolata.

CHAPTER 8

GASTROPODS, MONOPLACOPHORANS, AND ARCHINACELLIDS

The Platteville Formation contains a diverse, locally abundant, and well-preserved fauna of more than 40 species of gastropods, four monoplacophorans, and at least two archinacellids (Appendix 2). The gastropods, which were a major element of the Platteville faunal assemblage (Figure 8-1), include representatives of the subclasses Archaeobranchia, Neritimorpha, Caenogastropoda, and Amphigastropoda. Platteville monoplacophorans include the subclass Tergomya. The archinacellids are a group of mollusks whose position within the gastropods or monoplacophorans is presently considered uncertain. At the generic level, the overall fauna is typical of that in the eastern North American Late Ordovician carbonate succession.

Many of these mollusks are present throughout the upper Midwest outcrop belt but are most common in the Platteville dolostone facies of north-central Illinois and south-central Wisconsin, where they are typically preserved as steinkerns. In the limestone facies, the calcitic shells are sometimes preserved as "crack-out" specimens. One of the most remarkable occurrences is in the Mifflin Member marl beds at Dixon, Illinois, where the calcitic shells tend to weather free of matrix, revealing extraordinary morphologic details. However, the shells are completely replaced with sparite crystals.

Gastropods, commonly referred to as snails, are one of the most diverse living and fossil animal groups. They inhabit marine, freshwater, and terrestrial environments and have a fossil record extending back more than 500 million years. Platteville gastropods have a variety of shell forms (Figures 8-1 and 8-2-1 through 8-2-6), including (1) **conispiral**, asymmetrically spiraled cone-shaped shells with a protruding apex; (2) **planispiral**, shells that are coiled symmetrically about a median plane, and (3) **pseudoplanispiral**, shells that are coiled asymmetrically about a median plane. A **whorl** is the exposed surface of a single complete coil. In most cases, each coil conceals part of the preceding whorl along a contact called a **suture** (Figures 8-2-1 and 8-2-2). A **spire** is the **adapical** (toward the **apex**) part of all whorls except the last. **Abapical** refers to a direction away from the apex. Shells that are broadly coiled about the axis commonly form a conical cavity called the **umbilicus**. The axis of a coiled shell is the columella. An umbilical **carina** is a narrow, spirally wound ridge extending upward from the inner lip of the aperture into the umbilicus. Whorl profile, an important parameter in distinguishing species, can be convex, flat, or concave. Next to the suture, the whorl may have a sutural **shelf** or **ramp**, followed below by a spiral **keel** forming an obtuse angulation on the periphery of the whorl. The **peripheral band** (=selenizone) is a spiral band of concentric growth lines that are generated by a **slit**, **notch**, or **sinus** at the periphery of the **aperture** during shell growth. The **spire angle** is the angle of two lines drawn down opposite sides of the shell. The **height** is the distance between two planes set perpendicular to the axis and touching the extremities of the shell. The maximum **diameter** is the distance between two planes set parallel to the axis and touching the shell on opposite sides. The apex is **posterior** and the aperture **anterior**.

The bellerophontoid gastropods, monoplacophorans, and archinacellids have relatively simple shells that are bilaterally symmetrical. By convention, the aperture in the bellerophontoids is **ventral** and the reverse side is **dorsal**. Viewed from the dorsal surface, the aperture end is considered **anterior**. In the monoplacophorans and archinacellids, the apex is considered anterior in position relative to the center of the shell.

Despite their excellent fossil record, problems exist in the classification of Paleozoic gastropods, monoplacophorans, and archinacellids. Whereas living members are classified on the basis of soft-body characters, molecular data (in some cases), and shell morphology, fossils are limited to observations of hard parts only. As a result, phylogenetic relationships in the fossil record decrease with increasing geologic age because of the uncertain relationships with living groups and the effects of convergent evolution (Frýda and Rohr 2004). As a result, there are many unresolved classification problems in the fossil record, particularly at the higher taxonomic levels. Advances have been made in recent years (Wagner 2002; Bouchet et al. 2017), but classification above the superfamily level remains somewhat in flux. The classification scheme followed here is based largely on Bouchet et al. (2017). Parts of the Wagner (2002) scheme were employed for some subfamily–superfamily assignments.

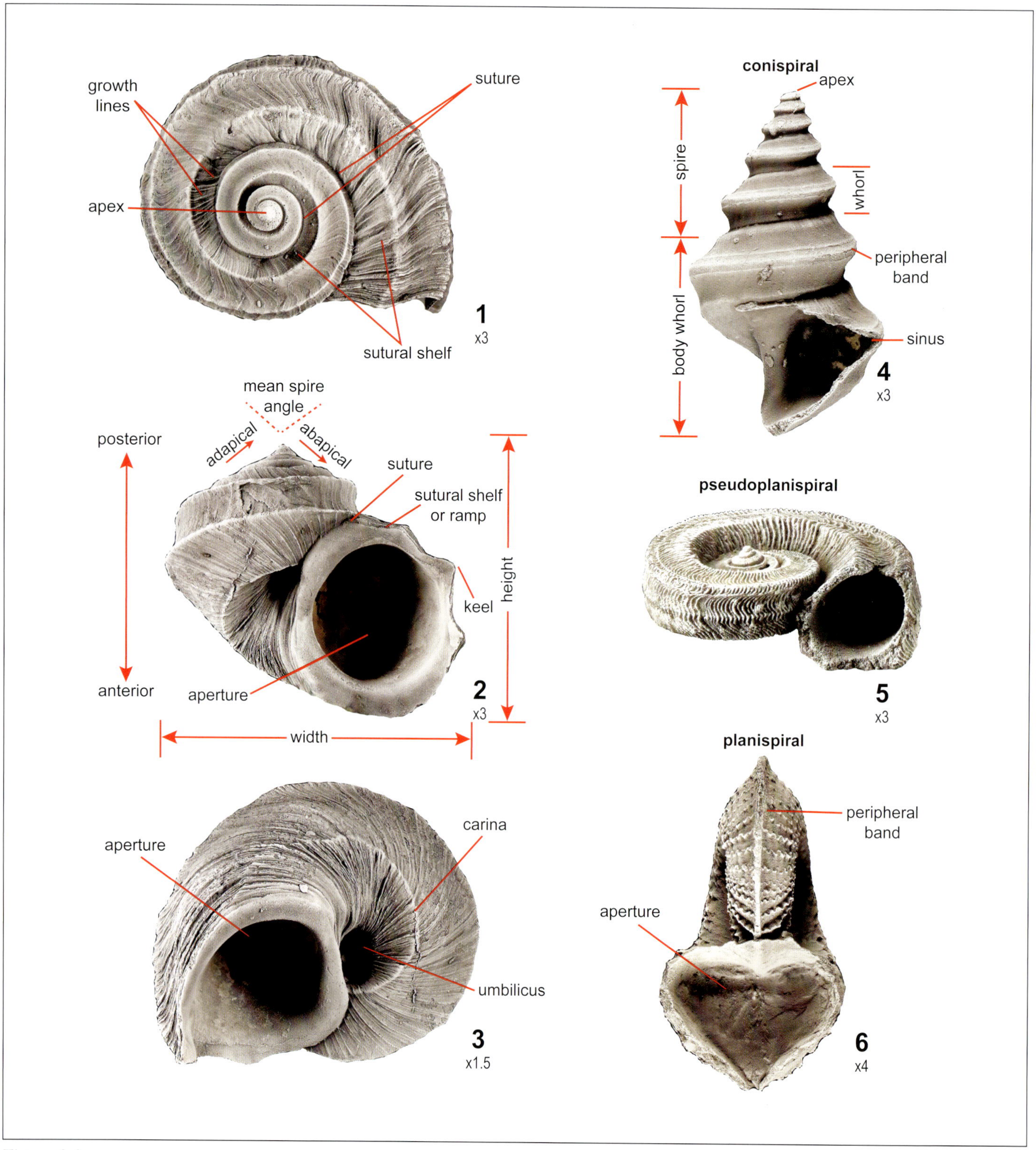

Figure 8-2

1–6. Selected specimens showing terms describing gastropod morphology and orientation. (**1**) *Trochonema umbilicata* top view; ISGS 992; (**2**) *Trochonema beloitense* apertural view; ISGS 930; (**3**) basal view of *Trochonema beloitense*; ISGS 930; (**4**) *Lophospira perangulata* lateral view of a typical conispiral shell; ISGS 920; (**5**) *Ophiletina angularis* apertural view of a typical pseudoplanispiral shell; ISGS 1029; (**6**) *Phragmolites fimbriatus* apertural view of a typical planispiral shell; ISGS 1575. All from the Mifflin Member, Dixon, Illinois (locality 2).

Primary sources used in the identification of these mollusks include Ulrich and Scofield (1897), Knight (1941), Knight et al. (1960), Wahlman (1992), and Wagner (2002).

SYSTEMATIC DESCRIPTIONS

Phylum MOLLUSCA

Class GASTROPODA

Subclass ARCHAEOBRANCHIA

Unassigned to Superfamily

Family RAPHISTOMATIDAE

Genus SCALITES

Scalites peracuta (Ulrich and Scofield, 1897)

Figures 8-3-12 through 8-3-14

Description. Shell height approximately 5 mm, diameter 10–20 mm, upper surface nearly flat, rounded below, three or four whorls, periphery very sharp and thin, umbilicus open, shell thin. V-shaped sinus in the outer lip of aperture culminating in a notch that generates an angular peripheral band; upper lip sigmoidal in outline corresponding to characteristic sigmoidal growth lines on upper surface. Growth lines on lower surface curve downward abapically then reverse and sweep adapically toward center of umbilicus.

Remarks. This species was long assigned to *Raphistoma*, but Wagner (2002) restricted the genus and reassigned it to *Scalites*. *S. peracuta* is distinguished from other raphistomatids by its nearly flat spire and sharp peripheral edge.

Occurrence. Common in the Mifflin and Grand Detour Members in northern Illinois and southern Wisconsin and the McGregor Member of southwestern Wisconsin, northeastern Iowa, and southeastern Minnesota.

Genus RAPHISTOMINA

***Raphistomina* sp.**

Figures 8-3-15 through 8-3-19

Description. Shell height approximately 10 mm, width 30 mm, lenticular, low spire, apical angle approximately 125°, umbilicus open and wide. Whorl profile very gently arched and nearly horizontal above, gently concave in a narrow band adjacent to peripheral carina, rounded on the outer two-thirds of the base and curving sharply into the umbilicus; sutures sharply incised beneath the periphery of the previous whorl but quite shallow; base more convex than the spire; abandoned early part of whorl closed off by septa. Peripheral carina increases in width abapically, edge of carina curved upward, hollow flattened tubular structures set tangentially to edge of carina at an angle of 25°. Shell thin except along carina; growth lines fine and regularly spaced.

Remarks. The wide, upturned peripheral carina, low spire, and unique tubular structures at the whorl periphery distinguish this undescribed species from other *Raphistomina*.

Occurrence. Common in the Mifflin and Grand Detour Members in northern Illinois and southern Wisconsin and the McGregor Member of southwestern Wisconsin, northeastern Iowa, and southeastern Minnesota.

Superfamily EUOMPHALOIDEA

Family EUOMPHALIDAE

Genus ASGARDASPIRA

Asgardaspira undulata (Hall, 1861)

Figures 8-4-1 through 8-4-7

Description. Shell discoidal and openly coiled, height 15 mm, some high-spiraled specimens reach 25 mm in height; one or two whorls; abandoned early part of whorl closed off by septa. Whorl subtriangular or somewhat ovate in cross section; upper side convex with a low, rounded keel extending throughout length; lower side flat to slightly convex; growth lines indicate a broad, deep sinus at crest of whorl. A narrow spiral ridge is positioned on the interior of the shell at the lower inner edge; seen as a depression on the exterior of steinkerns. Shell thick, particularly at periphery, which is marked by small, concave indentations that are smooth or cemented to shelly debris on the outer surface; fine transverse growth lines.

Remarks. This species was previously assigned to the genus *Ecculiomphalus* Hall, 1861, and accepted by Ulrich and Scofield (1897; =*Eccyliomphalus*) for specimens occurring in the Platteville Formation. The type *Ecculiomphalus* is distinguished by having a prominent high frill-like crest at the upper-outer edge of the whorl, a feature that is lacking in the species described here. This species is more appropriately assigned to the Ordovician–Silurian genus *Asgardaspira*, which lacks the pronounced frill-like crest. *A. undulata* has a more rapidly expanding shell but otherwise is most like *A. gerula* described by Rohr (1993) from the Upper Ordovician Telsitna Formation of west-central Alaska.

Occurrence. Common in the Mifflin and Grand Detour Members of northern Illinois and southern Wisconsin.

1
x5
2
x5
5
x3
3
x5
4
x5
6
x2
7
x2
8
x2
9
x2
12
x2
13
x2.5
10
x2
11
x2
14
x2.5
15
x2
16
x2
19
x2
x4
17
x2
18
x2
flattened
tubules
x5

Family HELICOTOMIDAE

Genus HELICOTOMA

Helicotoma umbilicata Ulrich and Scofield, 1897

Figures 8-4-12 through 8-4-15

Description. Shell low-spired, discoidal, with nearly horizontal ramp bordered at upper-outer edge of whorl by a carina with a short slit that appears to mark a peripheral band; diameter approximately 20–25 mm, height 10 mm; four or five whorls, sutures sharply incised and rather deep where the whorls are in contact with umbilicus. Whorl profile flat across upper ramp to carina, sides falling away from the carina with a slight outward slope, the sides gently concave above and roundly convex below, the roundness carrying across the base of the whorl into the umbilicus; umbilicus wide. Shell relatively thick, fine transverse growth lines, five to eight revolving, simple or bifurcated ridges at periphery and lower edge of whorl.

Remarks. *H. umbilicata* is based on specimens collected from the Platteville Formation at Minneapolis and St. Paul, Minnesota; Mineral Point and Beloit, Wisconsin; and La Salle and Rockton, Illinois (Ulrich and Scofield 1897).

Occurrence. Common in the Mifflin and Grand Detour Members of northern Illinois and southern Wisconsin.

Figure 8-3 (Opposite page)
1–5. ***Ophiletina angularis*** Ulrich and Scofield. (**1**) Top view of spire; (**2**) perspective view of top showing the aperture; (**3**) lateral view showing ornamentation; (**4**) perspective view of spire; (**5**) basal view showing the umbilicus. Mifflin Member, Dixon, Illinois (locality 2), **1–4**, ISGS 1029; **5**, ISGS 952.

6, 7. ***Ophiletina sublaxa*** Ulrich and Scofield. (**6**) Top view showing spire and wide flange on outer edge; (**7**) basal view showing umbilicus. Mifflin Member, Dixon, Illinois (locality 2), **6**, JAC 1064; **7**, ISGS 984.

8–11. ***Lesueurilla beloitensis*** (Ulrich and Scofield). (**8**) Perspective view of top showing spire; (**9**) top view showing spire; (**10**) lateral view showing aperture; (**11**) basal view showing umbilicus. Mifflin Member, Dixon, Illinois (locality 2), **8–10**, ISGS 978; **11**, ISGS 979.

12–14. ***Scalites peracuta*** Ulrich and Scofield. (**12**) Basal view showing umbilicus; (**13**) top view showing spire; (**14**) lateral view showing aperture and sharp peripheral edge. Mifflin Member, Dixon, Illinois (locality 2), **12**, ISGS 961; **13**, **14**, ISGS 962.

15–19. ***Raphistomina* sp.** (**15**) Perspective view of top showing spire; (**16**) lateral view showing aperture and wide, upturned peripheral carina; (**17**) top view of spire and photographic enlargement showing blunt apical end possibly indicating the presence of septa; (**18**) basal view showing umbilicus; (**19**) top view with photographic enlargement of spire showing diagnostic partly decapitated flattened tubules at edge of peripheral carina. Mifflin Member, Dixon, Illinois (locality 2), **15–17**, ISGS 1041; **18**, **19**, ISGS 1042.

Genus OPHILETINA

Ophiletina angularis Ulrich and Scofield, 1897

Figures 8-3-1 through 8-3-5

Description. Shell discoidal, with whorls coiled approximately in one plane; notch-keel or peripheral band just within the margin of the upper whorl face, diameter in incomplete specimens approximately 15–20 mm, height 5 mm; three to four whorls; sutures sharply incised and relatively deep; umbilicus wide. Shell relatively thick, with prominent fine, evenly spaced, transverse ridges aligned in sinuous patterns over upper, outer, and lower whorl surfaces.

Remarks. *O. angularis* is based on specimens collected from the *Phylloporina* bed, Black River group (=Decorah Formation), near Cannon Falls, Minnesota (Ulrich and Scofield 1897). *O. angularis* differs from *O. sublaxa* in having evenly spaced, sinuous ridges (lirae) on the upper, outer, and lower whorl surfaces.

Occurrence. Common in the Mifflin and Grand Detour Members of northern Illinois and southern Wisconsin.

Ophiletina sublaxa Ulrich and Scofield, 1897

Figures 8-3-6 and 8-3-7

Description. Shell discoidal with whorls coiled approximately in one plane, in contact with prominent flat peripheral flange of previous whorl; diameter in incomplete specimens approximately 20 mm, height 5 mm; three whorls, prominent circular nucleus, sutures sharply incised and rather deep where the whorls are in contact; peripheral band just within the margin of the upper whorl face, last whorl bearing wide flange near middle of the outer whorl face; umbilicus wide. Shell relatively thick, with fine, evenly spaced growth lines.

Remarks. *O. sublaxa* is based on specimens collected from the Platteville Formation at Minneapolis, Minnesota; Mineral Point, Wisconsin; and Dixon, Illinois (Ulrich and Scofield 1897). *O. sublaxa* differs from *O. angularis* in having a wide, flat flange near the middle of the outer whorl face that is in contact with the subsequent whorl.

Occurrence. Common in the Mifflin and Grand Detour Members in northern Illinois and southern Wisconsin and the McGregor Member of southwestern Wisconsin, northeastern Iowa, and southeastern Minnesota.

Family LESUEURILLIDAE

Genus LESUEURILLA

Lesueurilla beloitensis (Ulrich and Scofield, 1897)

Figures 8-3-8 through 8-3-11

Description. Shell discoidal with earlier whorls depressed below the last, diameter in incomplete specimens approximately 30–40 mm, height 10–15 mm; sutures sharply incised; earlier whorls with high, thin crest at outer-upper angle of the whorls; four or five whorls; disjuncture of abapical whorls not observed. Whorl profile concave on upper surface, gently convex on outer whorl face but slightly concave on upper half near the crest, rounded at base. Shell thin; growth lines fine, curving downward abapically from the thin crest, then gradually reversing to an adapical orientation in approach to the umbilicus, then reversing once more near the sutures.

Remarks. Ulrich and Scofield (1897) described this species from the Platteville Formation at Beloit, Wisconsin, and assigned it to the genus *Eccyliopterus*. Wagner (2002) proposed that the Beloit species *Eccyliopterus beloitensis* be reassigned to the genus *Lesueurilla* (new family Lesueurillidae) now restricted to lesueurilline species that are nearly planispiral, tightly coiled with rounded bases and lenticular apertures.

Occurrence. Locally common in the Mifflin and Grand Detour Members of northern Illinois and southern Wisconsin.

Family STRAPAROLLINIDAE

Genus STRAPAROLLINA

***Straparollina* sp.**

Figures 8-4-8 through 8-4-11

Description. Shell small, attained height of approximately 6 mm, 5 mm wide, moderately high-spired, apical angle 55°–60°; four to six whorls, shallow sutural indentations, umbilicus narrow. Whorl profile rounded on sides and base, lacking a peripheral band; umbilicus narrow with flat sides, open all the way to apex. Aperture round, height and lateral diameters approximately equal. A narrow, spirally wound carina extends upward from the base of the inner lip of the aperture into the umbilicus. Shell relatively thick, smooth, lacking growth lines.

Remarks. This Platteville species resembles *Straparollina pelagica* Billings, 1865, from the Whitrockian Catoche and Aguathuna Formations of western Newfoundland, Canada, in having a shell that is relatively small, rounded, moderately high-spired, with narrow umbilicus open to the apex; prominent umbilical carina, and lack of a peripheral band.

Occurrence. Locally very abundant in the Mifflin Member in northern Illinois.

Superfamily MACLURITOIDEA

Family MACLURITIDAE

Genus MACLURINA

Maclurina bigsbyi (Ulrich and Scofield, 1897)

Figures 8-5-6 through 8-5-12

Description. Shell discoidal, upper surface strongly convex with deep umbilicus instead of being raised into a spire, flat base, diameter approximately 80 mm, typical diameter 50–60 mm, height half to slightly less than half the diameter, three or four whorls; umbilicus deep, exposing part of inner whorls, width one-third to one-half the entire shell diameter. Whorl profile with rounded angulation and gently rounded keel surrounding the umbilicus; upper umbilical sutures sharply incised and moderately deep; basal sutures very shallow. Shell relatively thick, distinct, and with evenly spaced growth lines in adults and with 12 or more evenly spaced revolving ridges at periphery of whorl.

Remarks. *M. bigsbyi* was previously recognized in the Platteville Formation of the upper Midwest by Whitfield (1882) and Ulrich and Scofield (1897). This species was long assigned to *Maclurites* but reassigned to *Maclurina* by Rohr

Figure 8-4 (Opposite page)
1–7. ***Asgardaspira undulata*** (Hall). (**1**) Basal view; (**2**) upper surface; (**3**) side view showing peripheral indentations; (**4**) perspective view of upper surface showing the aperture; (**5**) lateral view showing relatively high spiral of dolomitized steinkern; (**6**) lower surface of steinkern revealing an impression of a spiral ridge on inner surface of shell; (**7**) apical end of steinkern showing a septal chamber. Mifflin Member, Dixon, Illinois (locality 2), **1–3**, ISGS 988; **4**, ISGS 998. Lee Center, Illinois (locality 31), **5–7**, JAC 108.

8–11. ***Straparollina* sp.** (**8**) Side view showing aperture; (**9**) side view; (**10**) top view of spire; (**11**) basal view of umbilicus showing the characteristic narrow, spirally wound ridge extending upward from the aperture into the umbilicus. Mifflin Member, Dixon, Illinois (locality 2), **8**, ISGS 993; **9**, ISGS 1055; **10**, ISGS 940; **11**, ISGS 1051.

12–15. ***Helicotoma umbilicata*** Ulrich and Scofield. (**12**) Lateral perspective showing revolving ridges; (**13**) lateral view showing outline of aperture; (**14**) upper surface showing spire; (**15**) lower surface showing umbilicus. Mifflin Member, Dixon, Illinois (locality 2), **12**, ISGS 986; **13–15**, ISGS 985.

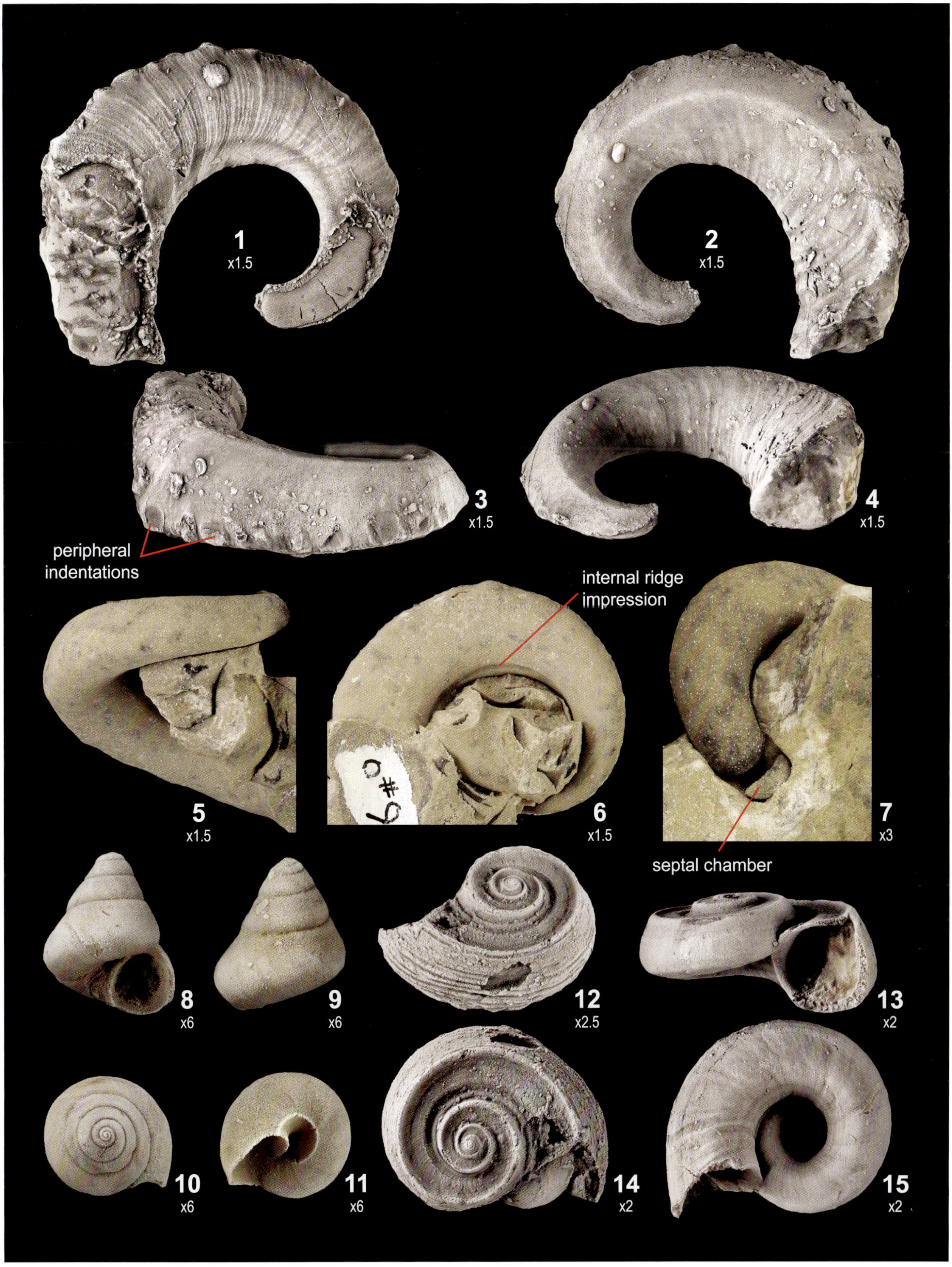
1
x1.5
2
x1.5
3
x1.5
4
x1.5
peripheral
indentations
internal ridge
impression
5
x1.5
6
x1.5
7
x3
septal chamber
8
x6
9
x6
12
x2.5
13
x2
10
x6
11
x6
14
x2
15
x2

et al. (1992) based on the revolving ridges and relatively narrow umbilicus.

Occurrence. Common in the Mifflin and Grand Detour Members in north-central Illinois and south-central Wisconsin and the McGregor Member of southwestern Wisconsin, northeastern Iowa, and southeastern Minnesota. Locally common in the Pecatonica Member in north-central Illinois.

Maclurites nitidus (Ulrich and Scofield, 1897)

Figures 8-5-1 through 8-5-5

Description. Shell discoidal, upper surface strongly convex, with deep umbilicus instead of being raised into a spire, flat base, diameter 20–30 mm, height at least half the shell diameter, three or four whorls; umbilicus deep, exposing part of inner whorls, width one-third to one-half the entire shell diameter, four or five whorls. Whorl profile with rounded angulation and gently rounded keel surrounding the umbilicus; upper umbilical sutures sharply incised and moderately deep; basal sutures very shallow. Shell moderately thick, subtle growth lines in immature specimens.

Remarks. *M. nitidus* was previously recognized in the Platteville Formation of the upper Midwest by Ulrich and Scofield (1897). It differs from *Maclurina bigsbyi* in having a relatively higher shell and a narrower umbilicus that largely conceals the inner whorls. In most other features, the two species are indistinguishable.

Occurrence. Common locally in the Mifflin and Grand Detour Members of northern Illinois and southern Wisconsin.

Maclurites operculum

Figures 8-5-13 and 8-5-14

Description. Operculum calcareous, very thick, growth layers originate from basal corner of inner aperture lip, sequentially stacked like a twisted horn, with a counterclockwise curvature when seen from the exterior. Inner surface with large roughened protuberance for attachment of left retractor muscle; smaller roughened area above for right retractor muscle. On rare occasions, a specimen shows the life association of shell and operculum (Figure 8-5-5). See Rohr and Yochelson (1999) for other examples.

Remarks. Isolated opercula likely belong to either *Maclurina bigsbyi* or *Maclurites nitidus*. In one rare example, the operculum is withdrawn within the shell, showing the aperture with concave margin to be larger in width than the operculum (Figure 8-5-5).

Occurrence. Common locally in the Mifflin and Grand Detour Members of northern Illinois and southern Wisconsin.

Superfamily TROCHONEMATOIDEA

Family TROCHONEMATIDAE

Genus TROCHONEMA

Trochonema umbilicata (Hall, 1847)

Figures 8-6-8 through 8-6-11

Description. Shell height approximately 35 mm, broadly conical spire, convex base, umbilicus open and relatively deep, apical angle approximately 95°, four or five whorls, disjuncture of abapical whorl in some specimens. Shell profile angular with three carinae; whorl surface concave between carinae; sutural shelf wide, concave, nearly horizontal, upper shoulder carina sharp, keel-like; most prominent carina at mid-whorl; third carina at lower edge of final whorl; below lower carina is a prominent umbilical carina that curves into the umbilicus. Shell relatively thick, thin along margins of the aperture. The three main spiral angulations extend forward of the outer edge of aperture as short digitations; aperture tangential to axis of coiling. Growth lines obscure on all but the last whorl, irregular but never coarse.

Remarks. *T. umbilicata* differs from *T. beloitense* in having less prominent growth lines, whorls that do not expand as rapidly, and a smaller apical angle.

Occurrence. Abundant locally in the Mifflin and Grand Detour Members in north-central Illinois and south-central

Figure 8–5 (Opposite page)
1–5. ***Maclurites nitidus*** (Ulrich and Scofield). (**1**) Basal view; (**2**) lateral view of aperture; (**3**) perspective view of base and aperture; (**4**) upper surface showing narrow apical depression; (**5**) lateral view of dolomitized steinkern showing edges of aperture with interior mold of intact operculum, indentation of left muscle platform exposed at bottom. Mifflin Member, Dixon, Illinois (locality 2), **1–4**, ISGS 935. Mifflin Member, Lee Center, Illinois (locality 31), **5**, JAC 109.

6–12. ***Maclurina bigsbyi*** (Ulrich and Scofield). (**6**) Upper surface of dolomitized steinkern; (**7**) lateral view; (**8**) lower surface of dolomitized steinkern; (**9**) perspective view of lower surface and side; (**10**) lateral view showing revolving ridges in photographic enlargement; (**11**) view of lower surface; (**12**) view of upper surface showing umbilicus. Mifflin Member, Lightsville, Illinois (locality 7), **6**, **7**, ISGS 1027. Mifflin Member, Byron, Illinois (locality 8), **8**, ISGS 1009. Mifflin Member, Dixon, Illinois (locality 2), **9–12**, JAC 110.

13, 14. ***Maclurites*** opercula. (**13**) Interior of operculum showing protruding left muscle platform; (**14**) exterior of operculum. Mifflin Member, Dixon, Illinois (locality 2), **13**, **14**, ISGS 1581.

1
x3
2
x3
3
x3
4
x3
5
x2
indentation of left muscle platform
6
x1
7
x1
8
x1
x2
9
x1
10
x1
left muscle platform
13
x1.5
11
x1
12
x1
14
x1.5

1
x1.5
2
x1.5
3
x1.5
4
x1.5
5
x1.5
6
x1.5
7
x1.5
8
x1.5
9
x1.5
10
x1.5
11
x1.5
12
x1

Wisconsin and the McGregor Member of southwestern Wisconsin, northeastern Iowa, and southeastern Minnesota.

Trochonema beloitense Whitfield, 1878

Figure 8-6-1 through 8-6-7

Description. Shell height approximately 35 mm, broadly conical spire, convex base, apical angle approximately 105°–110°, four or five rapidly expanding whorls, last whorl noticeably distended, disjuncture of abapical whorl in some specimens, umbilicus open. Shell profile angular with three carinae; whorl surface concave between carinae; sutural shelf wide, concave, nearly horizontal, upper shoulder carina sharp, keel-like; prominent carinae at mid-whorl; third carina at lower edge of final whorl; below the umbilical carina is a prominent basal angulation that curves into the umbilicus. Shell relatively thick, thin along margins of the aperture. The three main spiral angulations extend forward of outer edge of aperture as short digitations; aperture tangential to axis of coiling. Growth lines prominent, irregularly spaced, and coarse.

Remarks. *T. beloitense* differs from other Platteville trochonematids in having a prominent, coarse growth lines, rapidly expanding whorls, and a relatively large apical angle.

Occurrence. Abundant locally in the Mifflin and Grand Detour Members in north-central Illinois and south-central Wisconsin and the McGregor Member of southwestern Wisconsin, northeastern Iowa, and southeastern Minnesota.

Figure 8-6 (Opposite page)
1–7. ***Trochonema beloitense*** Whitfield. (**1**) Side view of shell showing disjuncture of body whorl and numerous nodular *Solenopora* encrustations; (**2**) side view showing thick, irregular growth lines; (**3**) side view of aperture; (**4**) top view showing prominent growth lines; (**5**) basal view showing umbilicus; (**6**) side view showing encrustation of the bryozoan *Diplotrypa*; (**7**) side view of dolomitized steinkern. Mifflin Member, Dixon, Illinois (locality 2), **1**, ISGS 929; **2**, **3**, **5**, ISGS 930; **4**, ISGS 1054; **6**, ISGS 1037. Grand Detour Member, Beloit, Wisconsin (locality 21), **7**, ISGS 1014.

8–11. ***Trochonema umbilicata*** (Hall). (**8**) Side view; (**9**) side view showing aperture; (**10**) basal view showing umbilicus; (**11**) top view showing spire. Mifflin Member, Dixon, Illinois (locality 2), **8–11**, ISGS 992.

12. ***Globonema niota*** (Hall). (**12**) Side view of large dolomitized steinkern. Grand Detour Member, South Beloit, Illinois (locality 20), **12**, JAC 112.

Globonema niota (Hall, 1861)

Figure 8-6-12

Description. Shell height approximately 65 mm, relatively high conical spire, umbilicus small, apical angle approximately 65°, five or six whorls, umbilicus open. Shell apparently thin, with subdued surface markings and lacking the prominent keel-like carinae present in other Platteville trochonematids. Upper shoulder slightly concave, mid-whorl flat to slightly convex, curving gently toward the umbilicus.

Remarks. *G. niota* is one of the largest of the Platteville gastropods. The relatively high spire of the shell and lack of carinae distinguish it from other Platteville trochonematids. This species was formerly assigned to *Trochonema* but was reassigned to *Globonema* of the same family by Wagner (2017).

Occurrence. Rare in the Grand Detour Member in north-central Illinois and south-central Wisconsin.

Family LOPHOSPIRIDAE

Genus LOPHOSPIRA

Description. The lophospirids have a dextral, moderately high-spired shell with an angular profile accentuated by whorls with three carinae, middle carina bearing a peripheral band, variable sinus or notch on middle carina at lip of the aperture, small umbilicus, lacking a solid columella. At least five lophospirid species occur in the Platteville Formation: *Eunema centralis*, *E. helicteres*, *Lophospira milleri*, *L. perangulata*, and *L. serrulata*.

Lophospira milleri (Hall, 1847)

Figures 8-7-8 through 8-7-10

Description. Shell height approximately 30 mm, moderately high-spired, apical angle 62°–68°; five or six whorls, closely coiled in the upper part of the spire, sutural indentation increases in depth abapically but disjuncture absent, umbilicus open. Shell profile angular with three carinae; whorl surface concave between carinae; sutural shelf narrow, concave, nearly horizontal, upper shoulder carina sharp, keel-like; most prominent carina at mid-whorl, bearing a trilineate peripheral band, middle thread ranges from equidistant between the upper and lower threads to slightly closer to lower thread; third carina at lower edge of final whorl is rounded. Plane of aperture nearly parallel to axis of coiling; main groove on interior of whorl corresponds to the peripheral band-bearing carina; inner lip thickened and reflexed outward, generating a small extension of the umbilicus along columellar lip. Narrow but deep umbilicus coils around axis of shell. Growth lines evenly spaced and parallel, convex forward above and below peripheral band, both curving backward toward the peripheral band; growth lines on peripheral band turn abruptly, forming a chevron-like pattern; turn gradually below the lower carina toward the umbilicus.

Remarks. Whitfield (1886) assigned this Platteville lophospirid to *L. milleri*. Ulrich and Scofield (1897) later assigned this same form to *Murchisonia bicincta*, which led to some nomenclatural confusion. In a taxonomic review, Tofel and Bretsky (1987) showed that *L. milleri* has precedence over *M. bicincta*. *L. milleri* is distinguished from other Platteville lophospirids in having a relatively high-spired shell with a relatively large apical angle, nondisjunct whorls, trilineate peripheral band with a carina that lacks serrations, and shell inflation of the lower part of the last whorl.

Occurrence. Abundant in the Mifflin and Grand Detour Members in north-central Illinois and south-central Wisconsin and the McGregor Member of southwestern Wisconsin, northeastern Iowa, and southeastern Minnesota.

Lophospira perangulata (Hall, 1847)

Figures 8-7-11 and 8-7-12

Description. Shell height approximately 20 mm, moderately high-spired, apical angle 48°–58°; six or seven whorls, closely coiled in the upper part of the spire, sutural indentation increases in depth abapically but disjuncture absent, umbilicus open. Shell profile angular with three carinae; whorl surface concave between carinae; sutural shelf narrow to absent, upper shoulder carina gently rounded; most prominent carina at mid-whorl, bearing a trilineate peripheral band, the middle thread being equidistant between the upper and lower threads; third carina at lower outer edge is gently rounded. Plane of aperture subparallel to axis of coiling; main groove on interior of whorl corresponds to the main carina; inner lip thickened and reflexed outward, generating a small extension of the umbilicus along columellar lip. Narrow but deep umbilicus coils around axis of shell. Growth lines parallel, convex forward above and below peripheral band and curving backward toward the peripheral band.

Remarks. The type specimens of *Lophospira perangulata* (Hall, 1847) are from the Black River Formation at Watertown, New York. *L. perangulata* also occurs in equivalently aged rocks at Murfreesboro, Tennessee, and in Mercer County, Kentucky (Ulrich and Scofield 1897). It is distinguished from other Platteville species of *Lophospira* in having a relatively small, high-spired shell with a relatively small apical angle.

Occurrence. Common in the Mifflin and Grand Detour Members in north-central Illinois and south-central Wisconsin and the McGregor Member of southwestern Wisconsin, northeastern Iowa, and southeastern Minnesota.

Lophospira serrulata (Salter, 1859)

Figures 8-7-13 through 8-7-17

Description. Shell height approximately 50 mm, moderately high-spired, apical angle 55°–65°; five or six whorls, closely coiled in the upper part of the spire, abrupt and wide disjuncture of abapical whorl, umbilicus open. Shell profile angular with three carinae plus an obtuse basal angulation; whorl surface concave between carinae; upper shoulder carina sharp; strongest carina at mid-whorl, bearing peripheral band bordered above and below by fine threads, with crenulations that produce a serrated edge that is particularly well developed on the adapical spires; third carina at lower outer edge; below carina is a broad basal angulation that curves into the umbilicus. Plane of aperture inclined to axis of coiling; three grooves on interior of whorl correspond to carinae; inner lip thickened and reflexed outward, generating a small

Figure 8-7 (Opposite page)
1, 2. ***Eunema centralis*** (Ulrich and Scofield). (**1**) Side view; (**2**) side view showing aperture. Mifflin Member, Dixon, Illinois (locality 2), **1**, **2**, ISGS 920.
3–7. ***Eunema helicteres*** (Salter). (**3**) Side view showing disjuncture of body whorl; (**4**) side view showing aperture; (**5**) side view of large dolomitized steinkern; (**6**) side view of large specimen showing typical disjuncture of the body whorl; (**7**) side view showing photographic enlargement of peripheral band. Mifflin Member, Dixon, Illinois (locality 2), **3**, **4**, ISGS 976; **6**, ISGS 977; **7**, ISGS 1040. Grand Detour Member, South Beloit, Illinois (locality 20), **5**, JAC 113.
8–10. ***Lophospira milleri*** (Hall). (**8**) Side view; (**9**) side view showing aperture; (**10**) side view of dolomitized steinkern. Mifflin Member, Dixon, Illinois (locality 2), **8**, **9**, ISGS 980. Grand Detour Member, Beloit, Wisconsin (locality 21), **10**, JAC 114.
11, 12. ***Lophospira perangulata*** (Hall). (**11**) Side view showing aperture; (**12**) side view. Mifflin Member, Dixon, Illinois (locality 2), **11**, **12**, ISGS 991.
13–17. ***Lophospira serrulata*** (Salter). (**13**) Side view showing prominent disjuncture of body whorl; (**14**) side view showing aperture; (**15**) side view showing prominent peripheral carina; (**16**) side view showing aperture; (**17**) perspective side view showing photographic enlargement of serrated peripheral carina. Mifflin Member, Dixon, Illinois (locality 2), **13**, **14**, ISGS 975; **15**, **16**, ISGS 973; **17**, ISGS 1038.

1
x2
2
x2
3
x1.5
4
x1.5
5
x1
6
x1.5
7
x3
peripheral band
x8
8
x1.5
9
x1.5
10
x1.5
11
x1.5
12
x1.5
13
x1.5
14
x1.5
15
x1.5
16
x1.5
17
x3
x5

extension of the umbilicus along columellar lip; aperture has a shallow sinus and notch. Narrow but deep umbilicus coils around axis of shell. Uncoiling of last whorl occurs in larger specimens. Growth lines parallel, convex forward above and below peripheral band and curving backward toward the peripheral band.

Remarks. This Platteville species most closely resembles *Lophospira serrulata* from the Black River Formation at Paquette Rapids in Ottawa, Ontario, Canada. Its occurrence in the Platteville was first reported by Ulrich and Scofield (1897) from Minneapolis and St. Paul, Minnesota, as well as from Mineral Point, Janesville, and Belleville, Wisconsin, and from Dixon, Illinois. *L. serrulata* differs from other Platteville species of *Lophospira* in having prominent serrations on the carina lying over the peripheral band.

Occurrence. One of the most common gastropods in the Platteville Formation throughout the upper Midwest outcrop belt, particularly in the Mifflin and Grand Detour Members of north-central Illinois and south-central Wisconsin.

Genus EUNEMA

Eunema helicteres (Salter, 1859)

Figures 8-7-3 through 8-7-7

Description. Shell height of approximately 70 mm, high-spired, apical angle 53°–63° in upper part of spire, fully grown shell 40°–45°; five or six whorls, closely coiled in the upper part of the spire, abrupt and wide disjuncture of abapical whorl, umbilicus open. Shell profile angular with three carinae; sutural shelf narrow and flat to slightly concave, upper shoulder carina sharp; upper ramp wide and concave; strongest carina at mid-whorl, bearing trilineate peripheral band bordered above and below by fine threads that are equidistant from medial thread, which is weakly developed or absent; third carina at lower outer edge; lower ramp relatively narrow and concave. Growth lines prominent, variably spaced; straight on sutural shelf; straight or swept away from aperture on upper ramp, forming concentric linear ridges on the peripheral band; straight or bifurcated on lower ramp and converging at the umbilicus.

Remarks. This species was formerly assigned to *Lophospira* but reassigned to *Eunema* of the same family by Wagner (1999). The type specimens are from the Black River Formation at Paquette Rapids in Ottawa, Ontario, Canada. Occurrence of this species in the Platteville Formation was first documented by Whitfield (1882) and later by Ulrich and Scofield (1897). *E. helicteres* differs from other Platteville lophospirids in having concentric linear growth lines on the peripheral band, widely disjunct abapical whorl, and greater size.

Occurrence. Common in the Mifflin and Grand Detour Members in north-central Illinois and south-central Wisconsin and the McGregor Member of southwestern Wisconsin, northeastern Iowa, and southeastern Minnesota.

Eunema centralis Ulrich and Scofield, 1897

Figures 8-7-1 and 8-7-2

Description. Shell height 30 mm, high-spired, apical angle 55°–60°; five or six whorls, closely coiled, no disjuncture, umbilicus open. Shell profile moderately angular with three carinae; sutural shelf narrow and flat to slightly concave, upper shoulder carina sharp; upper ramp wide and concave; strongest carina at mid-whorl, bearing trilineate peripheral band, middle thread slightly closer to the upper thread or equidistant from upper and lower threads; third carina at lower outer edge; lower ramp concave. Growth lines very fine, straight, and evenly spaced; intersecting peripheral band at nearly right angle. Plane of aperture inclined to axis of coiling.

Remarks. Formerly assigned to *Lophospira delicata* Tofel and Bretsky (1987), but reclassified by Wagner (2006) as a junior subjective synonym of *Eunema centralis*. *E. centralis* differs from other Platteville lophospirids in its moderately angular profile and growth lines that are very fine, straight, and evenly spaced and that have nearly right-angle intersections with peripheral band.

Occurrence. Common locally in the Pecatonica, Mifflin, and Grand Detour Members of north-central Illinois and south-central Wisconsin.

Genus GYRONEMA

Gyronema semicarinatum Salter, 1859

Figures 8-8-19 through 8-8-21

Description. Shell height approximately 40 mm, 35 mm wide, conical spire, convex base, umbilicus lacking, apical angle approximately 75°, four or five whorls, prominent spiraling carinae. Whorl profile with three principal carinate angulations and several minor ones; base slightly distended and marked by six or seven low-profile carinae; interspaces between angulations concave; sutures angular but not deep. Growth lines subdued and irregular.

Remarks. Formerly assigned to *Gyronema duplicata* Ulrich and Scofield (1897) but reclassified by Wagner (2017) as a junior subjective synonym of *G. semicarinatum* Salter, 1859. This species of *Gyronema* agrees in most aspects with *G. duplicata* as described by Ulrich and Scofield (1897) from the Platteville Formation at Beloit and other localities in southern Wisconsin as well as Dixon, Illinois.

Occurrence. Common locally in the Mifflin Member of north-central Illinois and south-central Wisconsin.

Order MIMOSPIRIDA

Family ONYCHOCHILIDAE

Subfamily CATALANISPIRINAE

Genus CATALANISPIRA

Catalanispira plattevillensis Ebbestad et al. (2021)

Figures 8-8-15 through 8-8-18

Description. Shell low-spired, discoidal, sinistrally coiled, base concave, wide umbilicus; attained height approximately 7 mm, width 12 mm, broadly conical spire, apical angle approximately 80°, three whorls, closely coiled, sutures shallow. Whorl profile with slightly convex sides, very angular at periphery, base concave with flat sides. Aperture unusually narrow and elongate. Shell moderately thick, growth lines subtle and curving downward adapically to the whorl periphery.

Remarks. This Platteville species is similar to *Catalanispira* (*Laeogyra*) *reinwaldti* Öpik, 1930, and discussed by Frisk and Ebbestad (2007) from the Upper Ordovician (Kukruse Stage) Viivikonna Formation of Estonia.

Occurrence. Common locally in the Mifflin Member at Dixon, Illinois.

Subclass VETIGASTROPODA

Paleozoic Taxa of Uncertain Position

Family HOLOPEIDAE

Genus HOLOPEA

Holopea ampla Ulrich and Scofield, 1897

Figures 8-8-1 through 8-8-6

Description. Shell height approximately 45 mm, 55 mm wide, low-spired, apical angle 110°; three or four whorls, closely coiled, deep sutural indentations, umbilicus open. Whorl profile strongly rounded on sides and base, distended; aperture rounded, height and lateral diameters approximately equal. Shell moderately thick for small size, surface smooth with surface undulations, growth lines present in some specimens.

Remarks. *H. ampla* is distinguished from the Platteville species *H.* cf. *insignis* in having a spire that rises higher.

Occurrence. Common locally in the Mifflin and Grand Detour Members of north-central Illinois and south-central Wisconsin.

Holopea* cf. *insignis Ulrich and Scofield, 1897

Figure 8-8-7

Description. Shell attained height approximately 45 mm, 55 mm wide, low-spired, apical angle 140°–145°; three or four whorls, closely coiled, deep sutural indentations, umbilicus open. Whorl profile strongly rounded on sides and base, distended; aperture elliptical with height slightly greater than lateral diameter. Shell thin, with surface undulations, growth lines present in some specimens.

Remarks. *H.* cf. *insignis* differs from other Platteville species in having a wider apical angle and an aperture that is subelliptic (higher than wide) in cross section.

Occurrence. Common locally in the Mifflin and Grand Detour Members of north-central Illinois and south-central Wisconsin.

Order PLEUROTOMARIIDA

Superfamily EOTOMARIOIDEA

Family EOTOMARIIDAE

Genus PARALIOSPIRA

Paraliospira supracingulata Billings, 1857

Figures 8-9-14 through 8-9-18

Description. Shell height 25–30 mm, width 35–45 mm, broadly conical spire with convex base, apical angle varies markedly from as little as 50° in early growth to approximately 95° at full size, six or seven whorls, closely coiled, no disjuncture, umbilicus open and large. V-shaped sinus in the outer lip of aperture culminating on the upper side of the periphery in a short slit or notch giving rise to a peripheral band; whorl profile convex above peripheral band, gently concave within peripheral band, periphery sharply angular, convex below peripheral band. Shell thin with variably spaced growth lines curving adapically downward from sutural shelf to the peripheral band, forming concentric linear ridges on the peripheral band, curving abapically downward from peripheral band then abruptly curving back

1
x3
2
x2
3
x3
4
x3
5
x1
6
x1
7
x1
8
x5
9
x10
10
x1
11
x2.5
12
x2.5
13
x2
14
x2
15
x3
16
x3
17
x3
18
x3
19
x2
20
x2
21
x2
22
x6
23
x6
24
x6
25
x6

in adapically before reversing direction again at opening to umbilicus.

Remarks. Formerly assigned to *Eotomaria semicarinatum* (Billings, 1857) but transferred to *Paraliospira* by Wagner (2002). *P. supracingulata* was reported previously from the Platteville Formation by Hall (1861), Whitfield (1895), and Ulrich and Scofield (1897).

Occurrence. Present in the Platteville Formation throughout the upper Midwest outcrop belt and particularly common in the Pecatonica, Mifflin, and Grand Detour Members of north-central Illinois and south-central Wisconsin.

Figure 8-8 (Opposite page)
1–6. ***Holopea ampla*** Ulrich and Scofield. (**1**) Side view showing rapidly expanding body whorl; (**2**) side view showing aperture; (**3**) top view of spire; (**4**) basal view of umbilicus; (**5**) top view of dolomitized steinkern; (**6**) side view of steinkern. Mifflin Member, Dixon, Illinois (locality 2), **1**, **3**, **4**, ISGS 990; **2**, ISGS 989. Grand Detour Member, Beloit, Wisconsin (locality 21), **5**, **6**, JAC 116.

7. ***Holopea* cf. *insignis*** Ulrich and Scofield. (**7**) Side view of limestone steinkern showing the aperture. Mifflin Member, Fennimore, Wisconsin (locality 26), **7**, ISGS 1015.

8–10. ***Omospira* cf. *laticincta*** Ulrich and Scofield. (**8**) Side view showing aperture and distinctive fine, closely spaced, and uniform threads; (**9**) photographic enlargement of whorl showing the distinctive threads; (**10**) side view of large specimen. Mifflin Member, Dixon, Illinois (locality 2), **8**, **9**, ISGS 950; **10**, ISGS 999.

11, 12. ***Hormotoma gracilis*** (Hall). (**11**) Side view showing aperture; (**12**) side view. Mifflin Member, Dixon, Illinois (locality 2), **11**, ISGS 994; **12**, ISGS 1022.

13, 14. ***Ectomaria prisca*** (Billings). (**13**) Side view showing aperture; (**14**) side view. Mifflin Member, Dixon, Illinois (locality 2), **13**, **14**, ISGS 993.

15–18. ***Catalanispira plattevillensis*** Ebbestad et al. (**15**) Top view showing characteristic left-hand spiral; (**16**) basal view showing umbilicus; (**17**) side view showing angular lower margin; (**18**) side view showing narrow aperture. Mifflin Member, Dixon, Illinois (locality 2), **15–18**, ISGS 1035.

19–21. ***Gyronema semicarinatum*** Salter. (**19**) Side view showing aperture; (**20**) side view showing prominent spiral carinae; (**21**) top view of spire. Mifflin Member, Dixon, Illinois (locality 2), **19–21**, ISGS 960.

22–25. ***Cataschisma convexum*** (Ulrich and Scofield). (**22**) Side view showing aperture; (**23**) side view showing faint trace of the peripheral band; (**24**) top view of spire; (**25**) basal view showing umbilicus. Mifflin Member, Dixon, Illinois (locality 2), **22**, ISGS 1057; **23**, ISGS 1062; **24**, ISGS 1063; **25**, ISGS 1056.

Genus EOTOMARIA

Eotomaria vicina Ulrich and Scofield, 1897

Figures 8-9-10 through 8-9-13

Description. Shell height 15–20 mm, width 20–25 mm, broadly conical spire with convex base, apical angle 105°, four or five whorls, closely coiled, no disjuncture, umbilicus open and large. V-shaped sinus in the outer lip of aperture culminating on the upper side of the periphery in a short slit or notch giving rise to a peripheral band; whorl profile sinuous from sutural shelf down to peripheral band, flat to gently concave within peripheral band, periphery sharply angular, convex below peripheral band. Shell moderately thick with fine, evenly spaced growth lines curving adapically downward from sutural shelf to the peripheral band, forming concentric linear ridges in peripheral band, curving abapically downward from peripheral band then abruptly curving back adapically before converging on umbilicus.

Remarks. *E. vicina* was proposed by Ulrich and Scofield (1897) for specimens collected from the Platteville Formation at Minneapolis, Minnesota, and Mineral Point, Wisconsin. It differs from *P. supracingulata* in its smaller size, wider apical angle, slightly higher spire, and configuration of growth lines.

Occurrence. Present in the Platteville Formation throughout the upper Midwest outcrop belt and particularly common in the Pecatonica, Mifflin, and Grand Detour Members of north-central Illinois and south-central Wisconsin.

Genus CLATHROSPIRA

Clathrospira subconica (Hall, 1847)

Figures 8-9-6 through 8-9-9

Description. Shell height approximately 45 mm, width 40 mm, broadly conical spire, apical angle approximately 75°, six or seven whorls, closely coiled, sutures shallow, no disjuncture, umbilicus open and narrow. Shell profile angular at periphery; above periphery, the profile is slightly convex in the upper part of the whorl and gently concave on the lower part. V-shaped sinus in the outer lip of the aperture, culminating on the upper side of the periphery in a short slit or notch, giving rise to a bilineate peripheral band. Shell thin; growth lines consist of fine threads, closely spaced and uniform, curving adapically downward from gently rounded sutural shelf to the peripheral band, forming concentric linear threads on the peripheral band, curving abapically downward from peripheral band then curving back adapically toward the umbilicus; a second series of fine threads spiral around the whorl parallel to the suture; together the two series form a prominent and distinct woven pattern.

Remarks. *C. subconica* was reported previously from the Platteville Formation by Ulrich and Scofield (1897). The broadly conical shell, shallow suture, and woven pattern of threads distinguish this species from other Platteville gastropods.

Occurrence. Present in the Platteville Formation throughout the upper Midwest outcrop belt and is particularly common in the Pecatonica, Mifflin, and Grand Detour Members of north-central Illinois and south-central Wisconsin.

Genus LIOSPIRA

Liospira obtusa Ulrich and Scofield, 1897

Figures 8-9-1 through 8-9-5

Description. Shell height approximately 15–20 mm, diameter 30–40 mm, spire low, apical angle approximately 135°, sutures very shallow, nearly indistinguishable on shell surface, four or five whorls, periphery narrowly rounded, upper surface flat to very slightly convex, umbilicus open. V-shaped sinus in the outer lip of aperture culminating in a notch that generates an angular peripheral band just above outer edge of whorl. Shell exterior smooth, relatively thick in the umbilicus region.

Remarks. This species is distinguished by its smooth shell, low spire, and sutures that are flush with the upper surface.

Occurrence. Present in the Platteville Formation throughout the upper Midwest outcrop belt and particularly common in the Pecatonica, Mifflin, and Grand Detour Members of north-central Illinois and south-central Wisconsin.

Family GOSSELITINIDAE

Genus CATASCHISMA

Cataschisma convexum (Ulrich and Scofield, 1897)

Figures 8-8-22 through 8-8-25

Description. Shell height approximately 10 mm, 9 mm wide, conical spired, apical angle 85°–100°; four or five whorls, shallow sutural indentations, umbilicus narrow. Whorl profile rounded on sides and base; bilineate peripheral band positioned low on the whorl, with growth lines outlining a shallow sinus; aperture elliptical, with height slightly greater than lateral diameter. Spirally wound narrow ridge extends upward from inner lip of aperture. Shell thin, mostly smooth, very faint peripheral band and growth lines; growth lines curve adapically downward from sutures to the peripheral band, forming concentric linear threads in the band, curving abapically downward from the band, then curving back adapically toward the umbilicus.

Remarks. The overall shell morphology, narrow umbilicus, bilineate peripheral band, and configuration of growth lines are very similar to *Clathrospira convexa* from the Platteville Formation in Calhoun County, Illinois (Ulrich and Scofield 1897). This species recently has been reassigned to the genus *Cataschisma* (Wagner 2002). *Cataschisma* is distinguished from *Clathrospira* by having a more rounded whorl profile and a broader but weakly developed peripheral band that is positioned in the middle of the aperture. The majority of shells found in northern Illinois are smooth and lack well-defined growth lines and a peripheral band.

Occurrence. Very abundant locally in the Mifflin Member in northern Illinois.

Superfamily MURCHISONIOIDEA

Family MURCHISONIIDAE

Genus HORMOTOMA

Hormotoma gracilis (Hall, 1847)

Figures 8-8-11 and 8-8-12

Description. Shell height approximately 30 mm, 7 mm wide, conical, high spire, umbilicus lacking, apical angle

Figure 8-9 (Opposite page)

1–5. ***Liospira obtusa*** Ulrich and Scofield. (**1**) Side view showing aperture; (**2**) basal view showing umbilicus; (**3**) upper surface; (**4**) upper surface of dolomitized steinkern; (**5**) side view. Mifflin Member, Dixon, Illinois (locality 2), **1**, **2**, ISGS 969; **3**, ISGS 971. Grand Detour Member, Beloit, Wisconsin (locality 21), **4**, **5**, ISGS 972.

6–9. ***Clathrospira subconica*** (Hall). (**6**) Side view of dolomitized steinkern; (**7**) side view of well-preserved specimen showing peripheral band and shell ornamentation in photographic enlargement; (**8**) side view showing the aperture and a part of the shell that was broken and healed during life; (**9**) top view of spire. Grand Detour Member, Lightsville, Illinois (locality 7), **6**, ISGS 963; **7–9**, ISGS 997.

10–13. ***Eotomaria vicina*** Ulrich and Scofield. (**10**) Side view; (**11**) top view of spire; (**12**) side view showing aperture; (**13**) basal view showing umbilicus. Mifflin Member, Dixon, Illinois (locality 2), **10–13**, ISGS 958.

14–19. ***Paraliospira supracingulata*** (Billings). (**14**) Top view showing spire and growth lines; (**15**) basal view showing umbilicus; (**16**) side view showing growth lines; (**17**) side view showing aperture; (**18**) side view of small specimen showing the abrupt change in apical angle in early growth stages; (**19**) perspective side view of a dolomitized steinkern. Mifflin Member, Dixon, Illinois (locality 2), **14**, **15**, **17**, ISGS 957; **16**, ISGS 956; **18**, ISGS 1052. Grand Detour Member, Beloit, Wisconsin (locality 21), **19**, JAC 115.

1
x2
2
x2
3
x2
4
x2
5
x2
6
x2
x10
peripheral band
7
x2
healed shell
8
x2
9
x2
10
x1.5
11
x1.5
12
x1.5
13
x1.5
14
x1.5
15
x1.5
16
x1.5
17
x1.5
18
x3
19
x1

approximately 18°–25°, 12 or 13 whorls. Whorl profile rounded on sides and base; deep sutural indentation; growth lines indicate outer lip of aperture has a deep V-shaped sinus culminating a little above mid-whorl height that gives rise to a peripheral band that is outlined by two weakly developed spiral ridges that are not developed in all specimens; aperture elliptical with height slightly greater than lateral diameter. Shell thin, growth lines subdued or absent.

Remarks. This Platteville species of *Hormotoma* agrees in most aspects with *H. gracilis* described by Ulrich and Scofield (1897) from the overlying Upper Ordovician Galena and Maquoketa Formations. The type specimens of *H. gracilis* are from the Trenton Formation at Watertown, New York.

Occurrence. Present in the Platteville Formation throughout the upper Midwest outcrop belt. Particularly abundant in the Mifflin and Grand Detour Members of north-central Illinois and south-central Wisconsin.

Genus ECTOMARIA

Ectomaria prisca (Billings, 1857)

Figures 8-8-13 and 8-8-14

Description. Shell height 40–50 mm, conical, high spire, umbilicus lacking, apical angle approximately 17°, 12 or 13 whorls. Whorl profile angular with two prominent carinae at periphery; upper shoulder flat to slightly convex, narrow concave band between the two carinae, gently concave between the lower carina and a low, rounded ridge adjacent to the lower suture, sutures sharply incised but not deep. Shell relatively thick, growth lines subdued or absent.

Remarks. The generic name *Ectomaria* (Koken 1896) takes precedence over *Solenospira* of Ulrich and Scofield (1897). This Platteville species of *Ectomaria* agrees in most aspects with those of *E. prisca* from the Ordovician Calciferous Sandrock, Mingan Islands, Québec, Canada.

Occurrence. Common in the Platteville Formation throughout the upper Midwest outcrop belt. Particularly abundant in the Mifflin and Grand Detour Members of north-central Illinois and south-central Wisconsin.

Superfamily PSEUDOPHOROIDEA

Family PSEUDOPHORIDAE

Pseudophorid sp.

Figures 8-10-6 through 8-10-12

Description. Shell height approximately 11 mm, 7 mm wide, conical spire, umbilicus lacking, apical angle ranging from 33° to 53°, seven or eight whorls. Whorl profile nearly flat on sides, shallow sutural indentation; a narrow convex band just below the upper suture followed below by a broad, slightly concave band that slopes gently outward to a subtle revolving ridge that occupies the whorl periphery and protrudes beyond the lower suture; a small rounded notch at the whorl periphery appears to give rise to a peripheral band; base broadly flat to slightly concave. Shell relatively thick, surface marked by evenly spaced sigmoidal-shaped growth lines that extend adapically downward from the suture to the rounded ridge at periphery of whorl, where they bend slightly abapically, then back adapically toward the lower suture.

Remarks. The trochiform shell, closed umbilicus, flat to slightly concave base, inclined aperture with sigmoidal outline, and notchlike sinus suggest a connection with the Silurian pseudophorid gastropods. This Platteville snail likely represents a yet to be named genus and species.

Occurrence. Common in the Mifflin Member, Platteville Formation (Sandbian) at Dixon, Illinois.

Superfamily SINUSPIROIDEA

Family NERITARIIDAE

Subfamily ONCOCHILINAE

Genus OMOSPIRA

Omospira* cf. *laticincta Ulrich and Scofield, 1897

Figures 8-8-8 through 8-8-10

Description. Shell height approximately 20 mm, moderately high-spired, apical angle 43°; six or seven whorls, closely coiled, deep sutural indentations, umbilicus very narrow or closed. Whorl profile somewhat flattened above, sloping outward from sutures at an angle of 50°–60° from the vertical to a relatively wide peripheral band at outer edge of the whorl; peripheral band marked above and below by a prominent sharp, revolving thread; whorl very gently rounded below the periphery of whorl. Outer lip of aperture with a broad, shallow sinus positioned just above the periphery. Shell thin; growth lines consist of fine threads, closely spaced, parallel and uniform, curving adapically downward from very narrow sutural shelf to the peripheral band, forming concentric linear threads in the peripheral band, curving abapically downward from peripheral band to intersect lower suture at a right angle and subsequently curve back adapically toward the umbilicus.

Remarks. This Platteville species of *Omospira* is most similar to *O. laticincta* from the Stones River Group near Lebanon, Tennessee.

Occurrence. Rare in the Platteville Formation and presently known only from the Mifflin and Grand Detour Members in north-central Illinois and south-central Wisconsin.

Subclass NERITIMORPHA

Paleozoic Taxa of Uncertain Position

Superfamily PLATYCERATOIDEA

Family PLATYCERATIDAE

Genus CYCLONEMA

Cyclonema textilis (Ulrich and Scofield, 1897)

Figures 8-10-13 through 8-10-16

Description. Shell height approximately 25 mm, 20 mm wide, conical spire, convex base, umbilicus lacking, apical angle approximately 70°, four or five whorls. Whorl profile rounded on sides, base distended; deep sutural indentation; commonly with widely separated, deep, oblique constrictions that parallel the growth lines; aperture rounded, height and lateral diameters approximately equal; columellar lip somewhat thickened, nearly straight in its inner (apertural) side, flaring with moderately well-arched outline on the outer (umbilical) side and with a gently concave surface. Shell moderately thick for small size; surface marked by fine, sharp, spiraling and obliquely transverse lines similar to coarsely woven cloth.

Remarks. This species of *Cyclonema* is most similar to *C. textilis* (Ulrich and Scofield, 1897) from the somewhat younger Ordovician Decorah and Galena Formations of Minnesota. It was originally assigned by Ulrich and Scofield (1897) to the genus *Strophostylus*.

Occurrence. Common locally in the Mifflin Member of north-central Illinois.

Subclass CAENOGASTROPODA

Paleozoic Taxa of Uncertain Position

Superfamily SUBULITOIDEA

Family SUBULITIDAE

Genus SUBULITES

Subulites dixonensis Ulrich and Scofield, 1897

Figures 8-10-2 and 8-10-3

Description. Shell elongate and fusiform, attained height approximately 80 mm, 20 mm wide, conical, high spire, umbilicus lacking, apical angle approximately 22°, six or seven whorls, slender extended base with an apparent siphonal fold and notch. Whorl profile flat to slightly convex between sutures; whorls relatively high; sutures very shallow and obscure on surface of shell; middle part of body whorl full and notably convex, causing a more rapid downward taper to the shell profile. Shell thin, smooth, lacking growth lines.

Remarks. The type specimens of *S. dixonensis* are from the Platteville Formation at Dixon, Illinois (Ulrich and Scofield 1897). This species differs from others in having a relatively large shell with a whorl profile that is flat to slightly convex and a body whorl that is distinctly full and convex.

Occurrence. Common in the Mifflin and Grand Detour Members in north-central Illinois and south-central Wisconsin and the McGregor Member of southwestern Wisconsin, northeastern Iowa, and southeastern Minnesota.

Subulites beloitensis Ulrich and Scofield, 1897

Figure 8-10-1

Description. Shell elongate and fusiform, attained height approximately 60 mm, 14 mm wide, conical, high spire, umbilicus lacking, apical angle approximately 19°, seven or eight whorls, slender extended base with an apparent siphonal fold and notch. Whorl profile flat between sutures; whorls relatively high; sutures shallow. Shell thin, smooth, lacking growth lines.

Remarks. *S. beloitensis* is distinguished by its flat whorl profile and relatively small apical angle.

Occurrence. Common in the Mifflin and Grand Detour Members in north-central Illinois and south-central Wisconsin and the McGregor Member of southwestern Wisconsin, northeastern Iowa, and southeastern Minnesota.

Genus FUSISPIRA

***Fusispira* sp.**

Figure 8-10-4

Description. Shell fusiform, attained height approximately 60 mm, 35 mm wide, conical, short spire, umbilicus lacking, apical angle approximately 55°, four or five whorls, extended base with an apparent siphonal fold and notch. Whorl profile convex between sutures; last whorl occupying more than half the total shell height. Shell thin, smooth, lacking growth lines.

Remarks. This Platteville species is distinguished by its rounded whorl profile, large body whorl, and relatively wide aperture.

1
x1
2
x1
3
x1
4
x1
5
x3
6
x8
7
x8
8
x8
9
x8
10
x8
11
x8
12
x8
columellar lip
13
x4
14
x4
15
x4
16
x9

Occurrence. Present but rare in the Mifflin and Grand Detour Members of north-central Illinois and south-central Wisconsin.

Genus CYRTOSPIRA

Cyrtospira tortilis Ulrich and Scofield, 1897

Figure 8-10-5

Description. Shell elongate and fusiform, axis of shell curved, attained height approximately 25–30 mm, 5 mm wide, conical, short spire, umbilicus lacking, apical angle approximately 40°, five or six whorls, slender extended base with an apparent siphonal fold and notch. Whorl profile convex between sutures; last whorl occupying more than half the total shell height. Shell thin, smooth, lacking growth lines.

Remarks. This Platteville species of *Cyrtospira* agrees in most aspects with *C. tortilis* described by Ulrich and Scofield (1897) from the lower part of the Stones River Group, Murfreesboro, Tennessee.

Occurrence. Common locally in the Mifflin Member of north-central Illinois.

Subclass AMPHIGASTROPODA

Order BELLEROPHONTIDA

Superfamily BELLEROPHONTOIDEA

Family BUCANIIDAE

Genus BUCANIA

Bucania* cf. *emmonsi Ulrich and Scofield, 1897

Figures 8-11-1 through 8-11-4

Description. Shell planispiral, wide umbilicus with sharp angular shoulders, shell width increases by about 2.5 times per whorl; maximum attained whorl width in incomplete specimens approximately 25 mm; three or four whorls. Whorl profile with gently arched and somewhat depressed dorsum, longitudinally oriented peripheral band raised slightly above middle of dorsum, sides of dorsum turning sharply into umbilicus, umbilical slopes very slightly convex to flat, sutures sharply incised and rather deep, ventral whorl gently concave, reflecting opposing shape of previous whorl; sutural indentations. Aperture height approximately half the width; growth lines indicate the presence of broad, shallow, V-shaped sinus. Shell relatively thin; revolving threads dominate shell ornament, consisting of about 12 dorsal threads on either side of peripheral band that angle very slightly toward the midline; corrugated growth lamellae spaced at intervals of approximately 1.5 mm form a prominent V-shaped pattern on the dorsum, with wavy anterior edges at intersection with revolving threads; the dorsal growth lamellae produce regularly spaced indentations on the ventral surface of the

Figure 8-10 (Opposite page)
1. ***Subulites beloitensis*** Ulrich and Scofield. (**1**) Side view showing aperture, slender profile, and nearly flat sides of dolomitized steinkern. Mifflin Member, Lee Center, Illinois (locality 31), **1**, ISGS 1074C.

2, 3. ***Subulites dixonensis*** Ulrich and Scofield. (**2**) Side view showing aperture and relatively convex whorl profile of dolomitized steinkern; (**3**) side view of specimen with calcitic shell. Mifflin Member, Lee Center, Illinois (locality 31), **2**, JAC 118; Dixon, Illinois (locality 2), **3**, ISGS 1004.

4. ***Fusispira* sp.** (**4**) Lateral view showing the aperture on lower right side. Mifflin Member, Lee Center, Illinois (locality 31), **4**, ISGS 1076C.

5. ***Cyrtospira tortilis*** Ulrich and Scofield. (**5**) Side view showing characteristic curve of shell. Mifflin Member, Dixon, Illinois (locality 2), **5**, ISGS 1586.

6–12. **Pseudophorid sp.** (**6**) Basal view showing umbilicus; (**7**) side view showing narrow aperture; (**8**) side view; (**9**) side view showing aperture; (**10**) side view; (**11**) side view showing regularly spaced threads marking the growth lines; (**12**) side view showing umbilicus and regularly spaced threads. Mifflin Member, Dixon, Illinois (locality 2), **6**, ISGS 1049; **7**, **8** ISGS 1585 **9**, **10**, ISGS 1048; **11**, **12**, ISGS 1050.

13–16. ***Cyclonema textilis*** (Ulrich and Scofield). (**13**) Side view showing aperture; (**14**) basal view showing columellar lip; (**15**) side view; (**16**) photographic enlargement of shell showing the characteristic woven-like ornamentation. Mifflin Member, Dixon, Illinois (locality 2), **13**, **14**, ISGS 1046; **15**, ISGS 1045; **16**, ISGS 1047.

succeeding whorl; prominent corrugated growth lamellae continue over angular shoulders into the umbilicus.

Remarks. This Platteville species is similar to *B. emmonsi* in having regularly spaced indentations on the ventral surface. In their description, Ulrich and Scofield (1897) do not mention the presence of dorsal threads or the angle of intersection with the peripheral band. Their illustration (figure 1, plate 66), however, does appear to show threads with a very slight angle toward the midline. *B.* cf. *emmonsi* is relatively large, with an attained whorl width of 25 mm.

Occurrence. Common in the Mifflin and Grand Detour Members in north-central Illinois and south-central Wisconsin.

Genus SALPINGOSTOMA

Salpingostoma buelli (Whitfield, 1878)

Figures 8-11-5 through 8-11-12

Description. Shell planispiral, wide umbilicus with rounded shoulders, adult shell expanded, with trumpetlike aperture extending anteriorly and laterally into a recurved apron; open slit extending from dorsum to just inside outer lip, where it terminates; four whorls in adult shells. Whorl profile with gently arched and somewhat depressed dorsum, longitudinally oriented peripheral band raised slightly above middle of dorsum, sides of dorsum gradually turning into umbilicus, umbilical slopes convex, sutures sharply incised and rather deep, ventral whorl gently concave, reflecting opposing shape of previous whorl. Aperture subcircular; abruptly expanded at frequent intervals, the expansion left behind forming transverse, imbricating, scalloped lamellae. Shell relatively thick; revolving threads dominate shell ornament, consisting of about 12 dorsal threads on either side of and parallel to the peripheral band; corrugated growth lamellae form prominent V-shaped pattern on the dorsum with wavy anterior edges intersecting with revolving threads, corrugated lamellae continuing onto umbilical whorls.

Remarks. *S. buelli* was originally described from the Platteville Formation of southern Wisconsin (Whitfield 1878). Ulrich and Scofield (1897) noted the occurrence of this species in the Platteville Formation at Minneapolis, Cannon Falls, and Old Concord, Minnesota; Beloit and Janesville, Wisconsin; and Rockton and Dixon, Illinois.

Occurrence. Common in the Mifflin and Grand Detour Members in north-central Illinois and south-central Wisconsin.

Genus TETRANOTA

Tetranota sexcarinata Ulrich and Scofield, 1897

Figures 8-12-7 through 8-12-12

Description. Shell planispiral, wide umbilicus, up to 25 mm long, broad dorsum with six revolving ridges, three or four whorls that are 2.5 times wider than high. Whorl profile with dorsal medial pair of ridges elevated and separated by slight concavity marking a peripheral band; shallow depressions separate the lateral ridges, outer ridges form the umbilical shoulders, umbilical slopes flat to very slightly concave, ventral whorl surface gently concave, reflecting opposing shape of previous whorl. Aperture wider than high, slight lateral flare of margins, wide U-shaped sinus. Shell moderately thick with faint growth lines.

Remarks. *T. sexcarinata* was originally described from the Platteville Formation at Minneapolis, St. Paul, and Cannon Falls, Minnesota; Janesville and Beloit, Wisconsin; and Dixon, Illinois (Ulrich and Scofield 1897). *T. sexcarinata* differs from *T. wisconsinensis* in having six revolving ridges and a wider umbilicus.

Occurrence. Common in the Mifflin and Grand Detour Members in north-central Illinois and south-central Wisconsin.

Figure 8-11 (Opposite page)

1–4. ***Bucania* cf. *emmonsi*** Ulrich and Scofield. (**1**) Left-lateral view of umbilicus showing prominent corrugated growth lamellae in the umbilicus; (**2**) posterior view showing longitudinally oriented peripheral band and corrugated growth lamellae forming a V-shaped pattern; (**3**) dorsal surface; (**4**) left-lateral view of dolomitized steinkern showing umbilicus. Mifflin Member, Dixon, Illinois (locality 2), **1–3**, ISGS 713. Mifflin Member, Lee Center, Illinois (locality 31), **4**, ISGS 1002.

5–12. ***Salpingostoma buelli*** (Whitfield). (**5**) Ventral view showing aperture with prominent bell-like expansions; (**6**) perspective view of left-lateral side showing encrustation of *Diplotrypa*; (7) perspective view of left-lateral side of dolomitized steinkern; (**8**) anterior view showing flared margins of aperture and prominent slit; (**9**) posterior view; (**10**) left-lateral view showing umbilicus; (**11**) dorsal view showing bell-like expansion of aperture and median slit; (**12**) impression of dorsal surface of dolomitized specimen showing the V-shaped pattern of the corrugated growth lamellae. Mifflin Member, Dixon, Illinois (locality 2), **5**, ISGS 934; **6**, ISGS 1000; **8–11**, JAC 121. Grand Detour Member, Durand, Illinois (locality 11), **7**, ISGS 1001. Grand Detour Member, Rochelle, Illinois (locality 35), **12**, JAC 122.

13–16. ***Pterotheca attenuata*** Hall. (**13**) Dorsal surface showing prominent keel; (**14**) ventral surface showing narrow aperture and median sinus at end of keel; (**15**) perspective view of posterior end showing the prominently elevated keel; (**16**) dorsal surface of dolomitized steinkern. Mifflin Member, Dixon, Illinois (locality 2), **13**, **15**, ISGS 995; **14**, ISGS 996. Mifflin Member, Byron, Illinois (locality 8), **16**, ISGS 1034.

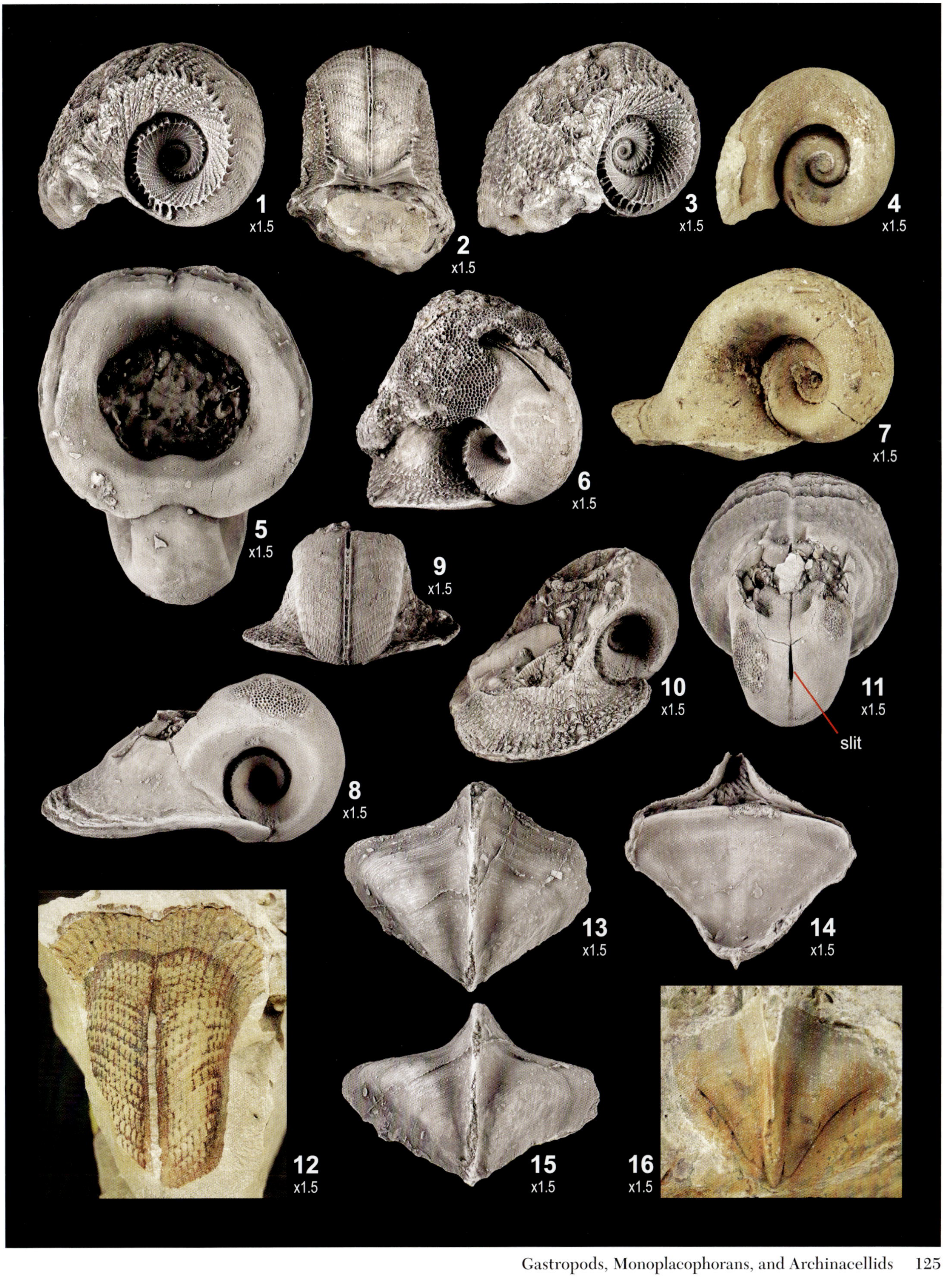
1
x1.5
2
x1.5
3
x1.5
4
x1.5
5
x1.5
6
x1.5
7
x1.5
8
x1.5
9
x1.5
10
x1.5
11
x1.5
slit
12
x1.5
13
x1.5
14
x1.5
15
x1.5
16
x1.5

Tetranota wisconsinensis Ulrich and Scofield, 1897

Figures 8-12-1 through 8-12-6

Description. Shell planispiral, narrow umbilicus, up to 25 mm long, two or three broad and depressed whorls. Whorl profile with dorsal medial band that is distinctly elevated and bears a peripheral band; dorsal lateral slopes broadly convex with two or four low, rounded, revolving ridges in juvenile whorls that gradually diminish in late-stage whorls; umbilical shoulders gently rounded, ventral whorl surface gently concave, reflecting opposing shape of previous whorl; sutures incised. Adult aperture wider than high, lateral and anterior margins prominently flared and flattened; broad, deep sinus. Shell moderately thick with faint growth lines.

Remarks. *T. wisconsinensis* was originally described from the Platteville Formation at Minneapolis and St. Paul, Minnesota; Janesville and Beloit, Wisconsin; and Dixon, Illinois (Ulrich and Scofield 1897). *T. wisconsinensis* differs from *T. sexcarinata* in having an adult aperture that is widely flared and in having fewer revolving ridges.

Occurrence. Common in the Mifflin and Grand Detour Members in north-central Illinois and south-central Wisconsin.

Family PTEROTHECIDAE

Genus PTEROTHECA

Pterotheca attenuata Hall, 1861

Figures 8-11-13 through 8-11-16

Description. Shell symmetrical, triangular in outline, without coiling, broadly expanded, apex at posterior margin and curved downward; hollow dorsomedial keel with sinus; average shell width at anterior lateral margins 45–50 mm, some specimens reach 120 mm in width; angle of posterior shell margins ranges from 95° to 110°. Shell profile narrowly concave between crest of triangular keel and lateral shell surfaces, which are gently convex, lateral margins flat; ventral surface concave. Shell thin, growth lines regularly spaced and subtle.

Remarks. *P. attenuata* was first described and illustrated from the Platteville Formation at Beloit, Wisconsin, by Hall (1861).

Occurrence. Common in the Mifflin and Grand Detour Members in north-central Illinois and south-central Wisconsin. Unusually large specimens occur in the Pecatonica Member at Dixon, Illinois, and the Grand Detour Member near Beloit, Wisconsin.

Family SINUITIDAE

Genus SINUITES

Sinuites rectangularis (Ulrich and Scofield, 1897)

Figures 8-12-13 through 8-12-15

Description. Shell planispiral, tightly coiled, umbilicus very small; maximum attained whorl width 25–30 mm. Whorl profile with sharply rounded dorsum, gently convex slope on each side, turning abruptly into umbilicus. Aperture triangular, wider than high; sinus deep, U-shaped, without a slit or peripheral band. Shell very thin, surface marked by obscure growth lines.

Remarks. *S. rectangularis* was originally described from the Platteville Formation at Minneapolis, St. Paul, and Cannon Falls, Minnesota; Janesville and Beloit, Wisconsin; and Dixon, Illinois (Ulrich and Scofield 1897).

Occurrence. Very abundant as steinkerns in the Mifflin Member in Grant County, Wisconsin; otherwise relatively rare in the Platteville Formation.

Family TROPIDODISCIDAE

Genus PHRAGMOLITES

Phragmolites fimbriatus (Ulrich and Scofield, 1897)

Figures 8-12-16 through 8-12-20

Description. Shell planispiral, laterally compressed, wide umbilicus with angular shoulders, maximum attained whorl width in incomplete specimens approximately 12 mm, 8 mm in height; three or four whorls; lacking a sinus, slit half a whorl deep. Whorl profile wider than high, with prominently elevated dorsal carina and long, narrow slit giving rise to a peripheral band with crescent-shaped ridges; gently convex between carina and sharp umbilical shoulder, sutures sharply incised, ventral whorl gently concave, reflecting opposing shape of previous whorl. Aperture subtriangular in outline with convex sides; abruptly expanded at frequent intervals, the expansion left behind forming transverse, imbricating, scalloped lamellae; anterior edges strongly serrated and projecting forward in the last whorl. Shell thin, growth increments lamellose, corrugated, scalloped, with very fine revolving threads.

Remarks. The type specimens are from the Platteville Formation at Minneapolis, Minnesota, and Dixon, Illinois (Ulrich and Scofield 1897). *P. fimbriatus* differs from *P. triangularis* in having a whorl that is wider than high and a shell with overlapping expansions with forward-projecting serrations.

Occurrence. Abundant in the Mifflin and Grand Detour Members in north-central Illinois and south-central Wisconsin and the McGregor Member of southwestern Wisconsin, northeastern Iowa, and southeastern Minnesota. Locally common in the Pecatonica Member in north-central Illinois.

Phragmolites triangularis (Ulrich and Scofield, 1897)

Figures 8-12-21 and 8-12-22

Description. Like *P. fimbriatus* but slightly smaller, with flat to slightly convex sides, height greater than width, whorl width approximately 6 mm, height 8 mm, and lacking overlapping shell expansions with forward-projecting serrations. Morphological variations somewhat gradational between these two Platteville species.

Remarks. *P. triangularis* was reported by Ulrich and Scofield (1897) from the Platteville Formation at Minneapolis, Minnesota; Janesville and Beloit, Wisconsin; Dixon, Illinois; and Lebanon, Tennessee.

Occurrence. Common locally in the Mifflin and Grand Detour Members in north-central Illinois and south-central Wisconsin and the McGregor Member of southwestern Wisconsin, northeastern Iowa, and southeastern Minnesota.

Class MONOPLACOPHORA

Subclass TERGOMYA

Order KIRENGELLIDA

Superfamily ARCHAEOPHIALOIDEA

Family ARCHAEOPHIALIDAE

Genus MICROPILEUS

Micropileus* cf. *variabilis Wahlman, 1992

Figures 8-13-1 through 8-13-3

Description. Shell high, cap shaped, typical length 20 mm, width 20 mm, height 12 mm; aperture in one plane, outline subcircular; apex blunt, slightly within anterior shell margin; anterior shell slope slightly concave to flat, posterior slope broadly convex. Shell surface smooth; apparently lacking muscle scars.

Remarks. The high, blunt apex positioned within the anterior margin is similar to that of *M. variabilis* from the Upper Ordovician Curdsville Limestone Member of the Lexington Limestone, Jessamine County, Kentucky.

Occurrence. Common in the Mifflin and Grand Detour Members in north-central Illinois and south-central Wisconsin.

Genus MACROSCENELLA

Macroscenella superba (Billings, 1865)

Figure 8-13-20

Description. Shell large, length approximately 80 mm, width 60 mm; aperture in one plane, outline ovoid-elliptical; apex subcentral to central, broadly rounded, well within the anterior margin. Shell largely smooth apically, prominent concentric growth lines in lower half of shell crossed by very fine radial lines; apparently lacking muscle scars.

Remarks. The size, shape, and ornamentation of this Platteville species are essentially the same as that in *M. superba* (Billings, 1865). Billings' type specimen is from the Black River Limestone at Paquette Rapids on the Ottawa River, Québec, Canada. The specimens described and illustrated by Yochelson (1994) are from the Platteville Formation and uppermost part of the Guttenberg Formation near Platteville, Wisconsin. Ulrich and Scofield (1897) also described specimens from the Platteville Formation near Cannon Falls, Minnesota.

Occurrence. Present but rare in the Mifflin and Grand Detour Members in north-central Illinois and south-central Wisconsin.

Genus SCENELLA

Scenella compressa Ulrich and Scofield, 1897

Figures 8-13-8 through 8-13-14

Description. Shell length approximately 17 mm, width 12 mm, height 20 mm; aperture in one plane, outline elliptical; shell profile strongly convex above, concave in some specimens where the aperture margins flare; longitudinal profile with a tendency for the posterior slope behind the apex to be concave; apex subcentral, well within the anterior margin, blunt to slightly pointed. Shell moderately thick; relatively coarse radial ridges added by intercalation during growth; fine concentric lines of growth, and somewhat irregular concentric wrinkles. Shell interiors and steinkerns lacking muscle scars.

Remarks. The size, shape, and shell ornamentation of the specimens figured here strongly suggest assignment to *S. compressa* Ulrich and Scofield, 1897, from the Platteville Formation at Minneapolis, Minnesota. Assignment of this monoplacophoran species to the Archaeophialidae is strongly suggested by the tall, cap-shaped shell that lacks obvious muscle scars.

Occurrence. Common locally in the Mifflin Member in northern Illinois and southern Wisconsin.

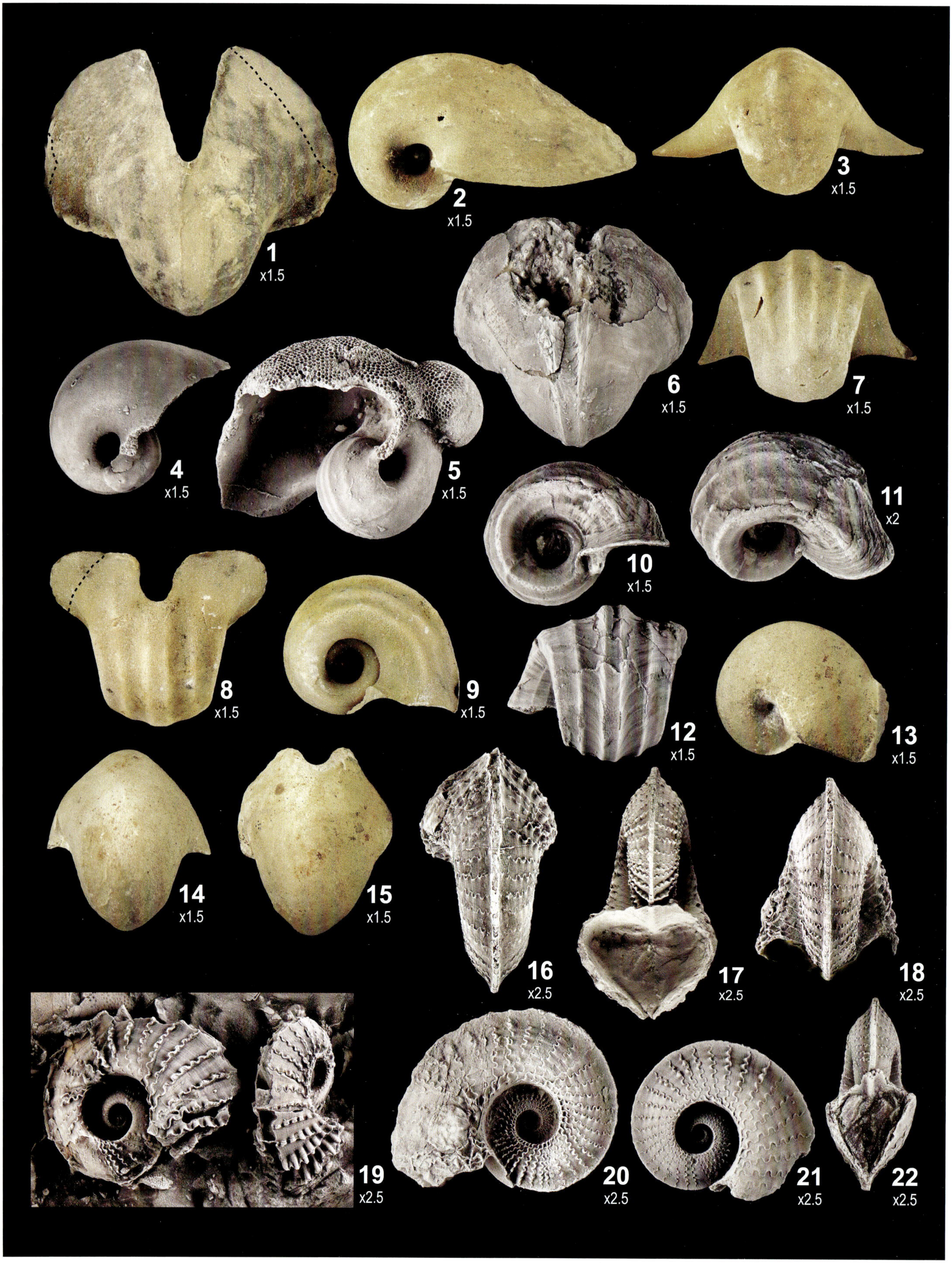
1
x1.5
2
x1.5
3
x1.5
6
x1.5
7
x1.5
4
x1.5
5
x1.5
11
x2
10
x1.5
8
x1.5
9
x1.5
12
x1.5
13
x1.5
14
x1.5
15
x1.5
16
x2.5
17
x2.5
18
x2.5
19
x2.5
20
x2.5
21
x2.5
22
x2.5

Order TRYBLIDIIDA

Superfamily TRYBLIDIOIDEA

Family PROPLINIDAE

Genus VALLATOTHECA

Vallatotheca* cf. *unguiformis (Ulrich and Scofield, 1897)

Figure 8-13-4 through 8-13-7

Description. Shell length approximately 13 mm, width 10 mm, height 7 mm; aperture in one plane, outline elliptical; shell profile strongly convex. Broadly rounded in posterior two-thirds, anterior third narrowing rapidly to blunt beak area; apex blunt, down-curved, becoming more so with growth, and extending far beyond the apertural margin; aperture outline elongate and oval, lateral apertural margin horizontal. Shell thin, growth lines closely and regularly spaced, radial threads fine, closely spaced, restricted to growth increments between growth lines.

Figure 8-12 (Opposite page)
1–6. ***Tetranota wisconsinensis*** Ulrich and Scofield. (**1**) Dorsal view of dolomitized steinkern showing deep sinus at anterior end, photographically restored at dashed lines, (**2**) right-lateral view of dolomitized steinkern showing umbilicus; (**3**) posterior view of dolomitized steinkern showing wide expansion of the aperture; (**4**) right-lateral view of umbilicus; (**5**) left-lateral perspective view showing encrustation of shell by bryozoan *Diplotrypa* and chaetetid sponge *Solenopora*; (**6**) dorsal view. Mifflin Member, Lee Center, Illinois (locality 31), **1**, JAC 123. Mifflin Member, Byron, Illinois (locality 8), **2**, **3**, ISGS 1006. Mifflin Member, Dixon, Illinois (locality 2), **4–6**, ISGS 1008, ISGS 1061, ISGS 1007.

7–12. ***Tetranota sexcarinata*** Ulrich and Scofield. (**7**) Posterior view of dolomitized steinkern showing the six prominent revolving ridges; (**8**) dorsal view of dolomitized steinkern showing the anterior sinus and expanded margins of the aperture, photographically restored at dashed line; (**9**) right-lateral view of dolomitized steinkern showing umbilicus; (**10**) right-lateral view of shell showing umbilicus; (**11**) right-lateral perspective view showing the expanded edge of the aperture; (**12**) posterior view. Mifflin Member, Lee Center, Illinois (locality 31), **7–9**, JAC 124. Mifflin Member, Dixon, Illinois (locality 2), **10**, **11**, ISGS 1060; **12**, ISGS 1059.

13–15. ***Sinuites rectangularis*** (Ulrich and Scofield). (**13–15**) Right-lateral, posterior, and dorsal view of limestone steinkern. Mifflin Member, Fennimore, Wisconsin (locality 26), **13–15**, ISGS 1010.

16–20. ***Phragmolites fimbriatus*** (Ulrich and Scofield). (**16–18**) Dorsal, anterior, and posterior views; (**19**) perspective view of two specimens in matrix showing typical corrugated and scalloped growth increments; (**20**) left-lateral view. Mifflin Member, Dixon, Illinois (locality 2), **16–18**, **20**, ISGS 1575; **19**, ISGS 1031.

21, 22. ***Phragmolites triangularis*** (Ulrich and Scofield). (**21**, **22**) Right-lateral and anterior views. Mifflin Member, Dixon, Illinois (locality 2), **21**, **22**, ISGS 1032.

Remarks. The overall shell morphology and ornamentation are comparable to *Vallatotheca* (=*Stenotheca*) *unguiformis* (Ulrich and Scofield, 1897), from the Lexington Limestone between Burgin and Danville, Kentucky.

Occurrence. Rare in the Mifflin Member at Dixon, Illinois.

Paleozoic Mollusks of Uncertain Position within Mollusca (Gastropoda or Monoplacophora)

Superfamily ARCHINACELLOIDEA

Family ARCHINACELLIDAE

Genus ARCHINACELLA

Archinacella* cf. *rugatina Ulrich and Scofield, 1897

Figures 8-13-17 through 8-13-19

Description. Shell length approximately 20 mm, width 18 mm, height 8 mm; dorsal outline round to slightly ovate, widest at mid-length of the shell; lateral profile convex; highest point of shell slightly posterior of apex; apex distinct, blunt, slightly curved anteriorly, not extending beyond anterior margin. Lateral aperture margin horizontal and straight. Muscle scar forming a continuous band parallel to shell periphery on the ventral surface. Shell thin, smooth, growth lines concentric and lamelliform in outer third of shell.

Remarks. In its size, dorsal outline, apex inset from the anterior margin, and lamelliform growth lines, this Platteville species is similar to *Archinacella rugatina* from the Bull Fork Formation (Upper Ordovician, Richmondian) at Waynesville, Ohio.

Occurrence. Common in the Mifflin and Grand Detour Members in north-central Illinois and south-central Wisconsin.

***Archinacella* sp.**

Figures 8-13-15 and 8-13-16

Description. Shell length approximately 6 mm, width 5 mm, height 2 mm; dorsal outline ovate, widest at anterior; lateral profile convex; highest point of shell slightly anterior of middle; subtle rounded dorsal median crest extends length of the shell, crest sharpest toward apex; apex pointed, extending well beyond anterior margin, incurved anteriorly one-half whorl. Shell thin, smooth, lacking well-defined growth lines.

Remarks. In its small size with apex extending beyond the anterior margin, this Platteville species is most like *A. instabilis* var. *incurva* (Ulrich and Scofield, 1897) from the *Ctenodonta* bed of the Guttenberg Formation near Cannon

Falls, Minnesota. It differs from the Guttenberg species in its slightly smaller size and being wider near the anterior end.

Occurrence. Common in the Mifflin and Grand Detour Members in north-central Illinois and south-central Wisconsin.

Figure 8-13 (Opposite page)
1–3. ***Micropileus* cf. *variabilis*** Wahlman. (**1–3**) Dorsal, left-lateral, and anterior view of dolomitized steinkern. Mifflin Member, Lightsville, Illinois (locality 7), **1–3**, JAC 125.

4–7. ***Vallatotheca* cf. *unguiformis*** (Ulrich and Scofield). (**4–6**) Anterior, right-lateral, and dorsal view; (**7**) photographic enlargement showing characteristic woven clothlike ornamentation formed by the intersection of radial lines and growth lines. Mifflin Member, Dixon, Illinois (locality 2), **4–7**, ISGS 968.

8–14. ***Scenella compressa*** Ulrich and Scofield. (**8–10**) Dorsal, right-lateral, and posterior view showing prominent radial ridges; (**11**, **12**) left-lateral and posterior view of dolomitized steinkern; (**13**) common association of *Scenella* attached to a gastropod shell; (**14**) encrustation of *Diplotrypa* over most of the shell but not overlapping the edges of the aperture. Mifflin Member, Dixon, Illinois (locality 2), **8**, ISGS 1082; **9**, **10**, ISGS 967; **13**, ISGS 1036; **14**, ISGS 1043; Lee Center, Illinois (locality 31), **11**, **12**, JAC 126.

15, 16. ***Archinacella* sp.** (**15**, **16**) Dorsal and left-lateral views. Mifflin Member, Dixon, Illinois (locality 2), **15**, **16**, ISGS 1012.

17–19. ***Archinacella* cf. *rugatina*** Ulrich and Scofield. (**17**) Dorsal view of dolomitized steinkern showing internal muscle scars fused in a ring; (**18**, **19**) dorsal and left-lateral views. Mifflin Member, Byron, Illinois (locality 8), **17**, JAC 127; **18**, **19**, Mifflin Member, Dixon, Illinois (locality 2), ISGS 608.

20. ***Macroscenella superba*** (Billings). (**20**) Posterior view of dolomitized steinkern. Grand Detour Member, South Beloit, Illinois (locality 20), **20**, JAC 145.

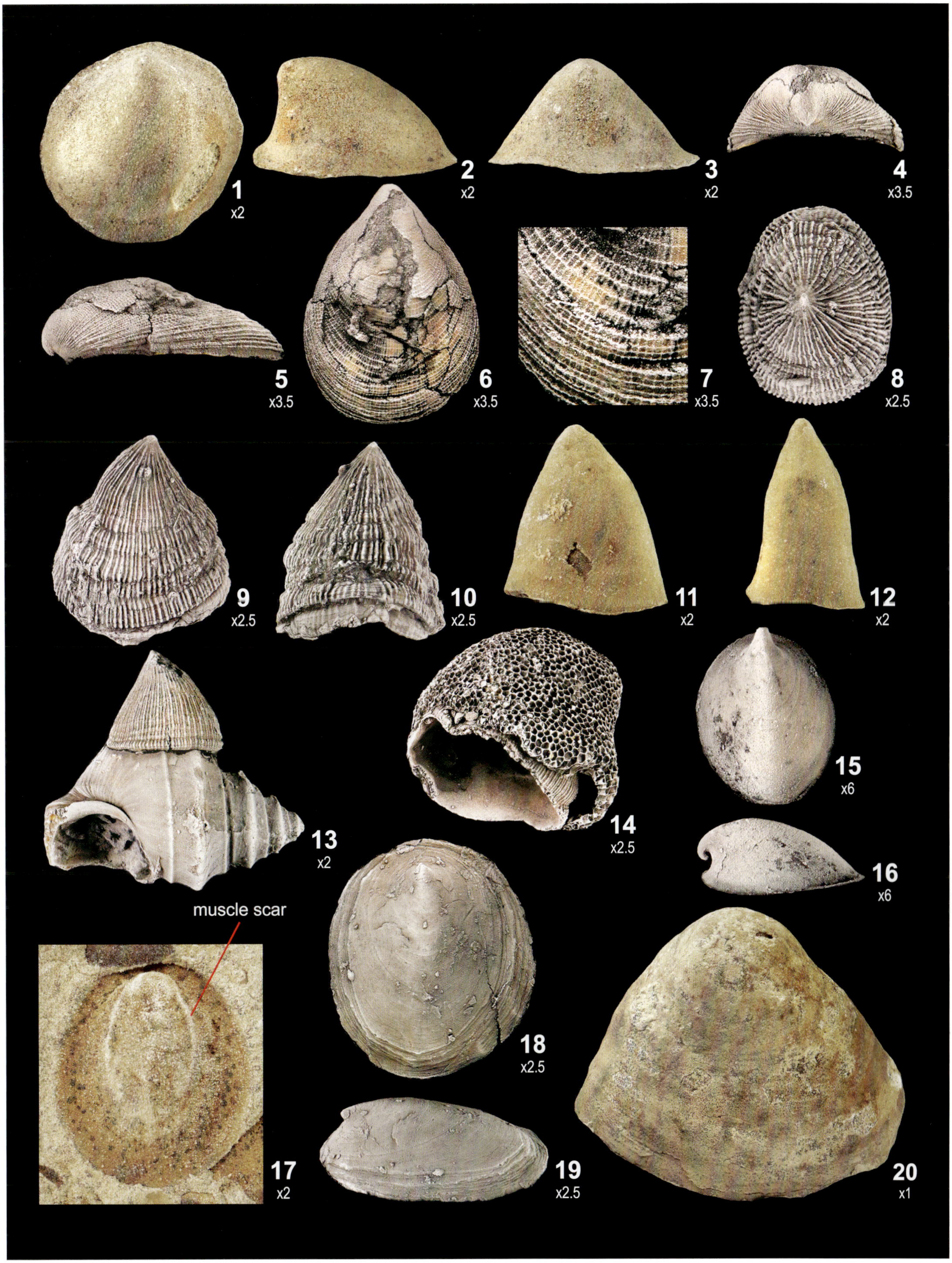
1
x2
2
x2
3
x2
4
x3.5
5
x3.5
6
x3.5
7
x3.5
8
x2.5
9
x2.5
10
x2.5
11
x2
12
x2
13
x2
14
x2.5
15
x6
16
x6
muscle scar
17
x2
18
x2.5
19
x2.5
20
x1

CHAPTER 9
BIVALVES

Bivalves are relatively diverse and locally abundant in the Platteville Formation of the upper Midwest (Appendix 2). The class is represented by subclass Protobranchia, including orders Nuculida, Solemyida, and Afghanodesmatida and subclass Autobranchia, including the Cyrtodontida, Myalinida, Ostreida, Colpomyida, Actinodontida, Modiomorphoida, and Solenida. Bivalves found in the Platteville dolostone facies typically are preserved as molds and casts, occasionally revealing muscle scars. These features may also be preserved in ""crack-out"" specimens found in the limestone facies. Some of the finest preservation occurs in the Mifflin Member marl beds at Dixon, Illinois, where shells tend to weather free of the limestone, revealing morphologic features, including muscle scars, hinge teeth, and surface ornamentation as well as various epibionts. Some lime wackestone beds are packed with articulated bivalves (Figure 9-1).

Bivalves (oysters, clams, mussels, and scallops) are a diverse class of marine and freshwater mollusks ranging in age from Early Cambrian to the present. They have a laterally compressed body typically enclosed by two opposing calcareous **valves**. Together the two valves make up the shell. The shell typically is bilaterally symmetrical, the plane of symmetry extending between the two valves. In most bivalves the two valves are of equal convexity but in some forms, including Platteville genera *Pterinea* and *Aristerella*, bilateral symmetry has been lost, perhaps resulting from cementation of one valve to the substrate. Platteville bivalves were adapted to a variety of environmental niches, including infaunal burrowing deposit feeders and epifaunal suspension feeders (Pojeta 1971).

The bivalve shell consists of two valves connected dorsally along a hinge that consists of a **hinge plate**, variably developed **hinge teeth**, and elastic **ligament** (Figures 9-2-1 through 9-2-4). In resting position, the valves are held open slightly by the ligament. The valves are closed by the action of **adductor muscles** connected to the inner surface of the valves, thus opposing the resistance of the ligament that acts as a spring to open the valves. The **ligament groove** is a narrow depression in the hinge area where fibers of the ligament are attached. A **commissure** marks the junction between valves. For directional orientation, the hinge area is regarded as **dorsal**, and the opposite side of the valves is **ventral**. In living bivalves, **anterior** is determined by the position of the mouth and **posterior** by the position of the siphon, if present. In some fossil specimens, determining the position of the soft parts can be challenging, thus making it difficult to determine anterior and posterior orientation. In most bivalves, the exterior of each valve culminates in an **apex** or **beak** on the dorsal margin above a convex region of the shell called the **umbo** or **umbones**. The beak and umbones tend to "point" toward the anterior end of the animal in most Platteville bivalves, thus making it possible to establish anterior and posterior orientation. Accordingly, the **right** and **left valves** are determined by orienting the shell with the hinge uppermost, commissure vertical, and anterior end pointing away from the observer (Figure 9-2-4). The **height** of the shell is the distance between two planes set parallel to the hinge axis and perpendicular to the plane of the commissure, one plane touching the dorsal and the other the ventral parts of the shell. If both shells are the same shape and size (mirror images), they are described as **equivalve**. **Inequilateral** refers to parts of the shell anterior and posterior to the beak that differ appreciably in length. The **pallial line** is a narrow groove on the interior of the valve close to the margin, marking the position of the marginal muscles that attach the mantle. The **pallial sinus** is an indentation in the pallial line marking the position of the siphon and attachment sites of the siphon retractor muscles. A more detailed discussion of bivalve morphology, although somewhat out of date, can be found in the *Treatise on Invertebrate Paleontology* (Cox et al. 1969). Also helpful is the online *Illustrated Glossary of the Bivalvia* (Carter et al. 2012).

Morphologic features used in the identification of bivalves include (1) overall shell shape; (2) hinge structure; (3) skeletal impressions of adductor muscle scars, pallial line, pallial sinus, or other parts of the mantle soft tissue; and (4) shell ornamentation. Primary literature sources for identification of Platteville bivalves include Ulrich (1890a, 1890b, 1890c, 1892a, 1892b, 1894a), Whitfield (1895), McAlester (1968), Pojeta (1971, 1978), and Cope (1997).

Figure 9-1 Vertically cut and polished section of lime wackestone from the Mifflin Member at Dixon, Illinois, showing densely packed, articulated shells of bivalve *Vanuxemia dixonensis*, many of which are filled with coarse sparite crystals. Arrow indicates upward direction of cut section.

SYSTEMATIC DESCRIPTIONS

Phylum MOLLUSCA

Class BIVALVIA

Remarks. There has been long-standing disagreement among malacologists regarding the preferred class name applied to these mollusks. In recent years, the Linnean name Bivalvia has gained favor over titles such as Pelecypoda, Lamellibranchiata, and Conchifera. The classification system used herein follows Carter et al. (2011).

Subclass PROTOBRANCHIA

Order NUCULIDA

Description. Nuculid bivalves are distinguished by small, nutlike shells with a hinge plate marked by numerous short hinge teeth, some or all transverse to the hinge margin, a type of dentition referred to as **taxodont**. Most Ordovician nuculids were infaunal burrowing deposit feeders (Pojeta 1971).

Family PRAENUCULIDAE

Genus HOMILODONTA

Homilodonta* cf. *scofieldi (Ulrich, 1894a)

Figures 9-3-1 through 9-3-6

Description. Shell 4–10 mm long, strongly convex, equivalve, anterior part of shell larger and more convex than the posterior part; anterior and posterior margins evenly rounded, posterior end tapered; ventral margin convex. Beaks small, prominent, incurved, positioned anterior of center. Hinge plate curved and narrow at beaks, widening on each side; approximately 15 posterior hinge teeth and 10 relatively large anterior teeth, gradually increasing in size away from the beaks (gradidentate). Adductor muscles very weakly impressed. Shell thin, smooth, or with fine concentric growth lines.

Remarks. A distinctive feature of this Platteville genus is the presence of numerous teeth that gradually increase in size away from the subumbonal area on both the anterior and posterior parts of the hinge plate. Pojeta (1971, 1978) assigned bivalves of this type to the genus *Deceptrix*; however, Cope (1997) pointed out that the type species of *Deceptrix* has a distinctly different dental pattern consisting of a straight row of posterior teeth and a small number of larger anterior teeth. Cope (1997) proposed the new genus *Homilodonta* for bivalves with numerous teeth that gradually increase in size away from the subumbonal area on both the anterior and posterior parts of the hinge plate (gradidentate). This Platteville species is similar in size and shape to *H. scofieldi* from the overlying Decorah Formation.

Occurrence. Abundant in the Mifflin Member of north-central Illinois.

Order SOLEMYIDA

Family CTENODONTIDAE

Genus CTENODONTA

Ctenodonta nasuta (Hall, 1847)

Figures 9-3-20 through 9-3-26

Description. Shell 20–45 mm long, transversely elongate, subovate, equivalve, inequilateral, moderately convex; anterior part of shell larger and thicker than the posterior part; anterior and posterior margins evenly rounded, posterior end tapered; ventral margin convex, anteriorly grading to straight or slightly concave posteriorly. Beaks positioned slightly anterior of center. Sharply defined ligament groove present on each side of the hinge line extending posteriorly from the beaks. Approximately 20–25 small hinge teeth are present; short and vertical beneath the beak, becoming larger, longer, and curved posteriorly and anteriorly to the beak. Pallial line and muscle scars very faintly impressed on the interior molds in some larger specimens. Shell thickness is thin, and outer surface ornamentation consists of fine growth lines and low, rounded ridges.

Remarks. Ulrich (1894a) documented the occurrence of *Ctenodonta nasuta* in the Platteville Formation (Trenton Limestone of Ulrich) at Minneapolis, Minnesota; Beloit and Janesville, Wisconsin; and Dixon, Illinois.

Occurrence. Locally common to abundant in the Pecatonica, Mifflin, and Grand Detour Members.

Genus TANCREDIOPSIS

Tancrediopsis gibberula (Salter, 1859)

Figures 9-3-7 through 9-3-10

Description. Shell 10–20 mm long, subovate, moderately convex, equivalve, equilateral; anterior end of shell larger and thicker than the posterior end; anterior and posterior margins evenly rounded, posterior end tapered; ventral margin convex. Beaks positioned near center. Hinge plate curved and narrow at beaks, widening on each side, terminating abruptly at the muscle scars; approximately 10 posterior and

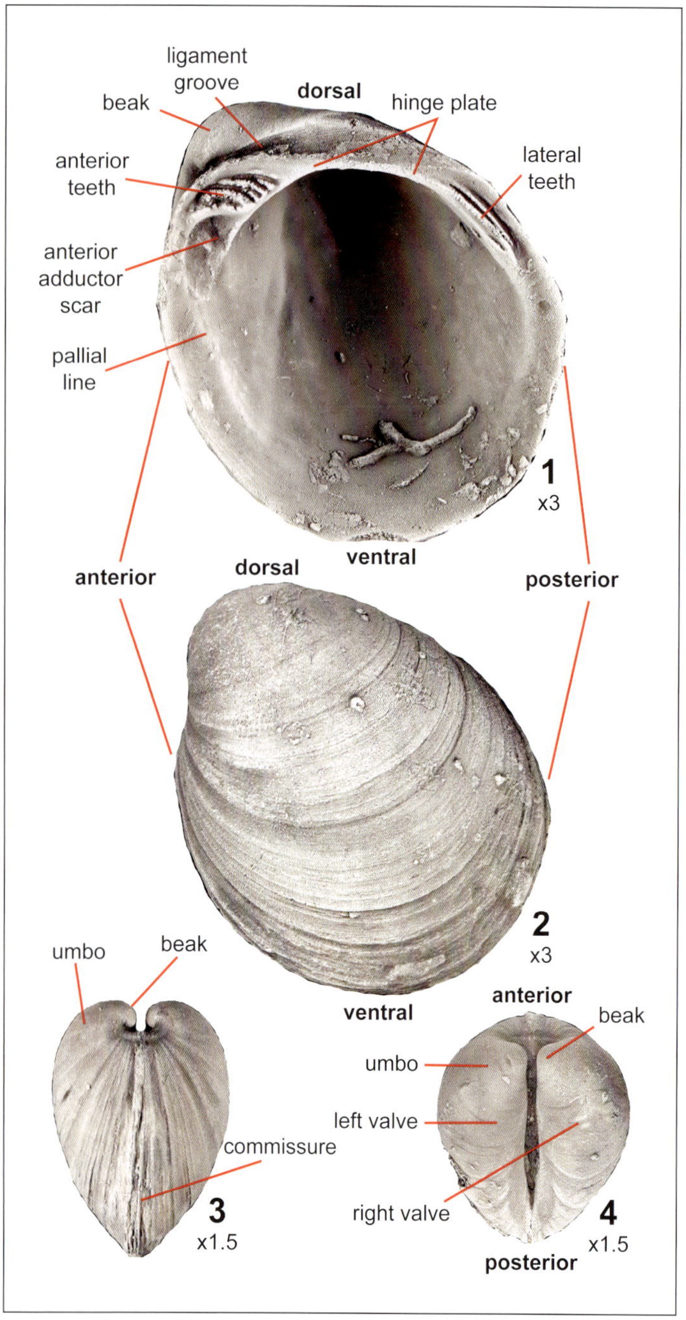

Figure 9-2
1–4. Specimens of *Vanuxemia dixonensis* showing typical bivalve morphologic terminology and orientation nomenclature. (**1**) Right valve interior showing the hinge plate, dentition, muscle scars, etc.; (**2**) left valve exterior; (**3**) anterior view of conjoined valves; (**4**) dorsal view of conjoined valves.

8 anterior hinge teeth, gradually becoming smaller at the beaks. Adductor muscles very deeply impressed. Shell very thick, particularly in the umbo and beak area; shell smooth or with fine growth lines.

Remarks. As pointed out by Ulrich (1894a), *T. gibberula* is one of the most striking fossils found in the Platteville Formation, particularly in the dolomite facies, where internal molds show the prominent adductor muscle scars and hinge teeth.

Occurrence. Present throughout the Platteville Formation but particularly common in the Grand Detour Member of north-central Illinois and south-central Wisconsin.

Order AFGHANODESMATIDA

Family SIMILODONTIDAE

Genus SIMILODONTA

Similodonta* cf. *similis (Ulrich, 1894a)

Figures 9-3-16 through 9-3-19

Description. Shell 5–15 mm long, equivalve, length and height approximately equal, umbones full, rounded; beaks curved anteriorly and projecting slightly above the hinge. Anterior and posterior rows of teeth meet at an angle of approximately 90° beneath the beaks. Each end of the hinge plate terminates at a shallow round-to-oval depression made by the adductor muscle scars. Outer surface of shell is smooth or with fine concentric ridges.

Remarks. In this publication, orientation terminology follows that of McAlester (1968) whereby the prominent curved beaks point anteriorly. This Platteville species of *Similodonta* is similar in size and shape to *S. similis* from the Late Ordovician Maquoketa Formation of Fillmore County, Minnesota. The Platteville species differs from the Maquoketa species in having a narrower hinge plate and smaller, more numerous teeth.

Occurrence. Common in the Mifflin Member of north-central Illinois.

Subclass AUTOBRANCHIA

Order CYRTODONTIDA

Description. Platteville cyrtodontid bivalves are represented by an abundant and diverse group, including *Cyrtodonta*, *Cyrtodontula*, and *Vanuxemia*. They typically possess (1) relatively thick ovoid shells that lack radial ornamentation, (2) prominent umbones, (3) groups of anterior and posterior teeth that are widely separated by a region of the hinge plate that lacks teeth, and (4) posterior lateral teeth that are parallel to the shell margin.

Family CYRTODONTIDAE

Genus CYRTODONTA

Cyrtodonta janesvillensis Ulrich, 1894a

Figures 9-4-6 through 9-4-10

Description. Shell 30–45 mm long, strongly convex, obliquely ovate; anterior end very short. Beaks incurved, projecting slightly above the hinge. Hinge plate moderately thick; three anterior teeth in each valve, curved and oblique; two or three posterior lateral teeth, curved and parallel to the shell margin. Adductor muscles very weakly impressed. Shell relatively thin, smooth, or with fine concentric growth lines.

Remarks. The type specimens of *C. janesvillensis* are from the Platteville Formation at Janesville and Beloit, Wisconsin.

Occurrence. Common in the Mifflin and Grand Detour Members of north-central Illinois and south-central Wisconsin.

Cyrtodonta billingsi Ulrich, 1894a

Figures 9-4-1 through 9-4-5

Description. Shell 20–40 mm long, strongly convex, obliquely ovate; valves prominently inflated in the umbone and central regions. Beaks small and incurved, projecting slightly above the hinge. Hinge line at least two-thirds the length of the shell, slightly arcuate, posteriorly declining and passing gradually into the broadly and uniformly curved posterior margin. Hinge plate moderately thick; three anterior teeth in each valve, curved and oblique; two or three posterior lateral teeth curved and parallel to the shell margin. Adductor muscles weakly impressed; anterior adductor scar small and oval; posterior adductor scar relatively large and oval. Shell relatively thin, smooth, or with fine concentric growth lines.

Remarks. The type specimens of *C. billingsi* are from the Platteville Formation at Beloit, Janesville, and Mineral Point, Wisconsin, and Cannon Falls and Minneapolis, Minnesota.

Occurrence. Common in the Platteville Formation throughout the outcrop belt, particularly in the Mifflin and Grand Detour Members in northern Illinois and southern Wisconsin.

Cyrtodonta* cf. *grandis Ulrich, 1894a

Figures 9-4-11 through 9-4-15

Description. Shell 30–60 mm long, equivalve, obliquely subovate, not greatly convex, broad anterior end, relatively short hinge line, small beaks. Hinge plate thick; three or four serrated teeth anterior to beaks; two or three serrated posterior lateral teeth parallel to shell margin. Shell thick, outer surface smooth or with fine concentric growth lines.

Remarks. The overall shape and outline of the shell are similar to *Cyrtodonta grandis* from the overlying Decorah Formation.

Occurrence. Common in the Mifflin Member in northern Illinois and southern Wisconsin.

Genus CYRTODONTULA

Cyrtodontula subcarinata (Ulrich, 1894a)

Figures 9-4-16 through 9-4-19

Description. Shell 25–70 mm long, equivalve, obliquely subovate, strongly convex, umbonal ridge and umbones prominent; beaks curved anteriorly and projecting slightly above the hinge. Hinge plate moderately thick; three or four folds or teeth in front of beaks; two or three posterior lateral teeth parallel to shell margin. Outer shell surface is smooth or with fine concentric growth lines.

Remarks. Recently collected specimens of this cyrtodontid compare closely to the figure and description of *Cyrtodontula* [*Whitella*] *subcarinata* documented by Ulrich (1894a) from the Platteville Formation in Jo Daviess County, Illinois. The most distinguishing characteristics of this species are its relatively large size and prominent umbones and umbonal ridge.

Occurrence. Common in the Platteville Formation, particularly in the Pecatonica, Mifflin, and Grand Detour Members of northern Illinois and southern Wisconsin.

Genus VANUXEMIA

Description. *Vanuxemia* differs from other cyrtodontid bivalves in having the anterior teeth immediately below the beaks and the anterior adductor muscle originating on a raised shell thickening or on the hinge plate. Many species have been named but few are distinctive enough for confident identification. The more obvious Platteville species are described below.

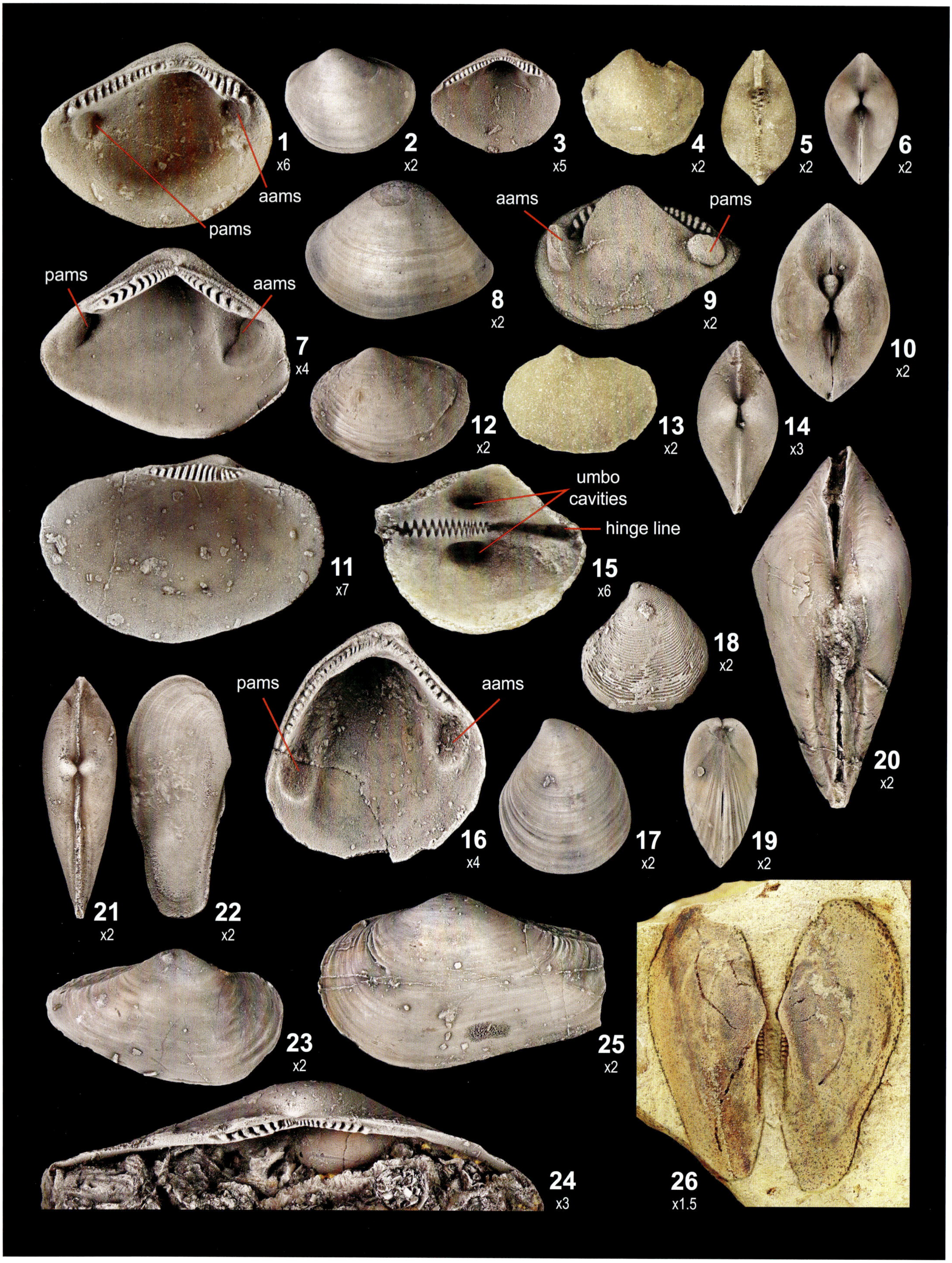
1
x6
aams
pams
2
x2
3
x5
4
x2
5
x2
6
x2
aams
pams
8
x2
9
x2
10
x2
pams
aams
7
x4
12
x2
13
x2
14
x3
umbo
cavities
hinge line
11
x7
15
x6
18
x2
pams
aams
20
x2
16
x4
17
x2
19
x2
21
x2
22
x2
23
x2
25
x2
24
x3
26
x1.5

Vanuxemia dixonensis Meek and Worthen, 1868

Figures 9-5-1 through 9-5-8

Description. Shell 10–30 mm long, equivalve, obliquely ovate, greatest convexity near center of valve; anterior end reduced, posterior broadly rounded; umbones inflated and prominent; beaks inflated, very oblique, and obtuse. Hinge plate thick and massive; three or four serrated folds or teeth beneath beaks; two or three serrated, posterior lateral teeth parallel to shell margin. Anterior adductor muscle scar deeply impressed on the anterior end of the hinge plate. Pallial line present on some larger valves. Shell thick, outer surface smooth or with fine concentric growth lines.

Figure 9-3 (Opposite page)
1–6. ***Homilodonta* cf. *scofieldi*** (Ulrich). (**1**) Left valve interior showing typical taxodont dentition as well as the anterior (aams) and posterior (pams) adductor muscle scars; (**2**) left valve exterior; (**3**) left valve interior; (**4**) dolomitic steinkern showing internal mold of left valve interior; (**5**) dorsal view of dolomitic steinkern showing internal mold of hinge teeth; (**6**) dorsal view of articulated valves. Mifflin Member, Dixon, Illinois (locality 2), **1**, ISGS 902; **2**, **6**, ISGS 871; **3**, ISGS 846. Mifflin Member, Lee Center, Illinois (locality 31), **4**, **5**, JAC 130.

7–10. ***Tancrediopsis gibberula*** (Salter). (**7**) Left valve interior showing tooth rows and anterior (aams) and posterior (pams) adductor muscle scars; (**8**) left valve exterior; (**9**) dolomitized steinkern showing impression of hinge teeth and anterior (aams) and posterior (pams) adductor muscle scars; (**10**) dorsal view of articulated valves. Mifflin Member, Dixon, Illinois (locality 2), **7**, ISGS 820; **8**, **10**, ISGS 876. Grand Detour Member, Beloit, Wisconsin (locality 21), **9**, ISGS 821.

11–15. ***Pseudocyrtodonta* sp.** (**11**) Right valve interior showing straight hinge plate with a single grouping of teeth; (**12**) right valve exterior showing smooth shell surface; (**13**) dolomitized steinkern showing internal mold of left valve; (**14**) internal mold of the dorsal surface showing the hinge line, umbo cavities, and dentition; (**15**) internal view of articulated but fragmented valves showing the hinge teeth. Mifflin Member, Dixon, Illinois (locality 2), **11**, ISGS 847; **12**, ISGS 854; **15**, ISGS 1044. Mifflin Member, Lee Center, Illinois (locality 31), **13**, **14**, JAC 129.

16–19. ***Similodonta* cf. *similis*** (Ulrich). (**16**) Left valve interior showing curved hinge plate with numerous teeth and anterior (aams) and posterior (pams) adductor muscle scars; (**17**) left valve exterior showing concentric growth lines; (**18**) left valve exterior; (**19**) dorsal view of articulated shells. Mifflin Member, Dixon, Illinois (locality 2), **16**, ISGS 901; **17**, **19**, ISGS 852; **18**, ISGS 899.

20–26. ***Ctenodonta nasuta*** (Hall). (**20**) Dorsal view of articulated shells; (**21**) dorsal view of small articulated shells; (**22**) left valve exterior showing elongate posterior end; (**23**) right valve exterior; (**24**) interior view of hinge line showing teeth beneath beak; (**25**) partial left valve exterior; (**26**) dolomitized steinkern showing conjoined right and left valves and hinge teeth. Mifflin Member, Dixon, Illinois (locality 2), **20**, ISGS 889; **21**, **22**, ISGS 857; **23**, ISGS 897; **24**, **25**, ISGS 890. Grand Detour Member, Beloit, Wisconsin (locality 21), **26**, ISGS 886.

Remarks. The type specimens of *V. dixonensis* are from the Platteville Formation at Dixon, Illinois (Meek and Worthen 1868).

Occurrence. Very abundant in the Pecatonica, Mifflin, and Grand Detour Members in northern Illinois and southern Wisconsin.

Vanuxemia obtusifrons Ulrich, 1894a

Figures 9-5-9 through 9-5-14

Description. Shell 20–45 mm long, equivalve, obliquely ovate, moderately convex; dorsoposterior margin relatively elongate and straight, posterior margin curved broadly to the ventral margin; greatest height is in the posterior half of the valve; anterior end obtuse, the upper part forming nearly a right angle with the hinge. Shell thick, outer surface is smooth or with fine concentric growth lines.

Remarks. The type specimens of *V. obtusifrons* are from the Platteville Formation at Minneapolis, Minnesota. *V. obtusifrons* differs from *V. dixonensis* in having a larger and wider shell with longer dorsal-posterior margin and less inflated valves.

Occurrence. Common in the Mifflin and Grand Detour Members in northern Illinois and southern Wisconsin.

Vanuxemia rectirostris (Hall, 1861)

Figures 9-5-18 through 9-5-20

Description. Shell 20–40 mm long, equivalve, somewhat elongate-ovate, outline uniformly rounded, very convex, beaks slightly incurved. Hinge line nearly straight, anterior end short. Hinge plate thick; three or four folds or teeth beneath beaks; two or three posterior lateral teeth parallel to shell margin. Shell thick, outer surface smooth or with fine concentric growth lines.

Remarks. *V. rectirostris* is one of the largest cyrtodontid bivalves in the Platteville Formation. The type specimens are from the Platteville Formation at Janesville, Wisconsin.

Occurrence. Common in the Mifflin and Grand Detour Members of northern Illinois.

Vanuxemia rotundata (Hall, 1861)

Figures 9-5-15 through 9-5-17

Description. Shell 15–20 mm long, equivalve, subcircular in outline, rotund, greatest convexity near center of valve; posterior broadly rounded. Hinge plate thick; three or four serrated folds or teeth beneath beaks; two or three serrated

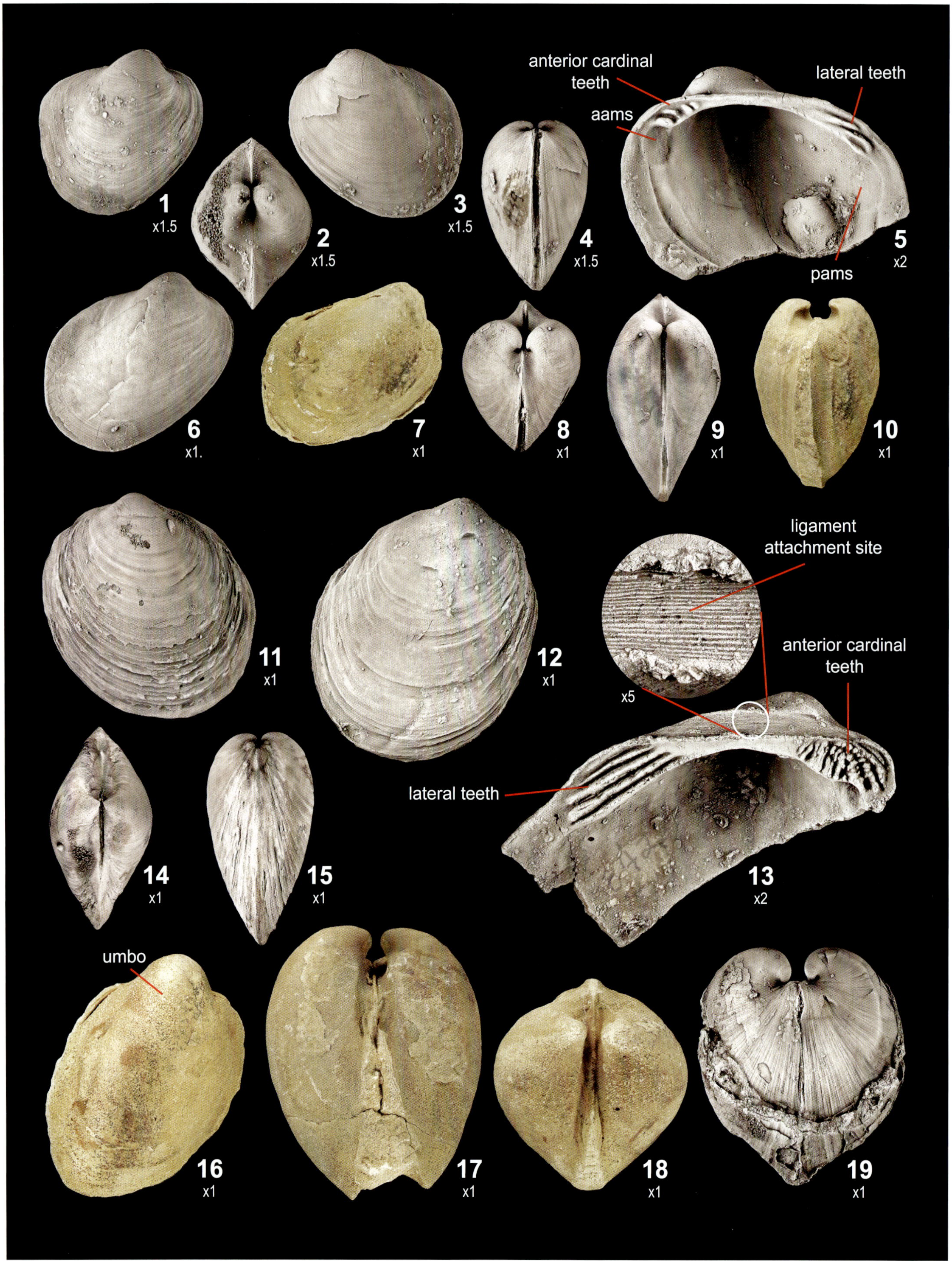
1
x1.5
2
x1.5
3
x1.5
4
x1.5
anterior cardinal teeth
aams
lateral teeth
pams
5
x2
6
x1.
7
x1
8
x1
9
x1
10
x1
11
x1
12
x1
ligament attachment site
x5
anterior cardinal teeth
lateral teeth
13
x2
14
x1
15
x1
umbo
16
x1
17
x1
18
x1
19
x1

posterior lateral teeth set parallel to shell margin. Anterior adductor muscle scar deeply impressed on the anterior end of the hinge plate. Internal casts reveal a prominent radial depression (sulcus) extending from the beak to the ventral margin. Pallial line present on some larger valves. Shell thick, outer surface smooth or with fine concentric growth lines.

Remarks. *V. rotundata* is similar to *V. dixonensis* but differs from the latter in having smaller beaks, a more rounded outline, and an internal cast that has a more deeply depressed sulcus that is more curved. The type specimens of *V. rotundata* are from the Platteville Formation at Janesville and Beloit, Wisconsin.

Occurrence. Common in the Mifflin and Grand Detour Members in north-central Illinois and south-central Wisconsin.

Vanuxemia terminalis Ulrich, 1894a

Figures 9-5-21 through 9-5-26

Description. Shell 15–50 mm long, equivalve, very convex, extremely oblique, small strongly incurved terminal beaks; umbones full and rounded. Beaks terminal. Hinge plate narrow, with three anterior teeth and two long posterior teeth set parallel to the shell margin. Anterior adductor muscle scars impressed on a relatively thin calcified shelf that curves down and inward from the hinge plate. Viewed from the anterior end of an internal cast (steinkern), the shelf is marked by a narrow slit that curves downward from the beak to the hinge plate (Figure 9-5-25). Pallial line distinct given the thinness of the shell. Shell thin, outer surface smooth or with fine concentric growth lines.

Remarks. The type specimens of *V. terminalis* are from the Platteville Formation at Minneapolis and Cannon Falls, Minnesota, and Beloit, Wisconsin.

Occurrence. Most common in the Mifflin and Grand Detour Members of northern Illinois and southern Wisconsin.

Order MYALINIDA

Remarks. Platteville myalinids include the ambonychiid *Cleionychia*. Morphologically, they appear to have been mostly epifaunal suspension feeders (Pojeta 1971).

Family AMBONYCHIIDAE

Genus CLEIONYCHIA

Cleionychia lamellosa (Hall, 1847)

Figures 9-6-1 through 9-6-5

Description. Shell 8–50 mm long, equivalve, obliquely subquadrangular in outline, hinge line straight, shell margin curved somewhat abruptly at posterior end, broadly curved along ventral margin; strongly convex, particularly in the umbones. Beaks terminal, small, curving slightly forward, projecting very little above the hinge. Hinge plate strong, without teeth, ligament area wide and faintly striated. Posterior adductor muscle scar relatively large and bilobed. Shell thin; concentric growth lines possess thin lamellae that are particularly prominent along the ventral margins.

Remarks. The type specimens are from the Platteville Formation at Beloit, Janesville, and Mineral Point, Wisconsin, as well as St. Paul and Minneapolis, Minnesota.

Occurrence. Common in the Mifflin Member of northern Illinois and southern Wisconsin.

Order OSTREIDA

Family PTERINEIDAE

Genus PTERINEA

***Pterinea* sp.**

Figures 9-6-6 through 9-6-9

Description. Shell small, inequivalve, inequilateral, subovate, beaks situated anteriorly, well-developed rounded

Figure 9-4 (Opposite page)

1–5. ***Cyrtodonta billingsi*** Ulrich. (**1**) Right valve exterior; (**2**) dorsal view showing convex articulated valves; (**3**) left valve exterior; (**4**) anterior view of articulated valves; (**5**) right valve interior showing the anterior teeth, lateral teeth, and anterior (aams) and posterior (pams) muscle scars. Mifflin Member, Dixon, Illinois (locality 2), **1**, **2**, ISGS 841; **3**, **4**, ISGS 839; **5**, ISGS 838.

6–10. ***Cyrtodonta janesvillensis*** Ulrich. (**6**) Right valve exterior; (**7**) dolomitized steinkern showing interior mold of right valve; (**8**) anterior view of articulated valves; (**9**) dorsal view of articulated valves; (**10**) dolomitized steinkern showing anterior view of articulated valves. Mifflin Member, Dixon, Illinois (locality 2), **6**, **8**, **9**, ISGS 840. Grand Detour Member, Rochelle, Illinois (locality 35), **7**, **10**, JAC 131.

11–15. ***Cyrtodonta* cf. *grandis*** Ulrich. (**11**) Left valve exterior; (**12**) right valve exterior; (**13**) left valve interior showing ligament attachment site, anterior teeth, and lateral teeth; (**14**) dorsal view of articulated valves; (**15**) anterior view of articulated valves. Mifflin Member, Dixon, Illinois (locality 2), **11**, **14**, **15**, ISGS 837; **12**, ISGS 843; **13**, ISGS 842.

16–19. ***Cyrtodontula subcarinata*** (Ulrich). (**16**) Dolomitized steinkern of right valve interior mold showing the prominent umbo; (**17**) anterior view of interior mold of articulated valves; (**18**) dorsal view of interior mold of articulated valves; (**19**) anterior view of articulated valves. Grand Detour Member, Beloit, Wisconsin (locality 21), **16**, **18**, JAC 132. Grand Detour Member, Durand, Illinois (locality 11), **17**, ISGS 884. Mifflin Member, Dixon, Illinois (locality 2), **19**, ISGS 885.

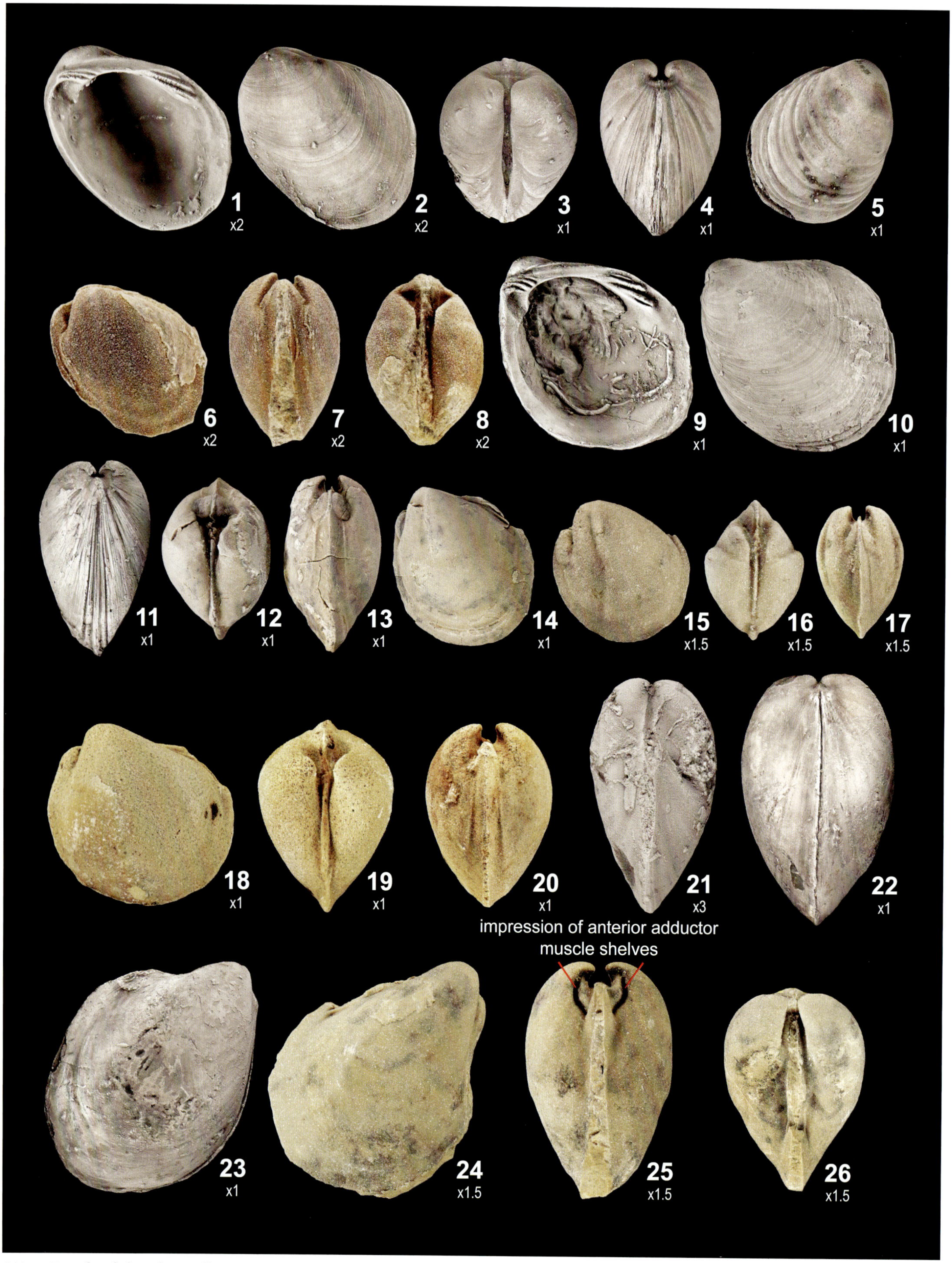
1
x2
2
x2
3
x1
4
x1
5
x1
6
x2
7
x2
8
x2
9
x1
10
x1
11
x1
12
x1
13
x1
14
x1
15
x1.5
16
x1.5
17
x1.5
18
x1
19
x1
20
x1
21
x3
22
x1
impression of anterior adductor muscle shelves
23
x1
24
x1.5
25
x1.5
26
x1.5

auricle and well-defined posterior wing. Right valve less convex than left valve, particularly obvious in larger specimens. Hinge and muscle scars unknown. Shell thin, outer surface marked by fine concentric growth lines and corresponding low, thin ridges.

Remarks. The oldest previously known and undoubted North American pterineids are from the Late Ordovician upper Lexington Limestone of Kentucky (Pojeta 1971). This Platteville pterineid represents an even earlier occurrence of these poorly known bivalves.

Occurrence. Rare in the Mifflin Member at Dixon, Illinois.

Figure 9-5 (Opposite page)

1–8. ***Vanuxemia dixonensis*** Meek and Worthen. (**1**) Right valve interior; (**2**) left valve exterior; (**3**) dorsal view of articulated valves; (**4**) anterior view of articulated valves; (**5**) right valve exterior; (**6**) dolomitized steinkern showing mold of left valve interior; (**7**) internal mold of anterior end; (**8**) interior mold of dorsal surface of articulated valves. Mifflin Member, Dixon, Illinois (locality 2), **1**, ISGS 829; **2**, ISGS 830; **3**, **4**, ISGS 827; **5**, ISGS 833. Mifflin Member, Byron, Illinois (locality 8), **6–8**, ISGS 906.

9–14. ***Vanuxemia obtusifrons*** Ulrich. (**9**) Right valve interior; (**10**) left valve exterior; (**11**) anterior view of articulated valves; (**12**) dolomitized steinkern showing internal mold of dorsal surface; (**13**) internal mold of anterior end of articulated valves; (**14**) left valve interior mold. Mifflin Member, Dixon, Illinois (locality 2), **9**, ISGS 1913; **10**, **11**, ISGS 834. Grand Detour Member, Rochelle, Illinois (locality 35), **12–14**, JAC 134.

15–17. ***Vanuxemia rotundata*** (Hall). (**15**) Dolomitized steinkern showing mold of left valve interior; (**16**) dorsal view interior mold of both valves; (**17**) interior mold of anterior end of articulated valves. Mifflin Member, Lee Center, Illinois (locality 31), **15–17**, JAC 134.

18–20. ***Vanuxemia rectirostris*** (Hall). (**18**) Dolomitized steinkern of left valve interior mold; (**19**) dorsal view of interior mold of articulated valves; (**20**) interior mold of anterior end of both valves. Grand Detour Member, Beloit, Wisconsin (locality 21), **18–20**, JAC 135.

21–26. ***Vanuxemia terminalis*** Ulrich. (**21**) Dorsal view of articulated valves; (**22**) anterior view of articulated valves; (**23**) right valve exterior; (**24**) dolomitized steinkern of right valve interior mold; (**25**) interior mold of anterior end showing impression of shelflike structures on which the anterior adductor muscles were attached; (**26**) dorsal view of interior mold of both valves. Mifflin Member, Dixon, Illinois (locality 2), **21**, ISGS 870; **22**, **23**; ISGS 828. Grand Detour Member, Rochelle, Illinois (locality 35), **24–26**, JAC 136.

Order COLPOMYIDA

Family COLPOMYIDAE

Genus COLPOMYA

Colpomya sp.

Figures 9-7-3 through 9-7-8

Description. Shell 5–10 mm long, subrhomboidal or ovate in outline, elongate, widest posteriorly; beaks anterior but not terminal; umbonal ridge strongly convex; broad lateral sulcus and prominent ventral sinus along shell margins. Hinge plate straight, thicker anterior to beaks, thinner posterior to beaks; right valve with a single tubercle, left valve with corresponding depression. Shell thin, outer surface smooth or marked by fine concentric growth lines.

Remarks. The overall shell outline with sulcus and prominent ventral sinus as well as the hinge morphology suggest assignment of this bivalve to *Colpomya*.

Occurrence. Abundant locally in the Mifflin Member of northern Illinois.

Order ACTINODONTIDA

Family CYCLOCONCHIDAE

Genus PSEUDOCYRTODONTA

Pseudocyrtodonta sp.

Figures 9-3-11 through 9-3-15

Description. Shell 4–10 mm long, subovate, equivalve, inequilateral, moderately convex; anterior and posterior margins evenly rounded; ventral margin convex. Beaks positioned anterior of center. Hinge plate narrows at beaks, widening posteriorly; a single series of approximately 14 interlocking, wedge-shaped hinge teeth are present. Four relatively small teeth curve anteriorly from the beak; the posterior 10 or so teeth are the longest and widest in the hinge. Pallial line and muscle scars very faintly impressed on the interior casts in some larger specimens. Shell is thin, and outer surface ornamentation consists of very fine growth lines.

Remarks. The valve size and shape as well as the actinodont-type dentition of this Platteville bivalve compare closely to that of *Pseudocyrtodonta ala* (Barrande, 1881) as figured by Polechová (2013, figure 6, A–C) and discussed by Steinová (2012) from the Middle Ordovician (early and middle Darriwillian) Sarka Formation of the Czech Republic.

Occurrence. Locally abundant in the Mifflin Member in northern Illinois and southern Wisconsin.

Order MODIOMORPHIDA

Remarks. The main unifying feature of this group of bivalves is the overall modiolid shape (in reference to the ear-shaped mussel *Modiolus*; Pojeta 1971). Aspects of the morphology, especially the hinge line, are poorly known.

Family MODIOMORPHIDAE

Genus ARISTERELLA

Aristerella cf. *nitidula* Ulrich, 1894a

Figures 9-6-15 through 9-6-21

Description. Shell 4–8 mm long, inequivalve, inequilateral, subovate, narrowest anteriorly; posterior margin slightly oblique, broadly rounded; umbonal ridge inconspicuous. Right valve less convex and slightly smaller than left valve, particularly obvious in larger, more mature specimens. Hinge line straight, relatively long, and without teeth. No obvious muscle scars. Shell thin, outer surface smooth or marked by fine concentric growth lines and corresponding low, thin ridges.

Remarks. This Platteville species is most similar to *Aristerella nitidula* from the Late Ordovician Decorah Formation of Minnesota. It differs slightly from *A. nitidula* in having ventral and posterior margins that are nearly parallel.

Occurrence. Common locally in the Mifflin Member of northern Illinois.

Genus GONIOPHORA

Goniophora sp.

Figures 9-7-1 through 9-7-2

Description. Shell 8–10 mm long, elongate, equivalve, rhomboidal to trapezoidal in outline; beaks anterior, curved inward and forward; prominent umbonal ridge (carina) from beaks to posterior ventral margin. Hinge line straight, dentition unknown. Shell thin, outer surface smooth.

Remarks. This Platteville species is assigned to *Goniophora* on the assumption that teeth are likely to be present. Lacking teeth, it is more appropriately assigned to *Goniophorina*.

Occurrence. Common in the Mifflin Member at Dixon, Illinois.

Order MYTILIDA

Family MODIOLOPSIDAE

Genus MODIOLOPSIS

Modiolopsis spp.

Figures 9-7-9 through 9-7-16

Description. Shell 10–30 mm long, equivalve, modioliform, expanding posteriorly; beaks anterior, curved inward and forward. Hinge plate straight; right valve with one tubercle, left valve with corresponding depression and strong process partly fitting into depression in front of tubercle. Multiple accessory muscle scars in front of the beaks; well-developed pallial line. Shell thin, outer surface smooth or marked by fine concentric growth lines.

Remarks. On the basis of variations in shell outline and the presence or absence of concentric growth lines, there appears to be more than one species of *Modiolopsis* in the Platteville Formation, but species are yet to be determined.

Figure 9-6 (Opposite page)

1–5. ***Cleionychia lamellosa*** (Hall). (**1**) Right valve exterior showing thin concentric lamellae; (**2**) exterior anterior view of articulated valves with prominent concentric lamellae; (**3**) exterior dorsal surface of articulated valves; (**4**) left valve interior showing details of the interior beak region; (**5**) dolomitic steinkern, left valve mold of interior showing prominent posterior adductor muscle scar. Mifflin Member, Dixon, Illinois (locality 2), **1–3**, ISGS 824; **4**, ISGS 900. Grand Detour Member, Durand, Illinois (locality 11), **5**, ISGS 823.

6–9. ***Pterinea* sp.** (**6**) Right valve exterior; (**7**) left valve exterior showing the characteristic auricle and posterior wing; (**8**) anterior view of articulated valves showing the concave profile of the right valve (left side); (**9**) right valve exterior. Mifflin Member, Dixon, Illinois (locality 2), **6**, ISGS 1441; **7**, **8**, JAC 138, **9**, ISGS 1592.

10–14. ***Cymatonota* sp.** (**10**) Left valve exterior; (**11**) dorsal surface of conjoined valves; (**12**) anterior end of conjoined valves; (**13**) left valve interior showing the hinge line to be lacking teeth; (**14**) dorsal surface of dolomitized steinkern showing conjoined valves. Mifflin Member, Dixon, Illinois (locality 2), **10–12**, ISGS 1567; **13**, ISGS 880. Grand Detour Formation, Beloit, Wisconsin (locality 21), **14**, JAC 139.

15–21. ***Aristerella* cf. *nitidula*** Ulrich. (**15**) Dorsal surface of conjoined valves showing the characteristically small and less convex left valve; (**16**) anterior end of conjoined valves; (**17**) left valve exterior; (**18**) dolomitized internal mold of right valve; (**19**) dolomitized internal mold of dorsal surface of conjoined valves; left valve is smaller and less convex than the right valve; (**20**) left valve interior showing small depression in hinge line that received tubercle from the right valve; (**21**) right valve interior showing small tubercle on hinge line. Mifflin Member, Dixon, Illinois (locality 2), **15**, ISGS 894; **16**, **17**, ISGS 893; **20**, ISGS 892; **21**, ISGS 813. Mifflin Member, Lee Center, Illinois (locality 31), **18**, JAC 140; **19**, JAC 141.

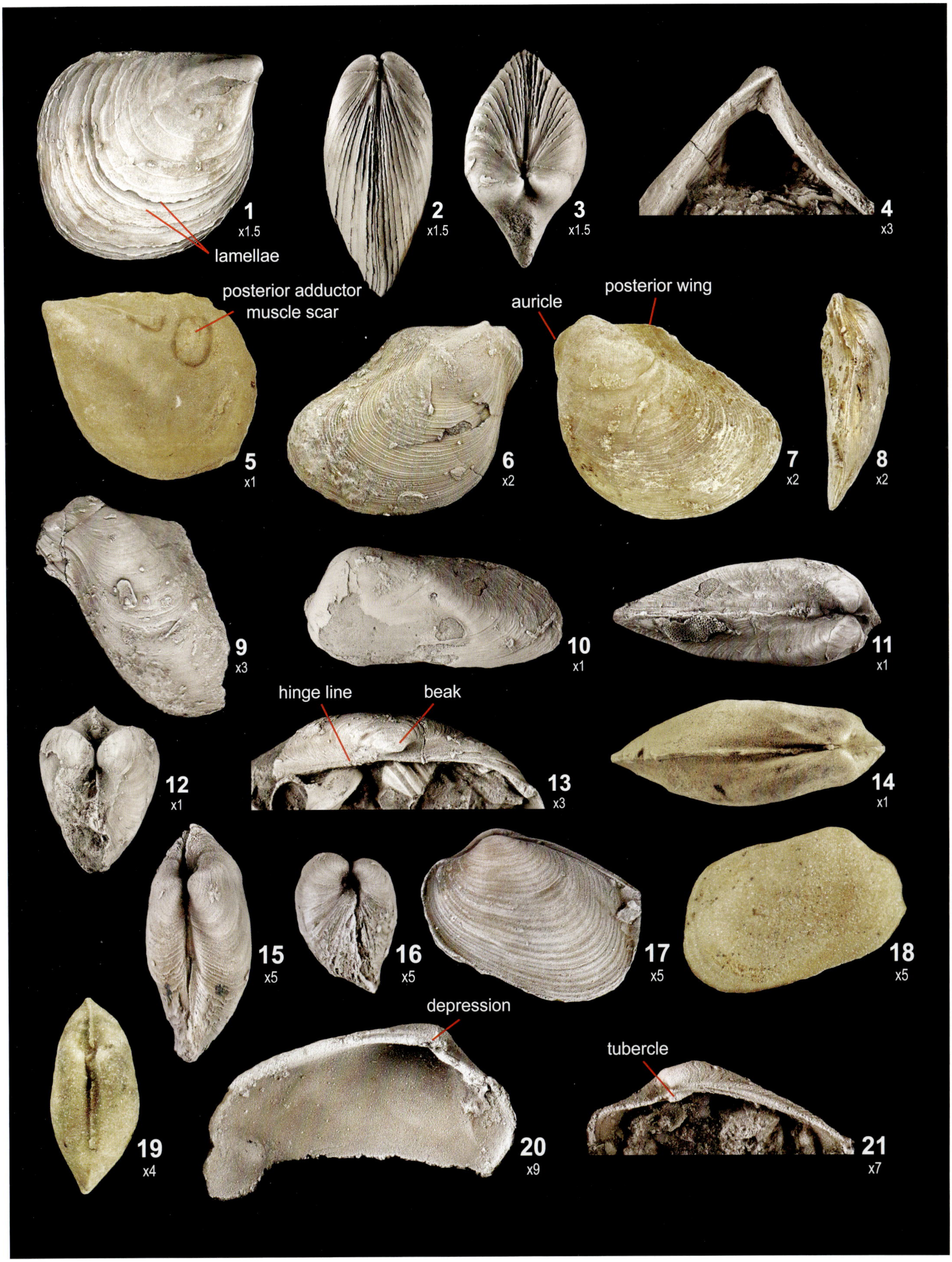
1
x1.5
lamellae
2
x1.5
3
x1.5
4
x3
posterior adductor
muscle scar
5
x1
6
x2
auricle
posterior wing
7
x2
8
x2
9
x3
10
x1
11
x1
hinge line
beak
12
x1
13
x3
14
x1
15
x5
16
x5
17
x5
18
x5
depression
19
x4
20
x9
tubercle
21
x7

Occurrence. Common in the Platteville Formation in northern Illinois and southern Wisconsin. Particularly common in the Mifflin Member in northern Grant County, Wisconsin.

Order SOLENIDA

Family ORTHONOTIDAE

Genus CYMATONOTA

Cymatonota sp.

Figures 9-6-10 through 9-6-14

Description. Shell 30–70 mm long, elongate, length nearly three times the height, very convex; subparallel dorsal and ventral margins; beaks about one-fourth distance from anterior end, ends rounded, umbonal ridge rounded; broad lateral sulcus and slight ventral sinus along shell margins. Hinge line lacking teeth. Shell thin, outer surface marked by fine concentric growth lines and corresponding low ridges.

Remarks. The relatively large, elongate shell is somewhat like that of *Cymatonota orthonotum* Meek and Worthen (1868), questionably from the "Platteville Formation" near Dunleith, Jo Daviess County, Illinois.

Occurrence. Common in the Mifflin and Grand Detour Members of northern Illinois and southern Wisconsin.

Figure 9-7 (Opposite page)
1, 2. ***Goniophora* sp.** (**1**) Dorsal surface exterior of conjoined valves; (**2**) right valve exterior showing the characteristic carina. Mifflin Member, Dixon, Illinois (locality 2), **1**, **2**, JAC 142.

3–8. ***Colpomya* sp.** (**3**) Left valve exterior; (**4**) dorsal surface exterior of conjoined valves; (**5**) right valve interior; (**6**) enlargement of same right valve; (**7**) left valve exterior; (**8**) left valve exterior showing the characteristic broad lateral sulcus (depression) and prominent ventral sinus (indentation) along shell margins. Mifflin Member, Dixon, Illinois (locality 2), **3**, **4**, ISGS 913; **5**, **6**, ISGS 818; **7**, ISGS 848; **8**, ISGS 815.

9–16. ***Modiolopsis* spp.** (**9**) Left valve exterior; (**10–12**) conjoined valves showing dorsal, anterior, and right valve exteriors, respectively; (**13**) steinkern of left valve; (**14**) right valve interior mold showing anterior and posterior adductor muscle scars (aams and pams) and pallial line; (**15**) same specimen as previous showing internal mold of dorsal surface and hinge line; (**16**) close-up of previous specimen showing anterior adductor muscle scar (aams), small accessory muscle scars, and right valve tubercle inserted in left depression along the hinge line. Mifflin Member, Dixon, Illinois (locality 2), **9**, ISGS 872; **10–12**, ISGS 911; **13**, ISGS 1597. Mifflin Member, Fennimore, Wisconsin (locality 26), **14–16**, ISGS 915.

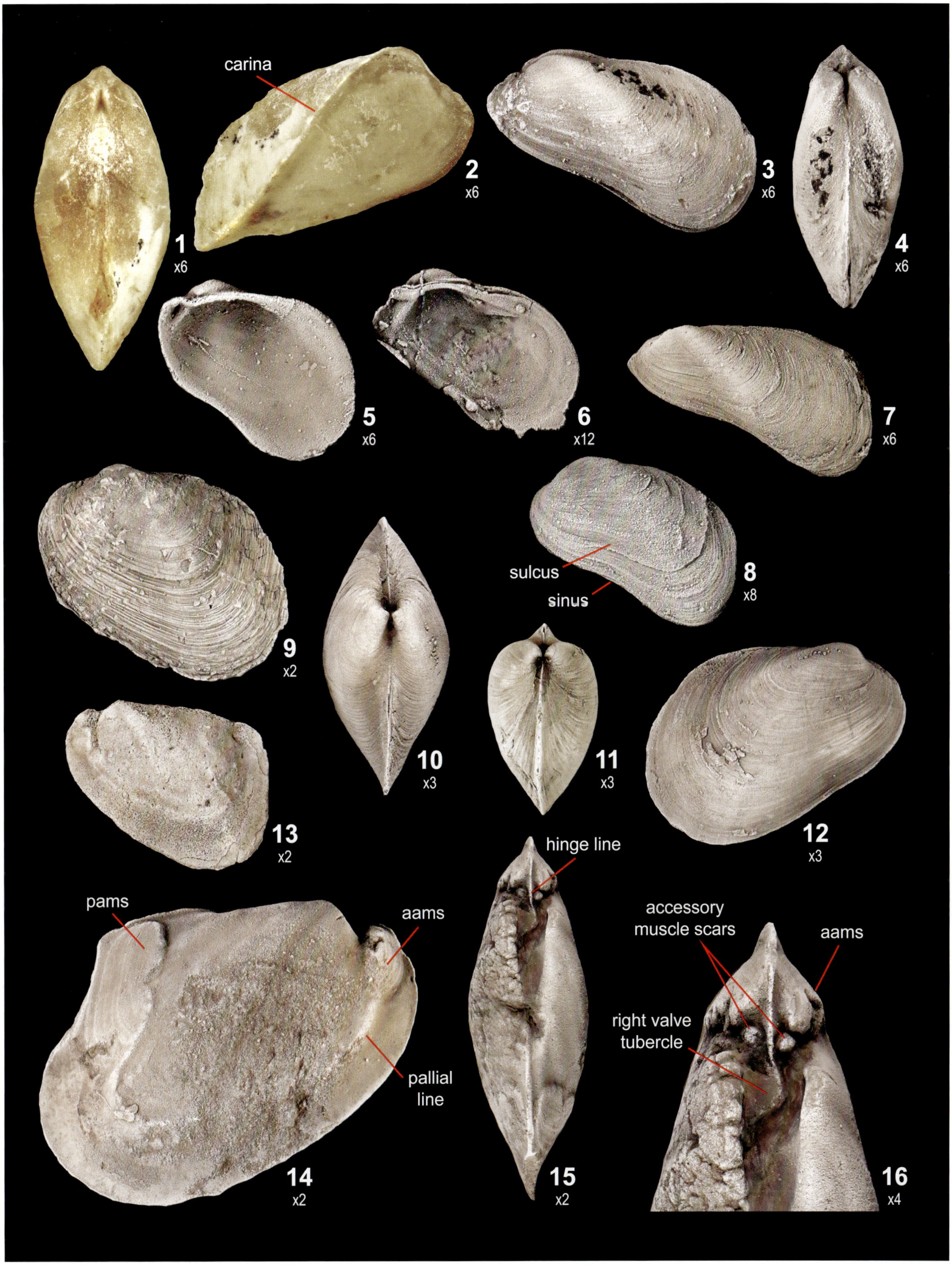

carina
1
x6
2
x6
3
x6
4
x6
5
x6
6
x12
7
x6
sulcus
sinus
8
x8
9
x2
10
x3
11
x3
12
x3
13
x2
hinge line
pams
aams
pallial
line
14
x2
15
x2
accessory
muscle scars
aams
right valve
tubercle
16
x4

CHAPTER 10
ROSTROCONCHS

At least two rostroconch genera are present in the Platteville Formation, *Eopteria* and *Hippocardia*, both of which are assigned to the order Conocardioida. These are small fossils in the range of 5–10 mm in length and generally quite rare. Some of the better specimens have been found in sieved bulk samples collected from the Mifflin Member marl beds at Dixon, Illinois. Here the rostroconchs tend to weather free from the matrix to reveal microscopic details of the shell.

Rostroconchs are an extinct class of mollusks that lived from the Early Cambrian to the Late Permian. Prior to the work of Pojeta et al. (1972), they were generally regarded as aberrant Paleozoic bivalves that gave rise to no other stock. Like bivalves, they are bilaterally symmetrical relative to the commissural plane that separates the shell into mirror images. Unlike bivalves, which have two opposing valves, rostroconchs have a single, pseudobivalved shell that lacks a functional hinge.

Little is known about the biology of rostroconchs because there are no living representatives. Therefore, some speculation is required to interpret aspects of functional morphology and orientation. By analogy with bivalves, the fused edge where valves connect in rostroconchs is regarded as **dorsal** (Pojeta and Runnegar 1976) and the opposite edge is **ventral**. One end of the shell has a relatively large **gape** that is interpreted as the opening for protrusion of the foot, thus is anterior in position. As with bivalves, the **right** and **left valves** are determined by orienting the shell with the dorsal side uppermost, commissure horizontal, and anterior end pointing away from the observer. The posterior end has a tubular **rostrum** (e.g., *Hippocardia*) and an opening called the **rostral orifice**. By analogy with bivalves and some gastropods, the rostrum would allow water and excretory products to enter and leave the mantle cavity (Pojeta and Runnegar 1976). The exterior of each valve culminates in an **apex** or **beak** on the dorsal margin. The **umbo** is the strongly convex part of the shell around the beak, extending to the beak and in some cases coinciding with it. The region of the shell between the umbones and anterior end is called the **snout**. Some rostroconchs (e.g., *Hippocardia*) have a prominent flange called a **hood** that projects in an anteroventral direction.

Morphologic features used in the identification of rostroconchs include (1) overall shell shape, (2) the presence or absence of a hood, (3) aspects of the rostral orifice and cleft, and (4) shell ornamentation. The primary literature sources used here for identification of Platteville rostroconchs includes Pojeta (1971) and Pojeta and Runnegar (1976).

SYSTEMATIC DESCRIPTIONS

Phylum MOLLUSCA

Subphylum DIASOMA

Class ROSTROCONCHIA

Order CONOCARDIOIDA

Description. The conocardioids have external and internal ribs expressed as marginal denticles on the inside edge of the commissure, an anterior gape, and dorsal clefts.

Superfamily CONOCARDIACEA

Family HIPPOCARDIIDAE

Genus HIPPOCARDIA

Hippocardia* cf. *cooperi Pojeta and Runnegar, 1976

Figures 10-1-1 through 10-1-7

Description. *Hippocardia* has a single hood forming prominent flanges on either side of the shell; external sculpture of snout reticulate, ventral part of anterior gape slitlike; length up to 5 mm.

Remarks. The Platteville species of *Hippocardia* is very similar in size, shape, and ornamentation to *H. cooperi* from the Late Ordovician lower Chambersburg Formation near Strasburg, Virginia.

Occurrence. Common in the Mifflin Member at Dixon, Illinois, and the dolostone facies of the Mifflin Member at Lee Center, Illinois.

Superfamily EOPTERIACEA

Family EOPTERIIDAE Miller

Genus EOPTERIA Billings

Eopteria* cf. *conocardiformis Pojeta and Runnegar, 1976

Figures 10-1-8 through 10-1-12

Description. A species of *Eopteria* with markedly elongated anterior snout so that the umbones are posterior to the center of the length of the shell; both anterior and posterior dorsal clefts present; anterior, ventral, and posterior shell gapes are continuous with one another; rostrum rudimentary; length up to 10 mm; external surface sculpture of raised concentric ridges anteriorly on the "snout" and faint radial ribs on the posterior, on the umbones and in the vicinity of the rostrum.

Remarks. This Platteville species of *Eopteria* with elongate anterior snout is most similar in overall shell shape to *E. conocardiformis* from the Upper Ordovician Little Oak Limestone of Alabama and High Bridge Group of Kentucky.

Occurrence. Common in the Mifflin Member at Dixon, Illinois, and the dolostone facies of the Mifflin Member at Lee Center, Illinois.

Figure 10-1 (Opposite page)
1–7. ***Hippocardia* cf. *cooperi*** Pojeta and Runnegar. (**1**) Right-lateral view showing morphological features and orientation nomenclature; (**2**) left-lateral view showing characteristic reticulate ornamentation; (**3**) posterior showing the rostral orifice and rostral face of the hood; (**4**) anteroventral surface showing the hood and anterior gape; (**5**) ventral surface showing the denticles; (**6**) anterior end; (**7**) dorsal surface. Mifflin Member, Dixon, Illinois (locality 2), **1**, ISGS 918; **2**, **7**, ISGS 919; **3**, **5**, ISGS 1589; **4**, **6**, ISGS 1590.

8–12. ***Eopteria* cf. *conocardiformis*** Pojeta and Runnegar. (**8**) Posterior end showing rostral cleft and commissure; (**9**) anterior end showing gape; (**10**) left-lateral view showing the snout, rostrum, and anterior gape; (**11**) ventral surface showing the denticles, umbo, and anterior gape; (**12**) dorsal surface showing cleft at anterior end. Mifflin Member, Dixon, Illinois (locality 2), **8–12**, JAC 184.

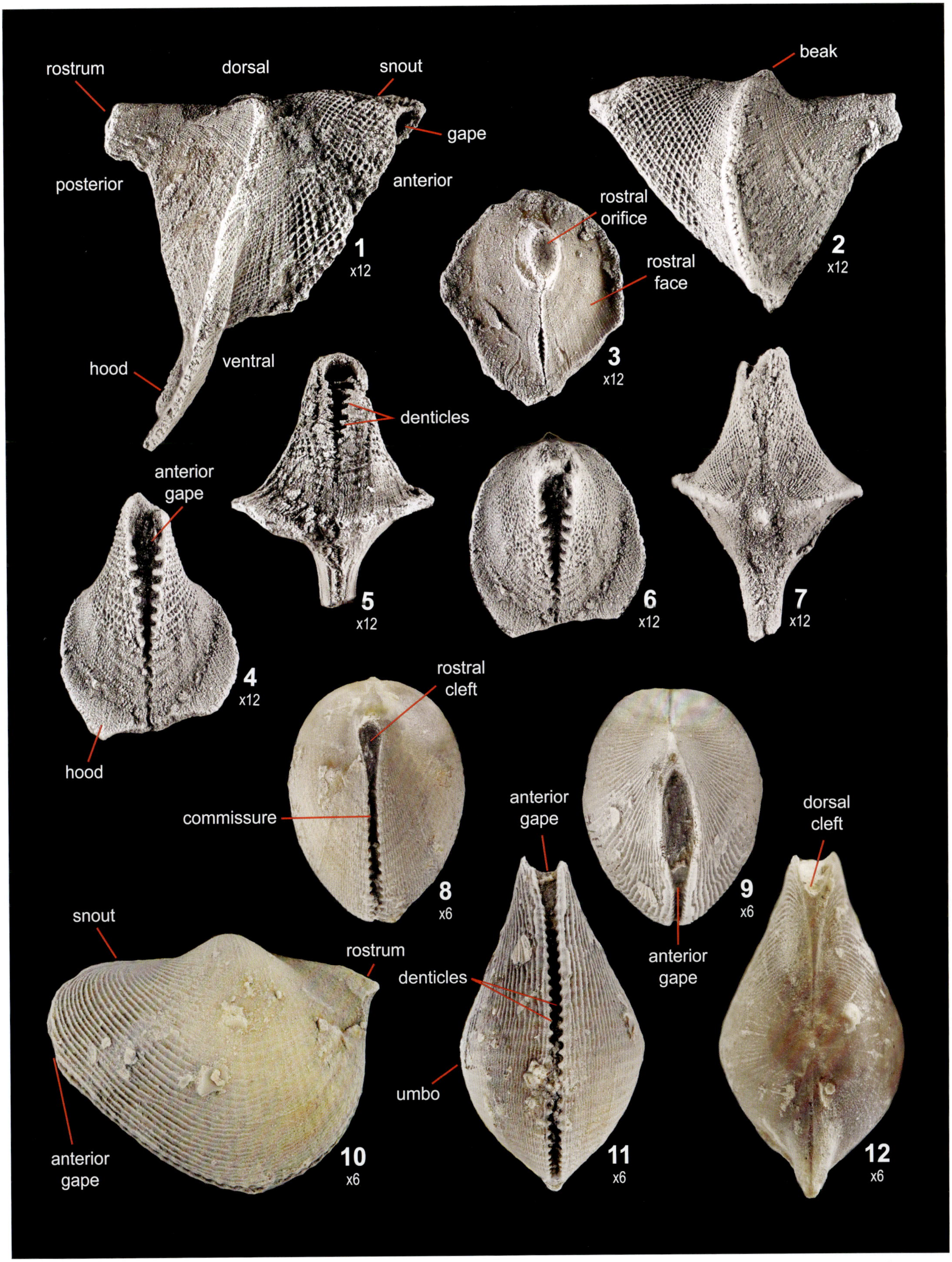
rostrum
dorsal
snout
gape
posterior
anterior
1
x12
hood
ventral
beak
rostral
orifice
rostral
face
2
x12
3
x12
denticles
anterior
gape
5
x12
6
x12
7
x12
4
x12
hood
rostral
cleft
commissure
anterior
gape
dorsal
cleft
8
x6
9
x6
snout
rostrum
denticles
anterior
gape
umbo
anterior
gape
10
x6
11
x6
12
x6

CHAPTER 11

NAUTILOID CEPHALOPODS

The Platteville Formation contains a diverse and relatively abundant nautiloid cephalopod fauna (Appendix 2) represented by (1) subclass Orthoceratoidea, including the orders Ellesmerocerida, Orthocerida, and Ascocerida; (2) subclass Endoceratoidea, order Endocerida; (3) subclass Actinoceratoidea, order Actinocerida; and (4) subclass Nautiloidea, including the orders Tarphycerida, Oncocerida, and Discosorida. The Formation contains one of the most diverse Late Ordovician cephalopod faunas in North America (Catalani 1987, 2010).

Cephalopods have a long geologic record, extending from the Late Cambrian to the present. They were particularly diverse and abundant during the Paleozoic and Mesozoic. The class is represented in modern marine environments by octopuses, squids, cuttlefishes, and *Nautilus*, the last being the only living form having an external shell. As such, *Nautilus* provides insight into the biology and functional morphology of all fossil cephalopods.

Cephalopods are agile, bilaterally symmetrical, predatory carnivores that have achieved a structural complexity unlike other unsegmented invertebrates (Sweet 1964). They were an important part of the benthic and nektonic marine communities during deposition of the Platteville Formation (Figure 11-1). Like all fossil cephalopods, the Platteville forms have an external shell called a **conch**. The **phragmocone** is the posterior part of the conch and consists of a series of partitions or **septa** (sing. **septum**) that divide the shell into chambers called **camerae** (Figure 11-2-1). **Sutures** are the line of juncture of the septa and the inner side of the phragmocone and are most readily observed on the surface of steinkerns. Platteville cephalopods are referred to as "nautiloid" cephalopods because they possess straight or slightly curved sutures as compared with the complexly convoluted sutures of "ammonoid" cephalopods. At the anterior (**adoral**) end of the conch is the **body chamber** containing most of the soft tissue, including the head, which is partly extended through the anterior **aperture** of the conch. The posterior end of the conch is referred to as the **adapical end**. The septa are perforated by an opening called the **septal foramen** (Figure 11-2-2) through which extends the siphuncular cord, a fleshy extension of the main body of the animal. Between septa, the siphuncular cord is enclosed by calcareous **septal necks** and **connecting rings**, which, taken together, are referred to as the **siphuncle**. The ventral side (**venter**) of the conch is conventionally interpreted to be the side nearest the axis of the siphuncle, which typically is the side on which the **hyponomic sinus** (concave sinus in the middle of the aperture) is positioned but rarely preserved. The dorsal side (**dorsum**) is opposite the venter. Platteville conchs may be straight (**orthoconic**), curved (**cyrtoconic**), loosely coiled (**gyroconic**), or tightly coiled with successive whorls in contact with each other (**tarphyconic**). In addition, the conch

Figure 11-1 (Opposite page) Depiction of a typical Platteville seafloor in the upper Midwest USA 450 million years ago showing a variety of cephalopods.

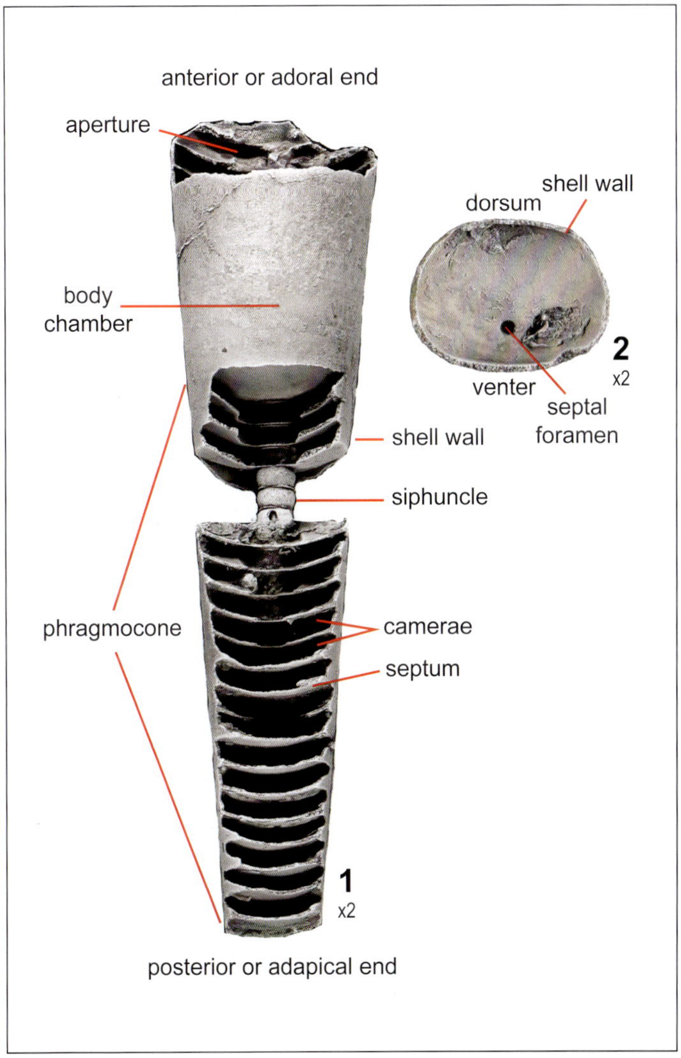

Figure 11-2 (**1**) Conch of *Cartersoceras shideleri* showing terms used to describe nautiloid morphology and orientation; (**2**) interior view of septa (*C. shideleri*) showing morphologic terms.

Figure 11-3
1–5. Variations in conch shape of typical Platteville nautiloids. (**1**) *Proteoceras*, longiconic orthocone; (**2**) *Gorbyoceras*, cyrtoconic orthocone; (**3**) *Beloitoceras*, cyrtoconic brevicone; (**4**) *Centrocyrtoceras*, gyrocone; (**5**) *Plectoceras*, tarphycone.

may be long and narrow (**longicone**) or short and rotund (**brevicone**; Figure 11-3). A primary key to the identification of fossil cephalopods is the position of the siphuncle relative to the venter. Coiled and curved conchs are considered **exogastric** if the siphuncle is positioned closer to the outer convex side and **endogastric** if it is closer to the inner concave side (Figure 11-4). Furthermore, a conch is considered **compressed** when it is taller than wide and **depressed** when wider than tall in transverse cross section. Ornamentation may be lacking entirely or consist of transverse ringlike expansions of the shell called **annuli** or very fine parallel ridges (**lirae**) that are transversely or longitudinally distributed, or both. Some orthoconic longicones (e.g., *Cartersoceras*) developed calcareous deposits within the camerae in the ventral part of the phragmocone to counterbalance the buoyant effect of gas-filled chambers, allowing the animal to maintain horizontal equilibrium (Frey 1995). A more detailed discussion of cephalopod morphology can be found in Teichert (1964).

Morphologic features used in the identification of cephalopods include (1) general shell shape (orthocone, longicone, brevicone, etc.), (2) maximum observed shell diameter and length, (3) apical angle or angle of shell expansion, (4) cross-sectional shape of the shell, (5) suture pattern, (6) cameral length, (7) features of the body chamber, (8) external shell ornamentation, and (9) siphuncle segment shape, diameter, and position (Frey 1995). Primary literature sources for the identification of Platteville cephalopods include Clarke (1897b), Foerste (1928, 1932), Flower (1946, 1952, 1955, 1957, 1963, 1964a, 1964b), Teichert (1964), and Frey (1995).

SYSTEMATIC DESCRIPTIONS

Phylum MOLLUSCA

Class CEPHALOPODA

Subclass ORTHOCERATOIDEA

Description. The Orthoceratoidea typically have orthoconic to slightly cyrtoconic (usually longiconic) shells, central to subcentral siphuncles, and cameral deposits.

Order ELLESMEROCERIDA

Description. The ellesmerocerids are an ancestral nautiloid stock characterized by wide morphologic variation. They are typically small, smooth, compressed cyrtoconic brevicones (e.g., *Cyrtocerina crenulata*) or slightly depressed orthoconic nautiloids (e.g., *Cartersoceras shideleri*) with closely spaced septa, a ventral tubular siphuncle, and structurally complex connecting rings.

Family BALTOCERATIDAE

Genus CARTERSOCERAS

Cartersoceras shideleri Flower, 1964a

Figures 11-5-1 through 11-5-4

Description. Depressed orthoconic longicone, gradually expanding at an angle of 8°–9°, reaching a maximum diameter of ~30 mm; circular in cross section; smooth exterior; mildly flattened venter, especially with maturity. Cameral lengths typically 2–3 mm, increasing slightly with maturity. Siphuncle subcentral adapically, becoming midventral with maturity; convex siphuncle segments. Siphuncular rod filling the siphuncle tube adapically and tapering to a point adorally (Figure 11-5-4). Calcareous cameral deposits common in the ventral part of the phragmocone.

Remarks. The Platteville species of *Cartersoceras* appears to be identical to *C. shideleri* from the essentially coeval Tyrone Limestone of Kentucky and Carters Limestone of Tennessee.

Occurrence. Common in the Mifflin and Grand Detour Members in northern Illinois and southern Wisconsin and the McGregor Member of southwestern Wisconsin, northeastern Iowa, and southeastern Minnesota. Locally common in the Pecatonica Member in north-central Illinois.

Figure 11-4
1, 2. Examples of exogastric and endogastric conchs. (**1**) *Cyrtocerina*, endogastric, dorsal surface shown as dashed line; (**2**) *Beloitoceras*, exogastric.

Family CYRTOCERINIDAE

Genus CYRTOCERINA

Cyrtocerina crenulata Flower, 1952

Figures 11-5-5 through 11-5-8

Description. Rapidly expanding cyrtoconic, endogastric, compressed brevicone; sutures crenulated, particularly on the ventral and lateral sides; siphuncle submarginal and diameter large relative to diameter of the shell; septal necks are vestigial or absent; connecting rings thick. Length of conch 40–50 mm; width at adoral end approximately 30 mm.

Remarks. *C. crenulata* was described by Flower (1952) from the "Beloit member" (probable Grand Detour Member) of the Platteville Formation at Beloit, Wisconsin. The recently collected specimens illustrated herein closely fit Flower's description.

Occurrence. Rare in the Mifflin and Grand Detour Members in northern Illinois and southern Wisconsin.

Order ORTHOCERIDA

Description. The Orthocerida are characterized by straight to slightly curved, gradually enlarging, longiconic shells that are circular to nearly circular in cross section. Typically, they have camerae of moderate to considerable length separated by septa that are symmetrically saucerlike in shape. The shell may be smooth (e.g., *Whitfieldoceras mumiaforme*) or distinctively annulated (e.g., *Anaspyroceras cylindricum*).

Family ORTHOCERATIDAE

Genus ANASPYROCERAS

Anaspyroceras cylindricum (Foerste, 1932)

Figures 11-5-9 through 11-5-12

Description. Small, slender, orthoconic longicone, very gradually expanding at an angle of ~3° or less to reach a diameter of ~7 mm; circular in cross section. Largest observed specimen is 80 mm long and 10 mm in diameter; most nearly complete body chamber observed is 36 mm long. Exterior marked by prominent raised annuli 1 mm wide separated by interannular depressions of equal width; annuli slope obliquely in an adapical direction from the dorsum to the venter, crossing from one camera to another; annuli on body chamber more subdued in adoral direction; annuli and interannular spaces overlain by fine transverse lirae. Septa are evenly spaced and 2 mm in length. Siphuncle small in

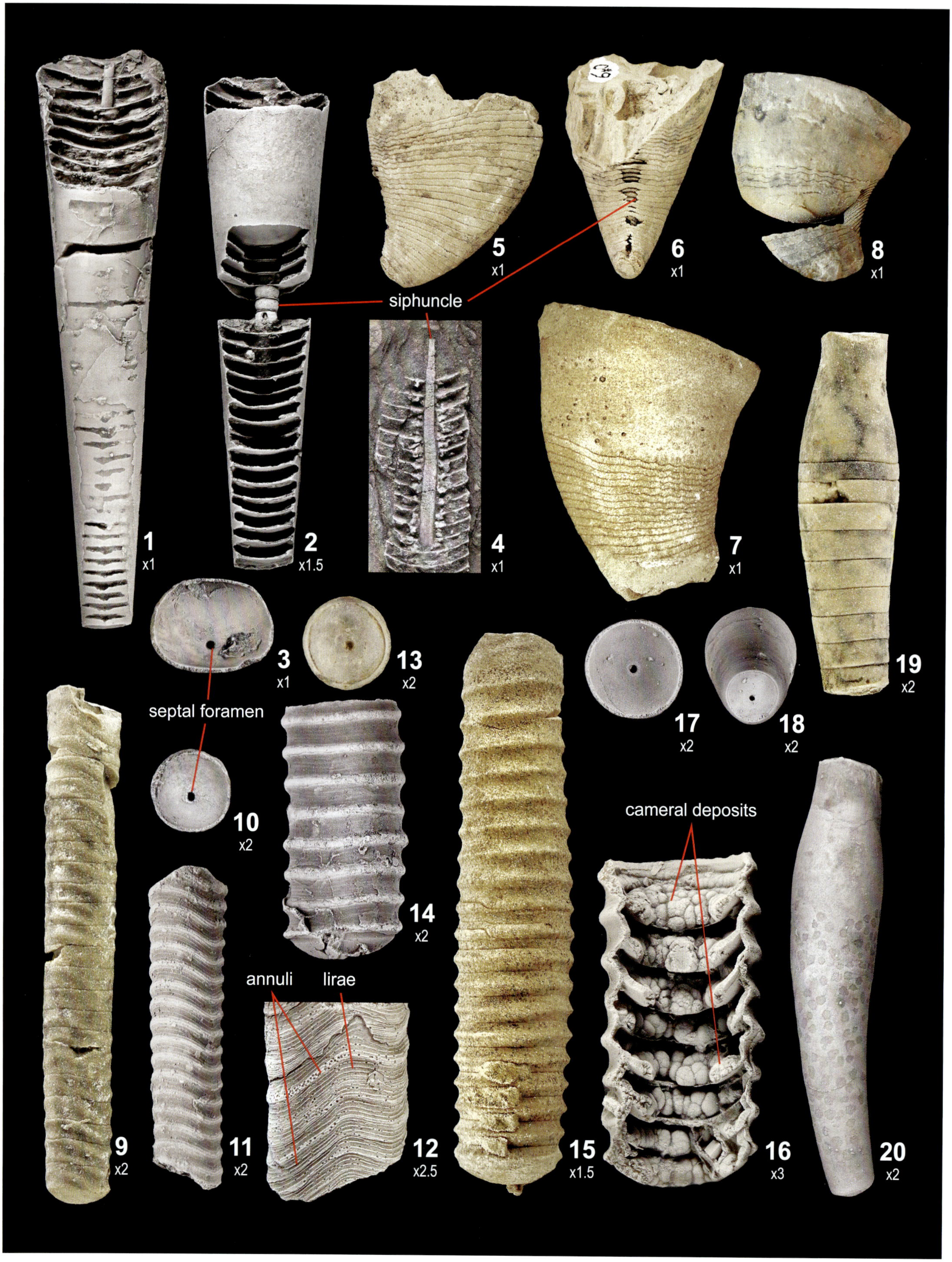

1
x1
2
x1.5
siphuncle
4
x1
5
x1
6
x1
8
x1
7
x1
19
x2
3
x1
13
x2
septal foramen
10
x2
17
x2
18
x2
14
x2
cameral deposits
annuli
lirae
9
x2
11
x2
12
x2.5
15
x1.5
16
x3
20
x2

diameter, tubular, and centrally positioned. Cameral deposits in the adapical parts of the shell, consisting of massive grapelike bodies in the ventrolateral parts of the camerae.

Remarks. The Platteville species of *Anaspyroceras* appears to be identical to *A. cylindricum* from the Trenton Limestone in New York (Foerste 1932).

Occurrence. Common in the Mifflin and Grand Detour Members in northern Illinois and southern Wisconsin and the McGregor Member of southwestern Wisconsin, northeastern Iowa, and southeastern Minnesota. Locally common in the Pecatonica Member in north-central Illinois.

Anaspyroceras* cf. *lesueuri (Clarke, 1897b)

Figures 11-5-13 through 11-5-16

Description. Small, slender, orthoconic longicone; very gradually expanding at an angle of ~3° or less to reach a maximum diameter of 16 mm at juncture with body chamber, then tapering to 13 mm at the aperture; body chamber 34 mm in length; circular in cross section. Exterior marked by prominent raised annuli 1.5 mm wide separated by interannular depressions of equal width; annuli mostly confined to a single camera (Figure 11-5-15); annuli on body chamber become more subdued and spacing increases to 4 mm in adoral direction. Septa are evenly spaced and 3 mm in length. Siphuncle small and centrally positioned.

Remarks. The type species of *Anaspyroceras lesueuri* was described by Clarke (1897b) from the "Platteville member of the Black River Formation" at Cannon Falls, Goodhue County, Minnesota. Clarke (1897b) described a second form of *A.* cf. *lesueuri* from the "Platteville member of the Black River Formation" at Dixon, Illinois, that appears to differ from the Minnesota specimens in having a perfectly straight shell and annulations that are not significantly curved transversely as in the Minnesota form. The specimens figured herein are from northern Illinois and appear to be most similar to *A.* cf. *lesueuri*; however, a more thorough study of both forms is needed to determine whether the differences are due to intraspecific variation.

Occurrence. Locally common in the Pecatonica, Mifflin, and Grand Detour Members in northern Illinois and southern Wisconsin.

Figure 11-5 (Opposite page)

1–4. ***Cartersoceras shideleri*** Flower. (**1**) Part of large phragmocone showing outer surface of shell and exposed parts of the siphuncle; (**2**) photographically restored specimen showing phragmocone, siphuncle, and body chamber; (**3**) interior surface of septum showing septal foramen; (**4**) tapered adoral end of siphuncle. Mifflin Member, Dixon, Illinois (locality 2), **1**, ISGS 717; **2**, ISGS 706; **3**, ISGS 16002; **4**, ISGS 716.

5–8. ***Cyrtocerina crenulata*** Flower. (**5**) Lateral view of conch; (**6**) venter of specimen in previous figure showing partially exposed siphuncle; (**7**) lateral view of conch; (**8**) lateral view of conch showing siphuncle. Mifflin Member, Lee Center, Illinois (locality 31), **5**, **6**, JAC 147; **8**, JAC 148. Grand Detour Member, Beloit, Wisconsin (locality 21), **7**, JAC 149.

9–12. ***Anaspyroceras cylindricum*** (Foerste). (**9**) Lateral view of dolomitized steinkern; (**10**) interior surface of a septum with septal foramen; (**11**) lateral view of shell; (**12**) enlargement of outer shell surface showing fine transverse lirae parallel to and covering the annuli and interannular spaces. Mifflin Member, Lee Center, Illinois (locality 31), **9**, JAC 150. Mifflin Member, Dixon, Illinois, (locality 2), **10**, ISGS 760; **11**, ISGS 695; **12**, ISGS 693.

13–16. ***Anaspyroceras* cf. *lesueuri*** (Clarke). (**13**) Interior surface of a septum showing the septal foramen; (**14**) part of a phragmocone showing the prominent annuli; (**15**) phragmocone and part of the body chamber showing the characteristic adoral contraction of the body chamber; (**16**) internal view of a phragmocone showing globular, grapelike, cameral deposits. Mifflin Member, Dixon, Illinois (locality 2), **13**, ISGS 759; **14**, **16**, ISGS 724. Grand Detour Member, Beloit, Wisconsin, (locality 21), **15**, JAC 151.

17–20. ***Whitfieldoceras mumiaforme*** Foerste. (**17**) Interior surface of a septum with septal foramen; (**18**) apical end of a phragmocone showing the septum and septal foramen; (**19**) dolomitized steinkern showing the tapered body chamber; (**20**) nearly complete conch. Mifflin Member, Dixon, (locality 2), **17**, ISGS 766; **18**, ISGS 765; **20**, ISGS 1530. Mifflin Member, Lee Center, Illinois (locality 31), **19**, JAC 152.

Family CLINOCERATIDAE

Genus WHITFIELDOCERAS

Whitfieldoceras mumiaforme Foerste, 1932

Figures 11-5-17 through 11-5-20

Description. Small, elongate, smooth-shelled, slightly curved cyrtoconic longicone. Conch gradually expands to the junction with the body chamber, then contracts to the aperture; cross section circular; cameral length 1–2 mm; apical end typically truncated. Largest specimens are approximately 50 mm in length and 12 mm maximum in diameter. Siphuncle small, positioned slightly ventral of center.

Remarks. The type species, *Oncoceras mumiaforme*, was described by Whitfield (1882) as being from the Lower Buff limestone of the Trenton group, below Carpenter's quarry, near Beloit, Wisconsin. Foerste (1932) reassigned the species to the genus *Whitfieldoceras*, noting that the type specimens were "from below Carpenter's quarry, at Beloit, Wisconsin, in the Lower Buff or Platteville member of the Black River" (p. 102).

Occurrence. Locally abundant in the Mifflin and Grand Detour Members in northern Illinois and southern Wisconsin and the McGregor Member of southwestern Wisconsin, northeastern Iowa, and southeastern Minnesota. Locally common in the Pecatonica Member in north-central Illinois.

Family PROTEOCERATIDAE

Genus GORBYOCERAS

***Gorbyoceras* sp.**

Figures 11-6-1 through 11-6-4

Description. Small- to medium-sized cyrtoconic longicone, gradually expanding at an angle of 8°–9° to reach a diameter of ~16 mm adorally; observed specimens reach lengths of 70 mm. Shell typically circular in cross section; exterior marked by prominent rounded annuli, well-developed longitudinal ridges, and microscopic transverse lirae (Figure 11-6-1). Sutures slope posteriorly from the dorsum to the venter. Cameral lengths typically 1–3 mm, increasing slightly with maturity. Siphuncle slightly subcentral with beadlike segments.

Remarks. The Platteville species of *Gorbyoceras* appears to be distinct in having a well-developed cyrtoconic longicone and prominent longitudinal ridges. Assignment to a particular species is herein deferred.

Occurrence. Common in the Mifflin and Grand Detour Members in northern Illinois and southern Wisconsin and the McGregor Member of southwestern Wisconsin, northeastern Iowa, and southeastern Minnesota. Locally common in the Pecatonica Member in north-central Illinois.

Genus MONOMUCHITES

***Monomuchites annularis* Frey, 1995**

Figures 11-6-7 through 11-6-9

Description. Moderate-sized, slightly endogastrically curved, orthoconic longicone; gradually expanding at an angle of 7°–8°. Circular in cross section with straight, transverse sutures. Largest observed specimen is 230 mm in length and adorally 30 mm in diameter. Some fragmentary specimens have maximum diameters of 60 mm or more. Cameral chambers decrease in relative length with maturity. Shell exterior has one prominent, raised, transverse annulus per cameral chamber separated by slightly wider interannular depressions that are marked by transverse growth lines. Siphuncle small with expanded beadlike segments; subcentrally positioned in shells less than 20 mm in diameter, becoming more centrally positioned as shells approach 30 mm or more in diameter. Cameral deposits present in adapical parts of the phragmocone.

Remarks. Frey (1995) described *M. annularis* as occurring in the Platteville Formation at Dixon, Illinois, and Mineral Point, Wisconsin, as well as in the coeval Tyrone Limestone of Jessamine County, Kentucky.

Occurrence. Common in the Mifflin and Grand Detour Members in northern Illinois and southern Wisconsin and the McGregor Member of southwestern Wisconsin, northeastern Iowa, and southeastern Minnesota. Locally common in the Pecatonica Member in north-central Illinois.

Figure 11-6 (Opposite page)

1–4. ***Gorbyoceras* sp.** (**1**) Typical shell showing the prominent annuli, longitudinal ridges, and lirae; (**2**) interior surface of a septum showing the centrally positioned septal foramen; (**3**) shell showing prominent annuli and longitudinal ridges and healed shell; (**4**) dolomitized steinkern with subdued annuli and longitudinal ridges. Mifflin Member, Dixon, Illinois (locality 2), **1**, ISGS 1532; **2**, ISGS 691; **3**, ISGS 1533. Mifflin Member, Lee Center, Illinois (locality 31), **4**, JAC 153.

5, 6. ***Monomuchites*? sp.** (**5**) Typical annulated phragmocone with a photographic enlargement showing very fine longitudinal lirae; (**6**) interior surface of a septum showing the centrally positioned septal foramen. Mifflin Member, Dixon, Illinois (locality 2), **5**, ISGS 729; **6**, ISGS 734.

7–9. ***Monomuchites annularis*** Frey, 1995. (**7**) Typical annulated phragmocone with a photographic enlargement showing the very fine transverse lirae; (**8**) interior surface of a septum showing the subcentrally positioned septal foramen; (**9**) dolomitized steinkern of a relatively complete phragmocone. Mifflin Member, Dixon, Illinois (locality 2), **7**, ISGS 728; **8**, ISGS 735. Pecatonica Member, Forreston, Illinois (locality 6), **9**, ISGS 723.

10, 11. ***Proteoceras* cf. *tyronensis*** (Foerste). (**10**) Phragmocone and body chamber; (**11**) interior surface of a septum showing the centrally positioned septal foramen. Mifflin Member, Lee Center, Illinois (locality 31), **10**, JAC 154. Mifflin Member, Dixon, Illinois (locality 2), **11**, ISGS 742.

12. ***Proteoceras* sp. 1.** (**12**) Small shell with photographic enlargement showing the very fine longitudinal lirae. Mifflin Member, Dixon, Illinois (locality 2), **12**, ISGS 732.

13. ***Proteoceras* sp. 2.** (**13**) Small shell showing very gradual expansion of the phragmocone. Mifflin Member, Dixon, Illinois (locality 2), **13**, ISGS 740.

14, 15. ***Redpathoceras* cf. *clarkei*** Flower. (**14**) View of dorsum; (**15**) lateral view showing a deeply rounded septum of truncation. Mifflin Member, Fennimore, Wisconsin (locality 26), **14**, **15**, ISGS 757.

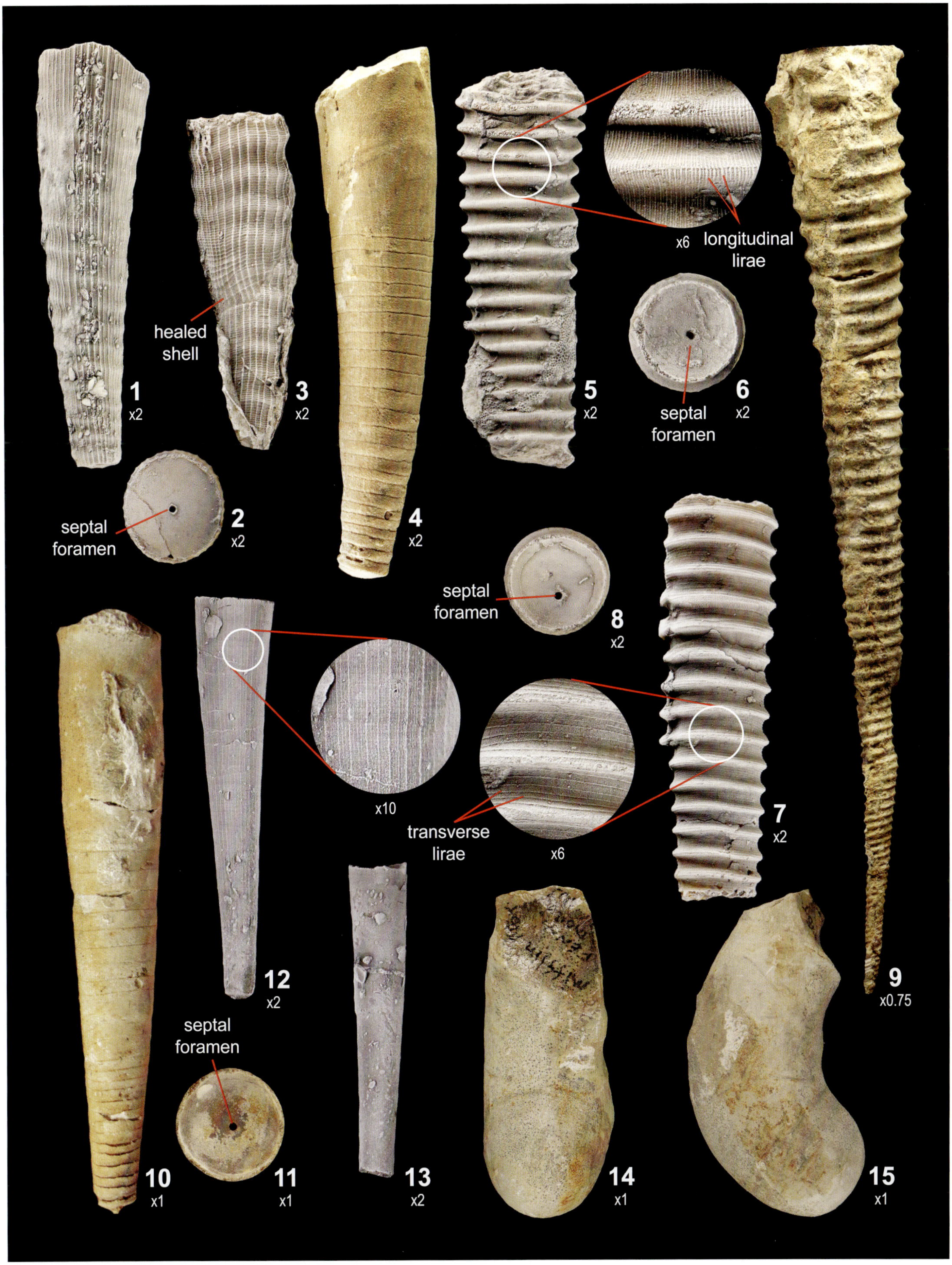
healed
shell
1
x2
3
x2
2
x2
septal
foramen
4
x2
5
x2
x6
longitudinal
lirae
6
x2
septal
foramen
9
x0.75
8
x2
septal
foramen
7
x2
x10
x6
transverse
lirae
12
x2
septal
foramen
10
x1
11
x1
13
x2
14
x1
15
x1

Monomuchites? sp.

Figures 11-6-5 and 11-6-6

Description. Moderate-sized, slightly curved, annulated, orthoconic longicone; gradually expanding at an angle of 5°–6°. Circular in cross section with straight, transverse sutures. Largest observed specimen is 70 mm in length and 16 mm in diameter at the oral end. Shell exterior has one prominent, raised, transverse annulus per cameral chamber separated by slightly wider interannular depressions; exterior of shell marked by fine longitudinal lirae that alternate in prominence. Siphuncle small; mostly centrally positioned; segments weakly expanded with small endosiphuncular annuli in septal foramina.

Remarks. Classification of this orthocerid is questionable, given the current understanding of the Ordovician annulated longicones. The somewhat obscure morphologic features of the siphuncle and typical poor preservation hinder accurate identification. This Platteville form is similar to *Monomuchites* in having a rapidly expanding shell with prominent annuli but differs in having longitudinal lirae. The expansion angle is greater than that of *Anaspyroceras*, and the annuli are more prominent than those of *Gorbyoceras*. The annulated longicone of *Metaspyroceras* (Foerste 1932) has similar longitudinal lirae but differs from this form in having annulations and sutures that are oblique rather than straight. On the whole, this Platteville form appears to be most like *Monomuchites*, but assignment is questionable and is herein deferred pending future research on the North American Late Ordovician annulated longicones.

Occurrence. Common locally in the Mifflin and Grand Detour Members in northern Illinois and southern Wisconsin and the McGregor Member of southwestern Wisconsin, northeastern Iowa, and southeastern Minnesota. Locally common in the Pecatonica Member in north-central Illinois.

Genus PROTEOCERAS

Remarks. *Proteoceras* is one of the more common cephalopods in the Platteville Formation and is represented by several morphological types, including (1) a small, smooth-shelled, slender, slightly curved, orthoconic longicone with an expansion angle of 6°–7° and beadlike siphuncle segments, most similar to *Proteoceras tyronensis* (Foerste, 1912); (2) a form that is similar in size and shape to the one above but has microscopic longitudinal lirae developed on the adapical parts of the shell (here referred to as *Proteoceras* sp. 1); (3) straight, smooth-shelled forms with expansion angles ranging from 4° to 5° (*Proteoceras* sp. 2); and (4) a variety of small to moderately large forms differing in expansion angles, cameral lengths, cameral deposits, and relative position of the siphuncle.

Proteoceras cf. *tyronensis* (Foerste, 1912)

Figures 11-6-10 and 11-6-11

Description. Shell slender, slightly endogastrically curved, orthoconic longicone; smooth exterior; gradually expanding at an angle of 6°–7°. Circular in cross section with straight, transverse sutures. Typical specimens are 20–30 mm in diameter at the body chamber. Cameral chambers decrease in relative length with maturity. Siphuncle small; subcentrally positioned in shells less than 15 mm in diameter, becoming more centrally positioned as shells approach 25 mm or more in diameter. Cameral deposits present in adapical parts of the phragmocone.

Remarks. This Platteville species has a small, slender, smooth shell with an expansion angle of 6°–7° and appears to be similar in most aspects to *Proteoceras tyronensis* (Foerste, 1912) from the coeval Tyrone Limestone of Kentucky.

Occurrence. Common in the Mifflin and Grand Detour Members in northern Illinois and southern Wisconsin and the McGregor Member of southwestern Wisconsin, northeastern Iowa, and southeastern Minnesota. Locally common in the Pecatonica Member in north-central Illinois.

Proteoceras sp. 1

Figure 11-6-12

Description. Shell slender, slightly endogastrically curved, orthoconic longicone; gradually expanding at an angle of 6°–7°; circular in cross section with straight, transverse sutures. Distinguished by having abundant longitudinally positioned microscopic lirae best developed on the adapical parts of the shell.

Remarks. Other than the presence of lirae, *Proteoceras* sp. 1 is essentially identical to the Platteville species *Proteoceras* cf. *tyronensis*. *Proteoceras* sp. 1 is less common than *Proteoceras* cf. *tyronensis*.

Occurrence. Rare locally in the Mifflin and Grand Detour Members in northern Illinois and southern Wisconsin.

Proteoceras sp. 2

Figure 11-6-13

Description. Shell, orthoconic longicone; smooth exterior; gradually expanding at an angle of 4°–5°. Circular in cross section with straight, transverse sutures. Largest observed specimen is 40 mm in length and 6 mm in diameter adorally. Siphuncle small; subcentrally positioned. Cameral deposits present.

Remarks. *Proteoceras* sp. 2 differs from other Platteville species of *Proteoceras* in having a small, slender, smooth shell with expansion angle of 4°–5°.

Occurrence. Common locally in the Mifflin Member in northern Illinois and southern Wisconsin.

Order ASCOCERIDA

Description. The ascocerids are small nautiloids characterized by unusually thin shells and septa as well as ontogenetic shell truncation. The conch consists of two distinct portions that are very rarely found together. The juvenile or deciduous part typically consisted of a slender cyrtocone that periodically was shed, leaving the more commonly found mature brevicone consisting of a living chamber and truncated phragmocone with a blunt end.

Genus REDPATHOCERAS

Redpathoceras* cf. *clarkei Flower, 1963

Figures 11-6-14 and 11-6-15

Description. Exogastric cyrtoconic brevicone; septum of truncation deeply rounded, with the edges so inclined forward that its suture is obscure. Reference specimen ISGS 757 is an internal mold of the living chamber and two attached camerae together measuring 60 mm in length. The shell expands no more than 2 mm in diameter from the basal living chamber septum, where it is 25 mm high and 26 mm wide to the aperture. The siphuncle is not exposed in the reference specimen.

Remarks. The overall size and unique shape of the shell with its deeply rounded septum of truncation are similar to *Redpathoceras clarkei* Flower, 1963, from the Black River Limestone of Québec.

Occurrence. Common in the Mifflin and Grand Detour Members in northern Illinois and southern Wisconsin.

Subclass ENDOCERATOIDEA

Description. The Endoceratoidea have medium-sized to very large shells that are straight and longiconic in most forms. They also have a large siphuncle containing a series of telescoped conelike structures called **endocones** occurring mainly in the posterior part of the phragmocone. Variations in the siphuncle, particularly the endocones, are important in distinguishing endoceratoid cephalopods (Flower 1955). Some Ordovician endoceratoids, such as *Cameroceras*, attained lengths of 6 m (Frey 1995), making these one of the largest invertebrate fossils in the Paleozoic rock record, if not the largest (Teichert 1964).

Order ENDOCERIDA

Description. Platteville endocerids are primarily large to giant longiconic shells with a large tubular siphuncle that is positioned at or near the venter. The siphuncles are filled apically with calcified, hollow endocones, each nested in the base of the next cone toward the apex of the shell. Camerae are very shallow, with closely spaced transverse sutures. Generally, the endocerids are difficult to separate into genera and species owing to the fragmentary nature of the fossils, modification of proportions by distortion, and uncertainty as to the extent of variation in proportions within a species (Flower 1958). *Endoceras* and *Cameroceras* are the most common endocerids in the Platteville Formation.

Family ENDOCERATIDAE

Genus ENDOCERAS

Endoceras annulatum Hall, 1847

Figure 11-7-3

Description. Large orthoconic longicone with annulations that are low and distinctly set apart; gradually expanding at an angle of 6°–7°; cross section slightly depressed; sutures straight and transverse. Siphuncle large, ventrally positioned but not intersecting the ventral shell.

Remarks. The large, straight shell with rounded annulations distinguishes *E. annulatum* from other Platteville endocerid cephalopods. Hall did not designate a type species for his new genus when it was first described in 1847. S. A. Miller (1889) later designated *E. annulatum* as the type species of *Endoceras* Hall. Thus, true *Endoceras* has an annulated shell exterior.

Occurrence. Common in the Mifflin and Grand Detour Members in northern Illinois and southern Wisconsin and the McGregor Member of southwestern Wisconsin, northeastern Iowa, and southeastern Minnesota. Locally common in the Pecatonica Member in north-central Illinois.

1
x0.33
siphuncle
2
x0.5
annuli
3
x0.75

Endoceras sp.

Figures 11-7-1 and 11-7-2

Description. Large, slender, orthoconic longicone lacking obvious annulations; very gradually expanding at an angle of 2°–3°; cross section slightly depressed; sutures straight and transverse. Estimates based on expansion angles of some fragmentary specimens indicate shell lengths of 3 m or more. Siphuncle large, ventrally positioned but not intersecting the ventral shell.

Remarks. This Platteville species of *Endoceras* is distinguished by its smooth, slender shell with an exceedingly small expansion angle of 2°–3°.

Occurrence. Common in the Mifflin and Grand Detour Members in northern Illinois and southern Wisconsin and the McGregor Member of southwestern Wisconsin, northeastern Iowa, and southeastern Minnesota. Locally common in the Pecatonica Member in north-central Illinois.

Genus CAMEROCERAS

Cameroceras cf. *alternatum* Flower, 1968

Figures 11-8-1 and 11-8-2

Description. Large to very large, straight conch with circular or slightly depressed cross section; gradually expanding at an angle of 5°–6°. Estimates based on expansion angles of some fragmentary specimens indicate shell lengths of 3 m or more. Sutures straight and transverse; septa rhythmically spaced with intervals of 3–5 mm in length separated by longer intervals in which camerae are 10–20 mm in length. Siphuncle as much as a third to half the diameter of the conch at maturity; siphuncle depressed, intersects ventral margin of shell; simple endocones.

Remarks. This Platteville *Cameroceras* is similar in most features to *C. alternatum* (Flower, 1968) from the coeval Black River Limestone at St.-Raymond, Québec, Canada.

Occurrence. Common in the Mifflin and Grand Detour Members in northern Illinois and southern Wisconsin and the McGregor Member of southwestern Wisconsin, northeastern Iowa, and southeastern Minnesota. Locally common in the Pecatonica Member in north-central Illinois.

Figure 11-7 (Opposite page)
1, 2. ***Endoceras* sp.** (**1**) Three-piece specimen (fits end to end, as shown with red arrows) showing the characteristic slender shell with an exceedingly small expansion angle of 2°–3°; photograph by John A. Catalani; used with permission; (**2**) interior surface of a septum showing the oval siphuncle positioned above the ventral shell wall. Mifflin Member, Dixon, Illinois (locality 3), **1**, JAC 158. **2**, ISGS 767.

3. ***Endoceras annulatum*** Hall. (**3**) Relatively large specimen showing the characteristic low, rounded annulations of this species. Mifflin Member, Magnolia, Wisconsin (locality 34), **3**, JAC 159.

Genus NANNO

Remarks. The genus *Nanno* was proposed by Clarke (1897a) and interpreted by him as preseptal apical ends of cephalopod siphuncles, citing the structural similarity to the siphuncles of *Cameroceras* and *Endoceras*. In an analysis of Chazyan and Mohawkian endocerids, Flower (1958) advanced the idea that *Nanno* type fossils are the free apical ends of endocerid siphuncles. At the present time, the general lack of definitive fossils showing the complete conch with intact apical ends makes it difficult to assert the generic identity of *Nanno*, or other free apical ends, with known endocerid genera. Apical ends of endocerid siphuncles found in the Platteville Formation are represented by at least two forms, one being the typical *Nanno* type and the other a straight, narrow cone. Assignment of these fragmentary fossils to the endocerids is based on their longiconic conch, large siphuncle positioned at or near the venter, and presence of endocones.

Nanno cf. *aulema* Clarke, 1897a

Figures 11-8-3 through 11-8-7

Description. Small conch with preseptal cone that continues forward as the siphuncle. The cone expands at an angle of about 30° for approximately 15 mm along the dorsal outline and then contracts sharply at the first septum, continuing forward as the ventrally positioned siphuncle. At the first septum, the expansion angle decreases to 7°–9°. Preserved camerae vary in number from four to six. Some specimens have as many as 10 camerae and a body chamber. The siphuncle is depressed and oval in cross section; the ventral surface is in contact with the outer shell wall.

Remarks. Platteville specimens are very similar to *Nanno aulema* (Clarke, 1897a) from the Decorah Formation at Minneapolis, Minnesota. This preseptal siphuncle is similar to that of *Endoceras* (Robert C. Frey, personal communications, 2019).

Occurrence. Common in the Mifflin and Grand Detour Members in northern Illinois and southern Wisconsin and the McGregor Member of southwestern Wisconsin, northeastern Iowa, and southeastern Minnesota. Locally common in the Pecatonica Member in north-central Illinois.

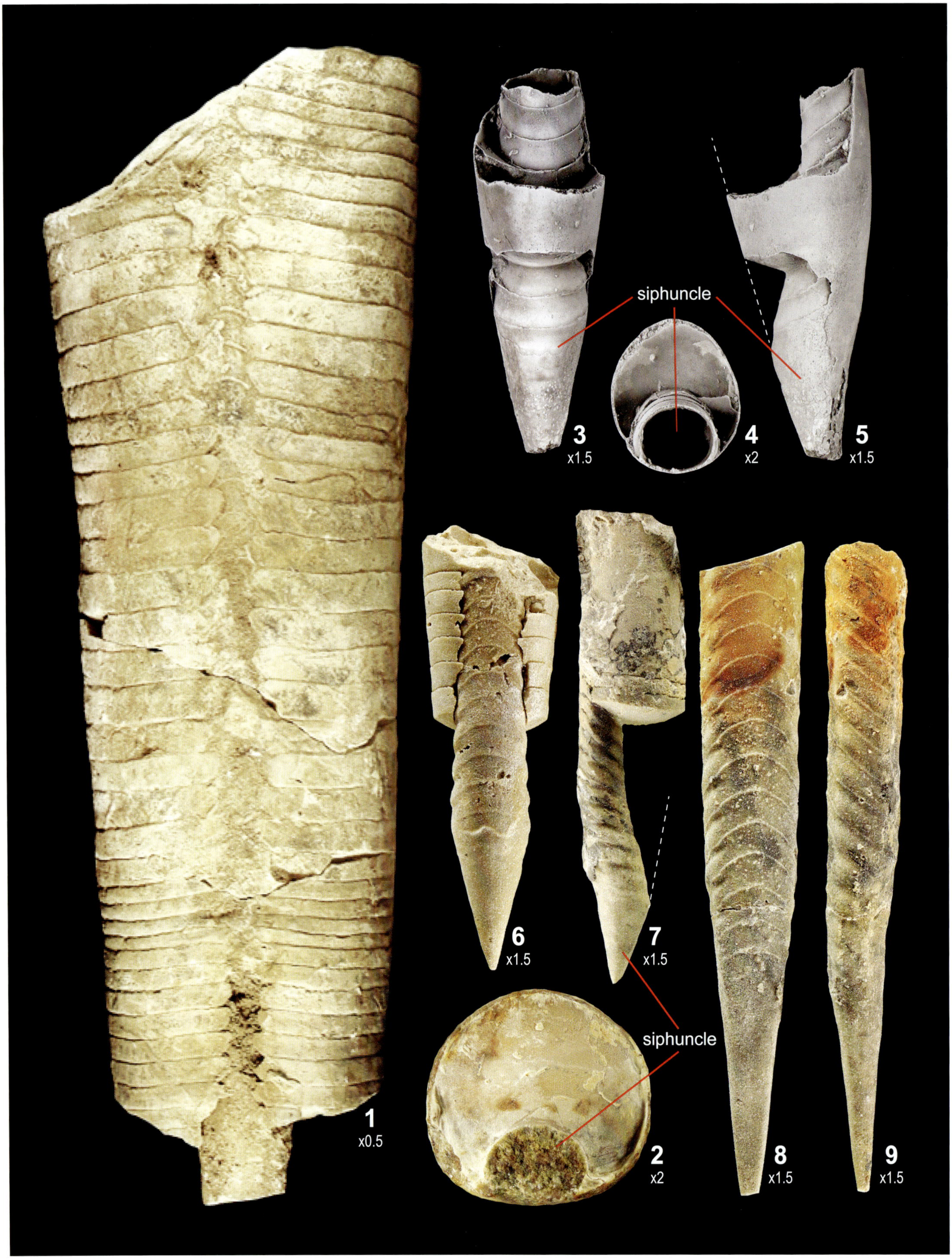
siphuncle
3
x1.5
4
x2
5
x1.5
6
x1.5
7
x1.5
siphuncle
1
x0.5
2
x2
8
x1.5
9
x1.5

Endocerid Apical Siphuncle sp. 1

Figures 11-8-8 and 11-8-9

Description. Straight, slender, conical, slightly depressed, apical siphuncle; apical end pointed; oval in cross section; apical angle of expansion 13°, rapidly diminishing to less than 3° within a distance of 35 mm of the apical tip. Length in excess of 75 mm, 13 mm in diameter. The ventral surface of the siphuncle has distinct forward-projecting lines marking the intersection with the septa in a pattern similar to other Ordovician endocerids.

Remarks. The characteristic slender siphuncle with camerae extending to the apical tip is most like those in *Cameroceras* (Robert C. Frey, personal communications, 2019).

Occurrence. Common in the Mifflin and Grand Detour Members in northern Illinois and southern Wisconsin and the McGregor Member of southwestern Wisconsin, northeastern Iowa, and southeastern Minnesota.

Subclass ACTINOCERATOIDEA

Description. Medium to large, straight conch; cross section broadly elliptical; some are strongly compressed, with a flat venter. Body chamber slightly contracted. Massive calcareous endosiphuncular deposits are well developed, particularly inside siphuncle segments toward the apex of the shell. Siphuncles typically large, subcentral to marginal in position; siphuncle segments broadly inflated between septa.

Figure 11-8 (Opposite page)

1, 2. ***Cameroceras* cf. *alternatum*** Flower. (**1**) Ventrum of large specimen showing integration of siphuncle with the ventral surface and alternating thick and thin cameral chambers; (**2**) interior surface of a septum showing oval siphuncle integrated with the ventral shell wall. Mifflin Member, Avon, Wisconsin (locality 32), **1**, JAC 155. Mifflin Member, Dixon, Illinois (locality 3), **2**, ISGS 770.

3–7. ***Nanno* cf. *aulema*** Clarke. (**3**) Dorsal view of apical end with several cameral chambers missing exposing the siphuncle; (**4**) interior surface of a septum showing the oval-shaped siphuncle integrated with the ventral shell wall; (**5**) lateral view of apical end; dashed lines show the projected surface of the shell; (**6**) dolomitized steinkern; ventral view of the apical end showing integration of the siphuncle and shell wall; (**7**) dolomitized steinkern; lateral view of the apical end; dashed line shows the projected surface of the shell. Mifflin Member, Dixon, Illinois (locality 2), **3–5**, ISGS 764. Mifflin Member, Lee Center, Illinois (locality 31), **6**, **7**, JAC 156.

8, 9. **Endocerid Apical Siphuncle sp. 1** (**8**) Dolomitized steinkern, ventral surface of siphuncle showing upward arch impressions of septal rings; (**9**) lateral view showing distinctly pointed apical end of siphuncle. Mifflin Member, Lee Center, Illinois (locality 31), **8**, **9**, JAC 157.

Order ACTINOCERIDA

Genus ACTINOCERAS

Remarks. Descriptions of species of *Actinoceras* are based mainly on the rate of shell expansion, cameral length, siphuncle position, siphuncle segment shape, and siphuncle diameter relative to the shell diameter (Frey 1995). Platteville actinocerids that have been described in the literature include *A. beloitense* (Whitfield, 1882) and *A. janesvillense* (Foerste, 1933), the most clearly recognizable species being *A. beloitense*. The Platteville species *A. janesvillense* is described by Foerste (1933) as being the most similar to *A. beloitense* but differing from the latter in having longer septal necks, smaller connecting rings, and a siphuncle that shows very little difference in its diameter within the length preserved. However, available specimens show considerable morphologic gradation between these two species, creating some uncertainty for the species concept of *A. janesvillense*. Flower (1957) noted that in addition to *A. beloitense* and *A. janesvillense*, the Platteville Formation has yielded *A. turinense* and at least two forms that are comparable to *A. bigsbyi*. A thorough analysis of the Platteville actinocerid species is needed but is beyond the scope of this report. Only *A. beloitense* is described here.

Actinoceras beloitense Whitfield, 1877

Figures 11-9-1 through 11-9-6

Description. Medium to large, smooth, fusiform shell; decreasing in diameter from anterior part of phragmocone toward aperture; complete conchs commonly in excess of 50 cm in length and 12 cm at widest part of conch; cross section circular, gradually becoming oval with growth. Approximately five camerae occur in a length equal to the lateral width of the conch at the contact of the body chamber and phragmocone; sutures on ventral surface curve toward the posterior end. Siphuncle large, positioned near the ventral surface but not intersecting the venter; siphuncle occupies the greater part of the initial conch, decreasing in diameter toward adult portion of the phragmocone; septal necks relatively long. Endosiphuncule deposits extensive in the apical portions of the conch; particularly heavy on the connecting rings, which can be greatly expanded between septa.

Remarks. A detailed comparison of *Actinoceras* species, including *A. beloitense*, can be found in Frey (1995).

Occurrence. Common in the Mifflin and Grand Detour Members in northern Illinois and southern Wisconsin and the McGregor Member of southwestern Wisconsin,

1
x0.5
2
x0.5
3
x0.5
siphuncle
4
x0.5
siphuncle
5
x0.5
6
x1

northeastern Iowa, and southeastern Minnesota. Locally common in the Pecatonica Member in north-central Illinois.

Gonioceras occidentale Hall, 1861

Figures 11-10-1 through 11-10-5

Description. Moderately large, straight, rapidly expanding smooth shell, strongly depressed in cross section with ventral side flat to slightly convex and dorsal side fully convex; both sides meet laterally at an acute angle; aperture contracted; septa deeply concave, closely arranged, septal sutures curve strongly downward toward the apex along the longitudinal axis of the venter and dorsum, then curve abruptly upward before reversing direction again along the flanks, producing an overall distinct trilobate pattern. At the large end of the growth spectrum, specimens may be as much as 20 cm long and 15 cm wide. Siphuncle close to the ventral wall of the conch; lateral diameter typically 8–10 mm; dorsoventral diameter 7–8 mm; connecting rings thin, with convex margins and cleft along the ventral axis.

Remarks. The type specimens of *G. occidentale* are from the Platteville Formation at Platteville, Wisconsin. Additional specimens were described by Foerste (1933) from the Platteville Formation at Mineral Point, Wisconsin; Dixon, Illinois; and Minneapolis, Minnesota.

Occurrence. Abundant in the Mifflin and Grand Detour Members and equivalent rocks throughout the outcrop belt; particularly common in the lower part of the Grand Detour Member in northern Illinois and southern Wisconsin.

Figure 11-9 (Opposite page)
1–6. ***Actinoceras beloitense*** Whitfield. (**1**) Dorsal surface of a specimen with most of the phragmocone and body chamber intact and illustrating the distinct fusiform body shape; (**2**) ventral surface of specimen showing part of the phragmocone and most of the body chamber; (**3**) lateral view of same specimen as **2** showing the depressed conch; (**4**) interior surface of a septum showing the oval-shaped siphuncle positioned above the ventral surface; (**5**) dolomitized steinkern of a partial phragmocone showing the characteristically large siphuncle; (**6**) well-preserved specimen showing a phragmocone with the dorsal shell removed exposing the thickened segments of the siphuncle. Mifflin Member, Dixon, Illinois (locality 2), **1**, JAC 160. Grand Detour Member, Beloit, Wisconsin (locality 21), **2**, **3**, JAC 161. Grand Detour Member, Dixon, Illinois (locality 3), **4**, ISGS 774. Grand Detour Member, Durand, Illinois (locality 11), **5**, ISGS 8703. Mifflin Member, Dixon, Illinois (locality 2), **6**, ISGS 771.

Subclass NAUTILOIDEA

Order TARPHYCERIDA

Description. The tarphycerids are mostly coiled conchs with a tendency to uncoil during late stages of conch formation. Primitive forms are ornamented by transverse annulations and fine lirae. Siphuncle is thin walled and tubular; central or subcentral position. Lacks cameral deposits.

Suborder BARRANDEOCERINA

Family BARRANDEOCERATIDAE

Genus CENTROCYRTOCERAS

Centrocyrtoceras cf. *annulatum* (Hall, 1847)

Figures 11-11-5 through 11-11-8

Description. Small, strongly curved gyrocone. Circular in cross section with straight, transverse sutures. The largest observed specimen is 90 mm long and has a dorsoventral diameter that enlarges from 2 mm at the apical end to 16 mm at the adoral end. Outer surface of conch crossed by transverse annulations with additional ribs intercalated on the ventral side; fine parallel ridges on the annulations and intcrannular depressions; annulations decrease in prominence adorally. Siphuncle small, subcentral, thin walled; segments cylindrical to subcylindrical.

Remarks. This Platteville species is similar in most respects to *C. annulatum* from the Trenton Limestone at Middleville, New York.

Occurrence. Common in the Mifflin and Grand Detour Members in northern Illinois and southern Wisconsin.

Centrocyrtoceras sp.

Figures 11-11-1 through 11-11-4

Description. Small, strongly curved, compressed gyrocone. Oval in cross section with straight, transverse sutures. Largest specimens exceed 150 mm in length; dorsoventral diameter enlarges from 2 mm at the apical end to 28 mm at the adoral end. Outer surface of conch crossed by transverse annulations, with additional ribs intercalated on the ventral side; fine parallel ridges on the annulations and interannular depressions; annulations decrease in prominence adorally. Siphuncle small, subcentral, thin walled; segments cylindrical to subcylindrical.

Remarks. *Centrocyrtoceras* sp. is similar in most respects to the Platteville species *C.* cf. *annulatum* but differs from the latter in having a distinctly compressed conch.

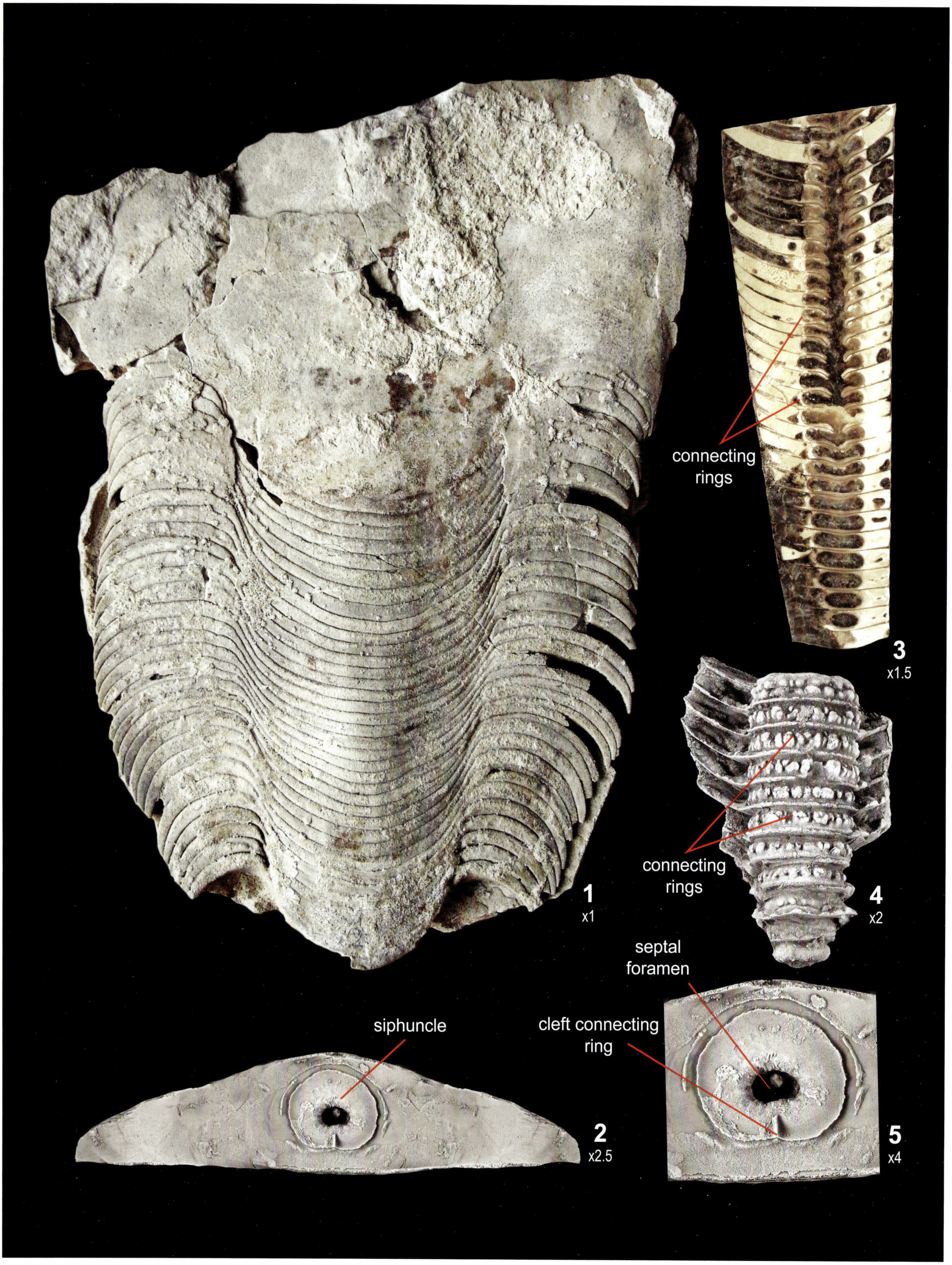
connecting
rings
3
x1.5
connecting
rings
4
x2
1
x1
septal
foramen
siphuncle
cleft connecting
ring
2
x2.5
5
x4

Occurrence. Common in the Pecatonica and Mifflin Members in northern Illinois.

Suborder TARPHYCERINA

Family PLECTOCERIDAE

Genus PLECTOCERAS

Plectoceras robertsoni (Hall, 1861)

Figures 11-11-9 through 11-11-11

Description. Conch tightly coiled (tarphyconic); whorls in contact but not deeply impressed; adorally phragmocone loses impressed zone, and mature body chamber is free; expansion angle 7°–8°. Venter flat to slightly convex and dorsum flat to slightly concave at impressed zone; lateral sides broadly convex; shell width greater than height. Outer surface of shell marked by faint undulating folds that arch backward in an apical direction from the dorsum to the venter. Growth lines parallel to the obscure folds. Septa transverse with downward curvature laterally. Siphuncle small, tubular, thin walled; less than 1 mm from the ventral surface.

Remarks. *P. robertsoni* differs from *P. occidentale* in having a smaller conch, slightly larger angle of expansion, more closely spaced septa, shallower impressed zone, more convex sides, and siphuncle positioned close to the venter.

Occurrence. The type specimens of *P. robertsoni* were described by Hall (1861) as being from the "Buff limestone of the Trenton limestone group at Beloit, Wisconsin, and Rockford, Illinois." In current stratigraphic classification, this would be approximately equal to the Mifflin and Grand Detour Members where *P. robertsoni* is common. Specimens are also common in the upper part of the Pecatonica Member at Dixon, Lee County, Illinois.

Figure 11-10 (Opposite page)
1–5. ***Gonioceras occidentale*** Hall. (**1**) Ventral surface of a well-preserved specimen showing the distinctive trilobed pattern created by the curved septa; photograph by John A. Catalani; used with permission; (**2**) interior surface of a septum showing the highly depressed shell and ventrally positioned siphuncle; oval siphuncle positioned above the ventral shell wall; (**3**) longitudinal polished section through the central axis of the siphuncle showing the thin connecting rings with convex margins; (**4**) phragmocone with dorsal shell removed exposing the connecting rings of the siphuncle; (**5**) interior surface of a septum showing the septal foramen and distinctive cleft ventral edge of a connecting ring. Grand Detour Member, South Beloit, Illinois (locality 20), **1**, JAC 162. Mifflin Member, Dixon, Illinois (locality 2), **2**, **5**, ISGS 775; **3**, ISGS 776; **4**, ISGS 777.

Plectoceras occidentale (Hall, 1861)

Figures 11-12-1 through 11-12-3

Description. Large, coiled conch; whorls in contact and impressed on the dorsal surface; expansion angle 6°–7°; commonly exceeding 25 cm in diameter. Venter flat to slightly convex and dorsum markedly concave at impressed zone; lateral sides slightly convex; shell height and width approximately equal. Outer surface of shell marked by faint undulating folds that arch backward in an apical direction from the dorsum to the venter. Growth lines parallel to the obscure folds. Septa concave, transverse with downward curvature laterally. Siphuncle small, tubular, thin walled; positioned above the venter at a distance of approximately 2.5 times the siphuncle diameter.

Remarks. Compared with *P. robertsoni*, *P. occidentale* has a larger conch, flatter sides, smaller expansion angle, more widely spaced septa, more deeply impressed dorsum, and siphuncle that is much closer to the center of the shell.

Occurrence. Particularly common in the Grand Detour Member of northern-central Illinois and south-central Wisconsin. Specimens are also common in the upper part of the Pecatonica Member at Dixon, Illinois.

Order ONCOCERIDA

Description. The oncocerids are compressed, exogastric cyrtocones and brevicones with short camerae and simple transverse or slightly curved sutures as well as relatively large body chambers. They are the most diverse cephalopods in the Platteville Formation. Particularly notable are *Oncoceras* and *Beloitoceras*. Cephalopod specialists have split these genera into numerous species, some of which are based on a small number of fragmentary specimens. Furthermore, the noted differences are relatively subtle, making it difficult confidently to assign specimens to a species, in some cases even at the generic level (Flower 1946). Only a few of the more distinct species are discussed and illustrated here.

Family ONCOCERATIDAE

Genus ONCOCERAS

***Oncoceras* spp.**

Figures 11-13-1 through 11-13-5

Description. Small (estimated length 4–6 cm, maximum shell diameter 2–3 cm), exogastric, slightly compressed, cyrtoconic brevicone; maximum shell convexity near contact between body chamber and phragmocone; from this

1
x2
2
x2.5
5
x2
lirae
annuli
x5
x5
3
x2
4
x2
6
x2
7
x2
annuli
8
x2
x5
lirae
9
x1
10
x1
11
x2

point, the shell gradually contracts toward the aperture; cross section of conch is only slightly compressed laterally along the phragmocone and basal part of body chamber but is distinctly compressed toward the aperture; septa straight and transverse. Shell exterior may be marked by longitudinal ridges. Siphuncle small, ventral; segments ovoid and longer than wide.

Remarks. *O. abruptum* from the Platteville Formation at Platteville and Beloit, Wisconsin, is the only documented species in the formation, but the brief description and illustrations provided by Whitfield (1895) are not sufficient to distinguish their fragmentary *O. abruptum* confidently from other species of *Oncoceras*. No attempt is made here to identify individual Platteville species of *Oncoceras*.

Occurrence. Common in the Grand Detour Member in northern Illinois and southern Wisconsin.

Genus BELOITOCERAS

Remarks. Some of the more identifiable species are described here, including *B. janesvillense*, *B. lycum*, *B. pandion*, and *B. plebeium*. Even in these species the differences are very minute, and *Beloitoceras* obviously needs a thorough investigation to understand what constitutes a species. Key features used in identification are the (1) degree of compression, (2) angle of expansion, (3) convexity of the shell at the juncture of the phragmocone and body chamber, and (4) overall shape of the body chamber and aperture. A significant part of the phragmocone and most of the body chamber are required for proper identification.

Figure 11-11 (Opposite page)
1–4. ***Centrocyrtoceras* sp.** (**1**) Interior surface of a septum showing the laterally compressed shell and subcentrally positioned siphuncle; (**2**) lateral view of a well-preserved specimen with intact apical tip of the phragmocone; photographically enlarged to show the low, rounded annuli and very fine lirae; (**3**) ventral surface showing the forward-arching annuli; (**4**) dorsal surface showing the interior surface of a septum and subcentrally positioned siphuncle. Mifflin Member, Dixon, Illinois (locality 2), **1**, ISGS 701; **2**, ISGS 681; **3**, ISGS 700; **4**, ISGS 725.

5–8. ***Centrocyrtoceras* cf. *annulatum*** (Hall). (**5**) Lateral view of a well-preserved specimen with phragmocone and most of body chamber intact; (**6**) ventral surface showing the annuli; (**7**) dorsal view showing the interior surface of the septum and near centrally positioned siphuncle; (**8**) lateral view and photographic enlargement of the annuli and lirae. Mifflin Member, Dixon, Illinois (locality 2), **5**, ISGS 675; **6**, **8**, ISGS 698; **7**, ISGS 699.

9–11. ***Plectoceras robertsoni*** (Hall). (**9**) Dolomitized steinkern of specimen with phragmocone and most of the body chamber; (**10**) well-preserved specimen with shell intact; (**11**) interior surface of a septum showing the ventrally positioned siphuncle and slightly concave dorsal surface. Grand Detour Member, South Beloit, Illinois (locality 20), **9**, JAC 163. Mifflin Member, Dixon, Illinois (locality 2), **10**, ISGS 779; **11**, ISGS 784.

Beloitoceras janesvillense Foerste, 1933

Figures 11-13-6 through 11-13-8

Description. Small, strongly curved, exogastric, compressed brevicone; ovate in cross section, becoming more compressed with ontogeny; maximum shell convexity just forward of the contact between body chamber and phragmocone; from this point, the shell gradually contracts toward the aperture and the lateral sides become flatter. Maximum lateral width of shell typically 20 mm. Dorsal outline concave for its entire length. Septa straight and transverse, forward-arching on the venter, sloping from the dorsum to venter in an adoral direction; cameral chambers short. Body chamber relatively short, laterally compressed, with well-developed hyponomic sinus ventrally. Shell smooth or with transverse growth lines. Siphuncle small, ventral; segments ovoid and longer than wide.

Remarks. The type specimen is from the Platteville Formation at Beloit, Wisconsin. *B. janesvillense* differs from other Platteville species in its greater compression, somewhat flattened sides of the body chamber, forward-arching septa on the venter, and more rapid rate of enlargement dorsoventrally.

Occurrence. Most abundant in the Grand Detour Member in northern Illinois and southern Wisconsin.

Beloitoceras lycum (Hall, 1861)

Figure 11-13-16 through 11-13-19

Description. Small, gently curved, exogastric, compressed brevicone; ovate in cross section, becoming more compressed with ontogeny; maximum shell convexity just forward of the contact between the body chamber and phragmocone; from this point, the shell gradually contracts toward the aperture. Dorsal outline concave for its entire length. Septa straight and transverse, sloping very gradually from the dorsum to venter in an adoral direction; cameral chambers short. Body chamber relatively short, laterally compressed, with well-developed hyponomic sinus ventrally. Shell smooth or with transverse growth lines. Siphuncle small, ventral; segments ovoid and longer than wide.

Remarks. *B. lycum* is described by Clarke (1897b) "to be quite common in the lower blue beds of the Trenton limestone at Janesville, Wisconsin. It also occurs in the Trenton

1
x2
siphuncle
2
x2
3
x1

at Preston and Minneapolis, Minnesota" (p. 801). By current stratigraphic nomenclature this would be equivalent to the Grand Detour Member of the Platteville Formation. Clarke (1897b) stated that *B. lycum* is distinguished from other species by its "subarcuate venter, broad but not ventricose [swollen] dorsal surface and very slight decrease in the diameter of the body chamber from its base to the constriction" (p. 801).

Occurrence. Most abundant in the Grand Detour Member in northern Illinois and southern Wisconsin.

Beloitoceras pandion (Hall, 1861)

Figures 11-13-9 through 11-13-11

Description. Small, robust, strongly curved, rapidly expanding, exogastric, compressed brevicone; broadly ovate in cross section, becoming more compressed with ontogeny; maximum shell convexity just forward of the contact between body chamber and phragmocone; from this point, the shell gradually contracts toward the aperture. Dorsum broad and slightly arched, venter significantly more curved. Dorsal outline concave for its entire length. Septa straight and transverse, slightly forward-arching on the venter, sloping from the dorsum to venter in an adoral direction; cameral chambers relatively long. Shell smooth or with transverse growth lines. Siphuncle small, ventral; segments ovoid and longer than wide.

Remarks. The differences in curvature of the outer and inner edges give the shell a distinctive protruding profile to the venter, particularly in the area where the phragmocone and body chamber connect. In addition, *B. pandion* differs from other Platteville species in having a robust and broad venter. The type specimens are from the Platteville Formation at Beloit, Wisconsin.

Occurrence. Most abundant in the Grand Detour Member in northern Illinois and southern Wisconsin.

Figure 11-12 (Opposite page)
1–3. ***Plectoceras occidentale*** (Hall). (**1**) Part of a phragmocone with siphuncle protruding from the apical segment; (**2**) interior surface of a septum showing the subcentrally positioned siphuncle and concave dorsal surface; (**3**) lateral view of a typical specimen. Grand Detour Member, Durand, Illinois (locality 11), **1**, **2**, ISGS 783. Everett Collection, Platteville Formation, Dixon, Illinois, **3**, ISGS 1531.

Beloitoceras plebeium (Hall, 1861)

Figures 11-13-12 through 11-13-15

Description. Small, curved, expanding, exogastric, compressed brevicone; broadly ovate in cross section, becoming more compressed with ontogeny; rapidly expanding in apical region, less rapidly in the middle, gradually contracting toward the aperture. Dorsal outline slightly concave for its entire length. Septa straight and transverse, slightly forward-arching on the venter, sloping from the dorsum to venter in an adoral direction. Shell smooth or with transverse growth lines. Siphuncle small, ventral; segments ovoid and longer than wide.

Remarks. The type specimens are from the Platteville Formation ("Buff limestone" = Pecatonica and Mifflin Members) at Beloit, Wisconsin. *B. plebeium* differs from other Platteville species in having a shell that expands abruptly at the apical end, less so in the middle, gradually contracting through the adoral part of the phragmocone and on through the body chamber.

Occurrence. Most abundant in the Grand Detour Member in northern Illinois and southern Wisconsin.

Genus CYRTORIZOCERAS

Cyrtorizoceras* cf. *minneapolis (Clarke, 1897b)

Figures 11-13-20 through 11-13-23

Description. Small, strongly curved, rapidly expanding, exogastric, slightly compressed brevicone; broadly ovate in cross section throughout ontogeny; lateral sides of phragmocone expand at a 30° angle; slight contraction of the body chamber. Dorsal outline deeply concave along its entire length. Septa straight and transverse, sloping from the dorsum to venter in an adoral direction. Shell exterior has fine parallel transverse lirae. Siphuncle small, ventral.

Remarks. The strongly curved, rapidly expanding, slightly compressed shell suggests that this oncocerid has affinities with *Cyrtorizoceras* and is comparable to *C. minneapolis* from the Platteville Formation of Minnesota.

Occurrence. Common in the Mifflin and Grand Detour Members in north-central Illinois and south-central Wisconsin and the McGregor Member of southwestern Wisconsin, northeastern Iowa, and southeastern Minnesota. Locally common in the Pecatonica Member in north-central Illinois.

Genus LOGANOCERAS

Loganoceras cf. *regulare* (Billings, 1857)

Figures 11-13-24 and 11-13-25

Description. Small, smooth, strongly curved, exogastric cyrtocone; not compressed; circular in cross section throughout ontogeny; no hyponomic sinus. Sutures straight or curving slightly downward laterally and rising at an increasing rate in a ventral direction. Siphuncle small, ventral.

Remarks. This Platteville species of *Loganoceras* is similar in most features to *L. regulare* from the Leray Member of the Black River Formation at Allumette Island on the Ottawa River, Canada.

Occurrence. Common in the Mifflin and Grand Detour Members in north-central Illinois and south-central Wisconsin and the McGregor Member of southwestern Wisconsin, northeastern Iowa, and southeastern Minnesota.

Genus RICHARDSONOCERAS

Richardsonoceras simplex (Billings, 1857)

Figures 11-14-1 through 11-14-4

Description. Gradually enlarging, strongly curved, compressed, exogastric cyrtocone; ovate in cross section; lacking a convex conch like *Beloitoceras* and enlarging less rapidly in a dorsoventral direction, thus more elongate; dorsal concave outline of phragmocone continues uninterrupted over the body chamber; aperture not contracted. Radius of curvature of convex outline equaling 30–40 mm. Adult conch as much as 120 mm long when measured along the ventral surface; dorsoventral diameter approximately 25 mm at mid-height of body chamber. Cross section more narrowly rounded ventrally than dorsally. At the contact between the body chamber and phragmocone where the dorsoventral diameter is 25 mm, seven camerae occur in a corresponding length when counted along the ventral outline. Septal sutures curve downward laterally about 2–3 mm; septa forward-arching on the venter, sloping from the dorsum to venter in an adoral direction. Body chamber lacks a well-developed hyponomic sinus ventrally. Shell smooth or with fine transverse growth lines. Siphuncle slightly larger than *Beloitoceras*, with ovoid segments that are longer than wide.

Remarks. The type specimens of *Richardsonoceras simplex* are from the Black River Limestone in Nepean Township, Carleton County, Ontario, Canada.

Occurrence. Locally abundant in the Grand Detour Member in northern Illinois and southern Wisconsin.

Richardsonoceras beloitense Foerste, 1933

Description. Essentially the same features as *R. simplex*, except in *R. beloitense* the lengthwise curvature of the conch is much less, the radius of curvature along the convex outline equaling approximately 70 mm.

Remarks. The type specimens were described by Foerste (1933) as being from the Platteville Formation at Samp's Quarry near Beloit, Wisconsin.

Occurrence. Locally common in the Grand Detour Member in northern Illinois and southern Wisconsin.

Figure 11-13 (Opposite page)

1–5. ***Oncoceras* spp.** (**1**) Dolomitized steinkern showing the ventral surface of a fragmentary phragmocone and body chamber; (**2**, **3**) dolomitized steinkern showing the ventral and lateral view, respectively; (**4**, **5**) dolomitized steinkern showing a photographically restored view of the dorsum and lateral view, respectively. Mifflin Member, Rochelle, Illinois (locality 35), **1**, JAC 164; **2**, **3**, JAC 165; **4**, **5**, JAC 166.

6–8. ***Beloitoceras janesvillense*** Foerste. (**6**) Lateral view of a dolomitized steinkern; (**7**) same specimen, view of the dorsum; (**8**) same specimen, view of the venter showing the forward-arching septal sutures and relatively great compression of the conch. Grand Detour Member, South Beloit, Illinois (locality 20), **6–8**, JAC 167.

9–11. ***Beloitoceras pandion*** (Hall). (**9**) Adoral view of a dolomitized steinkern showing the outline of the aperture and hyponomic sinus; (**10**) same specimen, lateral view; (**11**) same specimen, venter showing the characteristic robust conch. Mifflin Member, Lee Center, Illinois (locality 31), **9–11**, JAC 168.

12–15. ***Beloitoceras plebeium*** (Hall). (**12**) Lateral view of a dolomitized steinkern showing the typical slight curvature of the dorsum; (**13**) same specimen, view of the dorsum; (**14**) same specimen, view of the venter; (**15**), same specimen showing the apical end of a septum and ventral position of the siphuncle. Grand Detour Member, Beloit, Wisconsin (locality 21), **12–15**, JAC 169.

16–19. ***Beloitoceras lycum*** (Hall). (**16**) Dorsum of a dolomitized steinkern; (**17**) same specimen, lateral view; (**18**) same specimen, venter showing a partially exposed siphuncle and characteristic straight sutures; (**19**) same specimen showing the apical end of a septum. Grand Detour Member, Beloit, Wisconsin (locality 21), **16–19**, JAC 170.

20–23. ***Cyrtorizoceras* cf. *minneapolis*** (Clarke). (**20**) Lateral view showing the typical rapid expansion of the conch, particularly at the adoral end; (**21**) interior surface of a septum showing the oval outline of the conch and ventrally positioned septal foramen; (**22**) dorsal view showing the rapid expansion of the conch; (**23**) same specimen, lateral view showing the relatively short radius of the dorsal outline (left side). Mifflin Member, Dixon, Illinois (locality 2), **20**, ISGS 794; **21**, ISGS 799; **22**, **23**, ISGS 801.

24, 25. ***Loganoceras* cf. *regulare*** (Billings). (**24**) Adoral end of a dolomitized steinkern showing the nearly circular aperture; (**25**) same specimen, lateral view showing the typical strongly curved conch outline and straight sutures. Grand Detour Member, Beloit, Wisconsin (locality 21), **24**, **25**, JAC 171.

1
x1
2
x1
3
x1
4
x1
5
x1
6
x1
7
x1
8
x1
9
x1
hyponomic
sinus
10
x1
11
x1
12
x1
13
x1
14
x1
15
x1
16
x1
17
x1
18
x1
19
x1
20
x1.5
21
x1.5
22
x1.5
23
x1.5
24
x1
25
x1

Genus ZITTELLOCERAS

Zittelloceras beloitense Foerste, 1928

Figures 11-14-13 through 11-14-17

Description. Gradually enlarging, strongly curved, slightly depressed, exogastric cyrtocone; cross section circular except for very slightly depressed dorsum. Surface ornamentation with crenulate transverse frills; on the median part of the venter, the frills curve adapically to form a relatively narrow V-shaped hyponomic sinus. Septa faintly curved downward laterally, rising toward the venter. Siphuncle small, ventral; segments round.

Remarks. The type specimens are from the Platteville Formation at Beloit, Wisconsin (Foerste 1928).

Occurrence. Common locally in the Grand Detour and Mifflin Members of northern Illinois and southern Wisconsin.

Genus ZITTELLOCERAS

Zittelloceras clarkeanum Foerste, 1917

Figures 11-14-18 through 11-14-22

Description. Essentially the same features as *Z. beloitense*, except in *Z. clarkeanum* the conch is more strongly curved.

Remarks. The type specimens are from the Platteville Formation at Janesville, Wisconsin.

Occurrence. Common locally in the Grand Detour and Mifflin Members of northern Illinois and southern Wisconsin.

Family TRIPTEROCERATIDAE

Genus TRIPTEROCERAS

Tripteroceras planoconvexum (Hall, 1861)

Figures 11-14-9 through 11-14-12

Description. Small (estimated length approximately 12 cm, maximum shell width 4 cm), orthoconic to slightly cyrtoconic longicone; depressed triangle in cross section; venter broad and flat, lateral angles acute; dorsum divided by median ridge or keel; dorsolateral regions flat to slightly concave; shell exterior smooth except for fine growth lines; long body chamber; cameral chambers short; expansion angle 12°–14°. Siphuncle narrow, tubular, positioned at ventral shell margin.

Remarks. With considerable detail, Foerste (1932) described *T. planoconvexum* from the Platteville Formation at Beloit, Wisconsin. This species was also reported by Whitfield (1895) from the Trenton limestone at Mineral Point and Beloit, Wisconsin. A second species, *T. oweni*, based on a single fragmentary specimen from the Trenton limestone at Cannon Falls, Minnesota, was described by Clarke (1897b, p. 793) but is not well enough preserved to distinguish it confidently from *T. planoconvexum*.

Occurrence. Common in the Mifflin and Grand Detour Members in north-central Illinois and south-central Wisconsin and the McGregor Member of southwestern Wisconsin, northeastern Iowa, and southeastern Minnesota. Locally common in the Pecatonica Member in north-central Illinois.

Figure 11-14 (Opposite page)

1–4. ***Richardsonoceras simplex*** (Billings). (**1**) Dolomitized steinkern showing the venter and partially exposed segments of the siphuncle; (**2**) dolomitized steinkern showing the dorsum; (**3**) interior surface of a septum showing the relatively large siphuncle and overall oval shape of the conch; (**4**) lateral view of a dolomitized steinkern showing the characteristic 30–40 mm radius of curvature of the convex outline. Grand Detour Member, Beloit, Wisconsin (locality 21), **1**, JAC 172; **2**, JAC 173; **3**, ISGS 805; **4**, JAC 174.

5–8. ***Manitoulinoceras* cf. *neleum*** (Hall). (**5**) Dolomitized steinkern, view of the dorsum; (**6**) same specimen, view of the venter; (**7**) same specimen, interior surface of a septum showing the ventrally positioned siphuncle and overall circular conch with a slightly flattened dorsum; (**8**) lateral view of a phragmocone and most of the body chamber. Grand Detour Member, Durand, Illinois (locality 11), **5–7**, JAC 175. Grand Detour Member, Beloit, Wisconsin (locality 21), **8**, JAC 176.

9–12. ***Tripteroceras planoconvexum*** (Hall). (**9**) Interior surface of a septum showing a highly depressed triangular shell and ventrally positioned siphuncle; (**10**) dolomitized steinkern showing the venter; (**11**) dolomitized steinkern showing the long body chamber and central rounded keel; (**12**) same specimen as **11** showing lateral view of a depressed conch. Mifflin Member, Dixon, Illinois (locality 2), **9**, ISGS 719. Grand Detour Member, Beloit, Wisconsin (locality 21), **10**, ISGS 791. Grand Detour Member, Rochelle, Illinois (locality 36), **11**, **12**, ISGS 718.

13–17. ***Zittelloceras beloitense*** Foerste. (**13**) Lateral view showing crenulated transverse frills; (**14**) same specimen as **13** showing the venter with frills curved adapically, forming a relatively narrow V-shaped hyponomic sinus; (**15**) interior surface of a septum showing the ventrally positioned siphuncle and overall circular conch; (**16**) lateral view of a dolomitized steinkern; (**17**) same specimen as **16** showing the venter. Mifflin Member, Dixon, Illinois (locality 2), **13**, **14**, ISGS 696. Grand Detour Member, Beloit, Wisconsin (locality 21), **15**, ISGS 807; **16**, **17**, JAC 177.

18–22. ***Zittelloceras clarkeanum*** Foerste. (**18**) Venter of dolomitized steinkern; (**19**) same specimen as **18** showing lateral view of a partial phragmocone and body chamber; (**20**) venter of a well-preserved specimen showing crenulated transverse frills; (**21**) lateral view; (**22**) lateral view showing crenulated transverse frills and a strongly curved shell. Grand Detour Member, South Beloit, Illinois (locality 20), **18**, **19**, JAC 178. Mifflin Member, Dixon, Illinois (locality 2), **20**, **21**, ISGS 1026; **22**, ISGS 806.

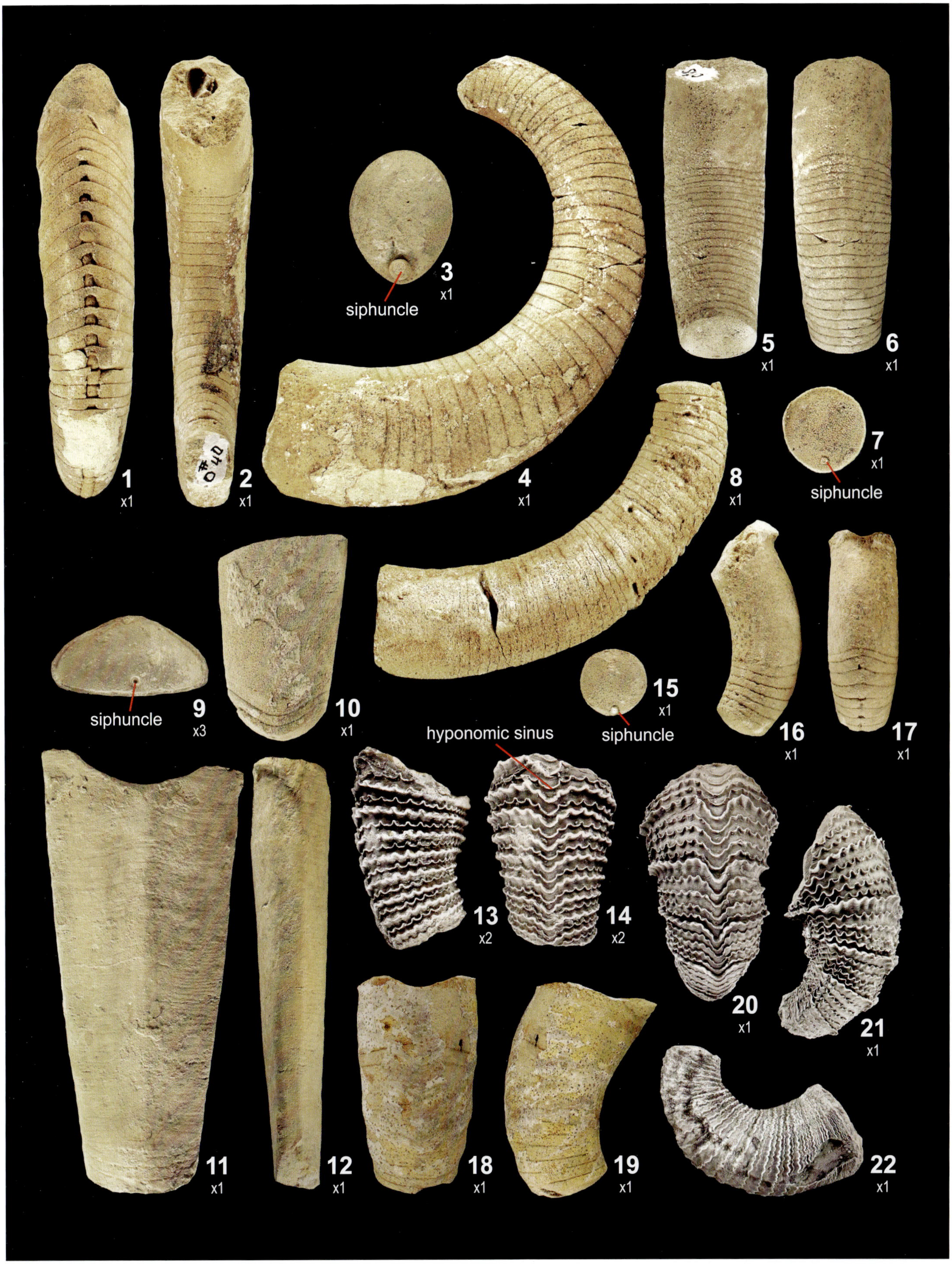
1
x1
2
x1
3
x1
siphuncle
4
x1
5
x1
6
x1
7
x1
siphuncle
8
x1
9
x3
siphuncle
10
x1
15
x1
siphuncle
16
x1
17
x1
hyponomic sinus
11
x1
12
x1
13
x2
14
x2
18
x1
19
x1
20
x1
21
x1
22
x1

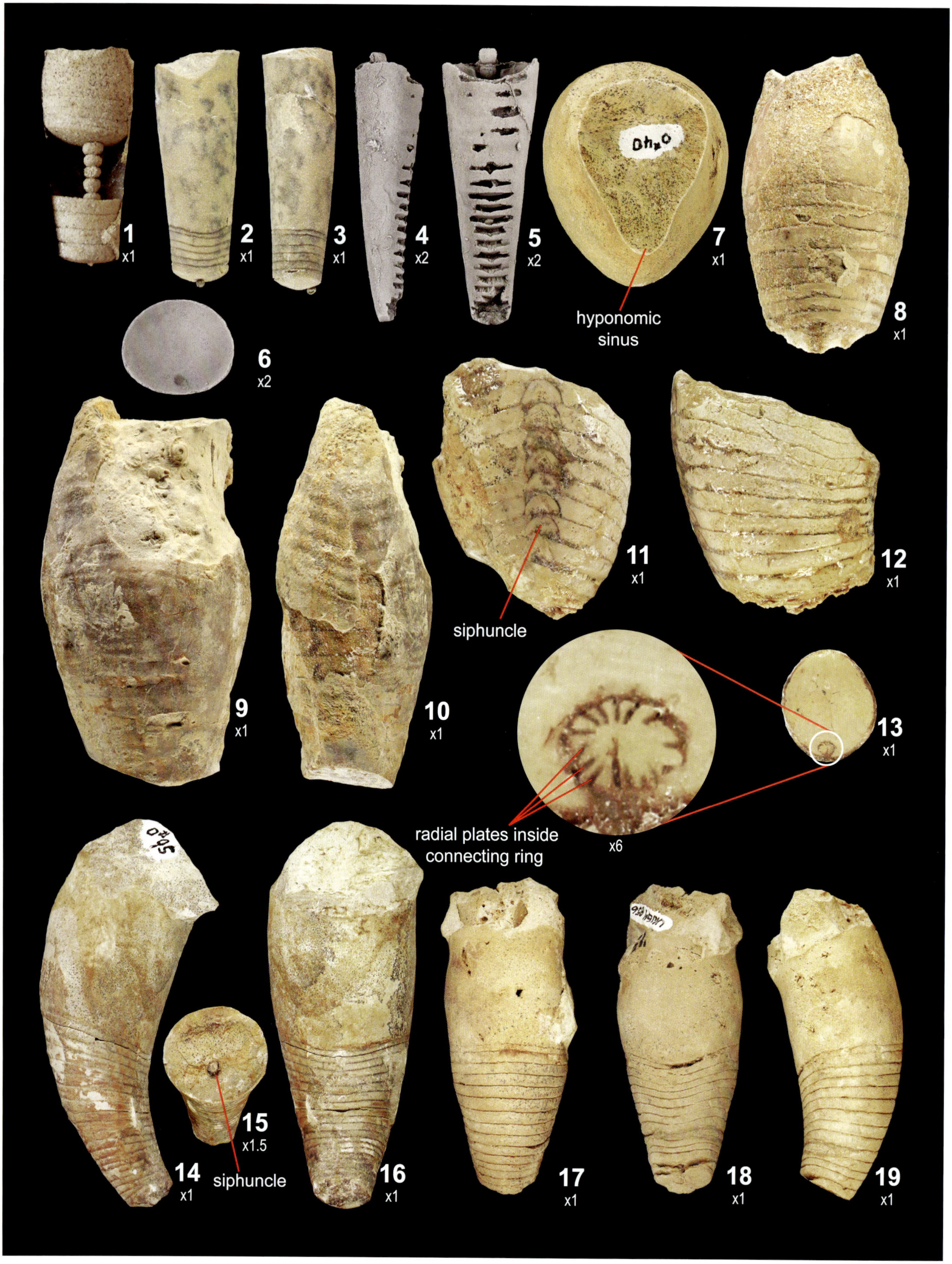
1
x1
2
x1
3
x1
4
x2
5
x2
7
x1
hyponomic
sinus
8
x1
6
x2
11
x1
12
x1
siphuncle
9
x1
10
x1
13
x1
radial plates inside
connecting ring
x6
15
x1.5
14
x1
siphuncle
16
x1
17
x1
18
x1
19
x1

Genus ALLUMETTOCERAS

Allumettoceras planodorsatum (Whitfield, 1882)

Figures 11-15-1 through 11-15-6

Description. Small (estimated length up to 10 cm, with maximum shell diameter of 2.5 cm), orthoconic to slightly cyrtoconic longicone; depressed ovate in cross section with the dorsum more convex than the venter; shell exterior smooth except for fine growth lines; long body chamber; cameral chambers short; expansion angle 8°. Siphuncle positioned at ventral shell margin; segments expanded, spherical in shape.

Remarks. Originally assigned to *Tripteroceras planodorsatum* by Whitfield (1882) and type locality described as found "three miles above Beloit, Wisconsin." (p. 231). The general location suggests the type specimens are from the Mifflin or Grand Detour Members.

Occurrence. Common in the Mifflin and Grand Detour Members of northern Illinois and southern Wisconsin.

Figure 11-15 (Opposite page)

1–6. ***Allumettoceras planodorsatum*** (Whitfield). (**1**) Dolomitized steinkern with rounded connecting rings of siphuncle exposed; (**2**) ventral surface and (**3**) lateral view of dolomitized steinkern of phragmocone and most of the body chamber; (**4**) lateral view and (**5**) dorsum showing small part of a siphuncle; (**6**) interior surface of a septum showing the ventrally positioned siphuncle, slightly flat venter, and rounded dorsum. Grand Detour Member, Beloit, Wisconsin (locality 21), **1**, JAC 179. Mifflin Member, Lee Center, Illinois (locality 31), **2**, **3**, JAC 180. Mifflin Member, Dixon, Illinois (locality 2), **4**, **5**, ISGS 788; **6**, ISGS 736.

7–13. ***Diestoceras*? sp.** (**7**) Adoral end showing a subtriangular aperture with prominent hyponomic sinus; (**8**) ventral surface showing part of a phragmocone and body chamber; (**9**, **10**), lateral view and ventral surface, respectively; (**11**) venter showing a siphuncle; (**12**) same as **11**, lateral view; (**13**) interior surface of a septum showing the ventrally positioned siphuncle; photographic enlargement showing distinctive calcareous deposits that form longitudinal plates in a radial orientation inside connecting rings and septal necks. Grand Detour Member, Beloit, Wisconsin (locality 21), **7**, JAC 181. Pecatonica Member, Dixon, Illinois (locality 1), **8**, ISGS 704; **9**, **10**, ISGS 702; **11**, **12**, ISGS 705; **13**, ISGS 703.

14–19. ***Ulrichoceras beloitense*** Foerste. (**14**) Lateral view of a well-preserved specimen showing the complete body chamber and most of the phragmocone; (**15**) same specimen with part of the phragmocone removed to show the interior surface of a septum and siphuncle; (**16**) same specimen showing the venter; (**17–19**) dolomitized steinkern showing the dorsum, venter, and lateral view, respectively. Mifflin Member, Fennimore, Wisconsin (locality 26), **14–16**, JAC 182. Mifflin Member, Beloit, Wisconsin, **17–19**, UWGM 856.

Family VALCOUROCERATIDAE

Genus MANITOULINOCERAS

Manitoulinoceras* cf. *neleum (Hall, 1861)

Figures 11-14-5 through 11-14-8

Description. Slender, strongly curved, exogastric cyrtocone; rate of enlargement of conch very small; somewhat flattened dorsum, leading to the development of a depressed cross section; body chamber not contracted at aperture. Dorsoventral diameter at adoral end of body chamber 20 mm; length more than 90 mm as measured along axis of venter. Sutures straight and transverse. Cameral chamber lengths short, approximately one-eighth the diameter of the shell. Siphuncle ventral; endosiphuncle (space inside the septal necks and connecting rings) contains distinctive calcareous deposits that form longitudinal plates in a radial orientation.

Remarks. According to Hall (1861) the type specimens are from "40 ft above the base of the Trenton limestone at Beloit, Wisconsin."

Occurrence. Common locally in the Grand Detour Member in northern Illinois and southern Wisconsin.

Family DIESTOCERATIDAE

Genus DIESTOCERAS

***Diestoceras*? sp.**

Figures 11-15-7 through 11-15-13

Description. Shell length 8–10 cm, robust, straight to faintly endogastric, compressed brevicone; broadly ovate in cross section; rapidly expanding with greatest height and width at adoral end of the phragmocone; body chamber large, inflated, contracting to aperture. Aperture subtriangular, open, and broadly rounded dorsally; constricted and narrowly rounded ventrally, forming a prominent hyponomic sinus. Dorsum convex throughout length of the shell. Shell exterior marked by low, rounded, transverse ridges that define a shallow V-shaped sinus. Cameral chamber lengths are approximately one-seventh to one-ninth the diameter of the shell. Sutures are straight and transverse. Siphuncle positioned close to the ventral shell margin. The endosiphuncle (space inside the septal necks and connecting rings) contains distinctive calcareous deposits that form longitudinal plates in a radial orientation (actinosiphonate type).

Remarks. The overall size and shape of the conch and distinctive siphuncle are suggestive of *Diestoceras*; however, in the Platteville species, the siphuncle is positioned closer to the ventral margin of the shell (Robert C. Frey, personal communications, 2019), causing some uncertainty about assignment to *Diestoceras*. The subtriangular aperture with constricted hyponomic sinus is similar to that of *D. indianense* from the Saluda Formation (Richmondian Stage) at Hamburg, Franklin County, Indiana.

Occurrence. Locally common in the Pecatonica and Grand Detour Members in northern Illinois and southern Wisconsin.

Order DISCOSORIDA

Remarks. The discosorids typically have robust breviconic or cyrtoconic conchs. They are also characterized by having central or subcentral siphuncle segments with thick, complex connecting rings or by having specialized endosiphuncular deposits, or both.

Family CYRTOGOMPHOCERATIDAE

Genus ULRICHOCERAS

Ulrichoceras beloitense Foerste, 1928

Figures 11-15-14 through 11-15-19

Discussion. Shell cyrtoconic, endogastric; rapidly expanding to middle of body chamber, then contracting toward the aperture; expansion angle approximately 25°; cross section slightly depressed; dorsum and ventrolateral sides strongly rounded; hyponomic sinus on the venter. Siphuncle positioned between the center and venter; segments short and broadly rounded.

Remarks. The type specimen is from the Platteville Formation at Beloit, Wisconsin.

Occurrence. Common locally in the Mifflin and Grand Detour Members in northern Illinois and southern Wisconsin.

CHAPTER 12

TRILOBITES

Trilobites are relatively diverse, locally abundant, and generally well preserved in the Platteville Formation. Approximately 30 species are known in the upper Midwest outcrop belt, including representatives of the orders Corynexochida, Lichida, Phacopida, Proetida, Asaphida, and Harpetida (Appendix 2). Overall, the fauna can be categorized as an illaenid–cheirurid–asaphid association with similarities to trilobite biofacies in coeval shallow, subtidal, level-bottom shelf lithofacies (Figure 12-1), including the Red River Formation of Manitoba (Ludvigsen and Westrop 1983). Analogous trilobite associations also occur in the Late Ordovician carbonate ramp to basin lithofacies of the lower Esbataottine Formation in the Sunblood Range of the Mackenzie District, northwestern Canada (Chatterton and Ludvigsen 1976), the Viola Group of south-central Oklahoma (Amati and Westrop 2006), and the Effna–Botetourt–Liberty Hall succession of the Taconic Foreland Basin of southwestern Virginia (Carlucci and Westrop 2012).

The terminology used herein to describe trilobite orientation and morphology follows Whittington (1997). Some of the more common trilobite terms and features are shown in Figure 12-2. Trilobites have a dorsal exoskeleton or cuticle made of chitin reinforced with calcite. The exoskeleton is a composite of articulated elements called **sclerites**. The length and width of the exoskeleton as well as any part of it are described with reference to either the longitudinal or transverse direction. **Sagittal** (abbreviated **sag.**) is the direction along the midline from front to back and **transverse** (abbreviated **tr.**) is the direction at right angles to sagittal (across the body from side to side). **Abaxial** is the direction away from sagittal and **adaxial** is toward sagittal.

The name trilobite is in reference to the three longitudinal lobes of the exoskeleton. A central **axial lobe** is marked on each side by an axial furrow that delineates two symmetrical, laterally positioned **pleural lobes**. The exoskeleton is also divided by transverse articulations into the **cephalon** at the **anterior** end, a central series of **segments** forming the **thorax**, and the **pygidium** at the **posterior** end. The margins of the exoskeleton curve around the edge and down onto the ventral surface, forming a **doublure** (Figure 12-3-8) of variable width. In some trilobites, the median portion of the doublure at the front of the head consists of a **rostral plate** (Figure 12-3-8). The **glabella** is a raised central part of the cephalon that commonly is outlined by the **axial** and **preglabellar furrows**. At the posterior end of the glabella is the **occipital lobe**, which may or may not be separated from the anterior part of the glabella by an **occipital furrow**. In many trilobites, the lateral margins of the glabella are intersected by pairs of **lateral glabellar furrows** designated **S0**, **S1**, **S2**, and **S3**. Positioned between the furrows are the **lateral glabellar lobes**, designated correspondingly **L1**, **L2**, and **L3**. Lichid trilobites have a pair of longitudinal furrows that divide the glabella into lateral **bullar lobes**. Unique to some illaenids are **lunettes** (Figure 12-3-3), which are small, paired, crescent-shaped swellings on the inner surface of the exoskeleton adjacent to the glabella.

Figure 12-1 (Opposite page) Depiction of Platteville seafloor showing life mode of several trilobites, including *Ceraurus*, *Dolichoharpes*, *Thaleops*, and *Walencrinuroides*. Created by Dennis R. Kolata.

Lateral to the glabella are the **cheeks** (syn. **genae**) that continue backward to the **genal angle**, where the cheeks meet the posterior margin. At this point, the cheeks may or may not terminate in a **genal spine**. The cheeks are divided into a fixed portion connected to the glabella called the **fixed cheek** (syn. **fixigena**) that is separated by a **facial suture** from the outer **free cheek** (syn. **librigena**). Three types of facial sutures are defined by where the suture ends relative to the genal angle: (1) **proparian** sutures end forward of the genal angle, as in *Ceraurus*; (2) **gonatoparian** sutures end at or near the genal angle, as in *Anapliomera*; and (3) **opisthoparian** sutures end along the posterior cephalic margin, as in *Isotelus*. The facial suture passes forward over the **eye lobe** between the **palpebral lobe** and the **visual surface** (Figure 12-2-3). The **cranidium** is the central part of the cephalon that includes the glabella and fixed cheeks, all bounded laterally by the facial sutures.

The **hypostome** (Figures 12-2-6 and 12-2-7), thought to be associated with the mouth, is part of the exoskeleton and is positioned on the ventral surface beneath the anterior part of the glabella. The **median body** of the hypostome is subdivided into a relatively large **anterior lobe** and a smaller **posterior lobe** by a **middle furrow** extending from the **lateral border furrow** inward and backward. The middle furrow dies out adaxially into a small flat area called the **macula** that commonly lacks surface sculpture. The anterolateral corners commonly project outward, forming short **anterior wings**.

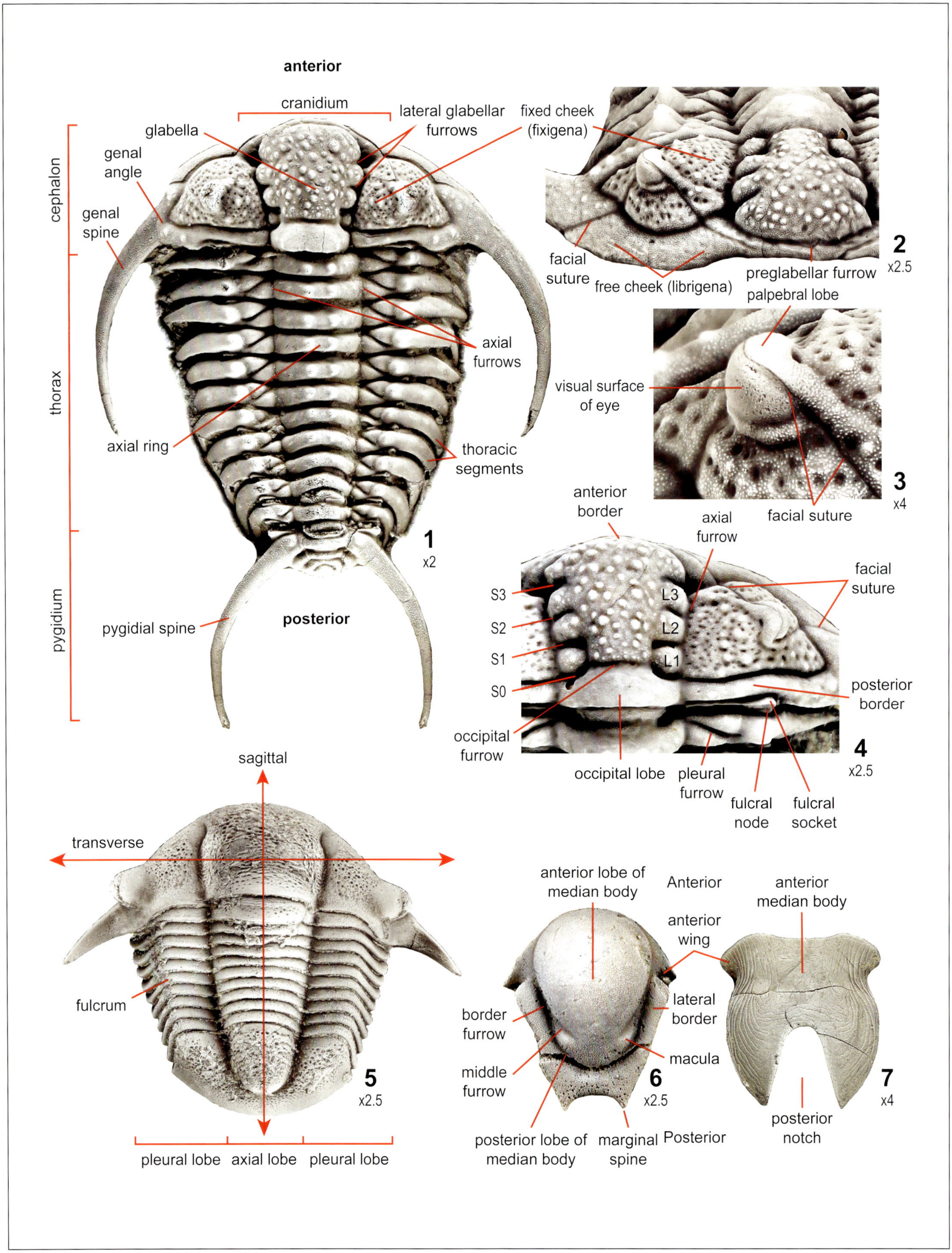
anterior
cranidium
glabella
lateral glabellar furrows
fixed cheek (fixigena)
genal angle
genal spine
cephalon
thorax
pygidium
axial furrows
axial ring
thoracic segments
1
x2
pygidial spine
posterior
facial suture
free cheek (librigena)
preglabellar furrow
2
x2.5
palpebral lobe
visual surface of eye
facial suture
3
x4
anterior border
axial furrow
facial suture
S3
S2
S1
S0
L3
L2
L1
posterior border
occipital furrow
occipital lobe
pleural furrow
fulcral node
fulcral socket
4
x2.5
sagittal
transverse
fulcrum
5
x2.5
pleural lobe
axial lobe
pleural lobe
anterior lobe of median body
Anterior
anterior wing
lateral border
border furrow
macula
middle furrow
posterior lobe of median body
marginal spine
Posterior
6
x2.5
anterior median body
posterior notch
7
x4

The thorax consists of multiple thoracic segments that articulate with one another and with the cephalon and pygidium. Each segment consists of an **axial ring** flanked on both sides by the **pleura**. The inner part of the pleura tends to be horizontal, whereas the outer part is bent down abruptly at a pivot point or **fulcrum**. The segments articulate with one another at a series of **fulcral nodes** on the anterior edges of the pleurae and are coupled with a **fulcral socket** on the posterior edge of the adjacent segment. Each segment has a pair of **pleural furrows** that extend from the axial furrow outward and slightly backward.

The pygidium is composed of fused segments and occupies the posterior part of the exoskeleton. Most pygidia have a convex axial lobe that is separated from lateral pleural lobes by an axial furrow. The two axial furrows merge at the posterior end of the axial lobe. The pleural area commonly extends around the terminal end of the axis. A border furrow and pygidial spines are present along the lateral and posterior margins in some species. The exoskeleton extends around and beneath the posterior edge of the pygidium, forming a doublure.

The dorsal and ventral exoskeleton surfaces are commonly sculptured with (1) knoblike prominences called **tubercles**, (2) small circular **pits** of various sizes or irregular-shaped **pockmarks** (syn. **scrobiculae**), and (3) narrow raised bands called **terrace ridges**, all collectively referred to as **surface sculpture**.

Primary sources used in the identification of the Platteville trilobites include Walcott (1877), Clarke (1897c), Delo (1940), Whittington (1949, 1953), DeMott (1963), Esker (1964), Darby and Stumm (1965), Shaw (1968), Chatterton and Ludvigsen (1976), Ludvigsen and Chatterton (1982), DeMott et al. (1987), Lespérance and Desbiens (1995), Brett et al. (1999), Amati and Westrop (2004), Carlucci et al. (2012), Congreve (2013), Amati (2014), Moss and Westrop (2014), and Swisher et al. (2015).

Figure 12-2 (Opposite page)
1–7. Morphologic and orientation terms applied to Platteville trilobites. (**1**) Dorsal view of *Ceraurus mifflinensis* (ISGS 1250) showing cephalon, thorax, pygidium, cranidium (includes glabella and fixed cheeks), lateral glabellar furrows, axial furrows, thoracic segments, pygidial spines, axial rings, and genal spines; (**2**) frontal view of *C. mifflinensis* (AWS 103) cephalon showing fixed and free cheeks, facial suture, and preglabellar furrow; (**3**) close-up of *C. mifflinensis* (AWS 103) eye showing palpebral lobe, visual surface of eye, and facial suture; (**4**) dorsal view of *C. mifflinensis* (AWS 103) showing anterior border, facial suture, posterior border, pleural furrow, occipital lobe, occipital furrow, axial furrow, lateral glabellar furrows (S0, S1, S2, and S3), and glabellar lobes (L1, L2, and L3); (**5**) dorsal view of *Thaleops ovata* (ISGS 1267) showing axial and pleural lobes, fulcrum, and sagittal and transverse positions; (**6**) ventral surface of *Ceraurinella templetoni* (AWS 135) hypostome showing anterior and posterior lobes of median body, anterior wings, lateral border, macula, marginal spines, middle furrow, and border furrow; (**7**) isoteline hypostome (ISGS 1318) showing anterior median body, anterior wings, and posterior notch.

SYSTEMATIC DESCRIPTIONS

Phylum ARTHROPODA

Order CORYNEXOCHIDA

Suborder ILLAENINA

Family ILLAENIDAE

Subfamily ILLAENINAE

Genus THALEOPS

Remarks. In a cladistic analysis of Middle and Late Ordovician illaenid trilobites of Laurentia, Amati and Westrop (2004) showed that *Thaleops* represents a large group of species characterized by long, articulating facets on the pygidium that are oriented at a high angle to the anterior margin of the pygidium and have posteriorly positioned palpebral lobes. Their results suggest that these illaenids display a wide range of morphologies, grading from strongly furrowed forms with well-developed genal spines and stalked eyes like *Thaleops ovata* to species like *Thaleops laurentiana* (formerly *Illaenus americanus*) that have comparatively weakly furrowed exoskeletons ("effaced"), no genal spines, and narrower palpebral lobes. These new observations suggest that several species, including some from the Platteville Formation that were previously assigned to *Illaenus* (DeMott 1963; DeMott et al. 1987), are more closely allied with *Thaleops*. These include the DeMotte specimens *Illaenus* sp. (cephalon A), *Illaenus* sp. (Pygidium 1), *Illaenus* sp. (cephalon B), and *Illaenus* sp. (Pygidium 2), none of which are illustrated in this report.

Thaleops ovata Conrad, 1843

Figures 12-3-1 through 12-3-8

Description. An illaenid trilobite that is approximately 30 mm in length, having 10 thoracic segments, well-defined axial furrows, a pygidium that is much smaller than the cephalon, and stout, curved genal spines. Cephalon broadly ovate at anterior end. Axial furrows diverge from posterior margin of cephalon then converge for a short distance before curving outward again and fading away in front of the lunettes. Glabella arched upward higher than the cheeks (genae). Rostral plate crescent shaped with a medial posterior projection. Eyes small, stalked, narrowly rounded, projecting laterally; palpebral lobes positioned near posterior margins of cranidium. Facial sutures form a symmetrical pair that extend from the posterior margin of the cephalon forward, forming a boundary with the adaxial side of the visual surface of the eye, then curving downward adaxially to the anterior margin of the cephalon. Occipital furrow is effaced. Free cheeks taper distally into prominent genal spines, the anterior and posterior surfaces of which converge to form a dorsal carina (Figure 12-3-1). Hypostome unknown. Thorax has well-defined axial furrows and a convex axis that is approximately one-third of the total width of the thorax. Thoracic segments gradually narrow (tr.) toward the posterior. Pygidium approximately half as long as wide, posterior margin rounded, well-defined axial furrows, posterior margin of axis well defined.

Surface sculpture consists of small, circular pits distributed over the cephalon, thorax, and pygidium. In addition, the cephalon and pygidium have clusters of pockmarks that are concentrated on the pleural and axial lobes. Some surface sculpture is aligned on the axial lobe of the pygidium, suggesting relict segmentation.

Remarks. The presence of a dorsal carina on the short, stout genal spines distinguishes *T. ovata* from other species of *Thaleops*. Amati and Westrop (2004) considered the relict segmentation on the pygidium unique to *T. ovata*.

Occurrence. Sclerites are abundant in the Mifflin and Grand Detour Members in north-central Illinois and south-central Wisconsin and the McGregor Member of southwestern Wisconsin, northeastern Iowa, and southeastern Minnesota. Complete specimens are common.

Thaleops cf. *longispina* Shaw, 1968

Figure 12-4-15

Description. An illaenid trilobite that is approximately 40 mm in length, having 10 thoracic segments, well-defined axial furrows, pronounced stalked eyes, and unusually long, thin, curved genal spines. Cephalon broadly ovate at anterior end. Axial furrows well defined, glabella arched upward higher than the cheeks (genae). Rostral plate crescent shaped with a medial posterior projection. Eyes markedly long for the genus, stalked, narrowly rounded, projecting upward and laterally with slight posterior tilt; palpebral lobes long and narrow, positioned near posterior margins of cranidium. Facial sutures form a symmetrical pair that extend from the posterior margin of the cephalon forward, bounding the adaxial side of the visual surface of the eye, then curving downward adaxially to the anterior margin of the cephalon. Free cheek below and slightly anterior of the eye lobe, tapering distally into a long, thin, prominent genal spine that is round in cross section, curving posteriorly to parallel the axis. Thorax has well-defined axial furrows and a convex axis that is somewhat wider than the pleural lobes. Thoracic segments gradually narrowing (tr.) toward the posterior. Pygidium approximately half as long as wide. Surface sculpture consists of small pits distributed over the cephalon, thorax, and pygidium. In addition, the cephalon is marked by clusters of pockmarks that are concentrated on the pleural and axial lobes.

Remarks. This Platteville species is similar to *T. longispina* Shaw, 1968, from the Upper Ordovician (Turinian Stage) Crown Point Limestone at Crown Point, New York. Like *T. longispina*, the Platteville specimens possess pronounced stalked eyes and long, thin, curved genal spines that are round in cross section.

Occurrence. Present but rare in the Grand Detour Member in Grant County, Wisconsin.

Thaleops (*Illaenus*) sp. 1

Figures 12-3-9 through 12-3-12

Description. An illaenid trilobite that is approximately 50 mm in length, having 10 thoracic segments, well-defined axial furrows, and a pygidium that is much smaller than the cephalon based on the projected margin of the thorax. Cephalon represented by a fragment of the glabella only; glabella appears to be arched upward higher than the pleural lobes. Thorax has well-defined axial furrows and a convex axis that is approximately one-third of the total width of the thorax.

Figure 12-3 (Opposite page)
1–8. ***Thaleops ovata*** Conrad. (**1**) Perspective view of dorsal surface showing dorsal carina on genal spine; (**2**) dorsal surface of partially rolled-up specimen; (**3**) anterior view of cephalon showing exfoliated exoskeleton and impression of lunettes on interior cast; (**4**) dorsal surface; (**5**) dorsal surface of pygidium showing scrobiculate surface sculpture, some of which suggests relict segmentation; (**6**) lateral view of partially rolled-up exoskeleton; (**7**) anterior view of cephalon; (**8**) ventral surface of cephalon showing the doublure and rostral plate. Mifflin Member, Dixon, Illinois (locality 2), **1**, ISGS 1271; **2**, **3**, **6**, ISGS 1266; **4**, **5**, ISGS 1267; **7**, ISGS 1272; **8**, ISGS 1286.

9–12. ***Thaleops* (*Illaenus*) sp. 1.** (**9**) Posterior view of pygidium; (**10**) lateral view showing the typical recurved pygidial profile; (**11**) dorsal surface; (**12**) photographic enlargement of terrace ridges on articulation surfaces of thoracic segments. Grand Detour Member, Dixon, Illinois (locality 4), **9–12**, ISGS 1294.

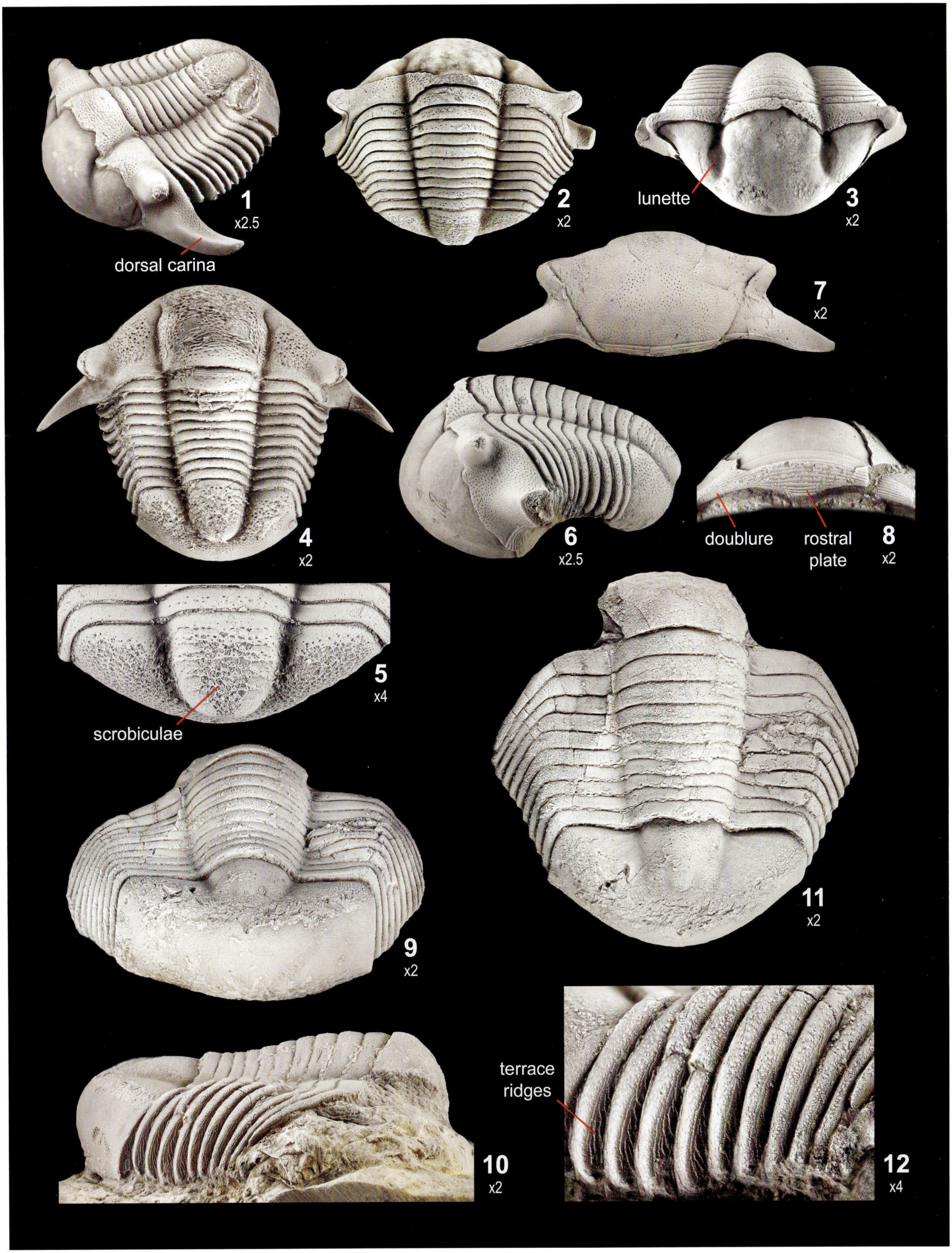
1
x2.5
dorsal carina
2
x2
lunette
3
x2
7
x2
4
x2
6
x2.5
doublure
rostral plate
8
x2
5
x4
scrobiculae
11
x2
9
x2
terrace ridges
10
x2
12
x4

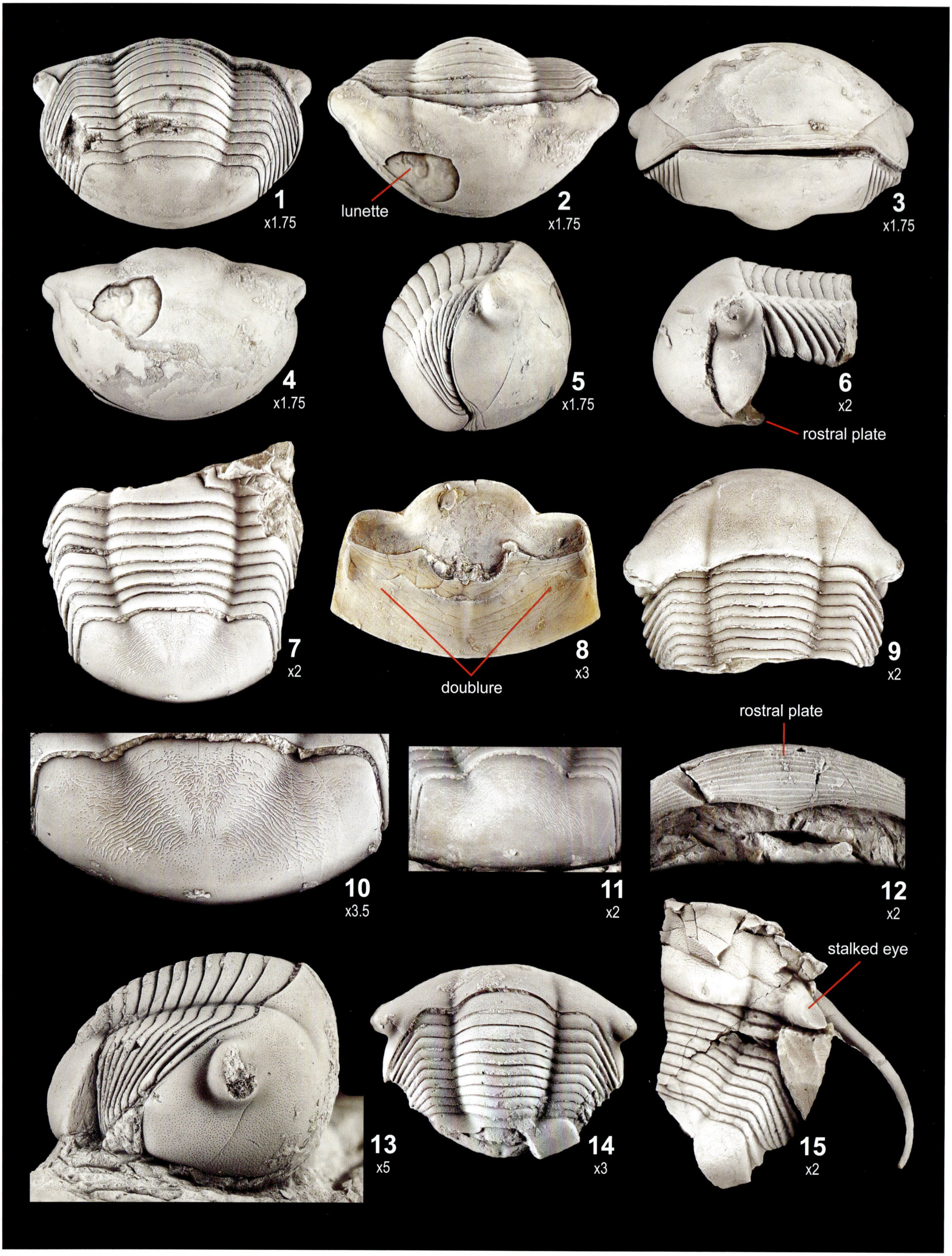
1
x1.75
lunette
2
x1.75
3
x1.75
4
x1.75
5
x1.75
6
x2
rostral plate
7
x2
doublure
8
x3
9
x2
rostral plate
10
x3.5
11
x2
12
x2
stalked eye
13
x5
14
x3
15
x2

Thoracic segments taper gradually toward the posterior. Segments of the axial lobe are marked on either side of the axis by a shallow transverse groove containing thin, closely spaced terrace ridges. Pygidium approximately half as long as wide; axis tapers posteriorly and merges with a relatively flat area of the pygidium that extends for a short distance posteriorly before curving abruptly downward in a slightly anterior direction, producing an overall pygidial profile that is unlike other known Platteville species of *Thaleops* (Figure 12-3-10). Relict segmentation is suggested by the presence of subtle, radiating ridges on either side of the pygidial axial lobe. Surface sculpture consists of pits and pockmarks on the cephalon and pygidium as well as very small tubercles on the outer parts of the thoracic segments and on the upper surface of the pygidium, particularly near the margin of the abrupt downward curve of the pygidium.

Remarks. DeMott et al. (1987) described a single pygidium (*Thaleops* sp. B) from the Grand Detour Member near Shirland, Illinois (locality 16 of this report) that closely matches the distinct pygidium of the specimen figured herein. *Thaleops* sp. 1 likely represents an undescribed species.

Occurrence. Present but rare in the Grand Detour Member of northern Illinois and southern Wisconsin.

Thaleops (Illaenus) sp. 2

Figures 12-4-1 through 12-4-14

Description. An illaenid trilobite with an observed length of approximately 60 mm, having 10 thoracic segments, defined axial furrows, and a pygidium that is smaller than the cephalon. Cephalon broadly ovate at the anterior end. Axial furrows well defined, straight to very slightly curved outward, converging somewhat from posterior margin of cephalon to the lunettes, then becoming effaced. Occipital furrow effaced. Glabella arched upward higher than the cheeks (genae). Rostral plate crescent shaped, with a medial posterior projection. Eyes projecting laterally at widest point of cephalon; palpebral lobes relatively large, rounded, and positioned near posterior margins of cranidium; visual surface convex laterally. Facial sutures form a symmetrical pair that extend from the posterior margin of the cephalon forward, forming a boundary with the adaxial side of the visual surface of the eye, then curving downward adaxially to the anterior margin of the cephalon. Free cheeks semicircular in outline and evenly rounded. Thorax with 10 segments, defined axial furrows, and a convex axial lobe that is wider than the pleural lobes. Thoracic segments gradually narrow (tr.) toward the posterior. Pygidium length approximately 60% that of the width, posterior margin evenly rounded, well-defined axial furrows. Surface sculpture consists of numerous small, circular pits, particularly on the cephalon and pygidium and less so on the thorax. In addition, the cephalon and pygidium have clusters of irregularly shaped pockmarks and terrace ridges that are concentrated on the pleural and axial lobes. Terrace ridges are also well developed on the distal ends of the thoracic segments.

Figure 12-4 (Opposite page)
1–14. ***Thaleops (Illaenus)* sp. 2** (**1**) Dorsal surface of rolled-up specimen; (**2**) dorsal view of cephalon showing exfoliated exoskeleton and impression of lunette on interior cast; (**3**) rolled-up specimen showing contact between cephalon and pygidium; (**4**) anterior view of cephalon; (**5**) lateral view of rolled-up specimen; (**6**) lateral view of cephalon and thorax showing recurved rostral plate; (**7**) outstretched thorax and pygidium; (**8**) ventral surface of pygidium showing the doublure; (**9**) outstretched cephalon and thorax; (**10**) photographic enlargement of pygidium showing array of terrace ridges; (**11**) posterior view showing outline of pygidium; (**12**) ventral margin of cephalon showing posterior medial projection of rostral plate; (**13**) lateral view of partially rolled-up specimen; (**14**) dorsal surface of partially rolled-up specimen. Mifflin Member, Dixon, Illinois (locality 3), **1–5**, ISGS 1261; **6**, **9**, ISGS 1269; **7**, **10**, **11**, ISGS 1268; **8**, ISGS 1290; **12**, ISGS 1287; **13**, **14**, ISGS 1275.

15. ***Thaleops* cf. *longispina*** Shaw. (**15**) Right side of exoskeleton showing the distinct long, thin genal spine and stalked eye. Grand Detour Member, Grant County, Wisconsin (precise locality unavailable), **15**, AWS 117.

Remarks. The Platteville specimens described and illustrated here appear to be conspecific with *Illaenus* sp. (Pygidium 1) of DeMott et al. (1987; plate 6, figures 16–18). *Thaleops* sp. 2 is similar to *T. mobydicki* (Amati and Westrop, 2004) from the approximately coeval Bromide Formation of Oklahoma but differs from the latter in having less distinct axial furrows on the cephalon. Furthermore, in *Thaleops* sp. 2, the angle measured from the axis of the pygidium to a line drawn parallel to the lateral edge is on average less than 10° as compared with 15° or greater in *T. mobydicki*. *Thaleops* sp. 2 likely represents an undescribed species.

Occurrence. Sclerites of *Thaleops* sp. 2 are common in the Mifflin and Grand Detour Members of northern Illinois and southern Wisconsin. Complete specimens are rare.

Thaleops (Illaenus) sp. 3

Figures 12-5-1 through 12-5-4

Description. An illaenid trilobite that is approximately 35 mm in length, with the greatest width at the outer angles of the free cheeks, having nine thoracic segments, weakly defined axial furrows, and a pygidium that is slightly smaller than the cephalon. Cephalon broadly ovate at anterior end. Axial furrows well defined, curved inward at posterior margin of cephalon, becoming straight before reaching the lunettes, then effaced. Occipital furrow effaced. Glabella arched upward slightly higher than the cheeks (genae). Eyes projecting laterally; palpebral lobes relatively large, rounded, and positioned near posterior margins of cranidium; visual surface crescentic, convex laterally. Facial sutures form a symmetrical pair that extend from the posterior margin of the cephalon forward, forming a boundary with the adaxial side of the visual surface of the eye, then curving downward adaxially to the anterior margin of the cephalon. Free cheeks relatively large, V-shaped, and projecting laterally to a blunt tip. Thorax has moderately defined axial furrows and a convex axial lobe that is wider than the pleural lobes. Thoracic segments approximately the same width throughout. Pygidium length approximately a third that of the width, posterior margin evenly rounded, weakly defined axial furrows. Surface sculpture consists of small, inconspicuous pits on the cephalon, thorax, and pygidium. Terrace ridges present on distal edges of the thoracic segments and on posterior edges of the free cheeks.

Remarks. The Platteville specimens described and illustrated here appear to be conspecific with *Illaenus* sp. (Pygidium 2) of DeMott et al. (1987; plate 6, figure 19). *Thaleops* sp. 3 differs from other species of *Thaleops* in having nine relatively wide thoracic segments that are of equal width throughout and a short but wide pygidium that lacks well-developed axial furrows. *Thaleops* sp. 3 likely represents an undescribed species.

Occurrence. Sclerites of *Thaleops* sp. 3 are common in the Mifflin and Grand Detour Members of northern Illinois and southern Wisconsin. Complete specimens are rare.

Genus BUMASTOIDES

Bumastoides milleri (Billings, 1859a)

Figures 12-6-1 through 12-6-9

Description. An illaenid trilobite that is approximately 30 mm in length, having nine thoracic segments, weakly defined axial furrows, axis comprising approximately 80% of the thoracic width. Cephalon broadly inflated, generally smooth and featureless; axial furrows very weak or absent; lunettes crescent shaped, positioned forward of the palpebral lobes; glabella not clearly differentiated. Front margin of cephalon with broad lateral indentations marked by prominent, transversely oriented terrace ridges. Rostral plate crescent shaped with a medial posterior projection. Eyes large, palpebral lobes positioned near posterior margins of cranidium; visual surface crescent shaped, convex laterally. Facial sutures form a symmetrical pair that extend from the posterior margin of the cephalon forward, forming a boundary with the adaxial side of the visual surface of the eye, then curving downward adaxially to the anterior margin of the cephalon. Free cheeks semicircular in outline and evenly rounded. Thorax has nine segments with very weakly defined axial furrows and a convex axis that is approximately 80% of the total width of the thorax. Thoracic segments gradually narrow toward the posterior; distal edges marked by pectinate (comblike) structures consisting of 10–20 closely spaced sinuous terrace ridges. Pygidium forms approximately one-third of the exoskeletal length; subtrapezoidal outline with rounded anterior and sharp lateral corners; length approximately 60% of width; lacking axial furrows; axis undifferentiated; rounded axial ridge present on interior surface. Surface sculpture consists of very fine elongate terrace ridges on the rostral and pygidial doublures as well as the distal parts of the thoracic segments.

Remarks. *Bumastoides milleri* has been a catchall for *Bumastoides*-type trilobites since its original description by Billings (1859a). DeMott et al. (1987) and Carlucci et al. (2012) argued that *B. milleri* should be restricted to *Bumastoides* species with nine thoracic segments.

Occurrence. Isolated sclerites are common in the Mifflin and Grand Detour Members in northern Illinois and southern Wisconsin and equivalent strata of the McGregor Member of southwestern Wisconsin, northeastern Iowa, and southeastern Minnesota. Sclerites locally common in the Pecatonica Member in northern Illinois and southern Wisconsin. Complete specimens are rare.

Figure 12-5 (Opposite page)
1–4. ***Thaleops (Illaenus)* sp. 3** (**1**) Anterior view of rolled-up specimen; (**2**) lateral view of the complete specimen; (**3**) dorsal view; (**4**) dorsal view of pygidium. Mifflin Member, Grant County, Wisconsin, **1**, ISGS 1274; **2**, **3**, AWS 109. Mifflin Member, Dixon, Illinois (locality 2), **4**, ISGS 1336.

5–9. ***Failleana indeterminatus*** (Walcott). (**5**) Ventral surface of hypostome; (**6**) lateral view; (**7**) dorsal view; (**8**) dorsal view of partial cephalon and thorax; (**9**) lateral perspective view of cephalon. Grand Detour Member, Dixon, Illinois (locality 4), **5**, ISGS 1255; **8**, **9**, ISGS 1292. Mifflin Member, Fennimore, Wisconsin (precise locality unavailable), **6**, **7**, AWS 115.

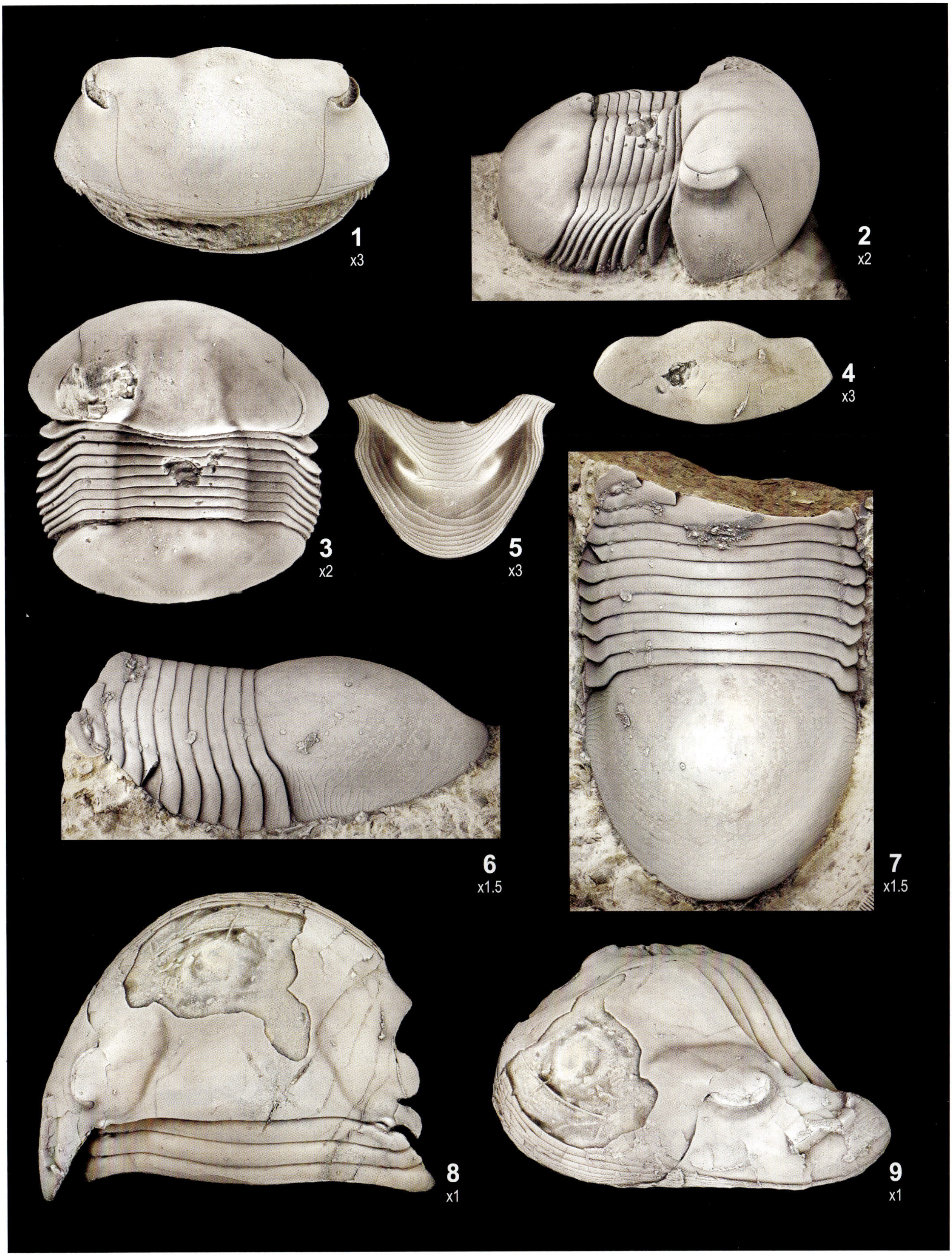
1
x3
2
x2
3
x2
4
x3
5
x3
6
x1.5
7
x1.5
8
x1
9
x1

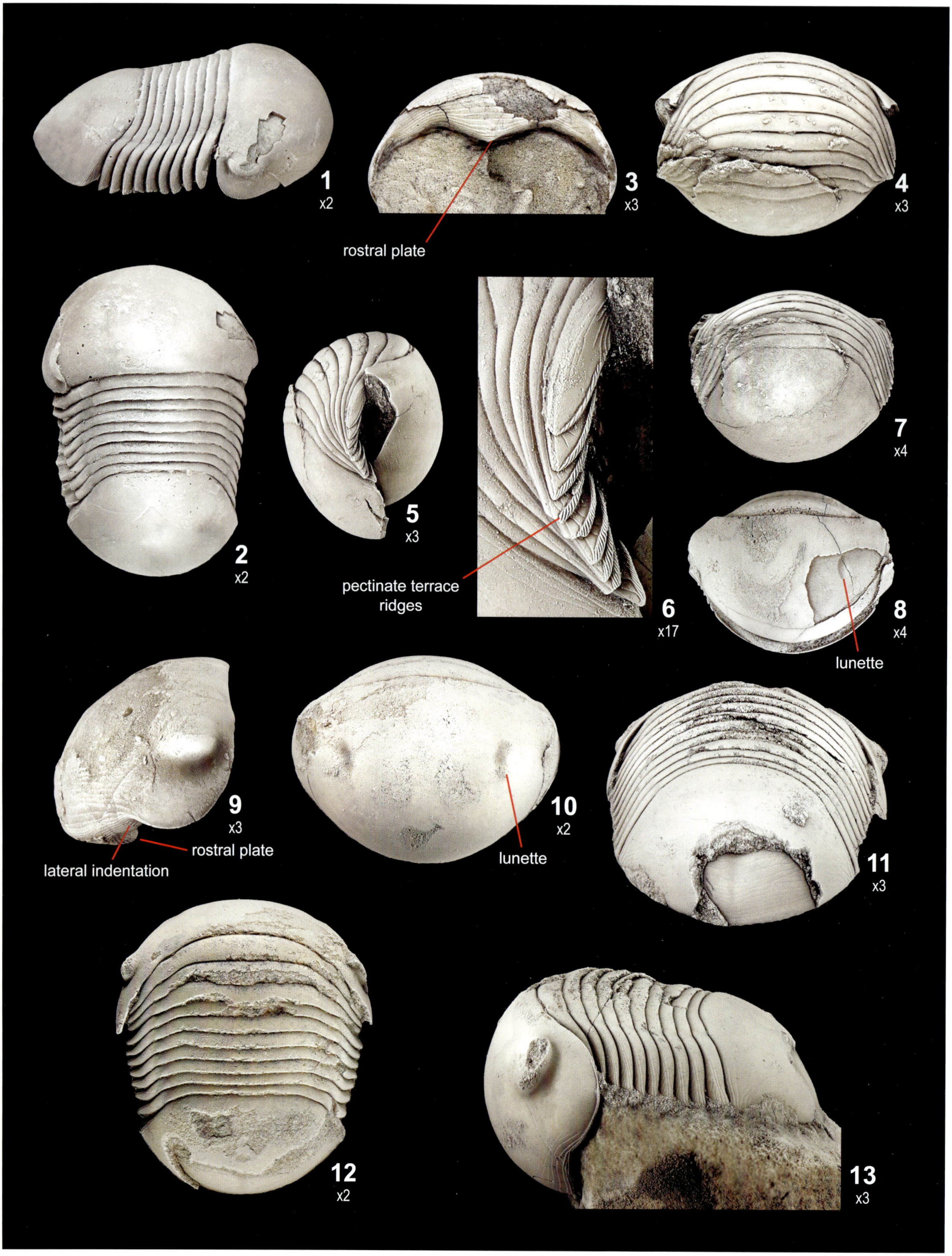
1
x2
3
x3
rostral plate
4
x3
2
x2
5
x3
pectinate terrace
ridges
6
x17
7
x4
8
x4
lunette
9
x3
rostral plate
lateral indentation
10
x2
lunette
11
x3
12
x2
13
x3

Bumastoides* cf. *graffhami Carlucci et al., 2012

Figures 12-6-10 through 12-6-13

Description. An illaenid trilobite that is approximately 35 mm in length, having 10 thoracic segments. Cephalon broadly inflated, generally smooth and featureless; axial furrows very weak or absent; lunettes oval, positioned forward of the palpebral lobes; glabella not clearly differentiated. Front margin of cephalon with broad lateral indentations marked by prominent transversely oriented terrace ridges. Rostral plate crescent shaped with a medial posterior projection. Eyes large, palpebral lobes positioned near posterior margins of cranidium; visual surface crescent shaped, convex laterally. Facial sutures form a symmetrical pair that extend from the posterior margin of the cephalon forward, bounding the adaxial side of the visual surface of the eye, then curving downward adaxially to the anterior margin of the cephalon. Free cheeks semicircular in outline and evenly rounded. Hypostome unknown. Thorax has 10 segments with very weakly defined axial furrows and a convex axis that is approximately 80% of the total width of the thorax. Thoracic segments gradually narrow toward the posterior; distal edges marked by sinuous terrace ridges. Pygidium forms approximately one-third of the exoskeletal length; subtrapezoidal outline with rounded anterior and sharp lateral corners; length approximately 60% of width; lacking axial furrows; axis undifferentiated; rounded axial ridge present on interior surface. Surface sculpture consists of very fine elongate terrace ridges on the rostral and pygidial doublures as well as on distal parts of the thoracic segments and on the lateral edges of the pygidium.

Figure 12-6 (Opposite page)
1–9. ***Bumastoides milleri*** (Billings). (**1**) Lateral view of the complete specimen; (**2**) dorsal view; (**3**) ventral surface of cephalon showing the posteriorly pointed rostral plate; (**4**) dorsal view of rolled-up specimen; (**5**) lateral view of rolled-up specimen; (**6**) photographic enlargement showing pectinate terrace ridges on outer edges of thoracic segments; (**7**) posterior view of rolled-up specimen; (**8**) anterior view of cephalon showing exfoliated exoskeleton and impression of lunette on interior cast; (**9**) lateral view of cephalon showing eye and posteriorly pointed rostral plate. Mifflin Member, Grant County, Wisconsin, **1**, **2**, AWS 138. Mifflin Member, Dixon, Illinois (locality 2), **3**, ISGS 1280; **4–8**, ISGS 1282; **9**, ISGS 1334.

10–13. ***Bumastoides* cf. *graffhami*** Carlucci et al. (**10**) Dorsal view of cephalon exoskeleton showing dark coloration of lunette; (**11**) posterior view of the complete specimen with partially exfoliated exoskeleton; (**12**) dorsal view of the complete specimen; (**13**) lateral view. Grand Detour Member, Beloit, Wisconsin (locality 21), **10**, **12**, JAC 100; **11**, **13**, JAC 101.

Remarks. This Platteville *Bumastoides* is very similar to and probably conspecific with *B. graffhami* Carlucci et al., 2012, from the Upper Ordovician (Sandbian) lower Pooleville Member of the Bromide Formation, Arbuckle Mountains region of Oklahoma. It is distinguished from *B. milleri* in having 10 rather 9 thoracic segments.

Occurrence. Isolated sclerites common in the Mifflin and Grand Detour Members and equivalent strata throughout the outcrop belt. Sclerites locally common in the Pecatonica Member in northern Illinois and southern Wisconsin. Complete specimens are uncommon.

Family STYGINIDAE

Genus FAILLEANA

Failleana indeterminatus (Walcott, 1877)

Figures 12-5-5 through 12-5-9

Description. An illaenid trilobite that is approximately 150 mm in length. Cephalon broad, semicircular, with short, wide, blunt genal spines. Cranidium subquadrate in outline, slightly inflated; axial furrows relatively well developed, converging strongly from fulcral nodes at posterior margin of cranidium adaxially for a short distance, then turns slightly anteriorly to a point in front of the palpebral lobes, then diverges forward before being effaced near anterior margin; front margin of cranidium curved down abruptly; occipital furrow indistinct. Eyes large, positioned far back on the cephalon; visual surface crescent shaped. Facial sutures form a symmetrical pair that extend from the posterior margin of the cephalon forward, bounding the adaxial side of the visual surface of the eye, then curving forward and downward abaxially to the lateral margin of cephalon. Hypostome width greater than length; anterior margin slightly curved backward, posterior broadly curved backward, crescent-shaped depression extends laterally between small anterior wings; maculae well defined on posterior lobe of median body; prominent terrace ridges aligned with lateral and posterior margins of hypostome. Thorax has 10 segments with weakly defined axial furrows and a convex axis that is approximately 80% of the total width of the thorax; transverse width of thoracic segments narrowest anteriorly, particularly segments 1 through 3; inner parts of pleurae with short, wide, fulcral nodes and corresponding fulcral sockets. Pygidium width slightly greater than length, inflated particularly in adult stages, broad border furrow. Surface sculpture consists of very small pits on dorsal surface of exoskeleton; prominent terrace ridges developed on outer margins of cephalon and pygidium as well as on hypostome.

Remarks. This Platteville species is most similar to *Failleana indeterminatus*, first described by Walcott (1877) from the Black River Limestone of Herkimer County, New York, and Platteville Formation at Platteville, Wisconsin. It was later reported from the Black River Limestone at the falls of La Petite Chaudiere at Ottawa, Canada, and by Clarke (1897c) from the Platteville Formation at Janesville, Wisconsin.

Occurrence. Isolated sclerites common in the Mifflin and Grand Detour Members in north-central Illinois and south-central Wisconsin and the McGregor Member of southwestern Wisconsin, northeastern Iowa, and southeastern Minnesota.

Order LICHIDA

Suborder LICHINA

Superfamily LICHOIDEA

Family LICHIDAE

Subfamily TETRALICHINAE

Genus AMPHILICHAS

Amphilichas* cf. *subpunctatus (Esker, 1964)

Figures 12-7-1 and 12-7-3

Description. A species of *Amphilichas* represented by several well-preserved cranidia. Cranidium subtriangular in outline, length 70% of cranidial width across palpebral lobes; median lobe strongly inflated and overhanging anterior border; adjacent bullar lobes separated from median lobe by narrow, depressed furrows that diverge forward to join well-defined axial furrows in front of and below the eyes, then merging with preglabellar furrow. Well-defined occipital lobe, furrow S0 transverse medially, with slight posterior deflection behind bullar lobes. Cranidium bears numerous tubercles of various sizes, some of which are perforated.

Remarks. Available cranidia are very similar in overall size, shape, and tuberculation to *Amphilichas subpunctatus* (Esker, 1964) from the Upper Ordovician (Turinian) Pooleville Member of the Bromide Formation, Arbuckle Mountains near Ardmore, Oklahoma.

Occurrence. Isolated sclerites common in the Mifflin and Grand Detour Members in north-central Illinois and south-central Wisconsin and the McGregor Member of southwestern Wisconsin, northeastern Iowa, and southeastern Minnesota. Complete specimens are very rare.

Genus PROBOLICHAS

Probolichas* cf. *robbinsi (Ulrich, 1892d)

Figures 12-7-8 and 12-7-9

Description. A species of *Probolichas* represented by several cranidia. Cranidium with elongate projection (proboscis) from median glabella; otherwise very similar to *Amphilichas*. Surface sculpture consists mainly of three tubercle sizes, evenly scattered and densely packed, some perforated.

Remarks. Available cranidia are very similar in overall size, shape, and tuberculation to *Probolichas robbinsi* (Ulrich, 1892d) from the middle beds of the Upper Ordovician (Chatfieldian Stage) Galena Formation near Wykoff, Minnesota. *P. kristiae* (Carlucci et al., 2010) from the coeval Pooleville Member of the Bromide Formation of Oklahoma differs in having elongate tubercles on the proboscis.

Occurrence. Isolated sclerites common in the Mifflin and Grand Detour Members in north-central Illinois and south-central Wisconsin and the McGregor Member of southwestern Wisconsin, northeastern Iowa, and southeastern Minnesota. Complete specimens are very rare.

Unidentified Tetralichine Sclerites

Remarks. Included here are isolated tetralichine-like free cheeks, hypostomes, thoracic segments, and pygidia from the Platteville Formation. They are found in association with *Amphilichas* cf. *subpunctatus* and *Probolichas* cf. *robbinsi* cranidia; however, the lack of articulated

Figure 12-7 (Opposite page)

1, 3. ***Amphilichas* cf. *subpunctatus*** (Esker). (**1**) Dorsal view of cephalon showing right side bullar lobe; (**3**) lateral view of cephalon. Mifflin Member, Grant County, Wisconsin, **1**, AWS 130; **3**, AWS 132.

2, 4–7. **Unidentified Tetralichine Sclerites.** (**2**) Dorsal view of free cheek; (**4**) dorsal view of thorax; (**5**) ventral view of hypostome; (**6**) dorsal view of Pygidium 2; (**7**) dorsal view of Pygidium 1. Mifflin Member, Grant County, Wisconsin, **2**, AWS 128; **4**, AWS 114; **5**, AWS 129; **6**, AWS 125; **7**, AWS 121.

8, 9. ***Probolichas* cf. *robbinsi*** (Ulrich). (**8**) Lateral view of crushed cephalon; (**9**) dorsal view of cephalon. Mifflin Member, Grant County, Wisconsin, **8**, **9**, AWS 124.

10–13. ***Hemiarges* cf. *paulianus*** (Clarke). (**10**) Dorsal view of cephalon showing longitudinal furrow; (**11**) dorsal view of pygidium; (**12**) dorsal view of nearly complete specimen; (**13**) ventral surface of hypostome. Mifflin Member, Dixon, Illinois (locality 2). **10**, ISGS 1304; **11**, ISGS 1338; **13**, ISGS 1232. Mifflin Member, Grant County, Wisconsin, **12**, AWS 110.

14. ***Hemiarges* sp.** Dorsal view of cephalon. Pecatonica Member, Green County, Wisconsin (locality 45), **14**, ISGS 520.

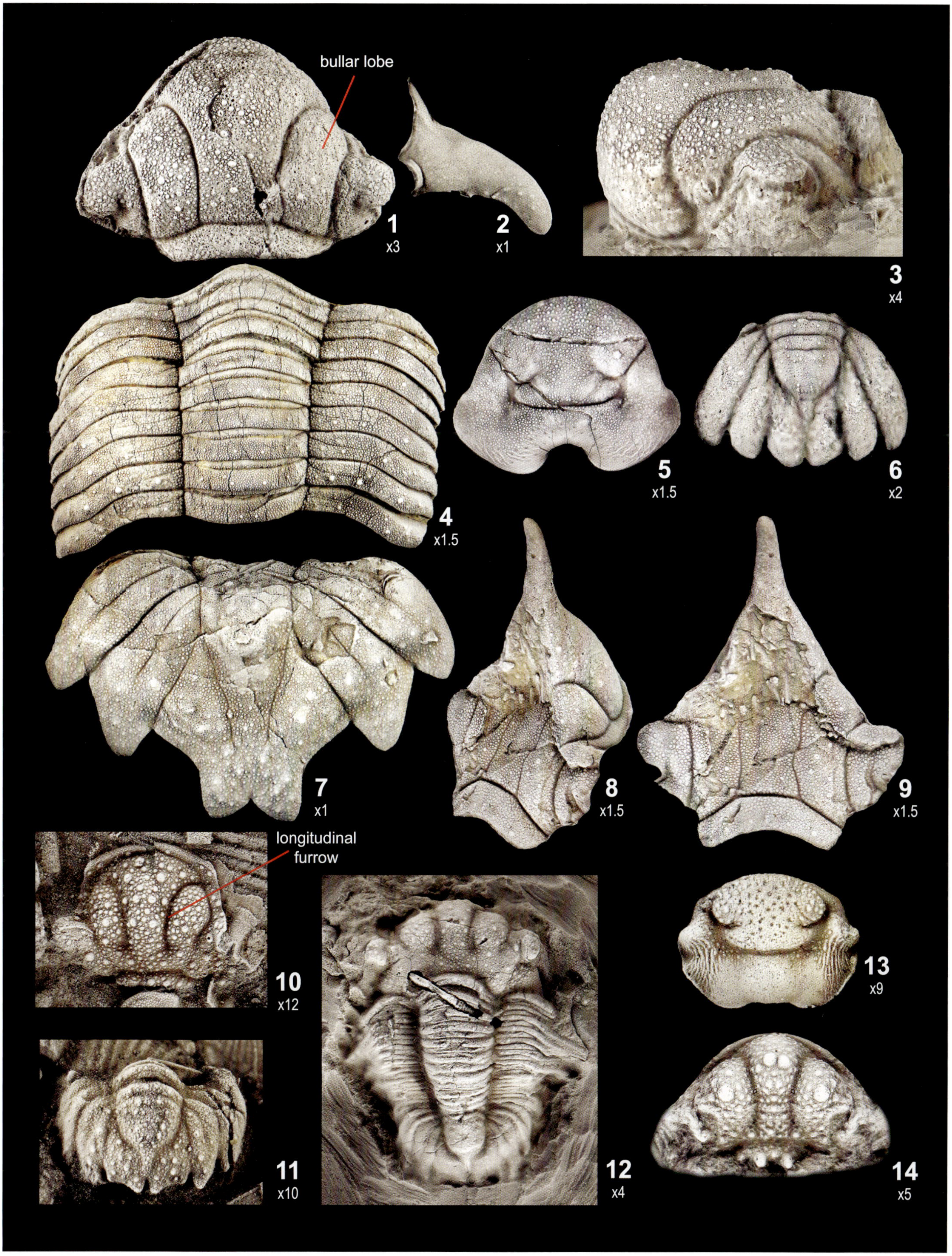
bullar lobe
1
x3
2
x1
3
x4
4
x1.5
5
x1.5
6
x2
7
x1
8
x1.5
9
x1.5
longitudinal furrow
10
x12
11
x10
12
x4
13
x9
14
x5

specimens precludes confident assignment to a particular genus or species.

Free Cheek

Figure 12-7-2

Description. Free cheek bearing a relatively large, curved, posterolaterally directed genal spine that is depressed in cross section and terminates in a blunt end. Visual surface of eye crescent shaped and convex outward. Surface sculpture consists of abundant tubercles of various sizes.

Occurrence. Present but rare in the Mifflin Member in Grant County, Wisconsin.

Thoracic Segments

Figure 12-7-4

Description. Articulated thoracic segments; axial ring rectangular in outline, width less than adjacent pleural lobes, rings narrowing (sag.) toward posterior end; axial furrows well defined. Pleurae horizontal adjacent to axial rings, flexed down and backward at fulcrum. Pleural furrows narrow and weakly developed, extending posterolaterally from axial furrow and terminating at fulcrum. Surface sculpture consists of abundant tubercles of various sizes.

Occurrence. Present but rare in the Mifflin Member in Grant County, Wisconsin.

Hypostome

Figure 12-7-5

Description. Hypostome relatively large, wider than long, anterior lobe extends to anterior border. Posterior lobe is shorter (sag.) and narrower than anterior lobe, bounded laterally by narrow, well-defined furrows that converge posteriorly; posterior border transverse. A pair of furrows extend for a short distance from the maculae, converging anteriorly for a short distance before branching and terminating. Posterior margin deeply notched medially. Surface sculpture consists of abundant tubercles of various sizes covering the ventral surface, except on the maculae.

Occurrence. Common locally in the Mifflin and Grand Detour Members in northern Illinois and southern Wisconsin and equivalent strata of the McGregor Member of southwestern Wisconsin, northeastern Iowa, and southeastern Minnesota.

Pygidium 1

Figure 12-7-7

Description. Pygidium large, somewhat depressed, width greater than sagittal length, with three pairs of spinose pleurae and narrow, well-defined pleural furrows; pleurae end in short, blunt spines that project beyond posterior margin. Third pygidial pleural segments terminate in adaxially curved spines separated by small V-shaped medial notch; posterior margin of pygidium concave between fixed segments 2 and 3. Complete furrow on first pleural segment only. Surface sculpture consists of abundant tubercles of various sizes over the dorsal surface.

Remarks. The adaxial convergence of spines on third pygidial pleural segment is similar to that of *Amphilichas hibernicus* (Portlock, 1843) from the Upper Ordovician Bardahessiagh beds of Northern Ireland.

Occurrence. Present but rare in the Mifflin Member, Grant County, Wisconsin.

Pygidium 2

Figure 12-7-6

Description. Pygidium width greater than length (sag.), with three pairs of closely spaced, spinose pleurae and short pleural furrows; pleurae end in short, blunt spines that project beyond posterior margin of pygidium and curve slightly inward. Pygidial axis with one complete ring furrow and partial furrow on second segment. Pygidium with well-developed axial lobe terminating in abruptly constricted axis. Surface sculpture consists of abundant tubercles of various sizes over the dorsal surface.

Remarks. This pygidium is similar to that of *Amphilichas conradi* Chatterton and Ludvigsen, 1976 (plate 18, figures 56 and 57) from the Upper Ordovician Esbataottine Formation, Sunblood Range, Mackenzie District, Canada.

Occurrence. Present but rare in the Mifflin Member in Grant County, Wisconsin.

Subfamily CERATARGINAE

Genus HEMIARGES

Hemiarges* cf. *paulianus (Clarke, 1897c)

Figures 12-7-10 through 12-7-13

Description. A species of *Hemiarges* that is approximately 12 mm in length. Cranidium convex, central lobe of glabella higher and expands forward in front of the adjacent bullar lobes; longitudinal furrows deep and relatively broad, convergent posteriorly, ending in a shallow depression in back of bullar lobes but not extending to occipital furrow. Occipital ring arched transversely, tapers abaxially and slopes anteriorly. Occipital furrow deep and broad abaxially, shallowing adaxially. Preglabellar furrow well defined. Anterior border moderately wide, expands slightly opposite axial furrow. Posterior border furrow relatively deep, widens abaxially. Palpebral lobes relatively large, visual surface crescent shaped. Facial suture opisthoparian type, sutures extend from posterolateral cephalic margin to base of eye, up around palpebral lobe and downward, intersecting anterior margin at a steep angle. Hypostome slightly inflated, subquadrate in outline, and wider than long; median body divided into slightly larger anterior lobe and smaller posterior lobe; prominent posterior and lateral border furrows; weakly developed maculae on sides of posterior lobe of median body; posterior margin concave backward. Thorax with 11 segments. Axial rings convex, axial furrows well defined. Pleural furrows narrow and well defined, extending from axial furrow to fulcrum. Pygidium with well-developed axial lobe, abruptly constricted axis extending posteriorly as a prominent postaxial ridge, with four pairs of marginal spines. First and second rings complete, third faint. First and second pleurae have pleural furrows that gradually terminate at base of short marginal spines. Third pleurae unfurrowed, with two pairs of posterior spines, the second being longer. Surface sculpture on dorsal exoskeleton consists of numerous small tubercles; fine parallel terrace ridges on anterior cephalic margin.

Remarks. In most features, this *Hemiarges* species is most like *H. paulianus* (Clarke, 1897c) from the lower Prosser (=Rivoli Beds, Dunleith Member, Galena Formation), St. Paul, Minnesota. It differs from the Galena species in having somewhat smaller tubercles. Associated with this species of *Hemiarges* is a small lichid hypostome very similar to *H. paulianus* from the Galena Formation of Minnesota in having a (1) length being 60% of width, (2) central body being broadly rounded in front, and (3) broad, shallow indentation on the posterior border.

Occurrence. Sclerites rare in the Mifflin Member of northern Illinois and southern Wisconsin. Complete specimens very rare.

Hemiarges sp.

Figure 12-7-14

Description. This species is represented by a single but distinctive cranidium. It is similar to *Hemiarges* cf. *paulianus* from the Platteville Formation but differs in having a single relatively large node (tubercle) on each bullar lobe, a pair on the axial lobe, and a pair of short, blunt spines on the occipital lobe. It also differs from *H.* cf. *paulianus* in having a glabella that does not expand forward of the bullar lobes.

Remarks. Overall shape and ornamentation are similar to *Hemiarges troedssoni* (Rudkin et al., 1994) from the Upper Ordovician Cape Calhoun Formation of northern Greenland.

Occurrence. Very rare in the Pecatonica Member in southern Wisconsin.

Order PHACOPIDA

Suborder CHEIRURINA

Family CHEIRURIDAE

Subfamily CHEIRURINAE

Genus CERAURINELLA

Ceraurinella scofieldi (Clarke, 1897c)

Figures 12-8-7 through 12-8-10

Description. A cheirurid trilobite that is approximately 35 mm in length, having 11 thoracic segments and well-defined axial furrows throughout. Cephalon moderately convex, crescent shaped, approximately three times wider than long (sag.), short, stout genal spine. Glabella extending to frontal margin with subparallel sides, rounded frontal lobe, and even inflation throughout; lobe L1 and frontal lobe slightly wider than occipital ring and lobes L2 and L3. Furrow S1 angles backward on approach to axis of glabella, producing a distinct triangular outline to lobe L1. Furrows S2 and S3 very lightly impressed, parallel, and curved convexly forward. Occipital ring and furrow bowed strongly forward. Cheeks triangular in outline with inclined flanks and prominent, elevated eyes. Facial sutures (proparian type) extend adaxially from lateral cephalic margin up the eye stalk between palpebral lobe and visual surface of eye, then gradually curve downward and forward adaxially to anterior cephalic margin. Lateral posterior margin of fixed cheek bears a short, backwardly curved genal spine. Hypostome subpentagonal in outline, width and length approximately equal, surface covered with very fine pustules; oval anterior end narrows and decreases in convexity backward, merging with small crescent-shaped posterior lobe; lateral borders flat, tilted inward, narrow posteriorly, terminating in a short pair of spines. Thorax with 11 segments, axis moderately high and convex, approximately equal in width to pleural lobes. Inner part of pleura inflated, crossed by narrow, diagonal, posteriorly trending pleural furrow; outer part of pleura extends into tapering spine that curves down and backward. Prominent articulation structures formed on inside lateral edges of pleurae, consisting of a fulcral node on anterior edge coupled with a fulcral socket on posterior edge of adjacent segment in a manner described and illustrated by Whittington and Evitt (1954, figure 3). Pygidium with anterior spines similar in size to genal spines; second segment with spines one-half the length and width of genal spines; third segment spineless and doubled on itself with a small median furrow. Surface sculpture consists of very fine tubercles distributed over the entire dorsal surface; small pits on fixed cheeks.

Remarks. The description given here is based on several nearly complete specimens. The type specimens are from the Platteville Formation at Minneapolis, Minnesota, and the Glades, Lebanon, Tennessee.

Occurrence. Isolated sclerites are abundant in the Mifflin and Grand Detour Members in north-central Illinois and south-central Wisconsin and the McGregor Member of southwestern Wisconsin, northeastern Iowa, and southeastern Minnesota. Complete specimens are rare.

Ceraurinella templetoni DeMott et al., 1987

Figures 12-8-11 and 12-8-12

Description. A *Ceraurinella* with a cephalon that is approximately 32 mm in width, measured at the tips of genal spines. Cephalon moderately convex, semicircular, approximately three times wider than long (sag.); short, slender genal spine. Glabella extending to frontal margin with subparallel sides, rounded frontal lobe, even inflation throughout; lobe L1 and frontal lobe slightly wider than occipital ring and lobes L2 and L3. Furrow S1 angles backward on approach to axis of glabella to produce a distinct triangular outline to lobe L1, with S1 becoming shallow and directed back to connect to the occipital furrow, isolating the two L1 lobes. Furrows S2 and S3 lightly impressed, parallel, and curved convexly forward. Occipital ring and furrow bowed strongly forward. Cheeks triangular in outline with inclined flanks and prominent elevated eyes. Facial sutures (proparian type) extend adaxially from lateral cephalic margin up the eye stalk between palpebral lobe and visual surface of eye, then gradually curve downward and inward adaxially to anterior cephalic margin. Lateral posterior margin of fixed cheek bears a short, slender, backwardly directed genal spine. Projected posterior cephalic margin intersects genal spine at nearly a right angle.

Remarks. *C. templetoni* differs from *C. scofieldi* in the isolation of the L1 lobes by longitudinal constriction, having a semicircular anterior cephalic margin, more elongate hypostome, and shorter genal spines. Available specimens are consistently twice as large as *C. scofieldi*.

Occurrence. The type specimens are from the Quimbys Mill Member in Lafayette County, Wisconsin. Isolated sclerites common in the Mifflin and Grand Detour Members in Grant County, Wisconsin. Complete specimens are very rare.

Genus CERAURUS

Ceraurus mifflinensis DeMott, 1963

Figures 12-9-7 through 12-9-11

Description. A species of *Ceraurus* that is approximately 50 mm in length as measured along the sagittal axis from anterior margin of cephalon to posterior end of pygidium; width of cephalon as measured from lateralmost edge of right and left fixed cheek border furrow approximately 30–40 mm. Glabella expanding anteriorly, does not reach front margin of cephalon, width of frontal lobe approximately 13% greater than distance between outside edges of L1 lobes. Glabella slightly wider between outer edges of L3 lobes. Furrow S1 extends adaxially a short distance, then turns back at a right angle to merge with the occipital furrow, isolating the two L1 lobes. Occipital ring broad and smooth. Cheeks triangular in outline, with inclined flanks and stalked eyes; prominently marked by deeply impressed posterior and lateral border furrows. Posterior margin of cephalon with a prominent swelling between thoracic articulating flange and base of genal spine. Lateral posterior margin of fixed cheek bears a

Figure 12-8 (Opposite page)

1–6. ***Eubufoceraurus scheerorum*** Congreve. (**1**) Dorsal view of cranidium and disarticulated left free cheek; (**2**) partial cranidium showing large tubercles on anterior part of glabella and occipital spine; (**3**) ventral surface of hypostome; (**4**) dorsal view of thoracic segment showing robust pleural lobe with prominent tubercles; (**5**) posterior end of same segment; (**6**) ventral surface of pygidium. Mifflin Member, Grant County, Wisconsin, **1**, AWS 141; **2**, AWS 142; **6**, AWS 144. Mifflin Member, Dixon, Illinois (locality 3), **3**, ISGS 927; **4**, **5**, ISGS 1328.

7–10. ***Ceraurinella scofieldi*** (Clarke). (**7**) Lateral view; (**8**) ventral surface of cephalon with hypostome intact; (**9**) dorsal surface; (**10**) perspective view of anterior end. Mifflin Member, Grant County, Wisconsin, **7**, **9**, **10**, AWS 108. Mifflin Member, Dixon, Illinois (locality 3), **8**, ISGS 1306.

11, 12. ***Ceraurinella templetoni*** DeMott et al. (**11**) Ventral surface of hypostome; (**12**) dorsal surface of partial cranidium and thorax. Mifflin Member, Grant County, Wisconsin, **11**, AWS 135; **12**, AWS 119.

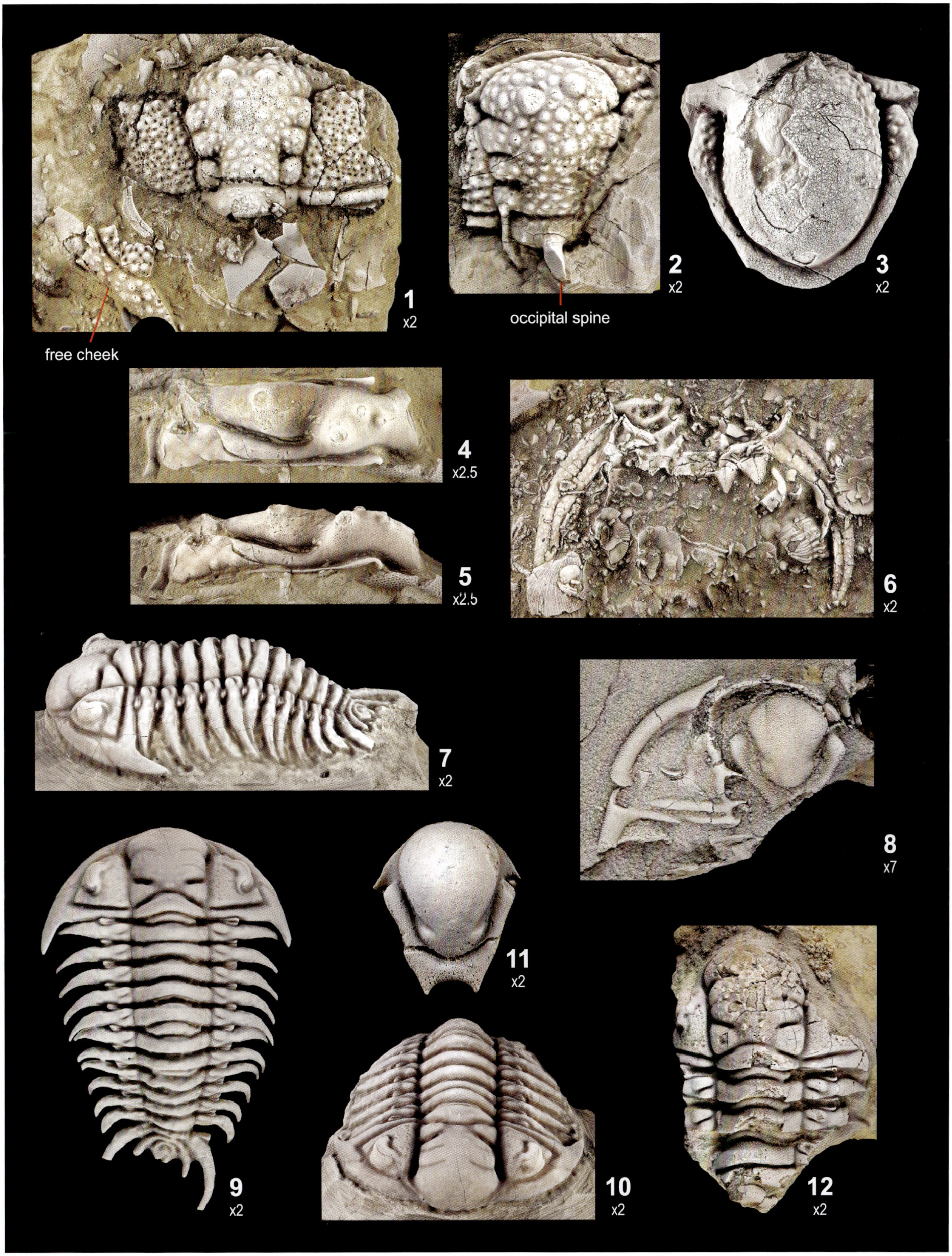
1
x2
free cheek
2
x2
occipital spine
3
x2
4
x2.5
5
x2.5
6
x2
7
x2
8
x7
9
x2
10
x2
11
x2
12
x2

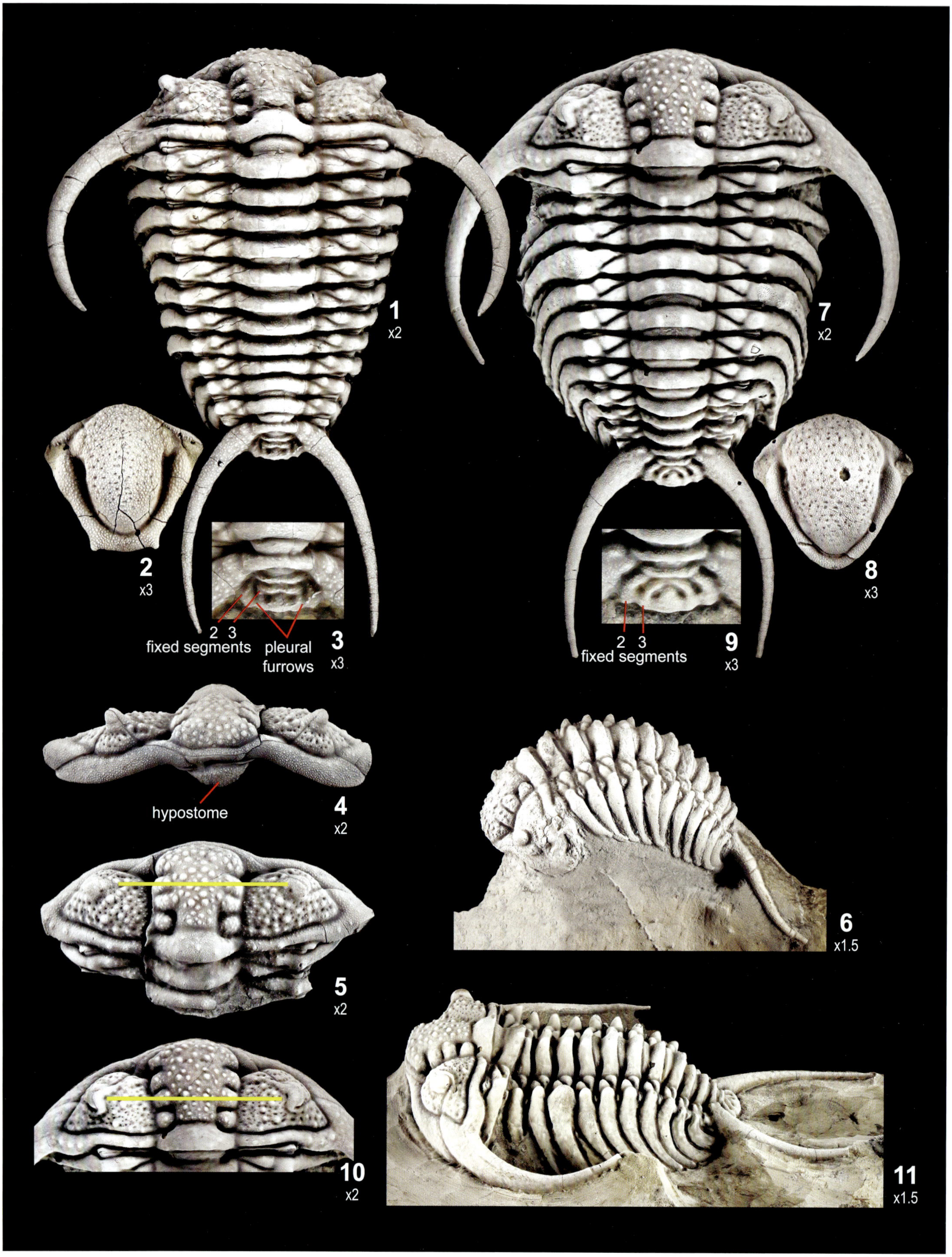
1
x2
7
x2
2
x3
8
x3
2 3
fixed segments
pleural
furrows
3
x3
2 3
fixed segments
9
x3
hypostome
4
x2
6
x1.5
5
x2
10
x2
11
x1.5

long, stout, backward-curved genal spine that distally curves inward. Facial sutures (proparian type) extend adaxially from lateral cephalic margin up around eye stalk between palpebral lobe and visual surface of eye, then gradually curve downward and forward adaxially to anterior cephalic margin. Eyes positioned well out on cheeks in line with the L2 lobes and at or slightly above glabella surface. Hypostome subpentagonal in outline; slightly longer (sag.) than wide, widest point between anterior wings; large, inflated, oval anterior lobe narrows and decreases in convexity posteriorly; deep, narrow border furrow extends around lateral and posterior margins; ventral surface with very fine granule-like tubercles except in lateral furrow and anterior wings; central ovoid area marked with numerous small pockmarks lacking granular ornament; maculae mostly effaced. Thorax consists of 11 segments with pairs of tubercles on axial lobes; very prominent tubercle on midportion of pleural lobes; fulcral nodes on edges of pleurae, consisting of a fulcral process on anterior edge coupled with fulcral socket on posterior edge of adjacent segment, all in line with prominent pleural lobe tubercles. Inner part of pleura inflated, crossed by diagonal, posteriorly trending pleural furrow; outer part of pleura tapering into spine that curves down and backward. Pygidium with anterior segment consisting of axial lobe flanked by long, curved spines that are slightly smaller than the genal spines; second fixed segment with small axial lobe bordered by shallow axial furrows and flanked by downward-curved pleural lobes that are rounded and protrude very slightly past the posterior pygidial margin; third fixed segment smaller, lacking axial lobe, consisting of downward-curved pleural lobes separated by shallow median furrow; third fixed segment doubled on itself with a small, shallow median furrow; pygidium distinguished by the presence of six teardrop-shaped pits, each formed between the pleural lobes. Surface sculpture consists of abundant, very small, granule-like tubercles evenly distributed over the cephalon, hypostome, thorax, and pygidium. In addition, relatively large tubercles randomly distributed on the glabella, cheeks, and margin of the cephalon, some occupying a constant position from specimen to specimen.

Remarks. *Ceraurus mifflinensis* was proposed by DeMott (1963) based on specimens from the Pecatonica, Mifflin, and Grand Detour Members of the Platteville Formation in northern Illinois, southern Wisconsin, and northeastern Iowa. It was subsequently reassigned to the genus *Gabriceraurus* by DeMott et al. (1987). More recently, Congreve (2013) showed by phylogenetic analysis that *Gabriceraurus* does not represent a monophyletic group and proposed that Platteville species be reassigned to *Ceraurus mifflinensis*.

Occurrence. Isolated sclerites are abundant in the Mifflin and Grand Detour Members in north-central Illinois and south-central Wisconsin and the McGregor Member of southwestern Wisconsin, northeastern Iowa, and southeastern Minnesota. Sclerites locally common in the Pecatonica Member in northern Illinois and southern Wisconsin. Complete specimens are uncommon.

Ceraurus jenlorum Congreve, 2013

Figures 12-9-1 through 12-9-6

Description. In overall aspect, *C. jenlorum* is similar to *C. mifflinensis*. It differs from the latter in having (1) relatively small eyes positioned farther forward on the cheeks in line with glabella furrow S2 and the posterior part of lobe L3, (2) a glabella that has nearly parallel sides, (3) lateral posterior corners of hypostome bearing a short, blunt spine, (4) pleural furrows between fixed segments 2 and 3 that project straight backward parallel to the sagittal line, (5) tubercles on the pygidium that are larger and more variable in size, and (6) clusters of small tubercles on the prominent thoracic pleural lobe tubercles.

Occurrence. Isolated sclerites are abundant in the Mifflin and Grand Detour Members in north-central Illinois and south-central Wisconsin and the McGregor Member of southwestern Wisconsin, northeastern Iowa, and southeastern Minnesota. Sclerites locally common in the Pecatonica Member in northern Illinois and southern Wisconsin. Complete specimens are uncommon.

Figure 12-9 (Opposite page)
1–6. ***Ceraurus jenlorum*** Congreve. (**1**) Dorsal surface; (**2**) ventral surface of hypostome; (**3**) enlargement of medial pygidium showing the relatively deep pleural furrows on fixed segments 2 and 3; (**4**) anterior view of cephalon showing intact hypostome; (**5**) dorsal surface of cephalon; yellow bar showing characteristic alignment of eyes with glabellar L3 lobes; (**6**) lateral view of specimen with flexed thoracic segments. Mifflin Member, Grant County, Wisconsin, **1**, **3**, AWS 104; **2**, AWS 136. Mifflin Member, Dixon, Illinois (locality 2), **4**, **5**, ISGS 1343; **6**, ISGS 1308.

7–11. ***Ceraurus mifflinensis*** DeMott. (**7**) Dorsal surface; (**8**) ventral surface of hypostome; (**9**) enlargement of medial pygidium showing characteristic features of fixed segments 2 and 3; (**10**) dorsal surface of cephalon; yellow bar showing characteristic alignment of eyes with glabellar L2 lobes; (**11**) lateral view. Mifflin Member, Grant County, Wisconsin, **7**, **9–11**, AWS 103; **8**, AWS 133.

Genus EUBUFOCERAURUS

Eubufoceraurus scheerorum Congreve, 2013

Figures 12-8-1 through 12-8-6

Description. A unique Platteville cheirurid trilobite of *Ceraurus* lineage presently known from fragments of the cephalon, hypostome, thorax, and pygidium. Glabella with prominent segmentation, forwardly expanding, widest at lobe L3. Occipital lobe bearing a long, stout, centrally positioned spine extending upward and curving posteriorly. Cheeks triangular in outline, with inclined flanks and stalked eyes. Small spine on each side of posterior cephalic margin between articulating flange and genal spine. Eyes aligned with furrow S3. Hypostome slightly longer (sag.) than wide, widest point between anterior wings; large, inflated, ovate anterior lobe narrows and decreases in convexity posteriorly; deep, narrow border furrow extends around lateral and posterior margins. Thoracic segment with very prominent tubercle on midportion of pleural lobe crowned with at least three relatively large tubercles; fulcral nodes on inside lateral edges of pleurae, consisting of a fulcral process on anterior edge coupled with fulcral socket on posterior edge of adjacent segment. Inner part of pleura inflated, crossed by diagonal, posteriorly trending pleural furrow. Pygidium with anterior segment consisting of axial lobe flanked by relatively large, long, curved spines; second fixed segment flanked by downward-curved pleural lobes that end in short spines that protrude past the posterior pygidial margin; third fixed segment smaller and protruding beyond posterior margin. Surface sculpture consists of relatively large, sharp-tipped tubercles that cover all the exoskeleton; three or four extra-large tubercles on anterior half of glabella; small pits distributed between tubercles.

Remarks. Available fragments show this species to be one of the largest cheirurid trilobites in the Platteville Formation.

Occurrence. Isolated sclerites are abundant in the Mifflin and Grand Detour Members in north-central Illinois and south-central Wisconsin and the McGregor Member of southwestern Wisconsin, northeastern Iowa, and southeastern Minnesota. Complete specimens very rare.

Subfamily ACANTHOPARYPHINAE

Genus ACANTHOPARYPHA

Acanthoparypha* cf. *evitti
Chatterton and Ludvigsen, 1976

Figures 12-10-1 and 12-10-2

Description. A species of *Acanthoparypha* represented by an incomplete cranidium. Axial furrows narrow and deeply impressed, meeting along anterior margin. Glabella large, oval in outline, inflated, widest across L1 lobes, narrowing forward, overhanging preglabellar furrow; glabellar furrows S1, S2, and S3 deeply impressed and becoming shorter (tr.) forward. Occipital ring well defined, sits below glabella, widest (sag.) along axis. Cheeks steeply inclined, triangular in outline, outlined by deeply impressed furrow that is continuous, with occipital furrow extending abaxially to near base of genal spine, then curving adaxially forward. Short, blunt spine at genal angle. Facial sutures (proparian type) extend adaxially from lateral cephalic margin up around eye stalk between palpebral lobe and visual surface of eye, then gradually curve downward adaxially to anterior cephalic margin. Palpebral lobes small, positioned opposite glabellar lobe L2. Surface sculpture consists of small, perforated tubercles covering all the cranidium except palpebral lobes; cheeks bear numerous small pits.

Remarks. In overall features, this Platteville *Acanthoparypha* is comparable to *A. evitti* from the Upper Ordovician Esbataottine Formation, Sunblood Range, Mackenzie District, Canada.

Occurrence. Present but rare in the Mifflin Member in Grant County, Wisconsin.

Subfamily CYBELINAE

Genus CYBELOIDES

Cybeloides cimelia Chatterton and Ludvigsen, 1976

Figures 12-10-3 through 12-10-7

Description. A cheirurid trilobite that is approximately 15 mm (excluding genal and thoracic spines) and having 12 thoracic segments. Cephalon crescent shaped with slight medial indentation at anterior margin; long genal spines. Glabella expanded anteriorly, axial furrows deeply impressed, preglabellar furrow lacking, axial lobe flanked by smaller and slightly lower ovate lobes. Occipital ring slightly narrower than maximum width of glabella and bearing a long, slender, centrally positioned spine extending upward and curving posteriorly. Cheeks triangular in outline with steeply inclined flanks and prominent stalked eyes. Facial sutures (proparian type) extend adaxially from lateral cephalic margin up the eye stalk between palpebral lobe and visual surface of eye, then downward adaxially to anterior cephalic margin. Lateral posterior margin of free cheek bears a short, backward-curved spine. Posterior margin of cephalon convex and relatively narrow from axial furrow laterally to area behind eye, then gradually widening and sloping downward to merge with base of genal spine. Posterior cephalic margin bears a pair of small, posteriorly projecting spines, one near base of genal spine and the second midway between the first spine and the level of the eye stalk. Genal spines long, tapered, slightly divergent backward, with upward curvature. Thorax of 12 segments, axis moderately high and convex, narrower than pleural lobes, well-defined axial furrows diminishing in width posteriorly except sixth segment, which is wider than adjacent segments and bears a long, gently curved spine. Axial lobe bears a pair of relatively large tubercles except on sixth segment; single tubercle on pleural lobes just adaxial to fulcrum. Pygidium small, slightly longer than wide, axis slightly convex, approximately one-third width of pygidium at anterior end, narrowing to blunt termination; axial ring furrows approximately 20 in number; four pairs of pleural ribs; single prominent tubercle positioned on middle to upper part of pleural lobes. Surface sculpture consists of small to relatively large tubercles on cephalon, thorax, and pygidium.

Remarks. Chatterton and Ludvigsen (1976) considered this Platteville species to be identical to *C. cimelia* from the Upper Ordovician Esbataottine Formation, Sunblood Range, Mackenzie District, Canada. Available material, including several nearly complete specimens, indicates that the Platteville *Cybeloides* differs slightly from the Canadian *C. cimelia* in having two short spines on the posterior margin of the cephalon close to the base of the genal spine.

Occurrence. Isolated sclerites are abundant in the Mifflin and Grand Detour Members in north-central Illinois and south-central Wisconsin and the McGregor Member of southwestern Wisconsin, northeastern Iowa, and southeastern Minnesota. Complete specimens are rare.

Family PLIOMERIDAE

Subfamily PLIOMERINAE

Genus ANAPLIOMERA

Anapliomera shirlandensis DeMott et al., 1987

Figures 12-10-8 through 12-10-14

Description. A pliomerid trilobite that is approximately 40 mm in length, having 12 thoracic segments and strongly defined axial furrows. Cephalon moderately convex, steeply dipping along margin, crescent shaped, nearly three times wider than long (sag.), genal angle rounded and projecting posteriorly. Glabella expanded anteriorly, lobe L1 and furrow S1 absent, composed of lobes L2 and L3 and furrows S2 and S3; axial furrows and sulci deeply impressed, anterior of glabella relatively wide, separated from the anterior margin by a deep, narrow furrow that terminates at the flanking axial furrows. Proximal end of furrow S3 bifurcates for a short distance, then merges with overall curvature of the medial glabella; short, shallow axial (sag.) furrow present at anterior margin of glabella. Occipital furrow relatively narrow and curved forward; occipital lobe low and flush with glabella. Palpebral lobe relatively small, semicircular; cylindrical eye projects above glabella and cheeks and is bounded by shallow palpebral furrow along posterior of base. Facial suture (gonatoparian type) extends from rounded genal corner forward across lateral border furrow, running parallel to border furrow in adaxial direction to eye stalk between palpebral lobe and visual surface of eye, then downward adaxially across anterior cephalic margin bordering a short axial rostral plate. Relatively large pits present on cheeks. Hypostome width and length approximately equal; center of ventral surface convex; posterior margin with short, centrally positioned spine flanked on each side by a pair of smaller projections. Thorax consists of 12 thoracic segments; axial lobe arched and approximately the same width as pleural lobes; pleurae with prominent rounded ridges, lacking pleural groove, axial furrows deeply impressed. Distal ends of segment 8 bear a prominent spine that bends back abruptly at thorax margin and extends posteriorly to full length of the pygidium. Pygidium subtriangular in outline, slightly longer than wide, with five pleural ribs and approximately 15 axial rings. Surface sculpture consists of very small tubercles evenly distributed over cephalon, thorax, and pygidium.

Remarks. The original description of *A. shirlandensis* was based on three cranidia and one associated segment (DeMott 1963, DeMott et al. 1987). The recent discovery of

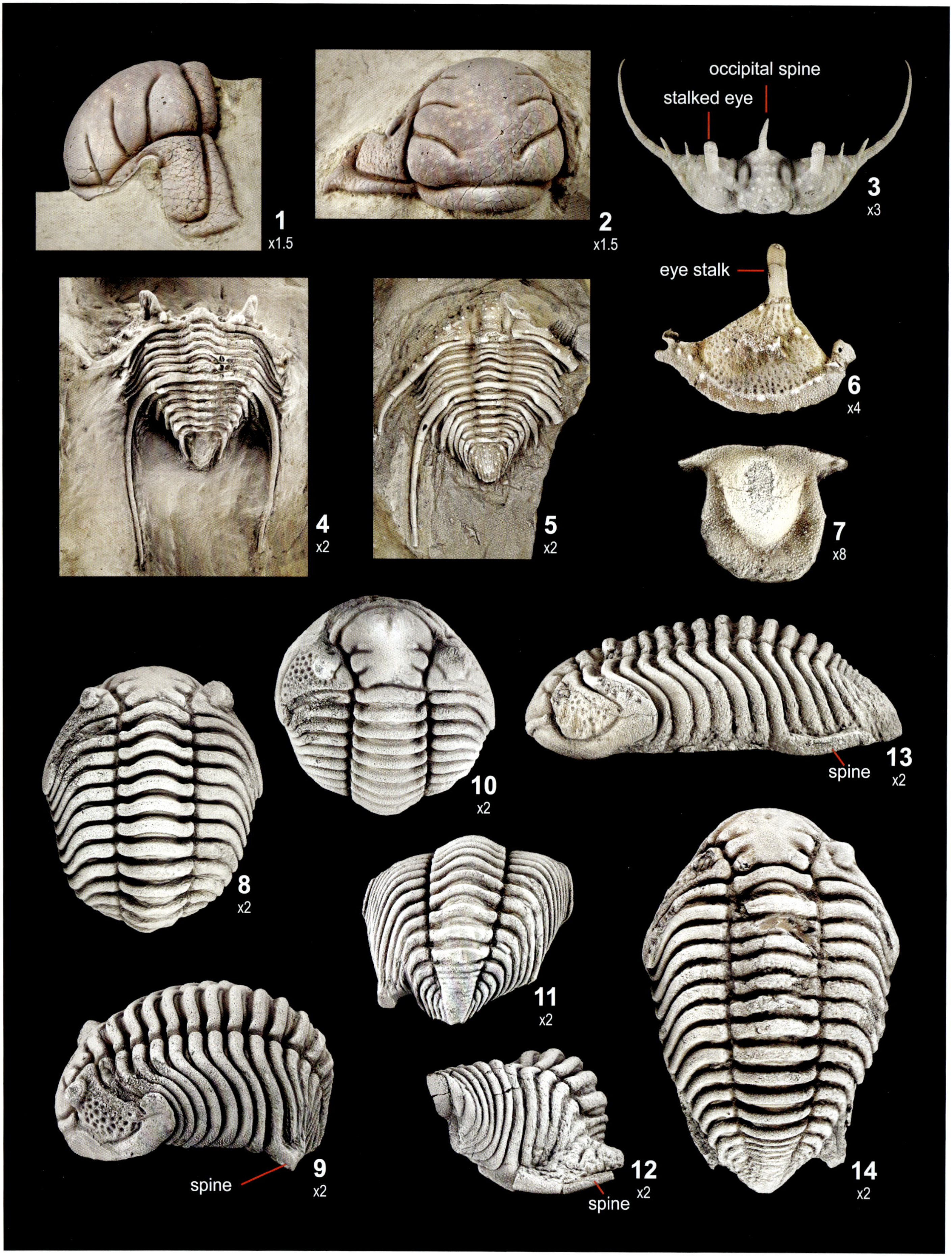
occipital spine
stalked eye
eye stalk
1
x1.5
2
x1.5
3
x3
4
x2
5
x2
6
x4
7
x8
8
x2
9
x2
spine
10
x2
11
x2
12
x2
spine
13
x2
spine
14
x2

several complete specimens (Figures 12-10-8 through 12-10-14) revealed new information about most of the exoskeleton.

Occurrence. Isolated sclerites locally common in the basal part of the Grand Detour Member in north-central Illinois and south-central Wisconsin. Complete specimens are rare.

Family ENCRINURIDAE

Subfamily ENCRINURINAE

Genus WALENCRINUROIDES

Walencrinuroides rarus (Walcott, 1877)

Figures 12-11-10 through 12-11-14

Description. An encrinurid trilobite that is approximately 20 mm in length, having 11 thoracic segments and strongly defined axial furrows. Cephalon wider than long and subtriangular in outline, genal angle rounded and projecting posteriorly. Glabella expanded anteriorly into mushroom-shaped outline, axial furrows deeply impressed and outwardly flaring forward, minimum width at lobe L2, axis flattened between furrow S0 and S3; furrows S1 through S3 deeply impressed; lobe L0 wider than L1. Furrow S0 extends laterally down to the facial suture on both sides of the cephalon, where it abruptly curves forward and terminates; the overall trend aligns with the border furrow on the free cheeks. Frontal lobe of glabella curved steeply downward to preglabellar furrow but not overhanging anterior edge of cranidium; small axial furrow present at anterior margin of glabella. Eyes relatively large, palpebral lobe positioned far out on cheek opposite lobe L2 and furrow S2; outlined adaxially by deep palpebral furrow that curves around posterior and anterior sides of visual surface; visual surface of eye bulbous and protruding. Facial sutures (gonatoparian type) form a symmetrical pair that extend from the posterior margin of the genal angle forward, bounding the adaxial side of the visual surface of the eye, then curving downward adaxially to the anterior margin of the cephalon. Thorax of 11 segments, axis moderately high and convex, slightly wider than pleural lobes, well-defined axial furrows. Profile of segments along axis evenly rounded, curving anteriorly downward to a shallow furrow, then slightly upward to form a flange that articulates with the posterior edge of the adjacent segment. Thoracic segments lack pleural furrows. Pygidium subtriangular in outline, slightly longer than wide, with five or six pleural ribs and approximately 15 axial rings.

Surface sculpture consists of relatively large, perforated tubercles on front of glabella and above border furrow on free cheeks; very small, nonperforated tubercles present on the anterior margin of the cephalon, on the anterior edge of thoracic segments, and on the pygidium; relatively large pits present on posterior parts of fixed cheeks.

Remarks. The holotype, *Ceraurus rarus*, was described by Walcott (1877) as being from the Grand Detour Member, Platteville Formation, at Beloit, Wisconsin. It was later assigned to *Encrinuroides* by DeMott (1963) and more recently to *Walencrinuroides* by Lespérance and Desbiens (1995). Additional material has been described from the Pecatonica, Mifflin, and Grand Detour Members in southern Wisconsin and northern Illinois (DeMott et al. 1987), the Upper Ordovician Esbataottine Formation, Sunblood Range, Canada (Chatterton and Ludvigsen 1976) and the Lebanon Formation of central Tennessee (Moss and Westrop 2014).

Occurrence. Isolated sclerites are abundant in the Mifflin and Grand Detour Members in north-central Illinois and south-central Wisconsin and the McGregor Member of southwestern Wisconsin, northeastern Iowa, and southeastern Minnesota. Complete specimens are common.

Figure 12-10 (Opposite page)

1, 2. ***Acanthoparypha* cf. *evitti*** Chatterton and Ludvigsen. (**1**) Lateral view of cranidium; (**2**) dorsal surface of cranidium. Mifflin Member, Grant County, Wisconsin, **1**, **2**, AWS 118.

3–7. ***Cybeloides cimelia*** Chatterton and Ludvigsen. (**3**) Photographic mosaic of cephalon showing stalked eyes and occipital spine; (**4**) dorsal surface; (**5**) dorsal surface; (**6**) free cheek showing eye stalk; (**7**) ventral surface of hypostome. Mifflin Member, Grant County, Wisconsin, **3**, AWS 137. Mifflin Member, Dixon, Illinois (locality 3), **4**, ISGS 1262; **6**, ISGS 1263; **7**, ISGS 1340. Grand Detour Member, Durand, Illinois (locality 11), **5**, BMNH 135; Thomas Baker, collector.

8–14. ***Anapliomera shirlandensis*** DeMott et al. (**8**) Dorsal surface; (**9**) lateral view showing partial spine at margin of thoracic segment 8; (**10**) dorsal surface of cephalon and part of the thorax; (**11**) posterior view of pygidium and part of thorax; (**12**) lateral view showing prominent spine at margin of thoracic segment 8; (**13**) lateral view showing prominent spine at margin of thoracic segment 8; (**14**) dorsal view. Basal 1 m (3.3 ft) of Grand Detour Member, Lightsville, Illinois (locality 7), **8–11**, ISGS 1265; **12**, ISGS 1270. Basal 1 m (3.3 ft) of Grand Detour Member, Durand, Illinois (locality 11), **13**, **14**, ISGS 1264.

Suborder PHACOPINA

Superfamily PHACOPACEA

Family PTERYGOMETOPIDAE

Subfamily EOMONORACHINAE

Genus SCEPTASPIS

Sceptaspis lincolnensis (Branson, 1909)

Figures 12-11-1 through 12-11-4

Description. A species of *Sceptaspis* that is approximately 25–30 mm in length. Cephalon wider than long, moderately convex, crescent shaped with a slight protuberance along the anterior margin, axial furrows moderately impressed, genal angle bearing long, slender spine. Glabella widest in middle of anterior lobe; expanded forward; occipital ring broad, delineated by occipital furrow and a short furrow that projects posteriorly from the occipital furrow, creating a small indentation in the anterolateral parts of the occipital ring; these lateral indentations are repeated on the axial rings of the thoracic segments. L1 lobes very small, well delineated, and inset. Lobes L2 and L3 fused to one another distally and somewhat inflated relative to median lobe; S2 furrows short and isolated from axial furrow; S3 furrows deep and extending posteromedially from axial furrow. Eyes large, palpebral lobes prominent with well-defined adaxially positioned palpebral furrows. Facial sutures (proparian type) extend adaxially from lateral cephalic margin up the eye stalk between palpebral lobe and visual surface of eye, then gradually curve downward adaxially to anterior cephalic margin. Thorax with 11 segments, well-defined axial furrows, axial lobe widest at fourth segment; pleura with prominent furrow extending abaxially from axial furrow to lateral margin of segment. Pygidium subtriangular, width slightly greater than length, approximately 20 distinct axial rings. Surface sculpture consists of relatively large tubercles on glabella; smaller tubercles on occipital ring, thorax, and pygidium; and fine granules over entire exoskeleton.

Remarks. This Platteville *Sceptaspis* is virtually identical to the type species from the coeval Upper Ordovician Plattin Limestone of Lincoln County, Missouri.

Occurrence. Isolated sclerites are abundant in the Mifflin and Grand Detour Members in north-central Illinois and south-central Wisconsin and the McGregor Member of southwestern Wisconsin, northeastern Iowa, and southeastern Minnesota. Complete specimens are rare.

Genus CALYPTAULAX

Calyptaulax plattevillensis (Delo, 1940)

Figures 12-11-7 through 12-11-9

Description. A species of *Calyptaulax* that is approximately 30 mm in length. Cephalon wider than long, moderately convex, crescent shaped, with slight upturned protuberance along medial anterior margin, axial furrows moderately impressed, genal angle rounded and projecting backward. Glabella widest in middle of anterior lobe, expanded forward; occipital ring delineated by occipital furrow and a short furrow that projects posteriorly from the occipital furrow, creating a small indentation in the anterolateral parts of the occipital ring; these lateral indentations are subtly repeated on the axial rings of the thoracic segments. L1 lobes very small, well delineated, and inset. Lobes L2 and L3 fused to one another distally. S2 furrows terminate before reaching the axial furrows; S3 furrows deep and extending posteromedially from axial furrow. Eyes large, with approximately 120 lenses; palpebral lobes prominent with well-defined palpebral furrows. Facial sutures (proparian type) extend adaxially from lateral cephalic margin up the eye stalk between palpebral lobe and visual surface of eye, then gradually curve downward adaxially to anterior cephalic margin. Thorax with 11 segments, well-defined axial furrows, axial lobe widest at fourth and fifth segments; pleura with prominent furrow extending abaxially from axial furrow to fulcrum. Pygidium subtriangular with rounded lateral margins, width greater than length, up to approximately 12 distinct axial rings. Surface sculpture consists of

Figure 12-11 (Opposite page)

1–4. ***Sceptaspis lincolnensis*** (Branson). (**1**) Dorsal surface; (**2**) lateral view; (**3**) dorsal view of pygidium and part of thorax; (**4**) dorsal surface. Mifflin Member, Grant County, Wisconsin, **1**, **2**, AWS 106; **3**, ISGS 1330. Mifflin Member, Dixon, Illinois (locality 2), **4**, JAC 102.

5, 6. **Pterygometopid hypostomes** associated with *Sceptaspis lincolnensis* and *Calyptaulax plattevillensis*. (**5**, **6**) Ventral surface. Mifflin Member, Dixon, Illinois (locality 2), **5**, ISGS 1342; **6**, ISGS 1298.

7–9. ***Calyptaulax plattevillensis*** (Delo). (**7**) Dorsal surface of cephalon; (**8**) lateral view of cephalon showing schizochroal-type eyes; (**9**) dorsal surface of complete specimen. Mifflin Member, Dixon, Illinois (locality 2), **7**, **8**, ISGS 1244. Mifflin Member, Grant County, Wisconsin, **9**, AWS 107.

10–14. ***Walencrinuroides rarus*** (Walcott). (**10**) Dorsal surface; (**11**) lateral view; (**12**) lateral view of rolled-up specimen; (**13**) view of cephalon, hypostome, and pygidium of rolled-up specimen; (**14**) dorsal view of rolled-up specimen. Mifflin Member, Dixon, Illinois (locality 2), **10**, **11**, ISGS 1534; **12–14**, ISGS 1252.

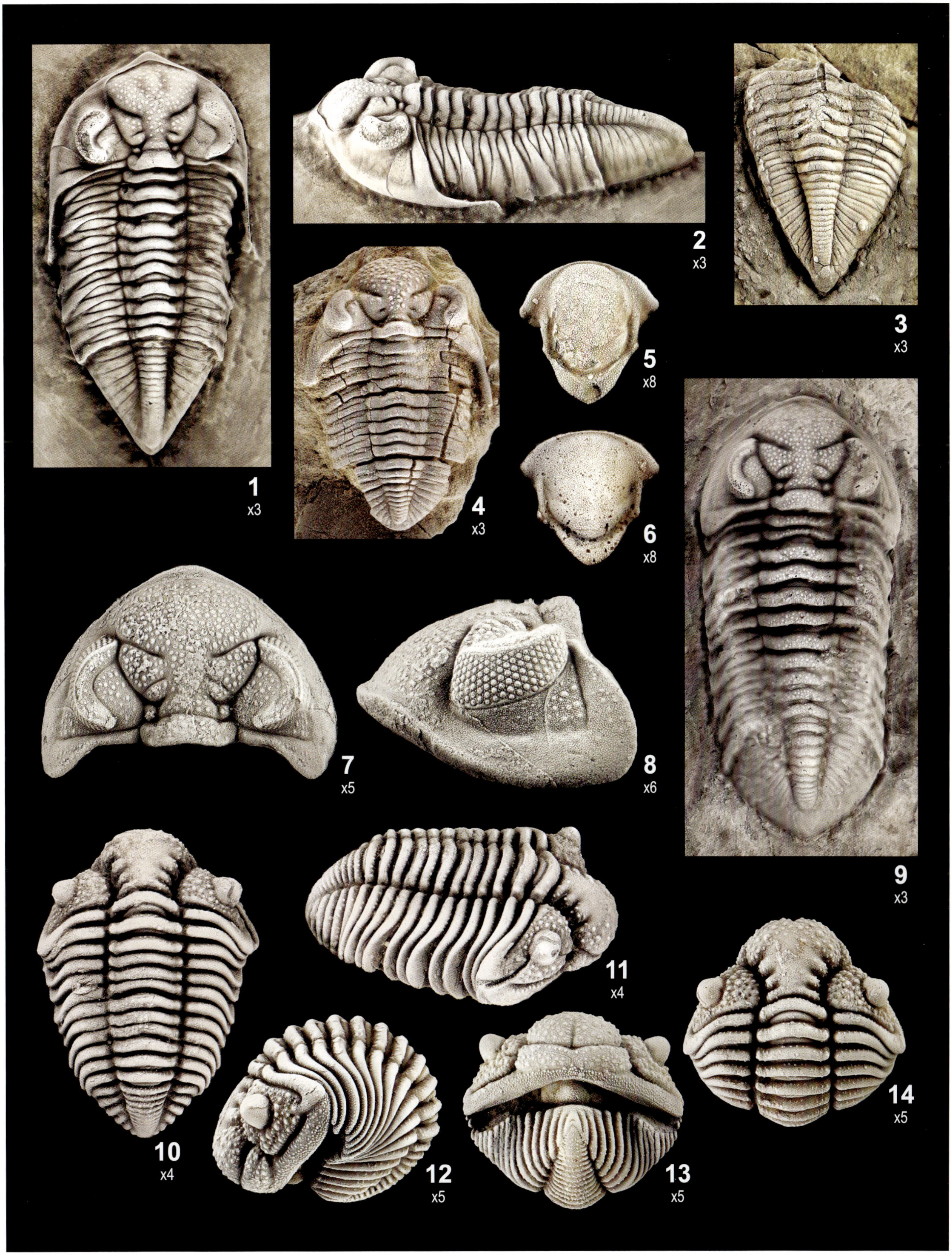
1
x3
2
x3
3
x3
4
x3
5
x8
6
x8
7
x5
8
x6
9
x3
10
x4
11
x4
12
x5
13
x5
14
x5

small tubercles on glabella, cheeks, and axial rings as well as fine granules over entire exoskeleton.

Remarks. This Platteville species was originally assigned to *Pterygometopus callicephalus* Clarke, 1897c, reassigned to *Calliops plattevillensis* Delo, 1940, and again reassigned to *Calyptaulax plattevillensis* by Ludvigsen and Chatterton (1982). It is similar to but easily distinguished from *Sceptaspis lincolnensis* in lacking genal spines.

Occurrence. Isolated sclerites are abundant in the Mifflin and Grand Detour Members in north-central Illinois and south-central Wisconsin and the McGregor Member of southwestern Wisconsin, northeastern Iowa, and southeastern Minnesota. Complete specimens are rare.

Pterygometopid Hypostomes

Figures 12-11-5 and 12-11-6

Description. Associated with specimens of *Calyptaulax* and *Sceptaspis* are isolated pterygometopid hypostomes that likely belong to one of these genera. Hypostome broadly curved along anterior margin, extending laterally into triangular anterior wings; median body ovoid in outline and defined laterally and posteriorly by narrow (tr.) border furrows; lateral border extends posteriorly, ending in poorly defined shoulders; posterior margin tongue shaped. Surface sculpture consists of very small granular tubercles; maculae indistinct.

Order PROETIDA

Family BATHYURIDAE

Genus BATHYURUS

Bathyurus (Raymondites) longispinus (Walcott, 1876)

Figures 12-12-1 through 12-12-4

Description. A species of *Bathyurus* with a cephalon that is approximately 45 mm wide as measured transversely along a line parallel with the second thoracic segment, and 55 mm long. Cephalon broadly inflated, elliptical, bearing large, long genal spines; marginal furrow moderately concave along anterior edge, becoming more deeply concave at widest point of cephalon. Cranidium widest across palpebral lobes. Glabella inflated, longer than wide, sides nearly parallel; well-defined axial furrows. Occipital lobe well developed, clearly defined anteriorly by moderately deep occipital furrow, posterior margin curved strongly backward, pointed tubercles present along posterior edge, lacking median occipital spine. Cheeks triangular in outline with inclined flanks and prominent, elevated eyes. Palpebral lobe crescent shaped, visual surface laterally convex. Facial sutures (opisthoparian type) extend from posterior cephalic margin slightly beyond outer edge of palpebral lobe adaxially around the eye between palpebral lobe and visual surface, then gradually curving downward abaxially to the marginal furrow, reversing to an adaxial orientation and intersecting the cephalic margin at an oblique angle. Free cheeks with very prominent genal spines that partly encircle the thorax and pygidium. Hypostome elongate, widest at anterior end, ventral side of median body convex, outlined at posterior and laterally by deeply impressed border furrow that disappears anteriorly; posterior lobe of central body defined by a pair of furrows that bifurcate from the border furrows but do not connect across the median body; maculae relatively smooth and distinct; anterior wings extend down sides and curve near the middle length of the median body; marginal spine on each posterior corner. Thorax with nine segments, axis moderately convex, well-defined axial furrows; prominent pleural furrows extend over half of pleural lobe width. Pygidium wider than long; axis convex, tapered and rounded at posterior end; anterior end has one well-defined axial ring furrow; five or six pleural ribs. Surface sculpture consists of fine, wavy terrace ridges on the cephalon, hypostome, thorax, and pygidium; prominent rounded tubercles on glabella and occipital ring.

Remarks. *B.* (*R.*) *longispinus* differs from all other members of the subgenus *Raymondites* in lacking an axial spine on the occipital ring and pygidium (Swisher et al. 2015).

Occurrence. According to Whittington (1953), the *B.* (*R.*) *longispinus* has been found in the "Upper Black River," Phelan and Jordan quarries, Dixon, Illinois; a similar horizon at Mineral Point, Wisconsin; and in the "Trenton" (Platteville Formation, McGregor Member, according to Stauffer

Figure 12-12 (Opposite page)
1–4. ***Bathyurus (Raymondites) longispinus*** (Walcott). (**1**) Dorsal surface of pygidium; (**2**) dorsal surface of cephalon and partial thorax; (**3**) ventral surface of hypostome; (**4**) dorsal surface of cranidium. Mifflin Member, Dixon, Illinois (locality 3), **1**, ISGS 1293; **2**, ISGS 1317. Mifflin Member, Grant County, Wisconsin, **3**, AWS 134. Grand Detour Member, Dixon, Illinois (locality 3), **4**, ISGS 1247.

5–10. ***Basiliella barrandi*** (Hall). (**5**) Dorsal surface of dolomitized specimen; (**6**) dorsal surface of dolomitized cranidium; (**7**) ventral surface of hypostome; (**8**) dorsal surface of complete specimen; (**9**) dorsal surface of pygidium; (**10**) perspective view of the complete specimen. Grand Detour Member, Lightsville, Illinois (locality 7), **5**, JAC 103. Grand Detour Member, Beloit, Illinois (locality 21), **6**, ISGS 1277; **9**, ISGS 1276. Mifflin Member, Dixon, Illinois (locality 2), **7**, ISGS 1279. Mifflin Member, Grant County, Wisconsin, **8**, **10**, AWS 105.

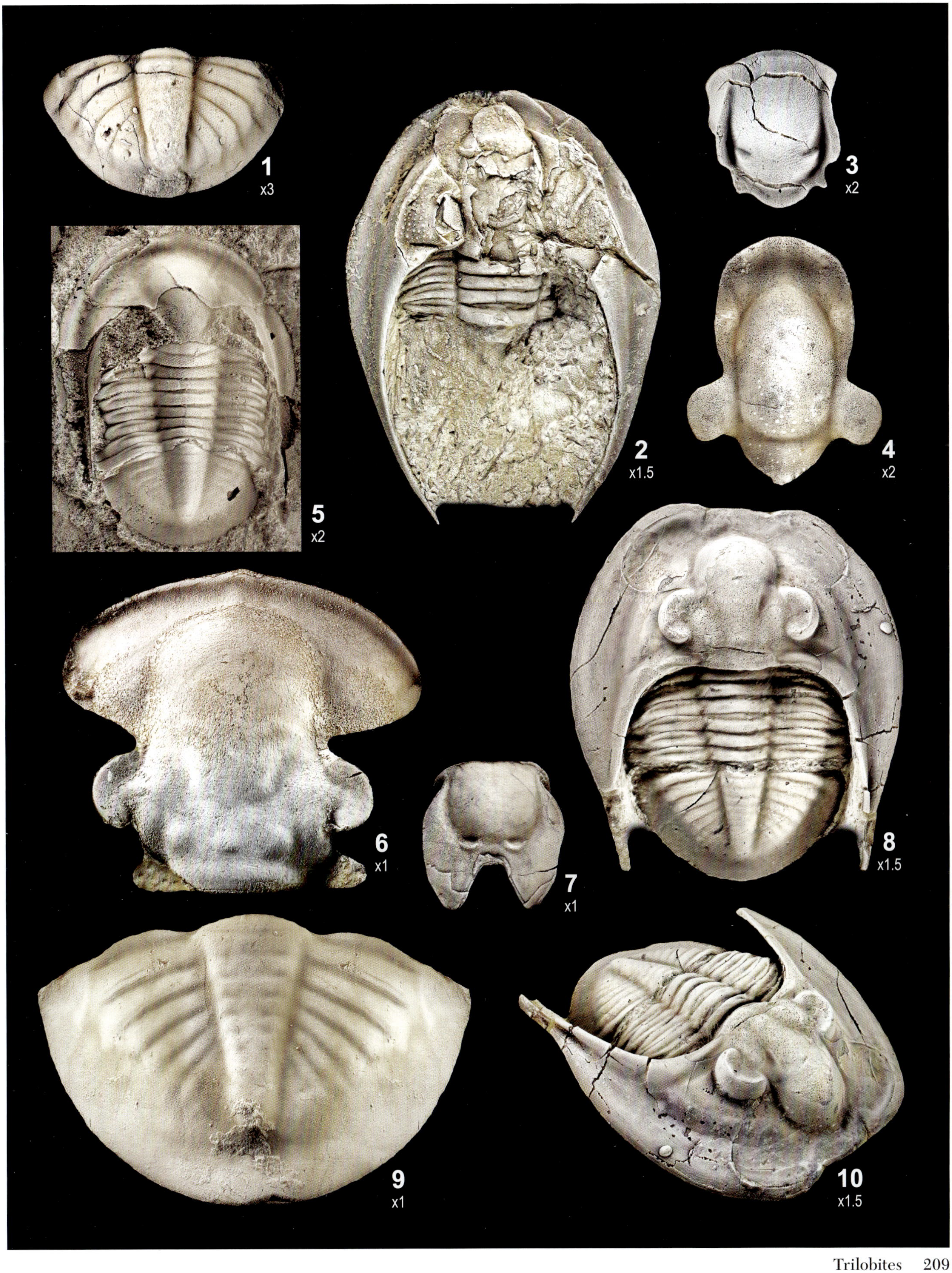
1
x3
2
x1.5
3
x2
4
x2
5
x2
6
x1
7
x1
8
x1.5
9
x1
10
x1.5

and Thiel 1941, p. 244) of Minneapolis, Minnesota. Isolated sclerites are common in the Mifflin and Grand Detour Members in north-central Illinois and south-central Wisconsin and the McGregor Member of southwestern Wisconsin, northeastern Iowa, and southeastern Minnesota. Relatively common as fragments in the Quimbys Mill, Mifflin, and Grand Detour Members in northern Illinois and southern Wisconsin. Complete specimens are very rare.

Order ASAPHIDA

Family ASAPHIDAE

Subfamily ASAPHINAE

Genus BASILIELLA

Basiliella barrandi (Hall, 1851)

Figures 12-12-5 through 12-12-10

Description. A species of *Basiliella* with a full-body length estimated to be approximately 200+ mm. Cephalon broadly but weakly inflated, subtriangular, bearing large, long genal spines; marginal furrow moderately concave along anterior edge, becoming more deeply concave along posterior half of free cheeks. Cranidium widest in front of palpebral lobes where facial sutures curve widely outward then back, meeting at anterior center line. Glabella expanding anteriorly very slightly; axial furrows weak, poorly defined posteriorly; frontal half of glabella elevated, rounded, and semiovate in outline. Weakly developed sigmoidal glabellar furrow extends adaxially from axial furrow near anterior part of palpebral lobes, curving posteriorly for short distance then back abaxially to form a very shallow furrow along a transverse line drawn through the centers of the palpebral lobes. Other glabellar furrows unclear. Glabellar tubercle positioned along axis in front of weakly developed occipital furrow and in larger specimens flanked by rounded, low-profile tubercles. Facial sutures (opisthoparian type) extend from posterior cephalic margin at a point slightly beyond outer edge of palpebral lobe, passing adaxially around the eye between palpebral lobe and visual surface, then gradually curving downward abaxially to the marginal furrow, reversing to an adaxial orientation and intersecting the cephalic margin at an oblique angle. Eyes positioned close to posterior margin of cranidium, with strong palpebral ridge and shallow furrow inside. Free cheeks with very prominent genal spines that partly encircle the thorax and most of pygidium. Hypostome length and width approximately equal; anterior lobe of median body rectangular in outline and inflated in front of prominent macular depressions; posterior lobes project back, forming fork with rounded edges; anterior wings small. Thorax with eight segments, axis moderately convex, well-defined axial furrows; prominent pleural furrows extend over half of the pleural lobe width. Pygidium wider than long; axis convex, tapered, and rounded at posterior end; approximately eight shallow axial ring furrows; seven or eight low, rounded pleural ribs. Surface sculpture consists of fine, wavy terrace ridges on the cephalon, hypostome, thorax, and pygidium.

Remarks. The type specimens, originally designated *Asaphus barrandi* Hall, 1851, are from the Turinian Bony Falls Formation exposed on the St. Mary's River in the upper peninsula of Michigan and the Platteville Formation near Platteville, Wisconsin.

Occurrence. Isolated sclerites are abundant in the Mifflin and Grand Detour Members in north-central Illinois and south-central Wisconsin and the McGregor Member of southwestern Wisconsin, northeastern Iowa, and southeastern Minnesota. Relatively common as fragments in the Quimbys Mill, Mifflin, and Grand Detour Members in northern Illinois and southern Wisconsin. Large cranidia and pygidia are somewhat common in the Grand Detour Member near Beloit, Wisconsin (locality 21). Complete specimens are rare.

Subfamily ISOTELINAE

Remarks. Isoteline trilobite fragments are relatively common in the Platteville Formation, but the lack of well-preserved, articulated specimens makes it difficult to determine precisely what genera and species are present. Further complicating identification is the current poor understanding of generic concepts and need for a formal phylogenetic analysis of the isotelines (Amati 2014). Identifiable genera include *Isotelus*, *Nahannia*, and *Vogdesia*.

Genus ISOTELUS

Isotelus platycephalus (Stokes, 1824)

Figures 12-13-9 through 12-13-13

Description. An isoteline trilobite that is approximately 80 mm wide; broad, concave cephalic and pygidial borders. Cephalon with smooth surface, anterior margin weakly pointed; convex centrally, with flat to slightly concave furrow at central anterior border, becoming more concave and distinct laterally, furrow gradually ending in front of genal angles; axial furrows lightly impressed and extending a short distance in front of the eyes; no apparent occipital furrow or median tubercle. Free cheeks bear small posteriorly directed spine in immature specimens but lack a genal spine in adults. Facial sutures (opisthoparian type) extend from posterior cephalic margin at a point well beyond outer edge of palpebral lobe; curved adaxially up the eye stalk between palpebral lobe and visual surface, then gradually curving downward abaxially, then reversing to an adaxial orientation and joining the opposing free cheek along a short medial suture; eyes prominent, set back behind mid-length of cephalon. Thorax with eight segments, all approximately the same shape, diminishing slightly in width posteriorly; axial lobe relatively narrow, axial furrows well developed; inner half of pleura crossed by well-developed pleural furrows. Pygidium approximately equal in size and shape to the cephalon; axial lobe tapering abruptly, terminal end falls short of posterior margin and is well defined; subtle surface expression of as many as 20 vestigial segments on axial lobe. Surface sculpture consists of numerous small pits covering entire exoskeleton, particularly evident in larger specimens; terrace ridges on cephalic and pygidial doublure, hypostome, and distal edges of thoracic segments.

Remarks. This Platteville species of *Isotelus* most closely resembles *I. platycephalus* from the Black River Limestone at St. Joseph Island in the northwest corner of Lake Huron in southernmost Ontario, Canada. These Black River rocks correlate with the Bony Falls Limestone of the Escanaba River section in the upper peninsula of Michigan (Darby and Stumm 1965) and upper part of the Platteville Formation in northern Illinois (Templeton and Willman 1963). *I. platycephalus* is similar to *I. gigas* but differs from the latter in having a more rounded and less triangular cephalon and pygidium.

Occurrence. Isolated sclerites are abundant in the Mifflin and Grand Detour Members in north-central Illinois and south-central Wisconsin and the McGregor Member of southwestern Wisconsin, northeastern Iowa, and southeastern Minnesota. Complete specimens are rare.

Isotelus aff. *homalonotoides* (Walcott, 1877)

Figures 12-14-1 through 12-14-4

Description. An isoteline trilobite presently known from isolated cranidia, free cheeks, and pygidia. Cranidium length and width approximately equal; anterior branch of facial suture in front of eyes less divergent than most isotelines; converging anteriorly along nearly straight flank to a central angle of 100°, where free cheeks meet. Glabella convex, bounded anteriorly by an abrupt flexure between it and the flat preglabellar field. Axial furrows moderately impressed, slight abaxial curve from posterior cephalic margin to base of palpebral lobes, reversing to adaxial orientation for a short distance before turning back abaxially and merging with prominent border furrow. The network of furrows outlines the fixed and free cheeks. The median tubercle is positioned in front of a weak occipital furrow and behind a posterior margin of palpebral lobes. Occipital ring weakly defined. Eye stalks stand well above the glabella. Pygidium subtriangular, slightly wider than sagittal length, axial furrows well defined, posterior margin of axial lobe blunt and prominent. Pleural regions slope gently outward, then abruptly turn down more steeply at border furrow; central posterior of edge of pygidium flared outward. The free cheek terminates in a long, slender spine. Surface sculpture consists of numerous pits and very small, perforated tubercles; tubercles on free cheeks are aligned end to end, forming unique terrace ridges.

Remarks. This Platteville isoteline is similar to *I. homalonotoides* as described by Walcott (1877) from the Upper Ordovician of New York and by DeMott et al. (1987) from the slightly younger Guttenberg Formation at Clayton County, Iowa. The Platteville specimens differ from *I. homalonotoides* in having a central angle of 100° instead 80° where the free cheeks intersect and in having unique terrace ridges formed by chains of small, perforated tubercles.

Occurrence. Presently known from isolated sclerites in the Mifflin Member in northern Illinois and southern Wisconsin. No known complete specimens.

Genus VOGDESIA

Vogdesia simplex (Raymond and Narraway, 1910)

Figures 12-13-1 through 12-13-7

Description. A species of *Vogdesia* that is approximately 70 mm long and 40 mm wide; flat cephalic and pygidial borders. Cephalon broadly inflated, longitudinal convexity even throughout; axial furrows effaced except for slight depression between posterior margin and palpebral lobes; occipital furrow effaced, no expression of occipital ring or median tubercle on shell exterior. Free cheeks somewhat triangular in outline, genal angle rounded, lacking spine. Facial sutures (opisthoparian type) extend from posterior cephalic margin at a point well beyond outer edge of palpebral lobe; curved abaxially and forward and passing adaxially up the eye stalk between palpebral lobe and visual surface, then gradually curving downward abaxially then reversing to an adaxial orientation and joining the opposing free cheek along a short medial suture; eyes prominent and relatively large. Thorax with eight segments, all approximately the same size and

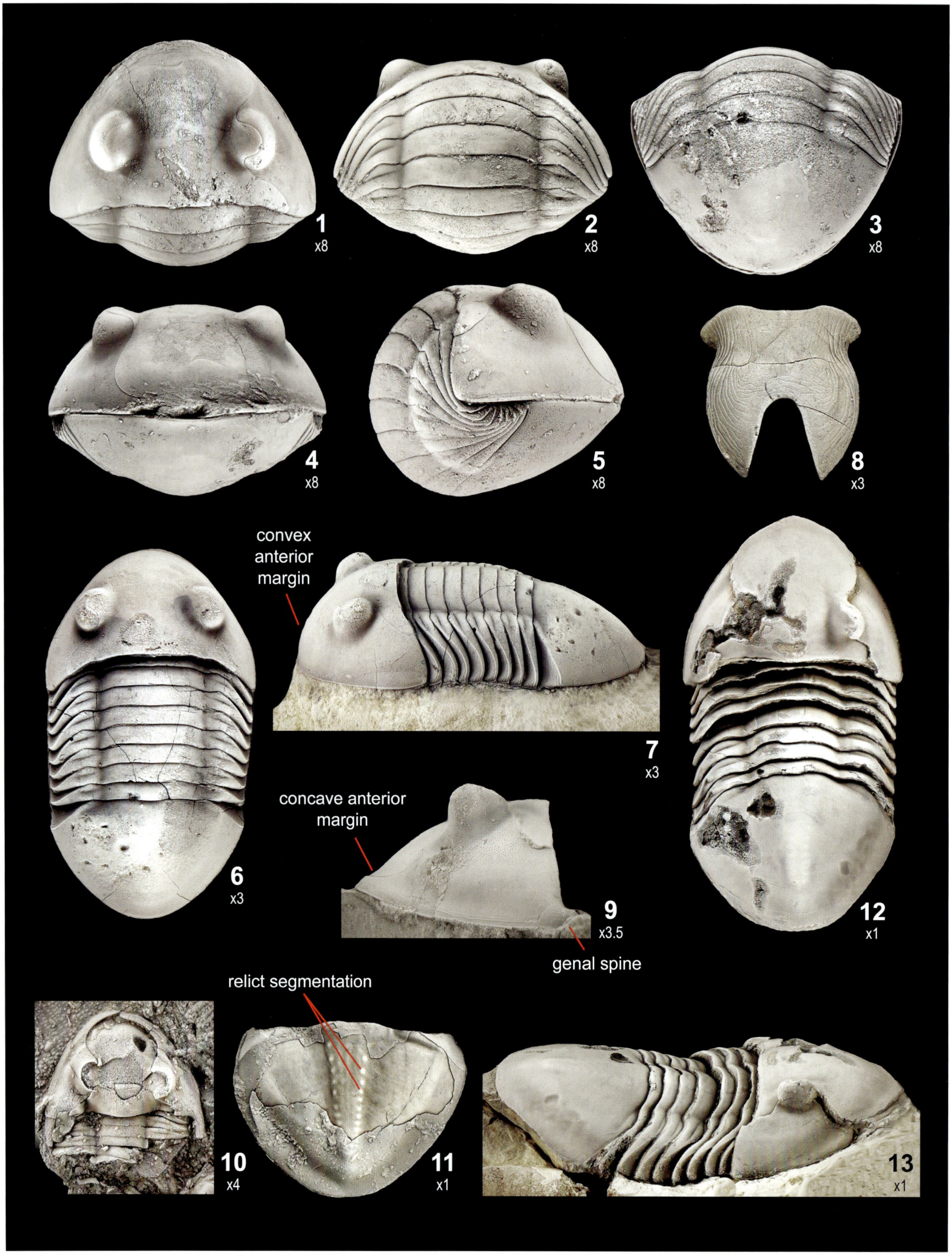
1
x8
2
x8
3
x8
4
x8
5
x8
8
x3
convex
anterior
margin
6
x3
7
x3
12
x1
concave anterior
margin
9
x3.5
genal spine
relict segmentation
10
x4
11
x1
13
x1

shape, axial furrows well developed, parallel and in line with center of palpebral lobes; inner half of pleura crossed by well-developed pleural furrows. Pygidium approximately equal in size and shape to the cephalon; axial furrows very weak. Surface sculpture consists of numerous very small pits covering entire exoskeleton, particularly evident in larger specimens; terrace ridges on doublure and distal edges of thoracic segments.

Remarks. The type specimens of *V. simplex*, from the Platteville Formation of Minnesota, were assigned originally to the genus *Onchometopus* by Raymond and Narraway (1910), later to *Homotelus* (Raymond 1920), and then to *Isotelus* (DeMott 1963; DeMott et al. 1987). More recently, Amati (2014) tentatively reassigned the species to *Vogdesia* based on the relatively elongate and posteriorly tapering pygidium and weak effacement of axial furrows on the posterior of the cephalon. As presently understood, *V. simplex* is distinguished from other Platteville isoteline trilobites in lacking concave borders on the cephalon and pygidium.

Occurrence. Described by Raymond and Narraway (1910) as being abundant in the "buff limestone, 20 feet above the St. Peter Sandstone on the Straight River, two miles south of Faribault, Minnesota" (p. 52). Isolated sclerites are abundant in the Mifflin and Grand Detour Members in north-central Illinois and south-central Wisconsin and the McGregor Member of southwestern Wisconsin, northeastern Iowa, and southeastern Minnesota. Complete specimens common.

Figure 12-13 (Opposite page)

1–7. ***Vogdesia simplex*** (Raymond and Narraway). (**1**) Cephalon of complete and rolled-up specimen; (**2**) thorax of rolled-up specimen; (**3**) pygidium of rolled-up specimen; (**4**) juncture of cephalon and pygidium in rolled-up specimen; (**5**) lateral view of rolled-up specimen; (**6**) dorsal surface; (**7**) lateral view showing convex profile of anterior margin. Mifflin Member, Dixon, Illinois (locality 2), **1–5**, ISGS 1256. Mifflin Member, Grant County, Wisconsin, **6**, **7**, AWS 139.

8. **Isoteline hypostome** associated with *Isotelus* and *Vogdesia*. Mifflin Member, Dixon, Illinois, **8**, ISGS 1318.

9–13. ***Isotelus platycephalus*** (Stokes). (**9**) Lateral view of cephalon showing small genal spine and relatively flat frontal profile with slight concavity of anterior margin; (**10**) small specimen with relatively large genal spines; (**11**) dorsal surface of pygidium showing relict segmentation on the interior cast; (**12**) dorsal surface; (**13**) lateral profile. Mifflin Member, Dixon, Illinois (locality 2), **9**, ISGS 1325; **10**, ISGS 1258; **11**, ISGS 1321. Grand Detour Member, Durand, Illinois (locality 11), **12**, **13**, BMNH K1958.

Isoteline Hypostome

Figure 12-13-8

Description. Associated with specimens of *Isotelus* and *Vogdesia* are isolated isoteline hypostomes that likely belong to one or both of these genera. Hypostome length slightly greater than width, typical asaphid type, forked with rounded V-shaped posterior notch, maculae and median body not well developed, anterior wings small. Surface sculpture consists of numerous well-defined terrace ridges that are particularly prominent adjacent to the lateral margins.

Genus NAHANNIA

Nahannia* cf. *mediocrisulcata
Chatterton and Ludvigsen, 1976

Figures 12-14-5 and 12-14-6

Description. An isoteline trilobite presently known from isolated cranidia and hypostomes. Cranidium subquadrate in outline and somewhat inflated; anterior margin broadly convex and very slightly pointed at medial edge; fixed cheeks very narrow. Glabella slightly inflated; axial furrows very shallow, converging forward to a point between the palpebral lobes, then diverging forward and downward in front of the eyes; occipital furrow very shallow; occipital lobe longer medially than laterally; median tubercle in front of occipital furrow. Hypostome typical isoteline form, width greater than sagittal length, posterior margin deeply notched; lateral margins indented near anterior margin, well-developed anterior wings; a pair of deeply impressed maculae on ventral surface opposite lateral indentations; ventral surface sculpture with prominent terrace ridges.

Remarks. The overall shapes of the cranidium and hypostome are most similar to *Nahannia mediocrisulcata* Chatterton and Ludvigsen (1976) from the Upper Ordovician Esbataottine Formation, Sunblood Range, Canada.

Occurrence. Present mainly as disarticulated parts in the Mifflin and Grand Detour Members of northern Illinois and southern Wisconsin. No known complete specimens.

Superfamily REMOPLEUROIDEA

Family REMOPLEUROIDIDAE

Subfamily HYPODICRANOTINAE

Genus HYPODICRANOTUS

Hypodicranotus* cf. *striatulus (Walcott, 1875)

Figures 12-14-7 through 12-14-9

Description. A relatively small remopleuroidid trilobite with posterolateral cephalic corners extended into a slender, curved spine that is in juxtaposition to a longer, crescent-shaped genal spine. Cephalon with smooth surface, anterior margin slightly tapered and evenly rounded; convex centrally. Eye lobes long, crescent shaped, broadly curved from anterolateral margin backward to near posterior cephalic border in line with axial furrows. Facial sutures (opisthoparian type) extend from posterior cephalic margin at a point well beyond outer edge of the palpebral lobe; curved adaxially up the eye stalk between the palpebral lobe and visual surface, then gradually curving downward adaxially to anterolateral margin. Hypostome approximately three times longer than wide and shaped like a tuning fork with two long, slender posterolateral projections that taper to sharp points. Thorax with 11 segments, wide axis, prominent articulating processes and sockets, diagonal pleural furrows, and backwardly curved and pointed pleurae. Pygidium small, oval, lacking furrows; a pair of spines on posterior margin. Surface sculpture of very small tubercles particularly well developed on anterior parts of the glabella.

Remarks. Most similar to *H. striatulus* from the Trenton Limestone, Trenton Falls, Oneida County, New York.

Occurrence. Isolated sclerites are abundant in the Mifflin and Grand Detour Members in north-central Illinois and south-central Wisconsin and the McGregor Member of southwestern Wisconsin, northeastern Iowa, and southeastern Minnesota. Complete specimens are rare.

Order HARPETIDA

Suborder HARPETINA

Family HARPETIDAE

Genus DOLICHOHARPES

Dolichoharpes* cf. *reticulata Whittington, 1949

Figures 12-14-10 through 12-14-12

Description. A harpetid trilobite that is approximately 40 mm long. Cephalon outline oval; genal elongations 1.5 times longer than sagittal length of cephalon and curved inward posteriorly, nearly encompassing the thorax and pygidium; maximum width transverse to occipital lobe. Axial furrows deeply impressed, joined at anterior end of glabella. Glabella convex, narrowing forward with well-defined occipital ring marking highest point of cephalon; subtle medial occipital tubercle; occipital furrow well defined and curved forward slightly. L1 lobes relatively large and subtriangular in outline, bearing a narrow ridge that curves forward and back, the pair meeting adaxially; S1 furrows deep and oriented at 45° angle to sagittal line. Lateral to L1 lobe is a rounded depression (ala) lacking pores or tubercles that narrows abruptly forward into the axial furrow (Figure 12-14-11). Furrows S2 and S3 very weakly impressed. Cheek lobes inflated and steeply sloping; preglabellar field slopes outward to upwardly concave bilaminar fringe. Marginal suture separates dorsal and ventral layers of fringe exoskeleton. Eyes small, positioned forward in line with L3 lobes. Thorax with at least 19 segments; axis narrow and arched; axial furrows weakly impressed; pleural furrows long, straight, and relatively deep, extending from axial lobe past downturned margin adjacent to articulating facets. Pygidium length approximately one-third of width, posterior margin rounded; consisting of at least four axial rings and corresponding pleural lobes. Surface sculpture on dorsal surface of cephalon consists of small pits and reticulate ridges covered with very small tubercles positioned

Figure 12-14 (Opposite page)

1–4. ***Isotelus* aff. *homalonotoides*** (Walcott). (**1**) Dorsal surface of pygidium; (**2**) dorsal surface of cranidium; (**3**) enlargement of free cheek showing characteristic small, perforated tubercles aligned to form terrace ridges; (**4**) free cheek bearing a genal spine and prominent eye stalk. Mifflin Member, Grant County, Wisconsin, **1**, AWS 122; **2**, AWS 126; **3**, **4**, AWS 123.

5–6. ***Nahannia* cf. *mediocrisulcata*** Chatterton and Ludvigsen. (**5**) Ventral surface of hypostome; (**6**) dorsal surface of cranidium. Mifflin Member, Grant County, Wisconsin, **5**, ISGS 1251; **6**, AWS 116.

7–9. ***Hypodicranotus* cf. *striatulus*** (Walcott). (**7**) Dorsal surface of cephalon and part of thorax; (**8**) lateral view showing prominent crescent-shaped eye and characteristic genal spine; (**9**) ventral surface of tuning fork-like hypostome. Mifflin Member, Grant County, Wisconsin, **7**, **8**, AWS 112; **9**, AWS 113.

10–12. ***Dolichoharpes* cf. *reticulata*** Whittington. (**10**) Complete specimen as found in the field before preparation; photographed by Allan Scheer; used with permission; (**11**) dorsal view of the complete specimen after preparation; inked and smoked with sublimate of ammonium chloride prior to being photographed by Tom Whiteley; used with permission; (**12**) interior view of dorsal fringe lamella; pustules are the interior wall of dorsal pits that align with similar pustules, projecting upward from the ventral fringe lamella. Mifflin Member, Grant County, Wisconsin, **10**, **11**, AWS 140. Mifflin Member, Dixon, Illinois (locality 2), **12**, ISGS 1257.

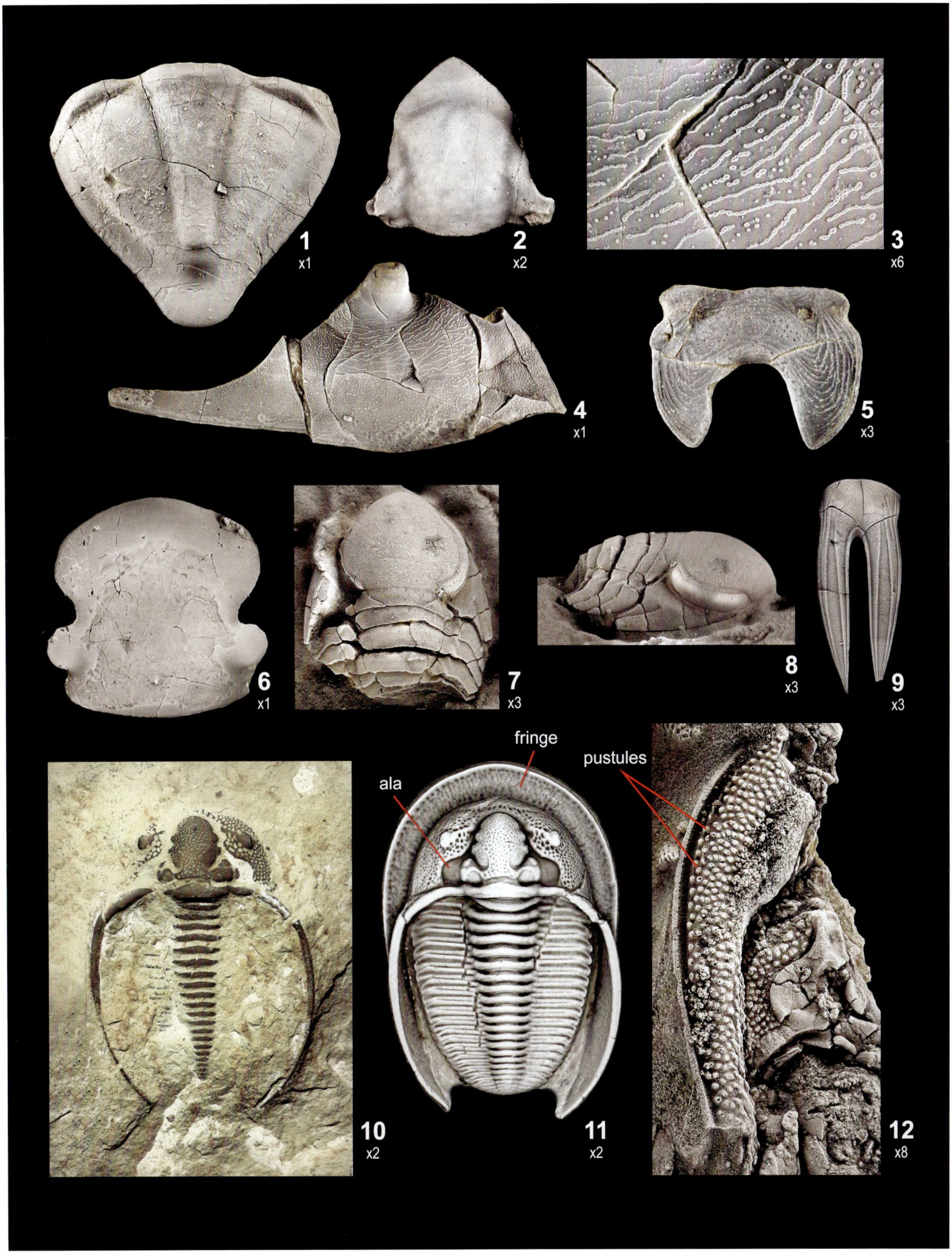
1
x1
2
x2
3
x6
4
x1
5
x3
6
x1
7
x3
8
x3
9
x3
fringe
pustules
ala
10
x2
11
x2
12
x8

mainly at junctions between ridges; pits on dorsal surface of fringe extend inward through upper lamella connecting with similar pits in lamella of ventral surface.

Remarks. This Platteville harpetid is very similar to *Dolichoharpes reticulata* Whittington, 1949, from the Upper Ordovician Edinburg Limestone of Virginia.

Occurrence. Isolated sclerites are abundant in the Mifflin and Grand Detour Members in north-central Illinois and south-central Wisconsin and the McGregor Member of southwestern Wisconsin, northeastern Iowa, and southeastern Minnesota. Some of the finest specimens of *D. reticulata* have been found in the Mifflin Member in Grant County, Wisconsin. Complete specimens are rare.

CHAPTER 13
OSTRACODES

The Platteville ostracodes are locally abundant but relatively low in diversity. Seven species representing the Orders Leperditicopida, Palaeocopida, and Podocopida are present (Appendix 2). In comparison, the overlying and slightly younger Upper Ordovician Decorah Formation in the upper Midwest contains 23 genera and 78 species (Swain et al. 1961; Swain and Cornell 1987). Platteville ostracodes are particularly abundant and collectable on shaley bedding planes in the Mifflin and Grand Detour Members and their lateral equivalents in the upper Midwest outcrop belt.

Ostracodes are primarily microscopic crustaceans with a long history extending from at least the Ordovician Period to the present. For much of their evolutionary history, they have occupied marine, brackish, and fresh water in depths ranging from shoreline to 2,800 m (Horne 2003). Although some ostracodes, such as the Platteville genus *Eoleperditia*, can be nearly a centimeter in length, most adult forms are 1 mm or less and require a hand lens or microscope to observe their morphologic details.

Ostracodes have a laterally compressed, bivalved **carapace** (**"shell"**) consisting of **right** and **left valves** that are calcified. The valves articulate along a **hinge** on the **dorsal** edge of the carapace, allowing the animal to extend its appendages along the **ventral** or free margin for feeding and locomotion. The surface of the carapace can be smooth and featureless or ornamented with **tubercles** (small, rounded protuberances), **reticulation** (network of polygonal to circular pits), or **punctae** (small pits), or it can be sculpted with **lobes** (elongate elevations) and **sulci** (elongated depressions), or both. The interior of the carapace is referred to as the **domicilium**. Some species have an **eye tubercle** with a transparent lens of calcite, indicating some level of vision. The **adductor** or closing muscles that extend between valves commonly leave an impression on the interior surface of the valves, forming an **adductor muscle scar**. In some ostracodes, the muscle scar may have some subtle expression on the exterior of the valves. **Height** is the maximum dimension of a valve from the dorsal to ventral margins measured perpendicular to the maximum **length**, which is parallel to the hinge line. The adductor muscle and eye are generally located in front of the carapace mid-length. **Anterior** is sometimes shown in the literature as a forward-pointing arrow.

Primary sources used in the identification of the Platteville ostracodes include Ulrich (1892c, 1894b), Kay (1934, 1940), Kesling (1960), Swain et al. (1961), Swain and Cornell (1987), and Williams and Vannier (1995). Morphologic terms follow Moore (1961), and suprageneric classification follows Whatley et al. (1993).

SYSTEMATIC DESCRIPTIONS

Phylum ARTHROPODA

Class CRUSTACEA

Subclass OSTRACODA

Order LEPERDITICOPIDA

Description. Ostracodes with a long, straight hinge, thick well-calcified shell, and large adductor muscle scar.

Remarks. Although it has traditionally been assumed that the leperditicopids (Ordovician-Devonian) are ostracodes, this assignment is considered tenuous because very little is known about their soft-part anatomy (Siveter 2008).

Family LEPERDITIIDAE

Genus EOLEPERDITIA

Eoleperditia fabulites (Conrad, 1843)

Figures 13-1-5 through 13-1-8

Description. Shell thick, smooth, approximately 12 mm long and 8 mm high; inequivalved, with sinuous margin in ventral view; right valve overlaps left valve along ventral margin; height greatest at posterior end behind mid-length of valve in lateral view. Hinge line straight, anterior and posterior margins evenly curved. Anterior cardinal angle approximately 125°, posterior cardinal angle approximately 135°. Adductor muscle scar relatively large. Eye tubercle low and inconspicuous, positioned between adductor muscle and hinge.

Remarks. Described originally by Conrad (1843) as being from the Platteville Formation at Mineral Point, Wisconsin.

Occurrence. Abundant in the Mifflin and Grand Detour Members in north-central Illinois and south-central Wisconsin and the McGregor Member of southwestern Wisconsin, northeastern Iowa, and southeastern Minnesota. Locally common in the Pecatonica and Quimbys Mill Members.

1
x20
frill
anterior node
S2 sulcus
2
x20
3
x20
4
x20
5
x4
adductor muscle
scar
6
x10
7
x4
eye tubercle
8
x4
9
x15
10
x15
11
x15
12
x25
13
x25
14
x25
15
x20
16
x20
17
x20
18
x20
19
x25
20
x25
21
x25
22
x25
23
x25
24
x20
25
x20
26
x20

Order PALAEOCOPIDA

Description. Ostracods with long and straight dorsal margin, smooth outer surface, or ornament; lobes, sulci, and adventral structure (ventral frill). Sexual dimorphism commonly reflected in the shell morphology.

Family EURYCHILINIDAE

Genus EURYCHILINA

Eurychilina subradiata Ulrich, 1894b

Figures 13-1-1 through 13-1-4

Description. Shell semicircular in outline; long, straight hinge, approximately 2 mm long and 1.5 mm high. Central part of shell composed of lobes and sulci, S2 (the main) sulcus wide and deep, posterior to a prominent preadductorial node. Frill wide and subtly radiate, set at an obtuse angle to the domicilium to form a sharp angulation along the ventral margin. In lateral view, a relatively deep, semicircular furrow extends from the anterior and posterior ends of the shell, outlining a thickened dorsal ridge to the frill; ventral side of the ridge gently concave or flat. Surface sculpture smooth or punctate. Dimorphic: heteromorphs (supposed females) have a broader and more convex ventral frill than tecnomorphs (juveniles and supposed males).

Remarks. Described originally by Ulrich (1894b) as being from the Platteville Formation at Dixon, Illinois, and Mineral Point, Wisconsin.

Occurrence. Abundant locally in the Mifflin Member of northern Illinois and southern Wisconsin.

Family APARCHITIDAE

Genus BULLATELLA

Bullatella granilabiata (Ulrich, 1892c)

Figures 13-1-12 through 13-1-14

Description. Subelliptic-subquadrate in outline, approximately 1.5 mm long and 1.0 mm high. Hinge margin straight and about two-thirds of shell length; anterior and posterior cardinal angles nearly equal at 135°; valves equal to slightly subequal in size, overlapping along free margin, swollen midventrally. Low, narrow ridge or row of tubercles adjacent to free margin of each valve. Surface sculpture finely punctate. A faint, smooth spot (marking the point of attachment of the adductor muscle) occurs just in front and above center.

Remarks. This species was originally assigned to *Aparchites* by Ulrich (1892c) based on specimens collected from the Decorah Formation of Minnesota but was reassigned to the new genus *Bullatella* by Swain et al. (1961) based on mid-central inflation of the valves.

Occurrence. Common on shaley bedding planes in the Mifflin and Grand Detour Members of northern Illinois and southern Wisconsin.

Family LEPERDITELLIDAE

Genus LEPERDITELLA

Leperditella* cf. *tumida (Ulrich, 1892c)

Figures 13-1-9 through 13-1-11

Description. Shell smooth, approximately 2.5 mm long, 2.0 mm high. Right valve overlaps left valve along ventral margin. Height greatest behind mid-length of valve in lateral view. Hinge line straight, anterior and posterior margins

Figure 13-1 (Opposite page) Arrow indicates toward anterior.
1–4. ***Eurychilina subradiata*** Ulrich. (**1**) Left valve interior showing the frill; (**2**) left valve exterior showing puncta on carapace in front of the anterior node and behind the S2 sulcus; (**3**) right valve exterior lacking puncta; (**4**) ventral view of articulated valves. Mifflin Member, Dixon, Illinois (locality 2), **1**, ISGS 1246; **2**, **4**, ISGS 1357; **3**, ISGS 1356.

5–8. ***Eoleperditia fabulites*** (Conrad). (**5**) Left valve interior; (**6**) photographic enlargement showing adductor muscle scar on interior of left valve; (**7**) ventral view showing overlap of right valve onto left valve; (**8**) left valve exterior showing eye tubercle. Mifflin Member, Dixon, Illinois (locality 2), **5**, **6**, ISGS 1369; **7**, **8**, ISGS 1353.

9–11. ***Leperditella* cf. *tumida*** (Ulrich). (**9**) Left valve exterior; (**10**) ventral view showing overlap of right valve onto left valve; (**11**) right valve exterior. Mifflin Member, Dixon, Illinois (locality 2), **9**, **10**, ISGS 1358; **11**, ISGS 1370.

12–14. ***Bullatella granilabiata*** (Ulrich). (**12**) Left valve exterior showing very fine puncta; (**13**) ventral view of articulated valves showing rows of small tubercles along edge of commissure; (**14**) anterior view of articulated valves. Mifflin Member, Dixon, Illinois (locality 2), **12–14**, ISGS 1355.

15–18. ***Krausella inaequalis*** Ulrich. (**15**) Oblique right valve exterior showing short posterior spine; (**16**) left valve exterior; (**17**) ventral margin of articulated valves showing overlap of right valve onto left valve; (**18**) dorsal margin of articulated valves. Mifflin Member, Dixon, Illinois (locality 2), **15**, ISGS 1376; **16**, ISGS 1377; **17**, ISGS 1373; **18**, ISGS 1375.

19–23. ***Schmidtella crassimarginata*** Ulrich. (**19**) Right valve interior; (**20**) left valve exterior showing slight overlap of right valve; (**21**) dorsal view of articulated valves; (**22**) left valve exterior showing slight overlap of right valve; (**23**) ventral margin of articulated valves. Mifflin Member, Dixon, Illinois (locality 2), **19**, ISGS 1367; **20**, **21**, ISGS 1354; **22**, **23**, ISGS 1366.

24–26. ***Leperditella canalis*** (Ulrich). (**24**) Left valve exterior; (**25**) ventral view of left valve; (**26**) left valve exterior. Mifflin Member, Dixon, Illinois (locality 2), **24**, **25**, ISGS 1368; **26**, ISGS 1372.

evenly curved, anterior and posterior cardinal angles obtuse.

Remarks. This Platteville species is similar in overall shape and dimensions to *L. tumida* from the approximately coeval Tyrone Limestone at High Bridge, Kentucky.

Occurrence. Common in the Mifflin and Grand Detour Members in north-central Illinois and south-central Wisconsin and the McGregor Member of southwestern Wisconsin, northeastern Iowa, and southeastern Minnesota.

Leperditella canalis (Ulrich, 1892c)

Figures 13-1-24 through 13-1-26

Description. Shell smooth, approximately 2.0 mm long, 1.5 mm high. Height greatest behind mid-length of valve in lateral view; anterior and posterior margins evenly curved; posterior more broadly curved than anterior, posterior cardinal angle rounded; anterodorsal sulcus subtle.

Remarks. Described by Ulrich (1892c) as being from the Platteville Formation at Mineral Point, Wisconsin. *L. canalis* differs from *L.* cf. *tumida* in being slightly smaller, having rounded cardinal angles, and having a more distinct anterodorsal sulcus.

Occurrence. Common in the Mifflin and Grand Detour Members in north-central Illinois and south-central Wisconsin and the McGregor Member of southwestern Wisconsin, northeastern Iowa, and southeastern Minnesota. Abundant locally in the Mifflin Member of northern Illinois and southern Wisconsin.

Genus SCHMIDTELLA

Schmidtella crassimarginata Ulrich, 1892c

Figure 13-1-19 through 13-1-23

Description. Shell ovate, approximately 1.5 mm long, 1.3 mm high. Inflated mostly in dorsal region, which tends to be umbonate. Hinge straight; right valve slightly larger than left valve and overlapping it ventrally. No sulcus; prominent furrow extends around the carapace ends and ventral margin to define a subtle ridge.

Remarks. Described by Ulrich (1892c) as being from the Platteville Formation at Mineral Point, Wisconsin.

Occurrence. Abundant in the Mifflin and Grand Detour Members in north-central Illinois and south-central Wisconsin and the McGregor Member of southwestern Wisconsin, northeastern Iowa, and southeastern Minnesota.

Order PODOCOPIDA

Description. Asymmetrical ostracods with curved dorsal margin.

Family KRAUSELLIDAE

Genus KRAUSELLA

Krausella inaequalis Ulrich, 1894b

Figure 13-1-15 through 13-1-18

Description. Shell subelliptic and elongate, approximately 2 mm long and 1 mm high; smooth exterior surface. Valves relatively thick and very unequal in shape. Left valve overlaps right valve; right valve moderately convex, with subparallel dorsal and ventral margins. Posterior edge drawn out into a short, blunt, spinelike process extending beyond edge of left valve.

Remarks. First reported by Ulrich (1894b) from the Platteville Formation at Dixon, Illinois.

Occurrence. Common on shaley bedding planes in the Mifflin and Grand Detour Members in north-central Illinois and south-central Wisconsin and the McGregor Member of southwestern Wisconsin, northeastern Iowa, and southeastern Minnesota.

CHAPTER 14
ECHINODERMS

At least 30 echinoderm species are present in the Platteville Formation, including members of the classes Crinoidea, Paracrinoidea, Rhombifera, Asteroidea, Cyclocystoidea, Echinoidea, and Holothuroidea. The most abundant and diverse echinoderms are monocyclic and dicyclic inadunate and camerate crinoids. Disarticulated echinoderm skeletal plates (ossicles) are locally very abundant, but complete specimens are rare owing to the rapid disintegration of connective tissues after death and quick dispersal of plates by physical and chemical processes. Only in rare instances of rapid but gentle burial was the whole articulated echinoderm skeleton preserved intact. The largest number of identifiable Platteville echinoderm specimens has been collected from the Mifflin and Grand Detour Members and their lateral equivalents within the upper Midwest region (Appendix 2).

The Echinodermata (Gr., *Echinos*, hedgehog + *derma*, skin or "spiny skinned") range in age from Early Cambrian to the present. Various authors have recognized as many as 20 classes, only five of which are living today: the Crinoidea (crinoids), Asteroidea (starfish), Ophiuroidea (brittle stars), Echinoidea (sea urchins), and Holothuroidea (sea cucumbers). Because of their intolerance to salinity variations, they are largely restricted to near-normal marine environments ranging from intertidal to abyssal depths. They are relatively advanced invertebrates with a moderately complex internal anatomy consisting of coelomic cavities, a digestive system, a water vascular system, and gonads, but strikingly, they lack a head and brain. Unique to the echinoderms is the water vascular system, a hydraulic system used for locomotion, food and waste transport, and respiration. Many echinoderms have specialized appendages used for feeding, protection, locomotion, and attachment. The fossil record of the Platteville Formation indicates they were adapted to a wide variety of benthic niches (Figure 14-1).

Echinoderms possess a rigid or flexible endoskeleton consisting of fused or unfused microporous calcite **plates** or **ossicles** that are derived from mesoderm tissue. Isolated fossil plates are easily identified in hand specimens because they consist of a single crystal with well-developed calcite crystal cleavage that reflects light from the whole plate when held at the proper angle to a direct light source. The skeleton commonly, but not always, is arranged in a radial pattern of pentameral symmetry. Most echinoderms have a saclike body called a **theca** (also called **calyx**, **test**, or **disk**) that can be made up of tightly sutured plates that form a rigid skeleton (e.g., some crinoids, paracrinoids, rhombiferans, and echinoids) or loosely organized plates or ossicles that form a flexible theca. For example, the relatively flexible body of holothurians (sea cucumbers) is commonly strengthened by tens of thousands of microscopic calcite **spicules** or **sclerites**. Attached to the exterior of the theca are appendages that function in protection (**spines**), food gathering (**arms** bearing **pinnules** or smaller unbranched **brachioles**), and attachment (a **stem** or **column** comprising numerous disk-shaped **columnals** and a distal **holdfast** to anchor the animal on the seafloor). The **gonopore** and **hydropore** are external openings of the reproductive and water vascular systems, respectively. **Ambulacra** are simple or branched structures on the adoral surface of the theca that house the tube feet of the water vascular system and lead to the mouth. Additional morphologic terms are explained in the following class-specific descriptions.

Primary sources used in the identification of the Platteville echinoderms include Hall (1847), Billings (1854, 1857, 1858a, 1858b, 1859b, 1887), Sardeson (1928), Gutschick (1954b), Kolata (1982), Brower and Veinus (1974, 1978, 1982), Sprinkle (1982a), Guensburg (1984), Brower (2007), and Blake (2018). Parts of this chapter are modified from Kolata (1975) and reprinted with permission from the *Journal of Paleontology*.

Figure 14-1 (Opposite page) Depiction of Platteville seafloor showing the life mode of several echinoderms, including the crinoids *Abludoglyptocrinus*, *Cupulocrinus*, and *Porocrinus* and the echinoid *Bothriocidaris*.

SYSTEMATIC DESCRIPTIONS

Phylum ECHINODERMATA

Class CRINOIDEA

Description. Crinoid terminology used here largely follows that of the *Treatise on Invertebrate Paleontology* (Ubaghs et al. 1978). A **crown** is the whole crinoid exclusive of the stem and holdfast. It comprises the **aboral cup** or **calyx** (Figure 14-2), an upper part called the **tegmen** with mouth and anus, and the arms. The aboral cup has five **rays** that lead to the arms, designated by the letters **A** through **E**

counterclockwise as viewed from the bottom of the aboral cup; **A** is **anterior** and the **CD interray** is **posterior**. Each ray consists of five **radial** and five **basal** plates. **Monocyclic** crinoids have only radial and basal circlets, whereas **dicyclic** crinoids have an additional circlet of plates called **infrabasals** at the base of the cup. The radial plates bear **facets** for the attachment of the arms. A radial can be divided into a lower plate called an **inferradial** and an upper plate called a **superradial** (Figures 14-3 and 14-4). The posterior or anal side of the cup can have a group of plates called the **anal series** that leads upward to the anal opening at the top of the tegmen. The **radianal** plate lies below or supports a radial plate, and the **anal X** plate sits above the radianal. **Interbrachials** are lower arm plates that are incorporated into the aboral cup. An arm is composed of cylindrical or wedge-shaped plates called **brachials** that can consist of a single row referred to as **uniserial** or double alternating **biserial** row. **Primibrachials** are the first series of brachials above the radial plates. Subsequent bifurcations lead to the **secundibrachials**, **tertibrachials**, and so on. **Pinnules** are slender, unbifurcated branches borne on alternate sides of the brachials. Arms that branch into two equal segments are referred to as **isotomous**; branches that are unequal in size or length are **heterotomous**. The **ambulacral grooves** on the pinnules, arms, and tegmen have numerous small biserial **cover plates** that protect the water vascular system.

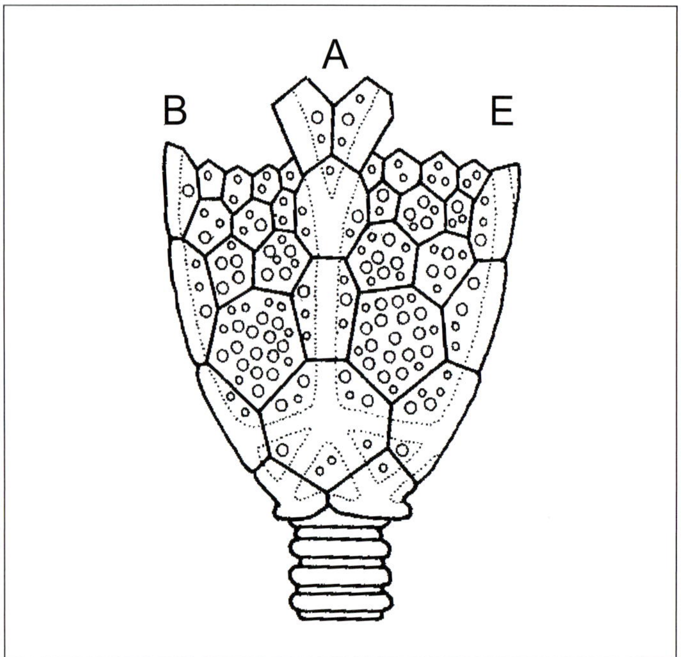

Figure 14-2 Diagram of the aboral cup on the anterior side of the monocyclic camerate *Abludoglyptocrinus pustulosis* showing the stem, basals, radials, interbrachials, two primibrachials, secundibrachials, and B, A, and E rays. Modified from Kolata (1975). Copyright 1975 Cambridge University Press. Used with permission.

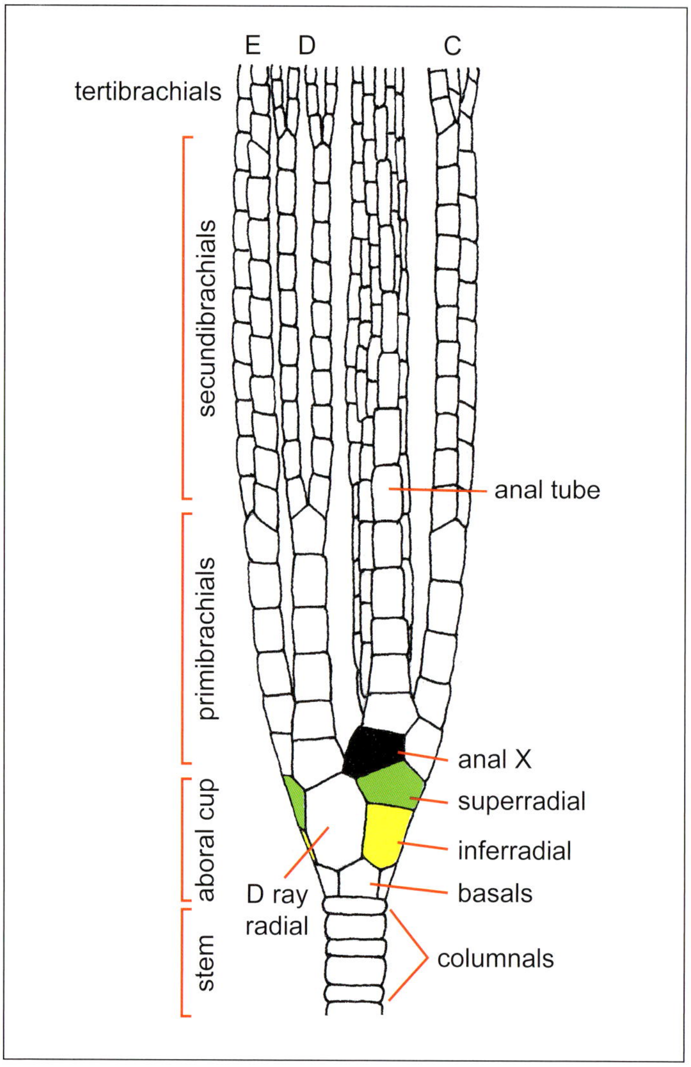

Figure 14-3 Diagram of the posterior side of the monocyclic inadunate *Isotomocrinus minutus* showing typical crinoid features, including the stem, columnals, aboral cup, basal circlet, inferradial, superradial, anal X, anal tube, primibrachials, secundibrachials, tertibrachials, and isotomously branching E, D, and C ray arms. Modified from Kolata (1975). Copyright 1975 Cambridge University Press. Used with permission.

Subclass CAMERATA

Description. Camerate crinoids have a monocyclic or dicyclic aboral cup with all plates in the calyx joined by rigid sutures; interbrachials commonly integrated into the upper part of the cup; tegmen consisting of numerous small plates that cover the mouth and food groove; uniserial or biserial arms that usually bear pinnules; and a stem with one-piece columnals.

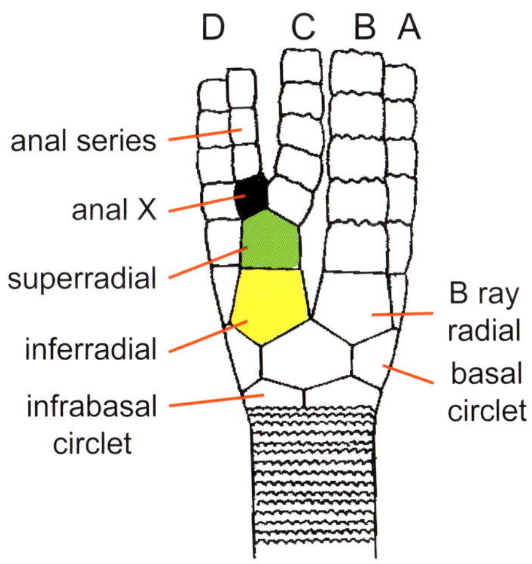

Figure 14-4 Diagram of the posterior side of the dicyclic inadunate *Merocrinus britonensis* showing proximal stem, infrabasals, basals, radials, inferradial, superradial, anal X, anal series, and D, C, B, and A ray arms. Modified from Kolata (1975). Copyright 1975 Cambridge University Press. Used with permission.

Order DIPLOBATHRA

Description. Camerate crinoids with a dicyclic aboral cup.

Suborder ZYGODIPLOBATHRIDA

Family CLEIOCRINIDAE

Genus CLEIOCRINUS

Cleiocrinus cf. *bromidensis* Kolata, 1982

Figures 14-5-1 and 14-5-2

Description. A relatively large dicyclic camerate crinoid with alternating basals and radials that form a circlet of 10 plates projecting down over the proximal columnals, concealing five infrabasals; lacking interbrachials; brachials in lateral contact except at anal side; anal series in a single column resting on the CD basal; cup plate sutures crenulate and firmly sutured, perforated by numerous sutural pores. Crown having an expansion angle of 65° from the horizontal up to primibrach 2, then abruptly bending outward to an angle of approximately 45° and continuing to top of the aboral cup. Cup plates smooth or with very fine ridges. Median groove on interior surface of brachials extends from base of cup, where five main branches radiate from a circular channel surrounding the cup axis upward in a radial position bifurcating on each axillary plate; small pores on each side of the median groove. Stem unknown.

Remarks. Species of *Cleiocrinus* are differentiated mainly on the size and shape of the aboral cup, number of arms, number and shape of sutural pores, and ornamentation (Kolata 1982). This Platteville *Cleiocrinus* is most similar to *C. bromidensis* Kolata, 1982, from the Pooleville Member of the Bromide Formation in southern Oklahoma.

Occurrence. Rare in the Grand Detour Member of northern Illinois and southern Wisconsin.

Family RETEOCRINIDAE

Genus RETEOCRINUS

Reteocrinus mahlburgi (Kolata, 1975)

Figure 14-5-3

Description. Crown large, approximately 10 cm high, steep straight sides, plates of circlets and rays with reticulate and nodular ornamentation. Infrabasal circlet missing from the single known specimen, but its presence is indicated by the truncated lower edges of the basal plates. Basals triaxial in outline, higher than wide, lower ray truncated in two places for reception of the adjacent infrabasal plates, upper two rays supporting adjacent radials; CD interray basal pentaxial, supporting primanal in center and two radials on upper sides; oval spaces between basals and radials contain 10–15 small, thick, star-shaped ossicles similar in size and shape to interradials; upper edge of basals in contact with interbrachials. Radials similar in size and shape to basals, separated by numerous small interbrachials, supporting five or six elongate primibrachials; upper radial facets half the width of the radial. First four or five primibrachials twice as high as wide, becoming progressively shorter distally; five or six elongate secundibrachs; interray areas wide and flat. Interbrachials small, irregular, very numerous, deeply sculpted, projecting above surface of theca to produce a spiny, rough appearance; there is some indication that small pores are present along the interbrachial margins. Arms free above first or second axillary brachial; very small interbrachials extend out under the free arms, forming a weblike structure between tegmen and arms on sides of the ambulacral grooves. Primanal higher than wide, followed above by 10 elongate anals that are equivalent in size to primibrachials, culminating in a protuberance (anal opening?) near apex of the tegmen; primibrachials, secundibrachials, and anals slightly swollen at joints. Tegmen prominent, domed, slightly inflated above free arms, consisting of small plates that are confluent with interprimibrachials and intersecundibrachials. Arms uniserial, long, branching isotomously at least five times, producing a total of 160 or more for the entire crown. Stem unknown.

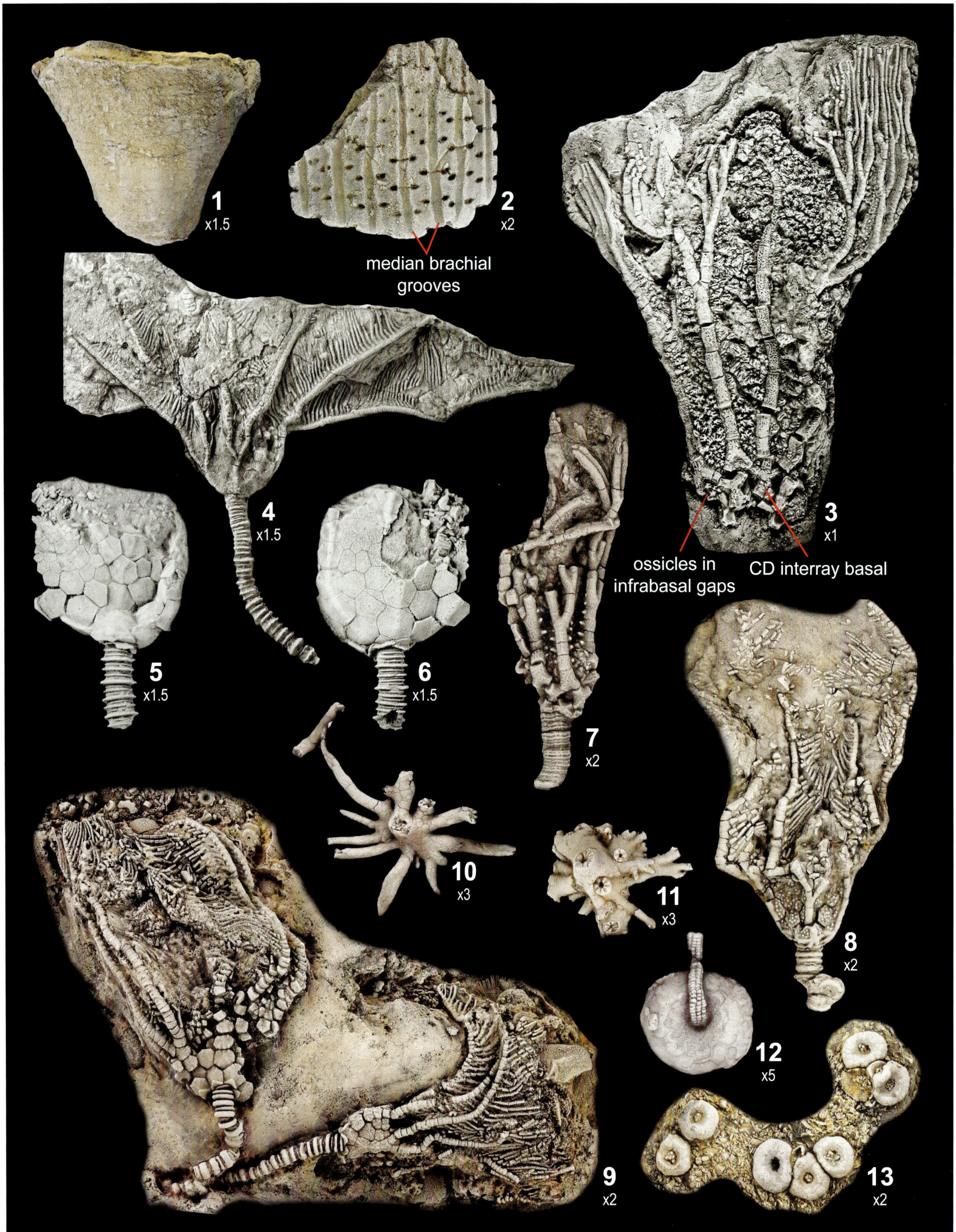
1
x1.5
2
x2
median brachial grooves
3
x1
ossicles in infrabasal gaps
CD interray basal
4
x1.5
5
x1.5
6
x1.5
7
x2
8
x2
9
x2
10
x3
11
x3
12
x5
13
x2

Remarks. The holotype, which is the only known specimen, consists of a slightly deformed crown with the posterior side exposed. The dorsal cup is complete except for the infrabasals, which broke off prior to discovery of the specimen, but the presence of which is indicated by the sharp angle on the lower broken edges of the basals. This specimen was assigned to the new reteocrinid genus *Traskocrinus* by Kolata (1975) mainly on the basis of having large spaces—between the deeply cleft sides of the infrabasals, basals, and radials—that are filled with numerous, small, irregular, deeply sculpted ossicles similar to the interbrachials and those on the tegmen. Subsequently, Guensburg (1984) pointed out that interbasal gaps filled with small ossicles are present in *Reteocrinus stellaris* Billings, 1859, the type species of *Reteocrinus*. This observation brings into question the validity of the genus *Traskocrinus*. Given the present understanding of known reteocrinid morphologies and the fact that the single known specimen of *T. mahlburgi* is incomplete (infrabasal not intact), this species is here reassigned to the genus *Reteocrinus*. The validity of the species concept awaits future discoveries and further study.

Occurrence. Present but very rare in the upper part of the Grand Detour Member near Durand, Illinois (locality 11).

Figure 14-5 (Opposite page)

1, 2. ***Cleiocrinus* cf. *bromidensis*** Kolata. (**1**) Lateral view of aboral cup; (**2**) interior of several aboral cup plates showing the median brachial grooves and flanking pairs of pore slits. Grand Detour Member, Durand, Illinois (locality 11), **1**, ISGS 1175. **2**, Grand Detour Member, Dixon, Illinois (locality 1), ISGS 1176.

3. ***Reteocrinus mahlburgi*** (Kolata). (**3**) Posterior side showing the CD interray with basal plates and small, irregularly shaped ossicles in the infrabasal gaps. Forreston Beds, Grand Detour Member, Durand, Illinois (locality 11), **3**, holotype BMNH PK-48.

4–6. ***Abludoglyptocrinus charltoni*** (Kolata). (**4**) Crown and stem of holotype; (**5**, **6**), anterior and posterior views of aboral cup, respectively. Grand Detour Member, Byron, Illinois (locality 42), **4**, ISGS 3951. Grand Detour Member, Durand, Illinois (locality 11), **5**, **6**, BMNH PK-40.

7. ***Reteocrinus spinosus*** Kolata. (**7**) Lateral view of crown and stem showing prominent spines on the A and B ray primibrachials. Lower part of Grand Detour Member, Durand, Illinois (locality 11), **7**, holotype UI X-4881.

8. ***Abludoglyptocrinus pustulosis*** (Kolata). (**8**) Anterior view of crown and proximal stem. Lower part of Grand Detour Member (locality 20), **8**, holotype UI X-4914.

9. ***Rhaphanocrinus buckleyi*** Kolata. (**9**) Anterior views of paratype UI X-4910 on left and holotype UI X-5095 on right; note difference in aboral cup ornamentation in the two specimens. Lower part of Grand Detour Member, Dixon, Illinois (locality 1).

10, 11. **Multiplated lobate- and digitate-encrusting holdfasts.** (**10**, **11**) Top view of holdfasts spread out on bedding plane. Mifflin Member, Dixon, Illinois (locality 2), **10**, ISGS 5135; **11**, ISGS 1164.

12, 13. **Multiplated discoidal (*Lichenocrinus*) holdfasts.** (**12**) A single holdfast with fragment of a stem that is longitudinally divided into five parts (pentamere); (**13**) multiple holdfasts encrusted on a shelly bedding plane. Mifflin Member, Dixon, Illinois (locality 2), **12**, UI X-5133; **13**, ISGS 1169.

Reteocrinus spinosus Kolata, 1975

Figure 14-5-7

Description. Dicyclic camerate, aboral cup small, steep-sided, plates of circlets and rays with fine reticulate ornamentation. Infrabasals small, pentagonal, wider than high, small spaces between adjacent plates; small, sharp ridges superimposed on larger median ridges and extending onto lower part of basals only; basals relatively large, height and width approximately equal, narrower below, large median ridge triaxial, truncated below on two sides at juncture with infrabasals; CD interray basal pentaxial, supporting the first anal plate in center and two radials on upper edges; small spaces between infrabasals, basals, and radials; radials similar in size and shape to basals; height and width equal, supporting four to six primibrachials; primibrachials possess a prominent spine on each side, height and width of primibrachials approximately equal; four elongate secundibrachials; radial width greater than width of radial facet; interray areas relatively narrow, consisting of small, irregular interbrachials. Primanal higher than wide, followed above by elongate anals approximately equal in width to primibrachials. Tegmen steeply conical and composed of numerous small irregular plates similar to interradials. Arms long and slender, branching isotomously three or possibly four times, apparently free above primaxials; distal brachials higher than wide, wedge shaped in transverse section, approximately four pairs of spear-shaped ambulacral cover plates per brachial. Stem subpentagonal proximally, not enlarged at base of dorsal cup, fine longitudinal striations proximally.

Remarks. *R. spinosus* differs from other *Reteocrinus* species by its slender crown with long narrow arms, narrow interray areas, and four to six primibrachials with prominent long lateral spines.

Occurrence. Present but very rare in the Grand Detour Member near Durand, Illinois (locality 11).

Family ARCHAEOCRINIDAE

Genus RHAPHANOCRINUS

Rhaphanocrinus buckleyi Kolata, 1975

Figure 14-5-9

Description. Crown small, aboral cup plates slightly convex, with fine ridges passing upward from basals converging on radials and continuing onto arms. Infrabasals apparently concealed beneath basals and proximal columnals. Basals large, hexagonal, width slightly greater than height; small, thick triangular process at juncture with stem. Radials pentagonal, wider than high, each separated from adjacent radials by a large interradial, supporting two primibrachials. Primibrachials wider than high; primibrachial 1 hexagonal; primibrachial 2 pentagonal, supporting first secundibrachials; interradial spaces slightly depressed. Ten arms, long and slender; brachials much wider than high, wedge shaped; pinnules long and delicate, two rows on alternate brachials beginning with secundibrachial 2; pinnules long and slender, keeled on dorsal side with deep groove on oral side. Stem round, consisting of dissimilar columnals, proximal portion consists of alternating prominent well-rounded columnals and smaller partially hidden columnals. Holdfast unknown.

Remarks. *R. buckleyi* most closely resembles *R. subnodosus* (Walcott, 1884) from the upper Trenton Limestone at Trenton Falls, New York. The latter differs in possessing shorter, more robust arms with stouter pinnules, heptagonal radials, and a proportionally larger stem.

Occurrence. Present but very rare in the Grand Detour Member at Dixon, Illinois (locality 1).

Order MONOBATHRIDA

Description. Camerate crinoids with a monocyclic aboral cup.

Family GLYPTOCRINIDA

Genus ABLUDOGLYPTOCRINUS

Abludoglyptocrinus charltoni (Kolata, 1975)

Figures 14-5-4 through 14-5-6

Description. A monocyclic camerate crinoid with aboral cup width approximately equal to height, plates either smooth or with small indistinct pustules, sutures flush. Basals pentagonal, small, with prominent sharp downward-projecting rim that forms a concavity at juncture of basals with stem, width greater than height (as measured from above projecting rim). Radials heptagonal, large, slightly convex, width greater than height, in full contact with adjacent radials, supporting two primibrachials. Primibrachial 1 hexagonal, wider than high; primibrachial 2 heptagonal, axillary, wider than high, supporting four secundibrachials per branch; secundibrachials 1 and 2 larger than 3 and 4; secundibrachial 2 is pinnule bearing. Interbrachials decreasing in size upward. Primanal heptagonal, slightly convex, anal plates consisting of numerous small irregular plates at the summit. Ten arms, long and slender, free above secundibrachial 4, pinnules long and slender; brachials twice as wide as high. Round stem, proximally consists of closely spaced columnals with wide, sharp edges alternating with smaller columnals; distally the wide columnals are spaced farther apart and possess small downward-projecting spines along the edges.

Remarks. The genus *Abludoglyptocrinus* (Kolata 1982) was proposed for glyptocrinids with 10 uniserial arms that are free above secundibrachials 3–5, possess median anal and ray ridges and have radials that are in full contact with adjacent radials.

Occurrence. Locally common in the Mifflin and Grand Detour Members of northern Illinois and southern Wisconsin. Also reported from the Mifflin Member in northeastern Iowa (Brower 2007).

Abludoglyptocrinus pustulosis (Kolata, 1975)

Figures 14-2 and 14-5-8

Description. A monocyclic camerate crinoid with small, relatively steep-sided aboral cup; plates thick and coarsely pustulose; strong ridges extend between basals, radials, and primibrachials; sutures flush. Basals small, pentagonal, height approximately equal to width, prominent V-shaped ridge extends to adjacent radials, indentation in lower rim at juncture with adjacent basals and stem. Radials relatively large, in full contact with adjacent radials, height equal to width, anterior three plates with five prominent radiating ridges, posterior two plates with one additional ridge extending to primanal, radials supporting two primibrachials. Primibrachial 1 higher than wide, hexagonal; primibrachial 2 higher than wide, heptagonal to decagonal, axillary, supporting two long, fixed secundibrachials per branch; three or four pustulose interbrachials between adjacent primibrachials; interray areas flat near lowest interbrachial, becoming slightly concave upward, depressions accentuated by strong ridges on primibrachials and radials. Interbrachials decrease in size but increase in number upward. Primanal not observed but apparently supported by two radials on posterior side; anal interray ridge apparently formed by the intersection of two ray ridges passing up to primanal from adjacent radials. Ten arms, long and slender, free above secundibrachial 3, uniserial; brachials wedge shaped, wider than high, bearing a pinnule alternately on each side; pinnules long, wedge shaped in transverse section. Stem round, consisting of alternating columnals of different diameters with rounded edges. Holdfast unknown.

Remarks. The prominent radiating ridges and pustulose ornament on the aboral cup plates distinguish *A. pustulosis* from other species of *Abludoglyptocrinus*. Note that *A. pustulosis* is based on a single slightly crushed crown that exposes only the anterior side. The basal circlet appears to rest directly on the proximal stem, with no sign of an infrabasal circlet. It is possible that the infrabasals are hidden from lateral view at the base of the aboral cup. If this is the case, then this species would be considered a dicyclic camerate with affinities to *Euptychocrinus* known from the overlying Decorah and Galena Formations of northeastern Iowa (Brower 1994).

Occurrence. Present but rare in the lower part of the Grand Detour Member near South Beloit, Illinois (locality 20).

Subclass INADUNATA

Remarks. Inadunate crinoids have a monocyclic or dicyclic aboral cup consisting of relatively few joined plates; a lightly plated tegmen with oral and anal plates; free arms without fixed brachials, uniserial or biserial in primitive forms and pinnulated in evolutionarily advanced forms; and a stem consisting of one-piece columnals.

Order DISPARIDA

Description. A monocyclic inadunate crinoid with weak to prominent bilateral symmetry developed in planes other than the A through CD interray; compound radials common; arms typically not pinnulated.

Superfamily CALCEOCRINACEA

Family CALCEOCRINIDAE

Genus CREMACRINUS

Cremacrinus arctus Sardeson, 1928

Figures 14-6-1 through 14-6-5

Description. Crown recumbent on stem with imperfect bilateral symmetry; crowns range from approximately 20–70 mm long and are characterized by punctate ornamentation, a long, slender, nonbifurcating E ray arm, normal calceocrinid pattern of heteromorphic branching; A, B, and D ray arms with two primibrachials, ramules (minor branch of arms) given off on alternate sides at every second brachial, and knuckle-like enlargements of brachials at bifurcation of ramules. The smaller, presumably immature specimens have long, narrow arms, elongate brachials with little swelling at the axillary joints, and nontumid plates in the aboral cup. The larger specimens have a robust cup with tumid inferradials in the B, C, and E rays and heavy arms with short nodose brachials that are swollen at the axillary joints to give the arms a knotty appearance. Specimens intermediate in size seem to show a gradation in features and thus support the assumption that all specimens are conspecific. Stem consists of nodose columnals. Holdfast unknown.

Remarks. This species was originally described from specimens collected in the upper Platteville Limestone at Minneapolis, Minnesota.

Occurrence. Abundant in the Mifflin and Grand Detour Members in north-central Illinois and south-central Wisconsin and the McGregor Member of southwestern Wisconsin, northeastern Iowa, and southeastern Minnesota.

Genus ISOTOMOCRINUS

Isotomocrinus minutus Kolata, 1975

Figures 14-3 and 14-6-6

Description. Crown small, slender, 14 mm or less in length. Aboral cup small, steeply conical, plates smooth. Basals small, height slightly greater than width. C ray inferradial pentagonal, height greater than width; C ray superradial hexagonal, laterally in contact with B and D ray radials, supporting a wide primanal and narrower primibrachial (Figure 14-3); E ray inferradial pentagonal, height greater than width, larger than superradial of the same ray; E ray superradial rectangular, width greater than height; A and B ray unknown; D ray radial hexagonal, height greater than width. Anal X pentagonal, laterally in contact with D ray radial and primibrachial 1 as well as primibrachial 1 of C ray; anal sac long and slender, reaching almost to the tips of arms, composed of elongate rectangular plates that diminish in size distally. Arms round, long and slender, isotomously branched; primibrachial 1 quadrangular, width slightly greater than height, succeeding primibrachials narrower and more elongate, first bifurcation on primibrachial 5 or 6, second bifurcation on secundibrachials 9–11. Stem with columnals divided into five equal parts, proximally round,

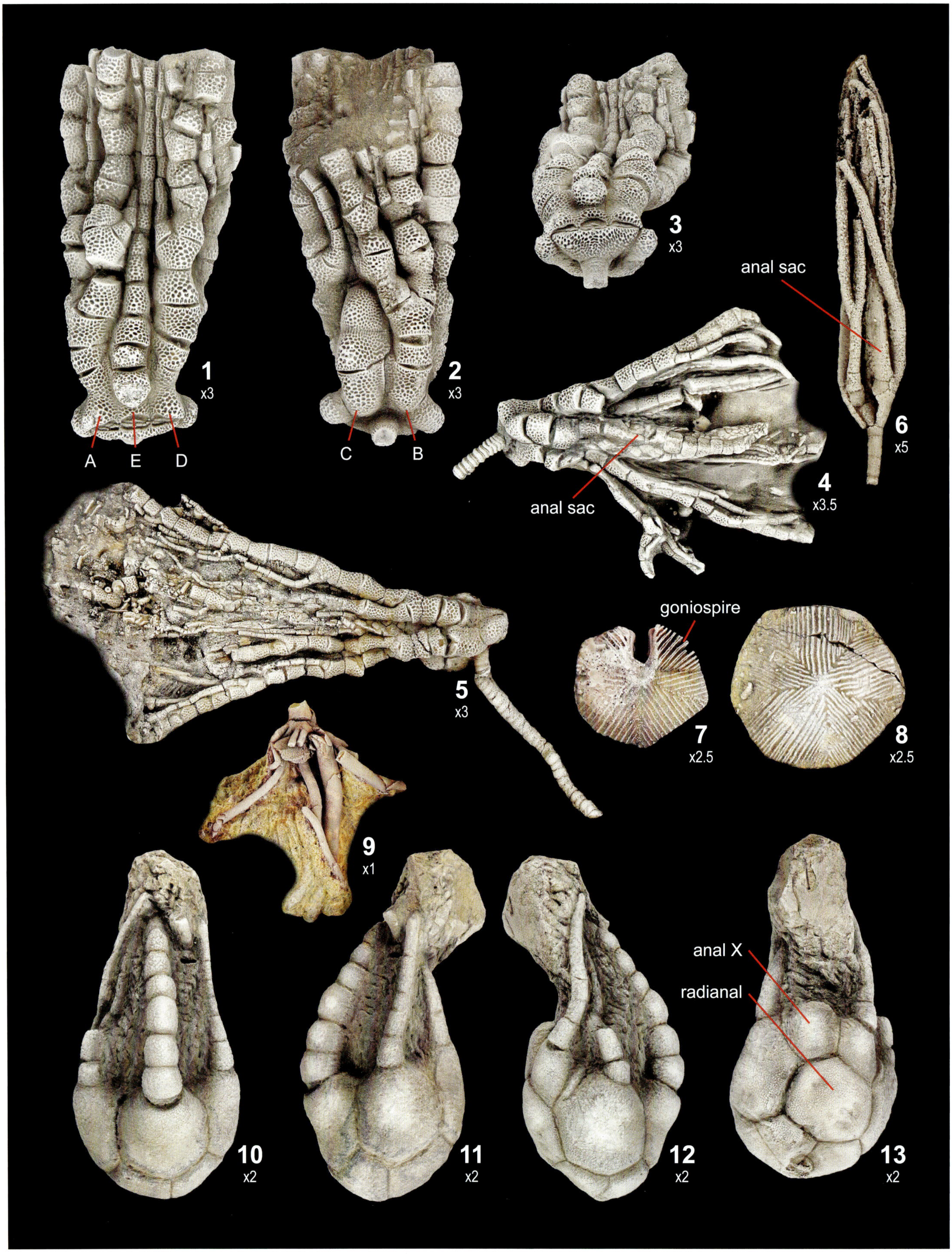
1
x3
A
E
D
2
x3
C
B
3
x3
anal sac
6
x5
4
x3.5
anal sac
goniospire
5
x3
7
x2.5
8
x2.5
9
x1
anal X
radianal
10
x2
11
x2
12
x2
13
x2

distally becoming pentagonal and rounded on outer edge; proximally composed of alternating short and long columnals, becoming equal in length distally; axial canal small, pentagonal.

Remarks. This species is assigned to *Isotomocrinus* based on the monocyclic aboral cup and isotomously branching arms. Although the small size of the available specimens suggests they may be immature, all plates appear sufficiently developed to reveal the distinctive features of the species. Similar small five-part columnals, possibly belonging to this species, occur in the Mifflin and Grand Detour Members. Some stems are attached to small, less than 10 mm in diameter, multiplated discs similar to that of *Lichenocrinus* Ulrich, 1925 (p. 84).

Occurrence. Present but rare in the Grand Detour Member of northern Illinois.

Order HYBOCRINIDA

Description. Monocyclic inadunate crinoids with five basals and radials, one or more anals, five orals, and numerous small plates in the tegmen; five uniserial, nonpinnulate arms.

Family HYBOCRINIDAE

Genus HYBOCRINUS

Hybocrinus cf. *bilateralis* Guensburg, 1984

Figures 14-6-10 through 14-6-13

Description. A species of *Hybocrinus* with a medium to large globular calyx and tumid plate surfaces; largest observed calyx 15 mm high. Basals five in number, pentagonal, asymmetrical, subequal in size, taller than wide. Radials five in number, largest in cup. Anal X slightly taller than wide, widest distally, upper margin convex, sits higher in the cup than adjacent radials, rests on upper left shoulder of radianal. Arms uniserial and unbranched, 20–25 mm long; A ray arm thicker and wider than others. Plate surfaces uniformly covered with round shallow pits approximately 0.2 mm in diameter. Column composed of dissimilar columnals.

Remarks. The overall size and shape of the calyx, tumid plates with very fine pits, and stout A ray arm are very similar and possibly conspecific with coeval *H. bilateralis* Guensburg, 1984, from the lower member of the Lebanon Formation of central Tennessee.

Occurrence. Rare in the Grand Detour Member of north-central Illinois. Also reported from the Mifflin Member in northeastern Iowa (Brower 2007).

Order CLADIDA

Description. Dicyclic inadunate crinoids with one to three anal plates in the dorsal cup.

Suborder CYATHOCRININA

Superfamily GASTEROCOMACEA

Family CARABOCRINIDAE

Genus CARABOCRINUS

Carabocrinus cf. *micropunctatus* Brower and Veinus, 1974

Figures 14-6-7 and 14-6-8

Description. Complete crown unknown, but based on isolated aboral cup plates, the calyx was relatively large and globular in shape. Basal plates subcircular in outline, unusually convex and marked by numerous distinctive ridges that radiate from the plate center outward in a chevron pattern. Radial plates marked by sharp ridges that meet at an angle of 60°, producing a V-shaped pattern radiating from the plate center; radial facet is narrow and ambulacral groove is

Figure 14-6 (Opposite page)

1–5. ***Cremacrinus arctus*** Sardeson. (**1**) Anterior view showing the A, E, and D rays and prominent punctate ornamentation; (**2**) posterior view showing C and B rays; (**3**) perspective view of aboral cup base; (**4**) side view of crown showing the A and C rays as well as the anal sac; (**5**) side view of crown showing the A and B rays. Mifflin Member, Dixon, Illinois (locality 2), **1–3**, ISGS 1162; **4**, ISGS 1161. Grand Detour Member, Durand, Illinois (locality 11), **5**, UI X-4923.

6. ***Isotomocrinus minutus*** Kolata. (**6**) Posterior view of crown showing the anal sac. Grand Detour Member, Dixon, Illinois (locality 1), **6**, holotype UI X-4886.

7, 8. ***Carabocrinus* cf. *micropunctatus*** Brower and Veinus. (**7**) Radial plate showing goniospires and radiating ridges; (**8**) basal plate with radiating ridges. Grand Detour Member, Dixon, Illinois (locality 1), **7**, UI X-4918. Pecatonica Member, Dixon, Illinois (locality 2), **8**, ISGS 1173.

9. **Type 2 Rootlike Holdfast.** (**9**) Lateral view of holdfast with long tubules that vertically penetrate the substrate. Grand Detour Member, Dixon, Illinois (locality 1), **9**, UI X-5132.

10–13. ***Hybocrinus* cf. *bilateralis*** Guensburg. (**10**) Anterior view showing the thick and wide A ray brachials; (**11**), lateral view showing E ray radial at center; (**12**) lateral view showing B ray radial at center; (**13**) posterior view showing the radianal plate and anal X. Grand Detour Member, South Beloit, Illinois (locality 20), **10–13**, ISGS 1172.

deeply cleft. Goniospires very large and consisting of about 10 folds that are slightly elevated above the plate surface.

Remarks. This Platteville species of *Carabocrinus* is very similar to *C. micropunctatus* Brower and Veinus, 1974, from the Upper Ordovician (Turinian Stage) Benbolt Formation at Rye Cove, Virginia, and *Carabocrinus* sp. Guensburg, 1984, from the Lebanon Formation (Turinian Stage) of central Tennessee.

Occurrence. Rare aboral cup plates are locally present in the Grand Detour Member in north-central Illinois and south-central Wisconsin.

Family POROCRINIDAE

Genus POROCRINUS

Porocrinus pentagonius Meek and Worthen, 1865

Figures 14-7-8 through 14-7-10

Description. A moderately sized (approximately 15 mm high and wide), globular, dicyclic aboral cup; posterior side protrudes slightly; calyx ridges sharp and prominent; sutures flush, surface finely punctate except on goniospires; punctae regularly arranged and radiating from plate corners; goniospires very shallow, medium to large, subcircular, 6–12 folds at adjacent corners of three cup plates. Infrabasals wider than high, much smaller than basals, juncture with stem crenulate. Basals large, height approximately equal to width, hexagonal except for CD and BC interray, which are heptagonal. Radials large, equal in size to basals, height and width approximately equal, radial facets narrow, horseshoe shaped. Radianal small, quadrate sutures with adjacent plates arcuate, upper left corner in contact with anal X. Anal X larger than radianal, radianal rests on upper edge of CD basal. Arms narrow, uniserial; brachials wedge shaped in transverse section, height equal to width; ambulacral grooves covered by small, biserial, rectangular covering plates. Oral area apparently small, oral plates not observed. Stem consists of dissimilar columnals; proximal stem pentagonal with rounded interradial angles, expanding abruptly at base of aboral cup, large central lumen, columnals telescopically imbricating, 15–20 in number, crenulations on outer surface form numerous longitudinal striations, distal stem more coarsely crenulate, composed of relatively thick columnals.

Remarks. Original description by Meek and Worthen (1865) is based on specimens collected from the Platteville Formation at Dixon, Illinois. *P. pentagonius* is distinguished from other species by having prominent sharp calyx ridges and goniospires that are flat and shallow.

Occurrence. Common locally in the Mifflin and Grand Detour Members of northern Illinois and southern Wisconsin. Also reported from the Mifflin Member in northeastern Iowa (Brower 2007).

Porocrinus* cf. *smithi Grant, 1881

Figure 14-7-7

Description. This species is structurally similar to *P. pentagonius* but differs in possessing smaller, more deeply inset goniospires and tumid punctate cup plates with more rounded ridges.

Remarks. This Platteville species most closely resembles *P. smithi* Grant from the Trentonian Hull beds of the Ottawa Formation at Belleville, Ontario. The Platteville species differs from *P. smithi* in possessing smaller, more deeply inset goniospires.

Figure 14-7 (Opposite page)

1. ***Cupulocrinus plattevillensis*** Kolata. (**1**) Posterior view of a large specimen showing the steeply conical aboral cup. Grand Detour Member, Durand, Illinois (locality 11), **1**, holotype UI X-4901.

2. ***Cupulocrinus gracilis*** (Hall). (**2**) Posterior view. Mifflin Member, Dixon, Illinois (locality 2), **2**, ISGS 1163.

3. ***Cupulocrinus plattevillensis*** Kolata. (**3**) Posterior view. Grand Detour Member, Davis, Illinois (locality 41), **3**, ISGS 1167.

4. ***Merocrinus britonensis*** Kolata. (**4**) Lateral view showing C and B rays of dorsal cup. Mifflin Member, Dixon, Illinois (locality 1), **4**, holotype UI X-4909.

5, 6. ***Quinquecaudex springeri*** (Kolata). (**5**) Lateral view of crown; (**6**) close-up of anal tube showing "pectinirhomb-like" endothecal folds along the ossicle margins. Grand Detour Member, Durand, Illinois (locality 11), **5**, **6**, paratype BMNH PK-31.

7. ***Porocrinus* cf. *smithi*** Grant. (**7**) Anterior view showing typical size, shape, and relief of goniospires. Forreston Beds, Grand Detour Member, Durand, Illinois (locality 11), **7**, BMNH PK-36.

8–10. ***Porocrinus pentagonius*** Meek and Worthen. (**8**) Base of aboral cup; (**9**) posterior of aboral cup showing goniospires; (**10**) anterior of cup. Grand Detour Member, Durand, Illinois (locality 11), **8–10**, BMNH PK-83.

11. **Type 1 Multiplated Rootlike Holdfast.** (**11**) Lateral view of five-part stem attached to three stout roots. Grand Detour Member, Durand, Illinois (locality 11), **11**, ISGS 1244.

12–15. ***Oklahomacystis* cf. *bibrachiatus*** (**12**) A very well-preserved theca exposing the anterior side with periproct and at least 10 pinnules; (**13**) enlargement of pinnules showing elongate ossicle with small cover plates; (**14**) enlargement of base of theca showing a circlet of four basal ossicles; (**15**) interior of theca showing pore slits along the margins of several plates. Grand Detour Member, Grant County, Wisconsin, **12–14**, PB-8686. Mifflin Member, Mifflin, Wisconsin, **15**, ISGS 1241.

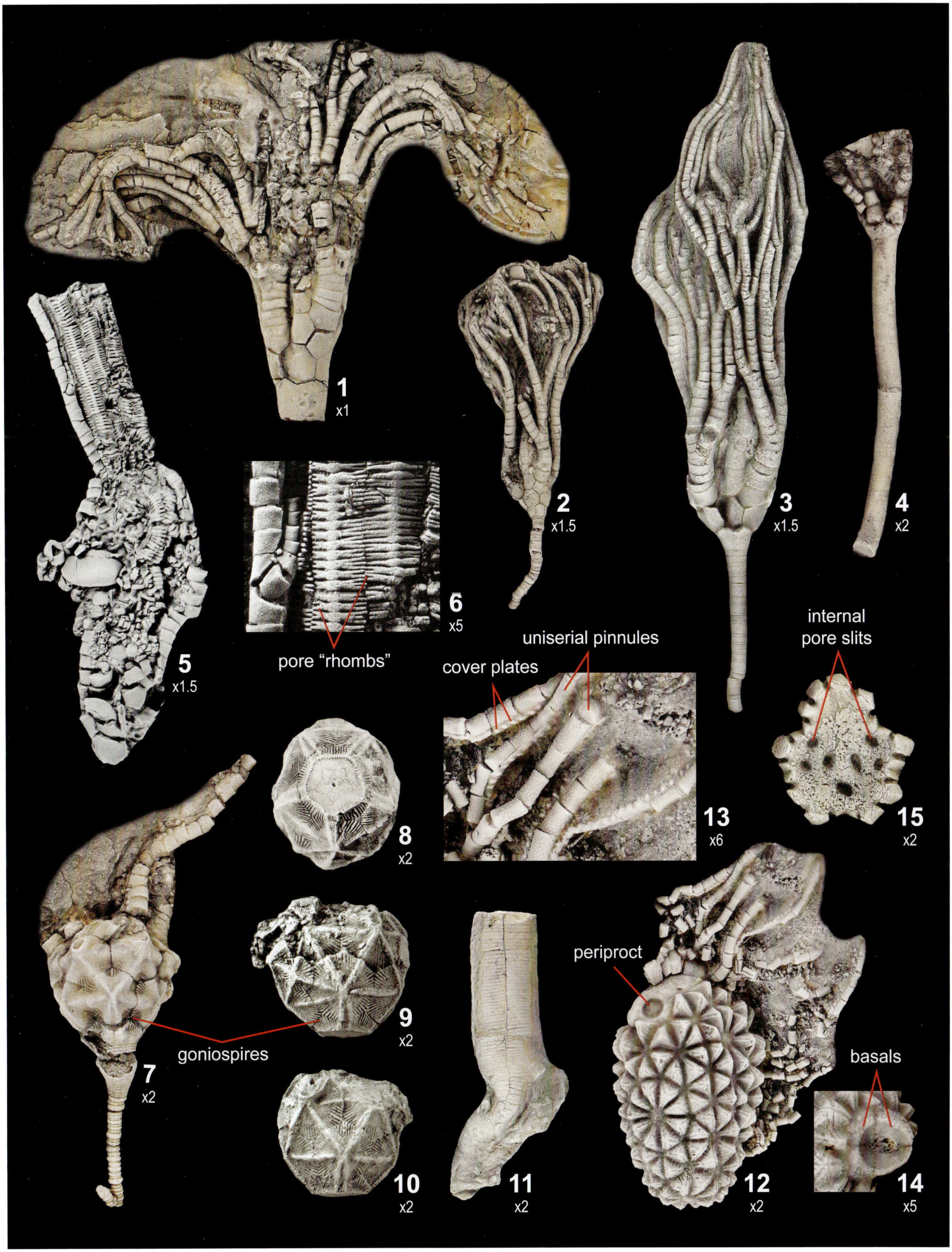
1
x1
2
x1.5
3
x1.5
4
x2
5
x1.5
6
x5
pore "rhombs"
uniserial pinnules
cover plates
internal
pore slits
13
x6
15
x2
8
x2
9
x2
goniospires
7
x2
periproct
basals
10
x2
11
x2
12
x2
14
x5

Occurrence. Disarticulated aboral cup plates are common in the Mifflin and Grand Detour Members in northern Illinois and southern Wisconsin.

Suborder DENDROCRININA

Superfamily DENDROCRINACEA

Family DENDROCRINIDAE

Genus QUINQUECAUDEX

Quinquecaudex springeri (Kolata, 1975)

Figures 14-7-5 and 14-7-6

Description. Crown relatively large, elongate, sides steep, plates smooth and flush. Infrabasals wider than high. Basal circlet high; height and width of basals approximately equal, larger than infrabasals. Radials wider than high. Radianal pentagonal, directly below and equivalent in size to C ray radial, laterally in contact with anal X and CD interray basal. Anal X hexagonal or heptagonal, suture with C ray radial short, distal edge above upper edge of D ray radial and below upper edge of C ray radial; size equivalent to radials; axillary, supporting two relatively large irregular plates, each of which supports a row of four or five plates that form the base of the anal sac. Anal sac very long and slender, six or seven times higher than aboral cup, consisting of seven or eight longitudinal rows of ossicles, each ossicle possessing two or three lateral folds that are sutured to folds in ossicles of adjacent rows similar to the pore rhombs of some cystoids. Arms long and slender, heterotomous branching after first bifurcation, which is on primibrachial 5 or 6; second bifurcation is random. Brachials wider than high throughout, dorsal surface broadly rounded (not keeled). Stem with columnals divided into five equal parts, sutures in radial position, stem does not expand at base of dorsal cup, diameter approximately equal from dorsal cup to holdfast, pentagonal in proximal two-thirds, becoming round near holdfast, crenulated articulation; holdfast rootlike, four or five hollow undivided branches consisting of thin, elongate plates.

Remarks. This species was assigned by Kolata (1975) to the genus *Dendrocrinus* with some hesitation. The structure of the dorsal cup and anal sac is typical for the genus, but the unusually wide radial facets are not. Subsequently, Brower and Veinus (1982) proposed the new genus *Quinquecaudex*, including *Q. glabellus* and the Platteville species *Q. springeri*. They stressed that the primary difference between the genera is that the columnals of *Dendrocrinus* are solid, whereas those in *Quinquecaudex* are divided into five parts (pentameres).

Occurrence. Rare in the Grand Detour Member in northern Illinois.

Superfamily MEROCRINACEA

Family MEROCRINIDAE

Genus MEROCRINUS

Merocrinus britonensis Kolata, 1975

Figures 14-4 and 14-7-4

Description. Aboral cup plates smooth and flush. Infrabasals approximately three times wider than high. Basals hexagonal, relatively large, more than twice as large as infrabasals. Radials pentagonal, height approximately equal to width, tapering slightly distally, equal in size to basals. C ray superradial pentagonal, equal in size to C ray radial, supporting primanal on left and C ray arm on right. Anal X higher than wide, followed above by elongate anal plates; anal series slightly narrower than adjacent C ray brachials. Arms narrow, branching unknown but apparently after the fifth primibrachial, brachials wider than high; primibrachial articulation surfaces crenulate. Stem round, relatively large, long (some greater than 185 mm), not expanding at base of aboral cup, composed of similar columnals throughout except distal end, which abruptly becomes smaller and is composed of proportionally thicker columnals; large axial canal proximally; crenulated articulation.

Remarks. *M. britonensis* is similar to *M. corroboratus* Walcott 1884, from the Trenton Limestone at Trenton Falls, New York, but the latter differs in having higher, narrower infrabasals, proportionally smaller basals, stem that expands at the base of the aboral cup, and thicker proximal columnals. *Merocrinus curtus* (Ulrich, 1879) from the Eden Shale of Cincinnati, Ohio, differs in possessing wider brachials and basals. *M. typus* Walcott, 1884, from the Trenton Limestone at Trenton Falls, New York, differs from *M. britonensis* in having wider radial, anal, and brachial plates.

Occurrence. Rare in the uppermost part of the Mifflin Member at Dixon, Illinois (locality 1) and northeastern Iowa (Brower 2007).

Family CUPULOCRINIDAE

Genus CUPULOCRINUS

Cupulocrinus gracilis (Hall, 1847)

Figure 14-7-2

Description. Aboral cup conical, small, steep-sided near base expanding upward, plates smooth and flush. Infrabasals higher than wide, vertical sutures nearly parallel. Basals slightly higher than wide, larger than infrabasals or radials. Radials wider than high; radial facets equal to width of radials, dorsal ligament fossa narrow; very small interbrachials adjacent to upper edge of radials; CD interray relatively wide. Radianal similar in size and shape to radials. Anal X hexagonal, supporting a median row of 8–10 relatively large rectangular anal plates; anal tube evenly rounded, diameter at base equivalent to that of the arms, curved inward above, small plates on lateral edges. Arms long and slender, dividing three or four times, isotomous, uniformly rounded; first bifurcation on primibrachials 4–6; primibrachials taper distally to axial 1 and are much narrower than radial facets; axial 1 only slightly wider than high; first secundibrachial pair well separated; second bifurcation on secundibrachial 8; brachials above axial 2 higher than wide; two or three paired covering plates per brachial. Tegmen consists of numerous small ossicles. Stem round, tapering distally, with longitudinal striations composed of alternating thick and thin columnals; epifacets occur on approximately every 16th columnal, outer edge curved in a distal direction and bearing short spines; weak, ill-defined columnal articulation; holdfast unknown.

Remarks. *C. gracilis* was first described by Hall (1847) from specimens collected at the base of the Trenton Limestone at Middleville, Herkimer County, New York.

Occurrence. Abundant in the Mifflin and Grand Detour Members in north-central Illinois and south-central Wisconsin and the McGregor Member of southwestern Wisconsin, northeastern Iowa, and southeastern Minnesota.

Cupulocrinus plattevillensis Kolata, 1975

Figures 14-7-1 and 14-7-3

Description. Aboral cup steeply conical, expanding uniformly from base, higher than wide, all plates smooth and flush. Infrabasal large, much higher than wide, height nearly equal to height of basals, vertical sutures nearly parallel. Basals large, height greater than width. Radials wider than high, radial facets extending full width of radials. Radianal pentagonal, located directly below C ray radial, approximately equivalent in size to anal X. Anal X pentagonal, supported below by truncated upper edge of CD basal and laterally in contact with C and D ray radials and upper left portion of radianal. Anal tube long and narrow, median row of quadrate plates higher than wide, small irregular lateral plates. Arms long, slender, branching isotomously, dividing four or five times; brachials wider than high, distally wedge shaped in transverse section, three pairs of cover plates per brachial; first bifurcation on primibrachs 3–6, second bifurcation on secundibrachials 5–8, first secundibrachial pair not well separated. Tegmen unknown. Stem round, not enlarged at base of dorsal cup; proximal columnals of variable thickness, sides straight and uniform, smooth and flush; distal columnals consisting of alternating thick and thin columnals with variable diameters.

Remarks. *C. plattevillensis* differs from *C. gracilis* in its larger size, more robust arms, larger infrabasals, long and straight anal tube, and smooth flush proximal columnals.

Occurrence. Locally common in the Grand Detour Member of northern Illinois and southern Wisconsin. Also reported from the Mifflin Member in northeastern Iowa (Brower 2007).

Class PARACRINOIDEA

Description. Paracrinoids are a relatively small group of echinoderms characterized by a lenticular or globular theca composed of numerous irregular plates with either recumbent ambulacra and erect uniserial pinnules or erect uniserial arm with pinnules, mouth offset at distal end of theca, and well-developed hydropore and gonopore.

Order COMAROCYSTITIDA

Family AMYGDALOCYSTITIDAE

Genus OKLAHOMACYSTIS

Oklahomacystis* cf. *bibrachiatus Parsley, 1982

Figures 14-7-12 through 14-7-15

Description. Theca globose, elongate, slightly compressed transversely, and oval in cross section. Thecal ornamentation consists of rosettes of raised triangular ridges; finely pustulose on the adoral thecal plates and pinnules. Theca consists of approximately 40 plates in a specimen 25 mm high. No orderly arrangement of plates is apparent. Plates four- to nine-sided, most commonly six-sided, centers slightly concave; sharply raised, keeled, distally expanding rays extend from plate center to plate corners; rays at juncture of three plates form raised trigonal pyramids. Paired, internally opening, transversely oriented, pore slits cross most thecal plates. Oral end of theca bears two unbranched, recumbent arms that lie on the thecal surface. Food groove faces the anterior in the left arm and posterior in the right. Pinnules approximately half the length of the theca, 10–12 in number, and consisting of approximately 12 gradually tapering ossicles each with three or four pairs of cover plates. In one exceptionally well-preserved specimen, there appear to be four near equally divided basal plates rather than the three that are typical in all known paracrinoids. Periproct (anal opening) round, positioned on the anterior side adjacent to the right arm.

Remarks. This Platteville species of *Oklahomacystis* is most similar to *O. bibrachiatus* Parsley, 1982. Both forms have two recumbent arms and a globose, elongate theca composed of numerous plates with distinctive rosettes of raised triangular ridges. The significance of what appears to be four basal plates in the extraordinarily well-preserved Platteville specimen figured herein is unclear because paracrinoids are generally thought to possess only three basals (Parsley and Mintz 1975).

Occurrence. Isolated thecal plates are common in the Grand Detour Member of northern Illinois and southern Wisconsin. Complete specimens are rare.

Class RHOMBIFERA

Description. Rhombiferans, a type of cystoid, have a flattened or globular theca with two to five **ambulacra** with **flooring plates** supporting erect **brachioles**. Body plates arranged into four or five circlets, **basals** (**B**), **infralaterals** (**IL**), **laterals** (**L**), **radials** (**R**), and **orals** (**O**), with associated small **distals** (**D**; Figure 14-8). The mouth is at the top of the theca opposite the stem, with the periproct (anus) at the side surrounded by numerous small plates. The **gonopore** and **hydropore** are positioned near the mouth.

Order DICHOPORITA

Description. The Dichoporita are characterized by long, narrow, parallel slits called **pore rhombs** that resemble the slots in a comb. Each slit of a pore rhomb is shared by two adjacent thecal plates. A **demirhomb** (Figure 14-9-12) is a half-rhomb that forms a chevron on a single plate that is more or less symmetrical with respect to the plate suture. A pore rhomb is **conjunct** if the slits extend between two adjacent plates and **disjunct** (Figure 14-9-12) if each plate has distinct and separate openings.

Superfamily GLYPTOCYSTITIDA

Family CHEIROCRINIDAE

Genus CORONOCYSTIS

Coronocystis durandensis Kolata, 1975

Figures 14-8 and 14-9-1 through 14-9-15

Description. Theca relatively large (as much as 40 mm long), ovoid, larger end aboral, with indentation at the periproct. Thecal plates with very prominent sharp axial ridges (except on oral plates) that divide the exterior surface into numerous triangular areas; few accessory ridges; numerous small pustules in rows that parallel plate boundaries. Basals deeply inset at juncture with stem; B2 pentagonal, higher than other basals but apparently not reaching L2; B1 and B3 pentagonal and B4 hexagonal; relatively long, slightly curved sutures between B2/IL1, B2/IL2, B3/IL3, and B3/IL4. Infralateral plates encircle the theca, all convex; IL4 and IL5 form the aboral boundary of the periproct. Circlet of laterals interrupted by IL2 and R2; L4 and L5 apparently form the oral boundary of the periproct; L3 largest in the lateral circlet. Radials large, forming a closed circlet, oral edges deeply cleft, R2 largest. Oral plates in interambulacral position. Ambulacra long, extending down over dome-shaped oral surface and over the radial plates; numerous rounded ambulacral flooring plates rest on elevated V-shaped radial facets; brachiole bases consist of two flooring plates with sutures through the center, brachioles set alternately on right and left sides of each ambulacrum; food grooves covered by numerous rectangular biserial cover plates that meet medially along a straight line and arch slightly. Pore rhombs disjunct; slit-spacing ranges from 2.5 to 3.0 mm depending on size of individual. Each pore rhomb and demirhomb generally has a ridge along the inner edge of the slits on one of the two plates. Pore rhombs occur along plate boundaries of B2/IL2, B2/IL1, L1/L2, L3/L4, L3/R3, R2/R3, R3/R4, and R4/R5; demirhombs occur on R5/R6 and L3/R3. The presence or absence of pore rhombs on IL1 and IL2, as well as on and radials 1 and 6 is not known. The periproct frame consists of at least four plates, including IL4 and IL5, as well as L4 and L5. The periproct was apparently covered by a flexible-plated periproctal membrane with a central opening surrounded by several stout plates. Brachioles, biserial, 9 or 10 per ambulacrum, elliptical in cross section, food groove covered by a biseries of small, long, erect covering plates. The stem is long (100 mm long in the holotype UI X-4961), tapering; typical glyptocystitid type, composed of a flexible proximal section that consists of telescopically imbricating columnals with serrated edges, large central lumen, and much longer distal section consisting of elongate columnals with smooth lateral edges and small central lumen. No holdfast observed.

Remarks. Currently known specimens are mostly disarticulated, making it difficult to determine the exact number and arrangement of thecal plates surrounding the periproct. Consequently, confident assignment to a particular genus is uncertain. Believing that at least five plates surround the periproct, Kolata (1975) assigned this species to the genus *Coronocystis* of the Family Cheirocrinidae. On the basis of recently discovered thecal plates, it now appears that as few as four could be involved, in which case this Platteville species could be assigned to the Family Glyptocystitidae as suggested by Sprinkle (1982b), who provisionally assigned it to *Tanaocystis*? *durandensis*. I choose to stay with assignment to *Coronocystis durandensis* Kolata (1975) until the thecal plate arrangement can be confirmed with new fossil material.

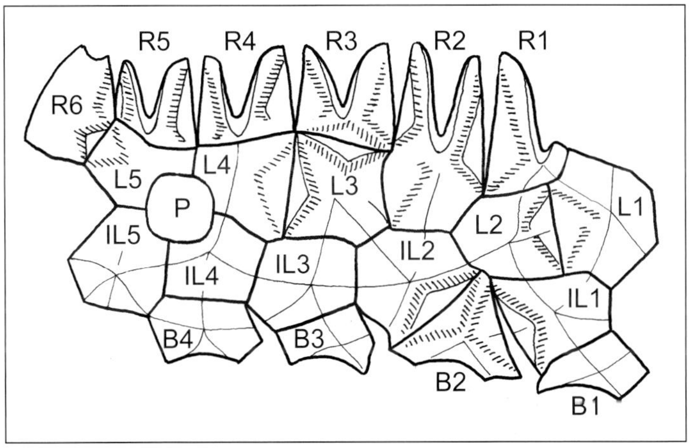

Figure 14-8 Thecal diagram of *Coronocystis durandensis* showing plate designations (R, radials; L, laterals; IL, infralaterals; B, basals; P, periproct) following that of Kesling (1967). Oral and distal plates not shown. The number of plates surrounding the periproct is not clearly known. Radial plates mark the top of the theca.

Occurrence. Disarticulated thecal plates are locally common in the Grand Detour Member of northern Illinois and southern Wisconsin.

Family PLEUROCYSTIDAE

Genus PRAEPLEUROCYSTIS

Praepleurocystis* cf. *ranaformis Guensburg, 1984

Figures 14-9-16 through 14-9-21 and 14-10

Description. Theca subtriangular in outline, nearly bilaterally symmetrical; largest specimens (excluding brachioles and stem) approximately 40 mm long and 35 mm wide; dorsal surface strongly arched and bearing three elevated, oval, disjunct pectinirhombs. Dorsal surface composed of relatively large basal (B), infralateral (IL), lateral (L), and radial (R) plates that are ornamented with prominent, radiating, rounded ridges; L2, L3, and IL2 are the largest plates on the dorsal surface; distal end of theca is formed by relatively small oral (O) and distal (D) plates. Ventral surface framed by B, IL, L, and R plates that surround a large, flat periproct consisting of numerous small, thin, irregularly hexagonal platelets. Hydropore and gonopore positioned along the suture of oral plates O1 and O6 on the ventral surface. Two long brachioles bearing biserial cover plates project from distal end of theca. Column tapers rapidly from proximal region of alternating wide to narrow columnals with rounded edges to distal columnals that are uniformly longer than wide.

Remarks. This Platteville species is most similar to and probably conspecific with *P. ranaformis* Guensburg, 1984, from the Lebanon Formation of central Tennessee. *P. nodosus* Westphal, 1974, known by a single specimen from the Platteville Formation of Grant County, Wisconsin, differs from *P. ranaformis* in having a dorsal surface with nodose radiating thecal plate ridges and in lacking contact between plates B2 and IL1 as well as B3 and IL3.

Occurrence. Rare in the Grand Detour Member of northern Illinois and southern Wisconsin.

Class Unknown

Holdfasts

Remarks. Holdfasts are structures at the distal end of an echinoderm column that fix the animal to the substrate. Various North American morphotypes of Ordovician age have been described and illustrated in the literature (Sardeson 1899, 1908; Kolata 1975; Warn and Strimple 1977; Brower and Veinus 1978; Lewis 1982; Guensburg 1984). The Platteville Formation contains at least four kinds of holdfasts: (1) multiplated disks (*Lichenocrinus*), (2) multiplated lobate- and digitate-encrusting forms, (3) multiplated rootlike forms that penetrate downward into the substrate, and (4) rootlike

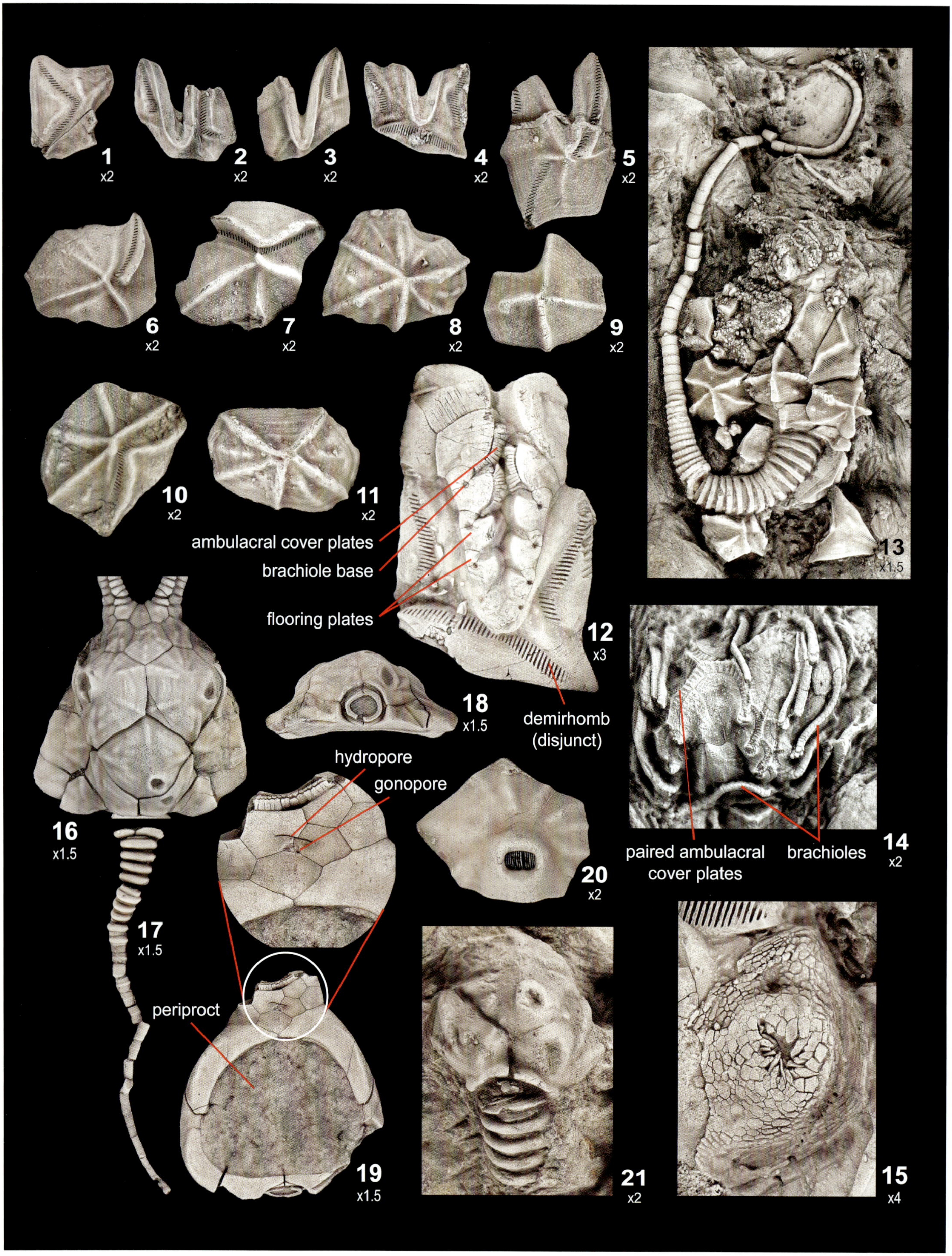
1
x2
2
x2
3
x2
4
x2
5
x2
6
x2
7
x2
8
x2
9
x2
10
x2
11
x2
ambulacral cover plates
brachiole base
flooring plates
12
x3
demirhomb
(disjunct)
13
x1.5
14
x2
paired ambulacral
cover plates
brachioles
15
x4
16
x1.5
17
x1.5
18
x1.5
hydropore
gonopore
19
x1.5
periproct
20
x2
21
x2

forms with long, slender branches that are not multiplated. Most of these holdfasts probably belong to crinoids, but other types of stalked echinoderms cannot be ruled out.

Multiplated Discoidal (*Lichenocrinus*) Holdfasts

Figures 14-5-12 and 14-5-13

Description. Disk (5–8 mm in diameter) with one-piece attachment plate with fine radiating ridges, multiplated upper surface that is depressed at base of stem and having a five-part stem. Typically encrusted on hardgrounds and a wide variety of shelly invertebrates, including brachiopods, gastropods, bryozoans, and trilobites (Kolata 1975).

Remarks. This distinctive morphotype is relatively common in Upper Ordovician limestones of eastern North America. Presently not associated with a particular echinoderm.

Occurrence. Abundant in the Mifflin and Grand Detour Members in north-central Illinois and south-central Wisconsin and the McGregor Member of southwestern Wisconsin, northeastern Iowa, and southeastern Minnesota.

Figure 14-9 (Opposite page)
1–15. ***Coronocystis durandensis*** Kolata. (**1–11**) Isolated thecal plates; (**1**) lateral plate L1; (**2**) R5 radial; (**3**) R4 radial; (**4**), R3 radial; (**5**) R2 radial; (**6**, **7**) exact position in theca unclear; (**8**) infralateral IL5; (**9**) infralateral IL4; (**10**) infralateral IL2; (**11**) infralateral IL3; (**12**) radial R4 with a well-preserved ambulacrum consisting of flooring plates, brachiole bases, ambulacral cover plates, and disjunct demirhombs; (**13**) partially disarticulated thecal plates and nearly complete stem; (**14**) well-preserved oral surface showing the paired ambulacral cover plates and brachioles; (**15**) suspected periproctal membrane showing imbricated plates near central opening. Grand Detour Member, Dixon, Illinois (locality 1), **1**, UI X-4967; **2**, UI X-4966; **3**, UI X-5081; **4**, UI X-4964; **5**, UI-4963; **6**, UI X-4965; **7**, UI X-5077; **8**, UI X-4968; **9**, UI X-5080; **10**, ISGS 1216; **11**, UI X-4969; **12**, UI X-5083; **13**, holotype UI X-4961; **15**, UI-X4936. Grand Detour Member, Durand, Illinois (locality 12), **14**, BMNH PK-82.

16–21. ***Praepleurocystis* cf. *ranaformis*** Guensburg. (**16**) Complete and well-preserved dorsal theca showing disjunct pectinirhombs, plate configurations, and ornamentation; (**17**) a nearly complete stem; (**18**) posterior end of theca showing opening at juncture with stem as well as a disjunct pectinirhomb; (**19**) ventral surface showing the large periproct and photographic enlargement of oral plates with hydropore and gonopore; (**20**) infralateral thecal plate IL2 showing the disjunct pectinirhomb; (**21**) an incomplete specimen showing the proximal stem, basals B2, and infralateral IL2. (**16–19**) Paratypes of *P. ranaformis* from the Lebanon Formation, central Tennessee, provided by Thomas E. Guensburg, **16**, PE82166; **17**, PE82168; **18**, **19**, UI X-5697; (**20**, **21**) Grand Detour Member, Dixon, Illinois (locality 1), **20**, UI X-5082; **21**, ISGS 1240.

Multiplated Lobate- and Digitate-Encrusting Holdfasts

Figures 14-5-10 and 14-5-11

Description. Holdfast multiplated, lobate, or digitate with 15–20 branches ranging from long and slender to short and stubby; longer branches up to 25 mm in length; typically encrusting shelly debris on and parallel to bedding planes.

Remarks. Presently not associated with a particular echinoderm.

Occurrence. Common in the Mifflin and Grand Detour Members of northern Illinois and southern Wisconsin.

Type 1 Rootlike Holdfast

Figure 14-7-11

Description. Holdfast multiplated, rootlike, with three to five branches extending into the substrate. Distal column round, 8 mm in diameter, composed of thin, straight-sided columnals divided into five-part circlets. Lumen is pentastellate, with star tips corresponding to the longitudinal pentamere articulations. Plates at top of each branch are wider than high; branches taper rather abruptly.

Remarks. This morphotype is very similar to "holdfast type B" described by Guensburg (1984) from the Upper Ordovician Lebanon Formation of central Tennessee as well as "type 1B" from the more or less coeval Bromide Formation of southern Oklahoma (Lewis 1982). The pentamere columnals and presence of five or fewer branches are also similar to the holdfast in the Platteville species *Quinquecaudex springeri* (Kolata, 1975).

Occurrence. Rare in the Grand Detour Member of northern Illinois.

Type 2 Rootlike Holdfast

Figure 14-6-9

Description. Holdfast consisting of 10–15 hollow nonbifurcating tubules that extend downward into the substrate in a rootlike manner. Tubules are round and maintain a consistent diameter of 2–4 mm.

Remarks. Presently not associated with a particular echinoderm.

Occurrence. Rare in the Grand Detour Member of northern Illinois and southern Wisconsin.

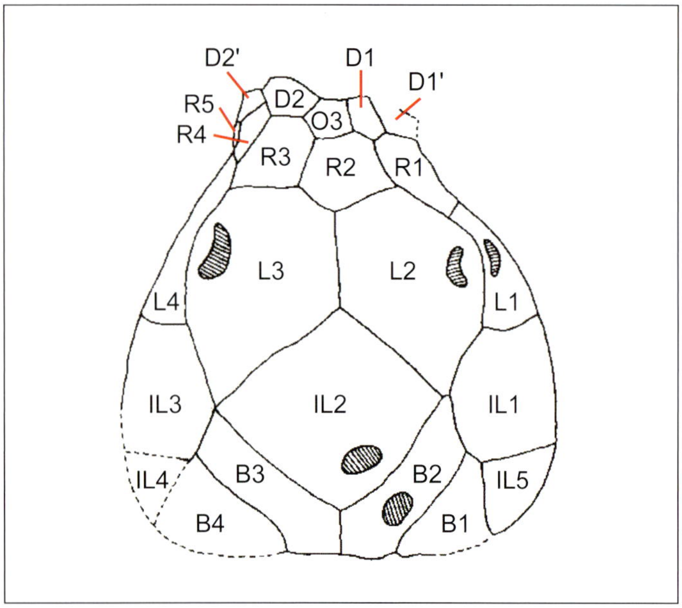

Figure 14-10 1. ***Praepleurocystis ranaformis*** Guensburg. (**1**) Drawing of theca showing basals (B), infralaterals (IL), laterals (L), radials (R), oral plate (O), and arm plates (D). Modified from Guensburg (1984, figure 13); reproduced with permission from the Paleontological Research Institution, Ithaca, New York.

Class ASTEROIDEA

Description. The skeletal ossicles of most asteroids are held together by connective tissue, providing substantial flexibility to the animal during life. However, after death the skeleton rapidly becomes disarticulated and the ossicles are scattered by biogenic activity and subsea currents. Consequently, complete asteroids are rarely preserved in the fossil record.

Asteroids have a pentagonal star-shaped ("starfish") body with **central disk**, open-water vascular system, commonly with an opening called the **madreporite**, and arms with ventrally positioned **ambulacral grooves** leading to a central mouth. The primary skeleton forms the body wall and consists of ventrally positioned **ambulacrals** and **adambulacrals**, marginal ossicles of the **inferomarginal series**, and the adjacent but more dorsally positioned **superomarginal series**. The dorsal surface includes a **primary circlet** consisting of **radials** at the head of the arms, **interradials**, and midarm ossicles of the **carinal series**. Terminology and classification follow those of Spencer and Wright (1966) and Blake (2018).

Order EUAXOSIDA

Family PALASTERINIDAE

Genus SCHUCHERTIA

Schuchertia sp.

Figure 14-11-8

Description. This Platteville species is known from a single specimen that exposes the dorsal surface of the central disk and five variously truncated arms. Distinguishing features include the relatively small, uniform, polygonal, marginal ossicles bearing numerous very small pustules; carinal plates arranged in two rows; apparently lacking superomarginals.

Remarks. Morphologically, this Platteville asteroid fits the emended diagnosis of the genus *Schuchertia* as described by Blake et al. (2016).

Occurrence. Articulated specimens are rare in the Mifflin Member at Dixon, Illinois (locality 1).

Kolataster perplexus Blake et al., 2020

Figures 14-11-6 and 14-11-7

Description. This Platteville species is known from two fragmentary specimens, both exposing the dorsal surface only, one an arm fragment and the other a disk with a nearly complete arm and four fragmentary arms. Dorsal surface composed of relatively large, irregularly arranged ossicles and lacking obvious carinal and marginal series, although multiple series reach the arm tips, suggesting that carinals and marginals are present. Paxilliform dorsal ossicles with slender elongate columns are present lateral to the midline; some are tilted outward, exposing the crescent-shaped bases with reentrants facing toward the midline; small depression present at tip of each column. Numerous tiny spines present on the dorsal surface. Arm radius approximately 15 mm. Madreporite small, interbrachial in position, upper surface marked by distinctive wavy ridges.

Remarks. The two known specimens are incomplete. Neither expose the ambulacral or adambulacral ossicles; therefore, the ordinal and familial assignments cannot be unequivocally demonstrated (Blake et al. 2020). These authors state that the form and configuration of dorsal ossicles suggest assignment of this asteroid to the euaxosidan Palasterinidae.

Occurrence. Very rare in the Mifflin Member, Dixon, Illinois.

Palasterinid sp.

Figures 14-11-1 and 14-11-2

Description. A relatively large Platteville asteroid known from a single incomplete specimen preserved with ventral surface exposed; characterized by elongate arms that are triangular in outline and possess curved interbrachial areas. Skeleton consists of four fragmentary arms and two interbrachia that outline a deeply weathered central disk. Arm furrows are flanked on each side by three rows of irregularly shaped ossicles. The inner row contains the largest ossicles. These are flanked by a row of relatively small ossicles, in turn flanked by medium-sized ossicles. Ossicles in all three rows support a short, blunt spine. It is difficult to determine whether the ossicular debris within the furrows represents pieces of the ambulacral and adambulacral ossicles or ossicles on the inside of the dorsal surface.

Remarks. The overall body form and ventral ossicles suggest assignment to the family Palasterinidae of the order Euaxosida (Daniel B. Blake, personal communication, 2018). Assignment to genus and species awaits the discovery of additional specimens.

Occurrence. Rare in the Mifflin Member at Dixon, Illinois (locality 1).

Genus and Species Undetermined

Figures 14-11-9 and 14-11-10

Description. This Platteville species is known from two specimens; both expose the ventral surface, one showing the central disc and five variously truncated arms and the other exposing three partial arms and a few plates of the central disk. Distinguishing features include a very small central disk lacking interbrachial arcs and long, slender arms with numerous relatively small, polygonal, paxilliform dorsal ossicles with slender, elongate columns. Ambulacral ossicles are fully dorsal to adambulacrals, are wider than long throughout the length of the arm, overlap each other proximally, and are approximately equal in number to the adambulacrals; ossicles of adjoining ambulacral columns appear to be offset slightly. Each ambulacral plate possesses a high, longitudinal ridge that turns at the adradial end and passes in a proximal direction outlining the V-shaped groove between the two columns of ambulacra and the ampullar basins between ambulacra of the same column. Ambulacral ossicles are wider than long throughout the length of the arm, overlap each other proximally, and are approximately equal in number to the adambulacrals; ossicles adjoining ambulacral columns appear to alternate slightly. Adambulacral ossicles form the margin of the ventral surface, are subquadrate in outline, are much wider than long, and bear four or five small spines on the ventral surface. Adorally, the adambulacrals decrease in width and terminate in pairs of triangular oral plates.

Remarks. One of the known specimens (UI X-4960) was described and figured by Kolata (1975) and questionably assigned to *Urasterella*? sp. Subsequently, this specimen was reexamined by Daniel B. Blake (personal communication, 2020), who determined that ambulacral form and position relative to the adambulacral ossicles is more aligned with species of the order Euaxosida.

Occurrence. Rare in the Grand Detour Member in north-central Illinois.

Family HUDSONASTERIDAE

Genus HUDSONASTER

Hudsonaster* cf. *rugosa (Billings, 1857)

Figures 14-11-3 through 14-11-5

Description. A small- to moderate-sized species of *Hudsonaster* measuring 12–15 mm from disk center to tip of full-length arm; dorsum arched. Dorsal disk composed of stellate ossicles that are larger than those on the arms. Arms relatively short and triangular in outline. Aboral arm skeleton composed of two rows of approximately 10 tumid, pustulose, relatively large superomarginals separated by a median row of smaller carinals; inferomarginals outset slightly from superomarginals; ossicles of marginal and carinal series decrease evenly in size away from disk. Axillary one per interray, significantly larger relative to adjacent inferomarginals; axillary indented at juncture with inferomarginals; distal end free at margin of disk. Adambulacrals approximately two per each adjacent inferomarginal. Ambulacrals vaulted and fully dorsal to adambulacrals.

Remarks. This Platteville *Hudsonaster* is similar to *H. rugosa* (Billings, 1857) from the Upper Ordovician (Richmondian Series) Charleton Formation at Charleton Point, Anticosti Island, Québec, Canada. The Canadian and northern Illinois specimens agree closely in size and shape, number of superomarginals and inferomarginals, and particularly in the stellate shapes of the aboral disk ossicles.

Occurrence. Rare in the Mifflin and Grand Detour Members of northern Illinois and southern Wisconsin.

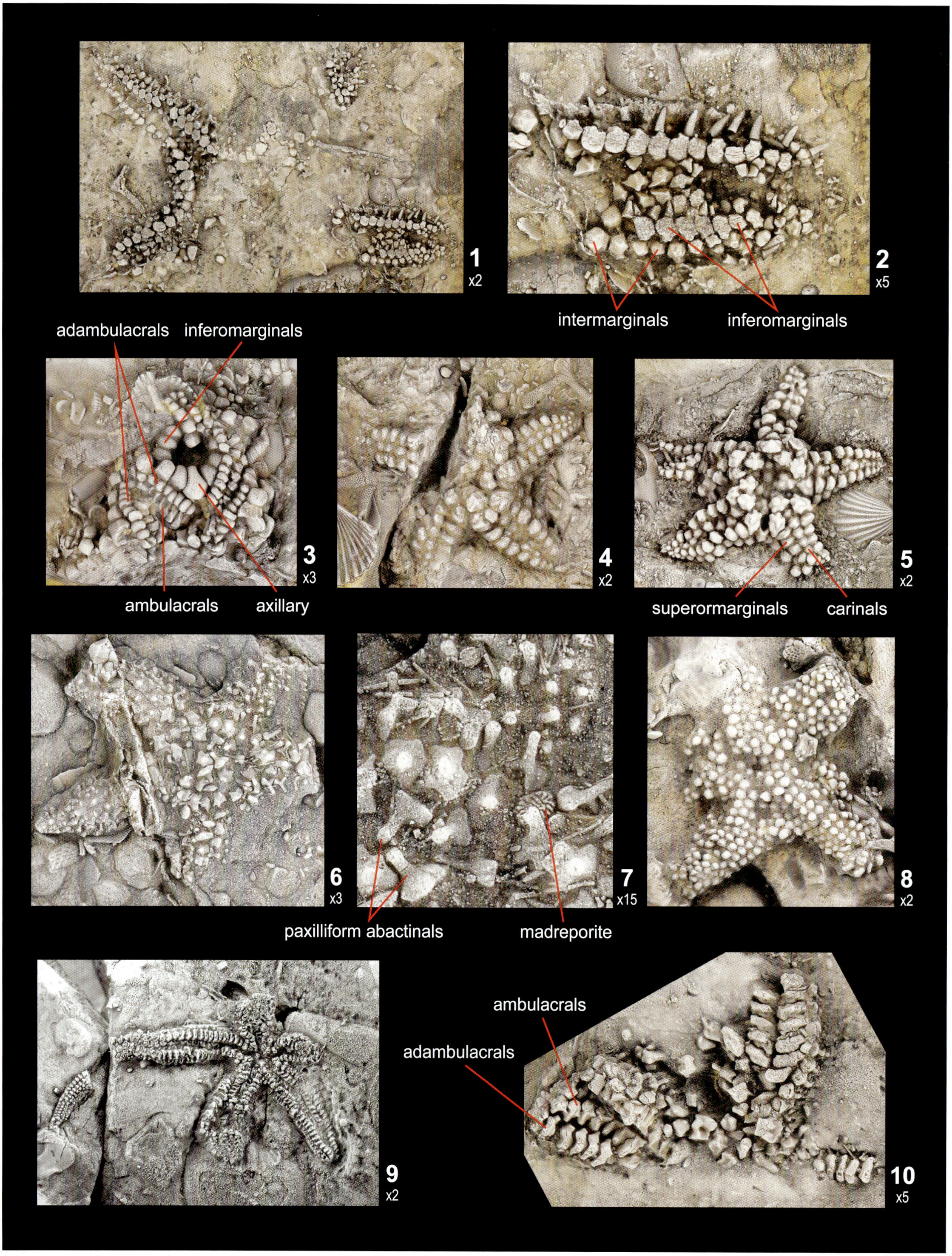
1
x2
2
x5
intermarginals
inferomarginals
adambulacrals
inferomarginals
3
x3
ambulacrals
axillary
4
x2
5
x2
superormarginals
carinals
6
x3
7
x15
paxilliform abactinals
madreporite
8
x2
9
x2
ambulacrals
adambulacrals
10
x5

Class CYCLOCYSTOIDEA

Description. Cyclocystoids are a small, relatively homogeneous group of echinoderms characterized by a low, disk-shaped theca consisting of a ring of stout **submarginal ossicles**, a peripheral ring of small **imbricated marginal plates**, and an upper and lower central disc composed of small, loosely articulated ossicles that apparently enclosed the main organs. The ring of submarginal ossicles is at times the only part of the organism preserved, but in exceptional specimens, the small, delicate marginals and disc plates are still intact. Unlike most other echinoderms, the mouth, anus, and food-gathering appendages are not readily apparent in cyclocystoids. Furthermore, the specialized ossicles are difficult to compare with those of other echinoderms. As a result, the functional morphology and mode of life are not well understood (Kolata 1975).

Family CYCLOCYSTOIDIDAE

Genus CYCLOCYSTOIDES

Cyclocystoides cf. *halli* Salter, 1858

Figures 14-12-10 and 14-12-11

Description. Complete specimens range from 15 to 22 mm in diameter and possess from 25 submarginal ossicles in the smaller specimens to 35 in the larger specimens; numerous ossicles of the lower or ventral disc are loosely articulated; in dorsal view (the surface exposed when submarginal cupules are facing up), these ossicles imbricate toward the center. Lateral extensions on individual ossicles meet adjacent ossicles on both sides, forming an irregular reticulate network with numerous pores of differing shapes and sizes passing through the disc. The center of the disc is marked by a pore containing four or five small ossicles surrounded by five or six larger ossicles that branch out from the center to successively smaller ossicles toward the submarginals. At the juncture with the submarginals, the radiating branches lie below and lateral to a series of relatively large spear-shaped plates that articulate with U-shaped prominences on the inside edge of the submarginals and project toward the center. A small furrow or channel lined with numerous very small triangular plates extends along the midline of each radiating branch. Skeletal elements of the upper disc are very small and reveal little information about the original structure. The disc apparently consisted of a flexible integument containing numerous irregular ossicles. The stout submarginal ossicles vary in number depending on the size of the individual, being more numerous in larger specimens. The distal part of each ossicle, on the dorsal side, contains one to three, but more commonly two, spoon-shaped depressions (cupules) leading to a pore that passes through the ossicle to the inside of the ring. The number of cupules per submarginal ossicle apparently is not arranged systematically about the ring. No cupule possesses a tubercle or facet in the center, as seen in most other cyclocystoid species. Each cupule is separated from adjacent cupules by a thin septum, with two septa separating cupules of adjacent submarginals. The distal edge of the cupule has from 10 to 15 small indentations at the line of articulation with the large proximal marginal plates. The proximal part of the submarginals is elevated above the disc, and the inside and outside edges overhang both sides of the ring. The surface of the submarginal ossicles is covered with prominent pustules, some of which are coalesced into ridges that are aligned in a radial direction corresponding with the septum between cupules. The lower (ventral) surface of the submarginals is slightly convex, is trapezoidal in outline, and has reticulate ornamentation. The periphery of the submarginal ring is enclosed by numerous imbricating plates that decrease in size distally. Additional skeletal structures are discussed in Kolata (1975).

Remarks. Platteville species of *Cyclocystoides* closely resemble *C. halli* Salter from the Trentonian Hull beds of the Ottawa Formation of Ottawa, Ontario, Canada, The

Figure 14-11 (Opposite page)
1, 2. **Palasterinid sp.** (**1**) Deeply weathered specimen exposing the aboral surface of four partially preserved arms; (**2**) magnified portion of a single arm showing the intermarginals, inferomarginals, and stout spines. Grand Detour Member, Dixon, Illinois (locality 1), **1**, **2**, ISGS 1214.

3–5. ***Hudsonaster* cf. *rugosa*** (Billings). (**3**) Adoral surface exposing two arms with adambulacrals, inferomarginals, ambulacrals, and an axillary plate; (**4**) aboral surface; (**5**) well-preserved specimen showing the superomarginals and carinals. Mifflin Member, Dixon, Illinois (locality 3), **3**, ISGS 1215; **4**, ISGS 1233. Mifflin Member, Grant County, Wisconsin, **5**, AWS-102.

6, 7. ***Kolataster perplexus*** Blake et al. (**6**) Well-preserved specimen showing the aboral surface; (**7**) photographic enlargement showing numerous paxilliform abactinal ossicles and madreporite. Mifflin Member, Dixon, Illinois (locality 2), **6**, **7**, PRIP 20027.

8. ***Schuchertia* sp.** (**8**) Dorsal (aboral) surface of central disk and five variously truncated arms. Mifflin Member, Dixon, Illinois (locality 2), **8**, ISGS 1213.

9, 10. **Genus and species undetermined.** (**9**) Ventral surface (adoral); (**10**) ventral surface showing adambulacral and ambulacral ossicles. Grand Detour Member, Shirland, Illinois (locality 16), **9**, UI X-4960. Mifflin Member, Dixon, Illinois (locality 2), **10**, ISGS 1234.

Canadian species differs in possessing a tubercle or facet within the submarginal ossicle cupules.

Occurrence. Isolated ossicles common in the Mifflin and Grand Detour Members in north-central Illinois and south-central Wisconsin and the McGregor Member of southwestern Wisconsin, northeastern Iowa, and southeastern Minnesota. Articulated specimens are rare.

Class ECHINOIDEA

Order BOTHRIOCIDAROIDA

Family BOTHRIOCIDARIDAE

Remarks. Bothriocidarid echinoids are a primitive family of sea urchins known only from Ordovician rocks of Europe and North America. At present, the family consists of *Bothriocidaris*, *Neobothriocidaris*, and *Unibothriocidaris*, all three of which are present in the Platteville Formation. They are characterized by a small, rigid **test** that is composed of thick, nonimbricated plates. In addition, they all possess **podial pores** that open externally in distinct **peripodia** surrounded by a rim on which one or more **perforate spine tubercles** are developed. The three genera differ markedly with respect to the position of the radial water vessels (internal or within the ambulacral plates), arrangement of ambulacra, position and shape of the podial pores, and morphology of the **interambulacra**. The three genera are easily differentiated by the position of the podial pores on the ambulacral plates. *Bothriocidaris* has a pair of pores positioned near the center of the plate, *Neobothriocidaris* has a single pore near the plate center and shares a pore with an adjacent plate, and *Unibothriocidaris* has a single large, elliptical pore corresponding to a pair of podia.

Genus BOTHRIOCIDARIS

Bothriocidaris solemi Kolata, 1975

Figures 14-12-14 and 14-12-15

Description. A species of *Bothriocidaris* possessing a spherical test approximately 15–20 mm in diameter. Ambulacra consist of two columns of alternating wide, hexagonal plates. Peripodia elliptical and lacking a conspicuous rim; two perforate primary tubercles on the peripodia, with the space between them equal to slightly less than their diameter; four to six additional secondary perforate tubercles ranging from 0.2 to 0.4 mm in diameter are closely spaced around the peripodium and primary tubercles; inside surface of the ambulacral plates smooth and slightly concave, with widely spaced podial pores that converge toward the exterior. Largest ambulacral plates approximately 2.3 mm wide and 1.3 mm high, with primary tubercles 0.6 mm in diameter. Interambulacra consist of five columns of hexagonal plates lacking podial pores, with a large, subcentral perforate tubercle equal in size to the primary tubercles of the ambulacrals; smaller perforate tubercles 0.2–0.4 mm in diameter are closely spaced over the slightly convex plate surface; largest interambulacrals approximately 1.9 mm wide and 1.6 mm high. Spines longitudinally striated, approximately 2.5 mm long and 0.4 mm in diameter, proximal ends slightly swollen, tips blunt.

Remarks. *Bothriocidaris solemi* most closely resembles *B. vulcani* Guensburg, 1984, from the approximately coeval Late Ordovician Lebanon Formation of central Tennessee, but *B. solemi* differs in having ambulacral plates with a width-to-height ratio less than 2 and four to six secondary tubercles. An excellent representation of a living *Bothriocidaris* (Figure 14-12-16) is illustrated in Durham (1966).

Occurrence. Isolated ossicles common in the Mifflin and Grand Detour Members in north-central Illinois and south-central Wisconsin and the McGregor Member of southwestern Wisconsin, northeastern Iowa, and southeastern Minnesota. Articulated specimens are rare.

Genus NEOBOTHRIOCIDARIS

***Neobothriocidaris* sp.**

Figures 14-12-1 through 14-12-5

Description. A species of *Neobothriocidaris* possessing a spherical test approximately 15–20 mm in diameter divided into five tracts of pore-bearing and pore-sharing plates that are divided by five nonporous sutures; pore pairs and internal lateral water vessels are aligned in a series of chevrons with four perforate plates in each limb; a single row of pore-sharing plates separates each limb. The perforate plates are typical for the genus, having one subcentral pore and a marginal pore shared with an adjacent plate. The larger plates are approximately 2.8 mm wide, 2.5 mm high, and 0.5 mm thick. One perforate tubercle 0.5 mm in diameter is present on the peripodium, and two or three imperforate tubercles occupy a constant position around the peripodium. In the larger well-preserved plates, the boundaries are commonly serrated, and the serrations are expressed on the outer surface as a row of small pustules adjacent to the sutures. Small pores, approximately 0.2 mm in diameter, opening into the internal system of lateral water vessels can be seen along the plate boundaries close to the inside surface of the plate.

Remarks. The first report of *Neobothriocidaris* in the Platteville Formation (Kolata 1975) was based on the occurrence of isolated ambulacral plates. Subsequently, Allan W. Scheer (personal communication, January 2019) discovered a complete and well-preserved specimen in the Mifflin Member in Grant County, Wisconsin, which is the primary source of information given in the above description. This Platteville species is most similar to *N. templetoni* Kolata, 1975, from the Dunleith Member of the Galena Formation in northern Illinois. *N. templetoni* differs in having three perforate plates in each chevron limb whereas the Platteville form has four, suggesting that at least two distinct species are present in the Upper Ordovician succession of the upper Midwest USA.

Occurrence. Isolated ambulacral plates of *Neobothriocidaris* are locally common in the Mifflin and Grand Detour Members of the upper Midwest, but complete specimens are extremely rare.

Genus UNIBOTHRIOCIDARIS

Unibothriocidaris* cf. *bromidensis Kier, 1982

Figures 14-12-6 through 14-12-9

Description. This Platteville species is presently known only from isolated coronal plates that are polygonal, range from 1.0 to 3.0 mm across, and are approximately 0.5 mm thick; coronal plates have a single, large, elliptical, subcentral pore, one perforate tubercle on the peripodium, and numerous imperforate tubercles randomly distributed over the plate surface. In well-preserved specimens there is a slight constriction on the sides of the elliptical pore, suggesting that one large pore formed from a coalesced pore pair. The circular peripodium has a low profile, and the space in the center is sunken slightly below the level of the plate surface. The single perforate tubercle is low and relatively large. The imperforate tubercles vary from 6 to 25 in number and are about 0.2 mm in diameter.

Remarks. The presence of apparent coalesced podial pores, a large perforate tubercle, secondary imperforate tubercles, and the peripodium is very similar to *Unibothriocidaris bromidensis* Kier, 1982, from the Upper Ordovician Bromide Formation of southern Oklahoma and *U. kieri* Guensburg, 1984, from the Upper Ordovician Lebanon Formation of central Tennessee.

Occurrence. Isolated ambulacral plates are rare in the Mifflin Member of northern Illinois.

Class HOLOTHUROIDEA

Remarks. Presumed holothurian sieve plates were described by Gutschick (1954b) from the Platteville Formation near Dixon, Illinois. These small, perforated plates were rejected by Frizzell and Exline (1955) as being unequivocal holothurians because similar plates are also known to develop in embryonic crinoids. Comparable echinoderm sclerites have been discovered in the Upper Ordovician Pooleville Member of the Bromide Formation in Carter County, Oklahoma (Reso and Wegner 1964). The holothurian origin is still under debate, but for the sake of completeness, the Platteville sieve plates are briefly described and illustrated here.

Family CALCLAMNIDAE

Genus THUROHOLIA

Thuroholia croneisi Gutschick, 1954b

Figure 14-12-12

Description. Thin calcareous, planar, or slightly domed sieve plates that are circular to oval or elliptical in outline; up to 2.5 mm in the longest dimension. Pores vary in size, shape, and number and form a hexagonal grid; commonly 8–10 perforations in a distance of 1.0 mm; approximately 100 perforations per plate; shapes vary from smoothly circular to oval or elliptical.

Remarks. The sieve plates figured here are indistinguishable from those illustrated by Gutschick (1954b).

Occurrence. Locally common on shaley bedding planes in the Mifflin Member of northern Illinois and southern Wisconsin.

Thuroholia cribriformis Gutschick, 1954b

Figure 14-12-13

Description. Like *T. croneisi* but distinguished by clusters of noticeably smaller pores that are eccentric or centrally positioned.

Remarks. The sieve plates figured here are indistinguishable from those illustrated by Gutschick (1954b).

Occurrence. Locally common on shaley bedding planes in the Mifflin Member of northern Illinois and southern Wisconsin.

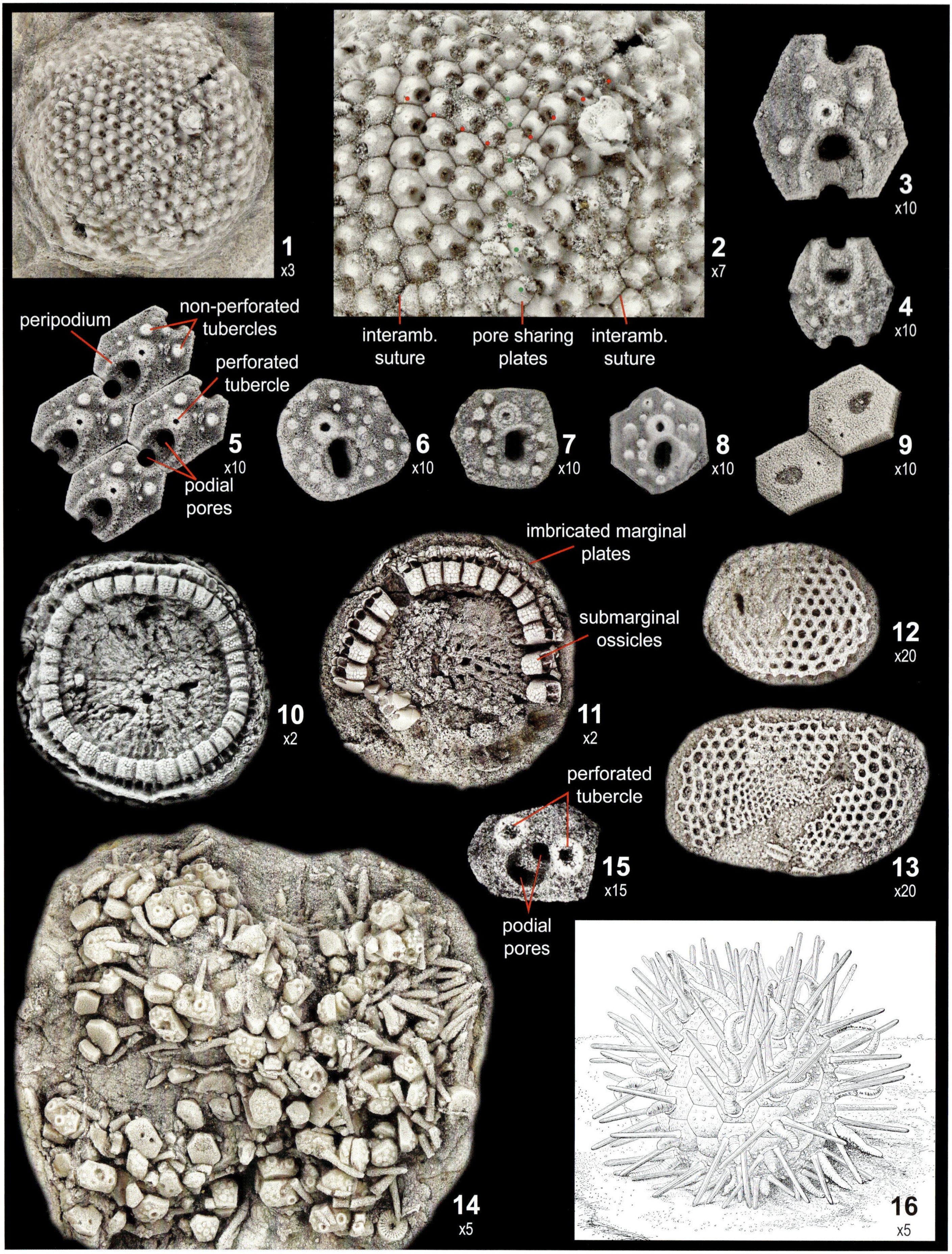
1
x3
2
x7
interamb. suture
pore sharing plates
interamb. suture
3
x10
4
x10
peripodium
non-perforated tubercles
perforated tubercle
podial pores
5
x10
6
x10
7
x10
8
x10
9
x10
imbricated marginal plates
submarginal ossicles
10
x2
11
x2
12
x20
13
x20
perforated tubercle
podial pores
15
x15
14
x5
16
x5

Figure 14-12 (Opposite page)
1–5. ***Neobothriocidaris* sp.** (**1**) An exceptionally well-preserved specimen showing the configuration of ossicles in the ambulacra; (**2**) photographic enlargement showing an ambulacral tract at the shell periphery; the chevron configuration of the ambulacra is revealed by red dots positioned over the large, perforate tubercles on the poriferous plates; each wing of a chevron is separated by a single row of pore-sharing plates marked by green dots; adjacent ambulacra meet along the interambulacral suture. (**3**) a single poriferous plate; (**4**) a single poriferous plate; (**5**) photographic reconstruction of several plates showing the peripodium, perforated tubercles at spine bases, nonperforated spine bases, and shared podial pores. Mifflin Member, Grant County, Wisconsin, **1**, **2**, AWS 101. Grand Detour Member, Durand, Illinois (locality 11), **3**, UI X-4946; **4**, UI X-4948; **5**, UI X-4947.

6–9. ***Unibothriocidaris* cf. *bromidensis*** Kier. (**6–8**) Isolated plates showing merged podial pore pairs, as well as perforated and nonperforated tubercles; (**9**) internal surface of two adjacent plates showing the merged podial pore pairs. Mifflin Member, Dixon, Illinois (locality 2), **6**, UI X-4950; **7**, ISGS 1179; **8**, UI X-4951; **9**, ISGS 1243.

10, 11. ***Cyclocystoides* cf. *halli*** Salter. (**10**) View of dorsal surface; (**11**) view of dorsal surface showing imbricated marginal plates and submarginal ossicles. Mifflin Member, Shirland, Illinois (locality 44), **10**, BMNH PK-69. Grand Detour Member, Dixon, Illinois (locality 1), **11**, UI X-4956.

12. ***Thuroholia croneisi*** Gutschick. (**12**) A single sieve plate. Mifflin Member, Dixon, Illinois (locality 2), **12**, ISGS 1177.

13. ***Thuroholia cribriformis*** Gutschick. (**13**) A single sieve plate showing the distinctive cluster of smaller pores. Mifflin Member, Dixon, Illinois (locality 2), **13**, ISGS 1178.

14, 15. ***Bothriocidaris solemi*** Kolata. (**14**) Disarticulated plates and spines of a single individual; (**15**) an isolated plate showing two perforated tubercles and podial pore pairs. Grand Detour Member, Dixon, Illinois (locality 1), **14**, holotype UI X-4882; **15**, UI X-4942.

16. ***Bothriocidaris eichwaldi*** Mannil. (**16**) Drawing of *Bothriocidaris* in living position showing spines and tube feet. From Durham (1966). Copyright 1966 John Wiley & Sons, Inc. Reprinted with permission.

CHAPTER 15

MISCELLANEOUS MACROFOSSILS (GRAPTOLITES, CORNULITIDS, HYOLITHIDS, MACROALGAE)

Included in this chapter are a small group of largely unrelated macrofossils, each represented in the Platteville Formation by a small number of species (Appendix 2).

GRAPTOLITES

Graptolites occur locally in the Mifflin and Grand Detour Members of the Platteville Formation in northern Illinois and southern Wisconsin, where they are preserved commonly as flattened organic films on shaley bedding planes. Well-preserved three-dimensional specimens have been found in both the limestone and dolomite facies, some of which are isolated from their matrix (Goldman et al. 2002).

Graptolites are colonial animals with a fossil record extending from the Middle Cambrian through the Mississippian, spanning a little more than 200 million years. Recent phylogenetic analyses suggest that fossil graptolites are most closely related to the extant colonial pterobranch *Rhabdopleura* and that this living example provides insight into the anatomy and life cycle of fossil graptolite zooids (Mitchell et al. 2013; Maletz 2014a). Planktonic graptolites (Graptoloidea) such as the Platteville species *Amplexograptus maxwelli* are believed to have lived unattached in marine environments, actively moving through the water column and filtering food particles (Maletz 2014a). The colony is made up of individual thecae, which house the sister **zooids** and are commonly arranged in single (uniserial) or double (biserial) branches (**stipes**) that constitute the overall colonial skeleton called the **rhabdosome** or **tubarium**, with the **periderm** being the outer wall. The graptoloid rhabdosomes formed from an initial zooid housed in a conical chitinous **sicula** that developed into a simple or branched colony. The classification used herein follows Maletz (2014b).

SYSTEMATIC DESCRIPTIONS

Phylum HEMICHORDATA

Class PTEROBRANCHIA

Subclass GRAPTOLITHINA

Order GRAPTOLOIDEA

Suborder AXONOPHORA

Family DIPLOGRAPTIDAE

Genus AMPLEXOGRAPTUS

Amplexograptus maxwelli Decker, 1935

Figures 15-1-5 through 15-1-7

Description. The essential characteristics of *A. maxwelli* were thoroughly described by Decker (1935) based on specimens from the Upper Ordovician Pooleville Limestone Member of the Bromide Formation in southern Oklahoma. *A. maxwelli* was first reported from the Platteville Formation in southern Wisconsin by Decker and Frederickson (1941), followed by a detailed description of rhabdosome development based on additional material from Grant County, Wisconsin (Walker 1953). More recently, three-dimensionally preserved specimens from the Mifflin Member near Lee Center, Illinois, were illustrated and discussed by Goldman et al. (2002). Characteristic features that distinguish *A. maxwelli* from other amplexograptids include thecae that are biform proximally, grading to straight tubes distally, and cuspate thecal apertures.

Remarks. As noted by Goldman et al. (2002), *A. maxwelli* is also present in the Mount Merino Shale ("Normanskill") of New York state and the Balclatchie Group, Glenkill and Lower Hartfell Shales of southern Scotland (Bulman 1945, 1962; Williams 1994). These correlate with the *Climacograptus peltifer* and *C. wilsoni* graptolite zones (=*Diplograptus foliaceous* Zone; Williams 1994).

Occurrence. Common locally on bedding planes in the Mifflin and Grand Detour Members in northern Illinois and southern Wisconsin.

commissure
aperture
1
x5
2
x5
rings
longitudinal lirae
3
x10
4
x10
5
x2
6
x5
7
x15

CORNULITIDS

Annulated, conical shells of *Cornulites* sp. are relatively common in the Platteville Formation. Typically, they occur singly or in clusters encrusted on various shelly invertebrates, including bryozoans, brachiopods, bivalves, gastropods, and cephalopods, or on bedding plane hardground surfaces.

Cornulitids range in age from the Middle Ordovician to the Late Carboniferous and are globally distributed largely in shallow marine sediments deposited on carbonate platforms (Vinn 2010).

They consist of straight, curved, or irregularly sinuous, conical, thick-walled, calcitic shells with prominent rings and longitudinal striae. Their shells exhibit cyclic changes, resulting in alternating growth of vesicular to compact lamellar structures in each annulation (Vinn and Mutvei 2005).

Their biological affinities are currently in debate. A wide range of possibilities have been proposed, including links to (1) lophotrochozoans (segmented worms, molluscs, brachiopods, and solitary bryozoans; Vinn and Zatoń 2012), (2) stem-group phoronids (Vinn 2005), and (3) aseptate cnidaria (Stolarski 2000). Cornulitids are commonly assigned to the phylum Insertae Sedis because of their uncertain affinities. A comprehensive revision of Ordovician cornulitids is needed to fully understand its species concepts.

Figure 15-1 (Opposite page)

1–4. ***Cornulites* sp.** (**1**) Multiple specimens of *Cornulites* encrusted on a shell of cephalopod *Loganoceras*; (**2**) multiple specimens of *Cornulites* encrusted on a shell of bivalve *Vanuxemia*; cornulitid apertures are aligned along margin of commissure in probable feeding position; (**3**) individual *Cornulites* specimen showing prominent rings and longitudinal lirae (**4**) individual specimen showing basal attachment surface. Mifflin Member, Dixon, Illinois (locality 2), **1**, ISGS 1416; **2**, ISGS 1418; **3**, ISGS 1417; **4**, ISGS 1499.

5–7. ***Amplexograptus maxwelli*** Decker. (**5**) Several rhabdosomes on dolostone bedding plane; (**6**) a single rhabdosome; (**7**) three-dimensionally preserved specimen isolated from matrix; photograph by Daniel Goldman; used with permission. Mifflin Member, Lee Center, Illinois (locality 31), **5**, JAC 148; **6**, ISGS 1420; **7**, Daniel Goldman collection.

Phylum INSERTAE SEDIS

Class TENTACULITIDA

Order CORNULITIDA

Family CORNULITIDAE

Genus CORNULITES

***Cornulites* sp.**

Figures 15-1-1 through 15-1-4

Description. Straight to curved, hollow, conical, thick-walled, calcitic shells entirely attached to substrate, consisting of shelly invertebrates or bedding plane hardground surfaces; tube length up to 20 mm, diameter at aperture approximately 4 mm; apical angle about 10°. Surface smooth in early growth stages, becoming annulated (rings) in intermediate stages and marked by prominent longitudinal lirae in advanced stages. Growth lines subparallel, forming approximately six prominent rings in a distance of 1 mm, with numerous very fine intervening annulations. Prominent external longitudinal lirae, approximately six or seven per 1 mm width near aperture. Shell wall with septa and small vesicles. Shell interior marked by the projecting edges of invaginated cones with apices directed away from the aperture.

Remarks. Many specimens of this Platteville species of *Cornulites* are attached to shelly invertebrates, showing evidence of commensalism during the life history of the host organism. A common example is seen in specimens of the bivalve *Vanuxemia*, where numerous cornulitids are lined up on both valves with apertures facing the bivalve commissure (Figure 15-1-2). The generation of currents by the bivalve during its respiration and feeding could have provided a source of nutrients for the awaiting cornulitids.

Occurrence. Common locally in the Mifflin and Grand Detour Members of northern Illinois and southern Wisconsin.

HYOLITHIDS

Disarticulated conical shells, opercula, and helens (see below) belonging to the hyolithid species *Solenotheca baconi* (Whitfield, 1878) are present locally in the Platteville Formation in the upper Midwest USA. Preservation ranges from exceptionally good in the limestone facies to relatively poor in the dolostone.

Hyolithids are globally distributed fossils that occur in rocks ranging in age from Early Cambrian to the end of Permian time. The skeleton is composed of a depressed **conical shell** with a broad **aperture**, **operculum**, and in many

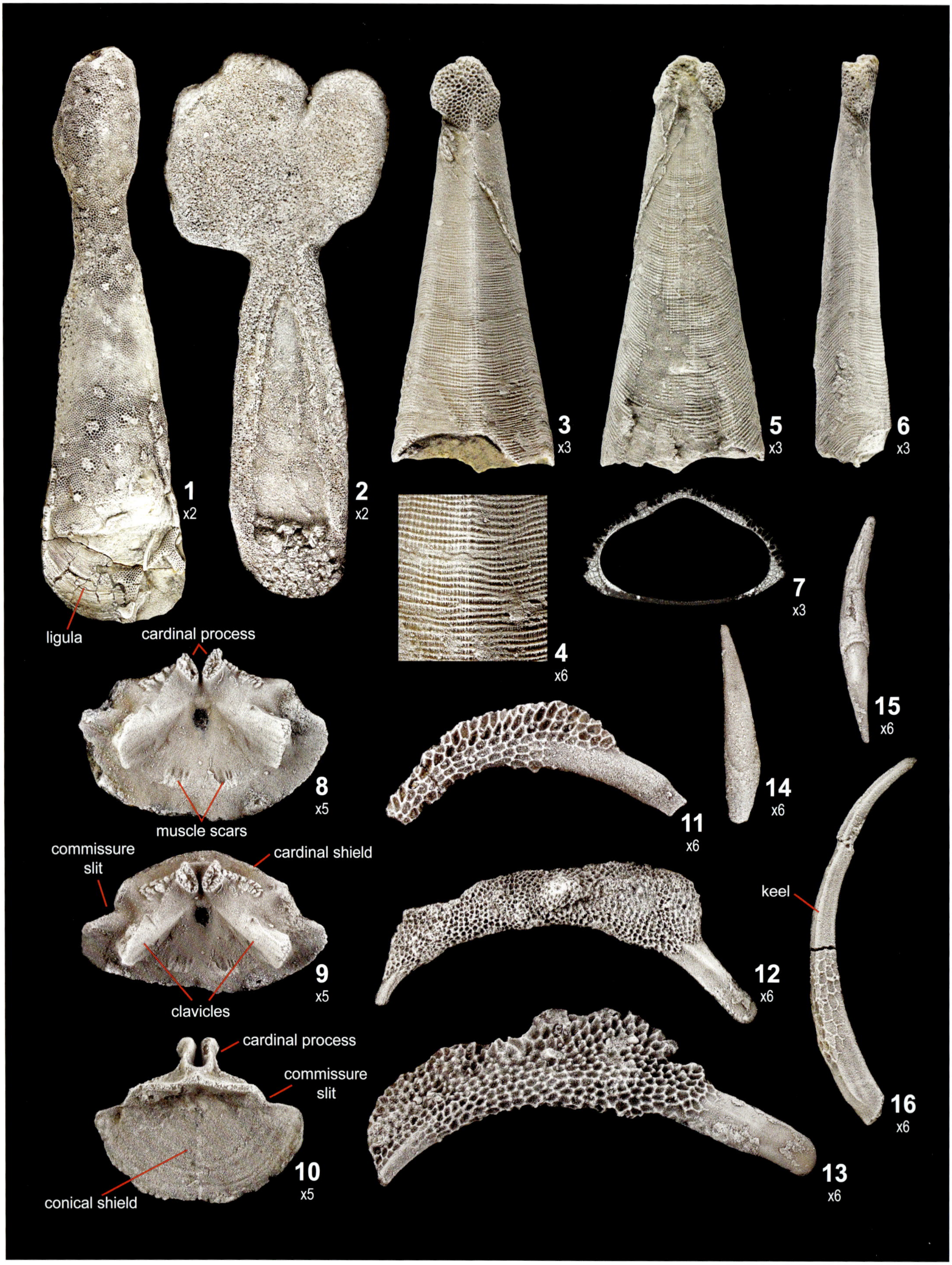
1
x2
2
x2
3
x3
5
x3
6
x3
ligula
cardinal process
4
x6
7
x3
15
x6
14
x6
8
x5
11
x6
muscle scars
commissure
slit
cardinal shield
keel
9
x5
12
x6
clavicles
cardinal process
commissure
slit
16
x6
10
x5
13
x6
conical shield

species, a pair of long, spirally curved spines called **helens**. The conical shell has a flattened ventral surface (**venter**), inflated dorsal surface (**dorsum**), broad oval-shaped aperture, flat shelf at the ventral edge of the aperture called the **ligula**, and a pair of lateral sinuses (**commissure slits**) that accommodate the externally projecting helens. The operculum is folded into a flattened **cardinal shield** that fits the dorsum of the shell and an inflated **conical shield** that fits the ligula. A **furrow** separates the two shields. Internally, the operculum has a dorsally projecting **cardinal process**, a series of muscle scars, and a stout pair of **clavicles** that diverge from the fold between the two shields.

Remarkably well-preserved hyolithid specimens have been described recently from the Middle Cambrian Burgess Shale of British Columbia, Canada, and the Spence Shale of Utah, USA (Moysiuk et al. 2017). These have well-preserved soft tissues that include an extendable tentacle-bearing organ surrounding a central mouth and a U-shaped digestive tract ending in a dorsolateral anus. These features suggest hyolithids have an affinity with the lophophorates (brachiopods, phoronids, and possibly bryozoans; Moysiuk et al. 2017).

Figure 15-2 (Opposite page)
1–16. ***Solenotheca baconi*** (Whitfield). (**1**) Dorsum of conical shell encrusted with *Leptotrypa hexagonalis*; note inflation of apical end due to bryozoan overgrowth, ligula exposed on interior surface of venter; (**2**) venter encrusted with *Leptotrypa hexagonalis*; note massive encrustation on apical end; (**3**) dorsum of conical shell with small bryozoan encrustation on apical end; (**4**) photographic enlargement of dorsum showing transverse and longitudinal lirae; (**5**) venter; (**6**) lateral view of conical shell; (**7**) transverse section of shell encrusted with *Leptotrypa hexagonalis* showing characteristic narrowly rounded dorsum with concavo-convex slopes; (**8**) operculum interior showing cardinal process and muscle scars; (**9**) operculum interior showing indentation of commissure slit, cardinal shield, and clavicles; (**10**) operculum exterior cardinal process, indentation of commissure slit and conical shield; (**11**) left helen with trepostome bryozoan encrustation; (**12**) right helen with *Leptotrypa hexagonalis* encrustation; (**13**) right helen with trepostome bryozoan encrustation; (**14**) helen of immature individual; (**15**) helen of immature individual; (**16**) left helen with small bryozoan encrustation and showing keel. Mifflin Member, Dixon, Illinois (locality 4), **1**, ISGS 1437; **2**, ISGS 1440; **3–6**, ISGS 1438; **7**, ISGS 379; **8**, **9**, ISGS 1430; **10**, ISGS 1439; **11**, ISGS 1433; **12**, ISGS 1436; **13**, ISGS 1434; **14**, ISGS 1431; **15**, ISGS 1432; **16**, ISGS 1435.

Phylum INSERTAE SEDIS

Class HYOLITHA

Order HYOLITHIDA

Family HYOLITHIDAE

Genus SOLENOTHECA

Solenotheca (=Hyolithes) baconi (Whitfield, 1878)

Figures 15-2-1 through 15-2-16

Description. A hyolithid with a depressed conical shell; narrowly rounded medial dorsum flanked by concave slopes, rounded lateral margins with shallow sinus on each side but not on the dorsal margin; venter flat to slightly convex; apical end in lateral view curves slightly upward; ranging up to approximately 60 mm in length. Ligula forms short, flat shelf at ventral edge of aperture. Surface sculpture consists of very fine transverse and longitudinal lirae; transverse lirae parallel outline of aperture, curving down from crest of dorsum to lateral margins following outline of apertural sinuses, ventral lirae follow anterior edge of ligula; longitudinal lirae superimposed on transverse lirae, forming a latticelike network. Operculum tri-oval in outline, up to 10 mm wide and 7 mm long, cardinal shield fits into dorsum of aperture and conical shield conforms with ligula; interior surface of operculum has a pair of dorsally positioned cardinal processes that project inward and a pair of stout, centrally positioned clavicles projecting in a posterolateral direction; upper lateral edge of operculum marked by indentations to accommodate helens; small tubercles positioned along dorsal edge on either side of cardinal processes. Helens paired, slender, broadly curved, ranging up to 22 mm in length and 2 mm wide proximally; tri-oval in cross section, with subtle keel on anterior edge, posterior edge flat to slightly convex; tapering distally, slightly narrower at proximal end.

Remarks. The earliest reported Ordovician hyolithid from the upper Midwest is *Hyolithes baconi* Whitfield, 1878, from "the hard, bluish layers of the Trenton group, below Carpenter's quarry, near Beloit Wisconsin" (p. 225). Given the current understanding of bedrock geology in the Beloit area, the type specimens most likely came from the dolostone facies of either the Mifflin or Grand Detour Member of the Platteville Formation. These specimens are preserved as dolomitized steinkerns lacking many details of the shell but revealing distinctive concavo-convex slopes on the dorsal surface. Since Whitfield's work, many well-preserved specimens have been found in the Platteville Formation at numerous locations in the upper Midwest USA. Some of these were described and illustrated by Malinky (1990), who proposed the new genus *Solenotheca*, noting it differs from other hyolithids in having a "narrowly rounded dorsum with concavo-convex slopes, short ligula and orthogonal aperture" (p. 269). Malinky (1990) assigned a suite of well-preserved specimens from the Platteville Formation of Illinois, Minnesota, and Wisconsin to his *Solenotheca bakerae* n. gen., n. sp. and stated that the known specimens of *Hyolithes baconi* Whitfield, 1878, were too poorly preserved to be confidently assigned to *S. bakerae*. Numerous well-preserved specimens are now available, yet only one hyolithid morphotype is known from the Platteville. The specimens figured in this report are provisionally assigned to *S. baconi* based on the rounded dorsum with concavo-convex slopes.

Platteville specimens of *S. baconi* are commonly encrusted with the trepostome bryozoan *Leptotrypa hexagonalis*. Encrustations typically cover the dorsal and ventral surface of the conical shell and are particularly thick on the apical end, where they produce a distended profile. The helens too are often encrusted with trepostome bryozoans. These hyolithid epibionts have been described and interpreted by Galle and Parsley (2005).

Occurrence. Abundant locally in the Mifflin and Grand Detour Members, particularly near Fennimore, Grant County, Wisconsin, and Dixon, Lee County, Illinois.

MACROALGAE

A small but significant assemblage of macroalgae, including bryopsidalean and dasycladalean green algae, occurs locally in the Platteville Formation. Most are preserved as carbonaceous compressions on shaley bedding planes, but they are also present as partially calcified facets of the dasyclad *Cyclocrinites*. The noncalcified macroalgae were first described by Hall (1861), who interpreted several forms as graptolites; these were later reinterpreted by Whitfield (1894) as algae. LoDuca (1997, 2019) reexamined the type specimens and reported the discovery of additional species in the Platteville Formation. The brief descriptions offered in this review closely follow those of LoDuca.

Phylum CHLOROPHYTA

Class ULVOPHYCEAE

Order BRYOPSIDALES

Genus BUTHOGRAPTUS

Description. Thallus erect, noncalcified, flattened and featherlike, consisting of an unbranched, flexuous, cylindrical, central axis bearing slender, unbranched pinnules arranged in opposite or near opposite position at nearly right angles to the central axis, becoming wider distally; external surface without pores or other openings. The overall featherlike morphology closely resembles the living bryopsidalean green macroalgae *Caulerpa* (LoDuca 2019).

Buthograptus laxus Hall, 1861

Figures 15-3-1 and 15-3-2

Description. Thallus with slender pinnules that widen gradually distally, typically 5–6 mm long and 0.3–0.4 mm in width and with a length-to-width ratio of at least 12 for longer pinnules; interval between adjacent pinnules typically less than two times the proximal pinnule width (LoDuca 2019).

Occurrence. Common locally. The type specimens are from the Platteville Formation near Platteville, Wisconsin (Hall 1861; Whitfield 1894). Additional localities noted by LoDuca (2019) include ravine exposures of the Quimbys Mill Member of the Platteville Formation near Platteville and a small quarry near Dickeyville, Wisconsin, as well as the lower part of the Pecatonica Member near Sun Prairie and Columbus in south-central Wisconsin. Specimens are also known to occur in the shaley dolostone of the upper part of the Forreston Beds, Grand Detour Member at several localities in north-central Illinois, including South Beloit and Durand (localities 20 and 11, respectively).

Figure 15-3 (Opposite page)

1, 2. ***Buthograptus laxus*** Hall. (**1**) Large thallus showing central axis and pinnules; (**2**) small thallus. Grand Detour Member, Dixon, Illinois (locality 4), **1**, ISGS 1160. Forreston Beds, Grand Detour Member, South Beloit, Illinois (locality 20), **2**, UI X-5117.

3, 4. ***Buthograptus gundersoni*** LoDuca. (**3**) Thallus showing central axis and stout pinnules; (**4**) thallus with central axis and pinnules. Forreston Beds, Grand Detour Member, South Beloit, Illinois (locality 20), **3**, UI X-5116; **4**, UI X-5118.

5. ***Buthograptus meyeri*** LoDuca. (**5**) Two thalli, holotype on right; photograph by Steven T. LoDuca. Pecatonica Member, Columbus, Wisconsin, **5**, UWGM 2408.

6. ***Callithamnopsis fruticosa*** (Hall). (**6**) Type specimen; photograph by Steven T. LoDuca. Platteville Formation near Platteville, Wisconsin, **6**, USNM P 41089.

7–10. ***Cyclocrinites* sp.** (**7**) Outer surface of calcified facets; (**8**) perspective view of interior of calcified facets; (**9**) interior view of calcified facets; (**10**) polished transverse section of outer cortex of calcified facets. Mifflin Member, Dixon, Illinois (locality 2), **7**, ISGS 1485; **8**, **9**, ISGS 1487; **10**, ISGS 1489.

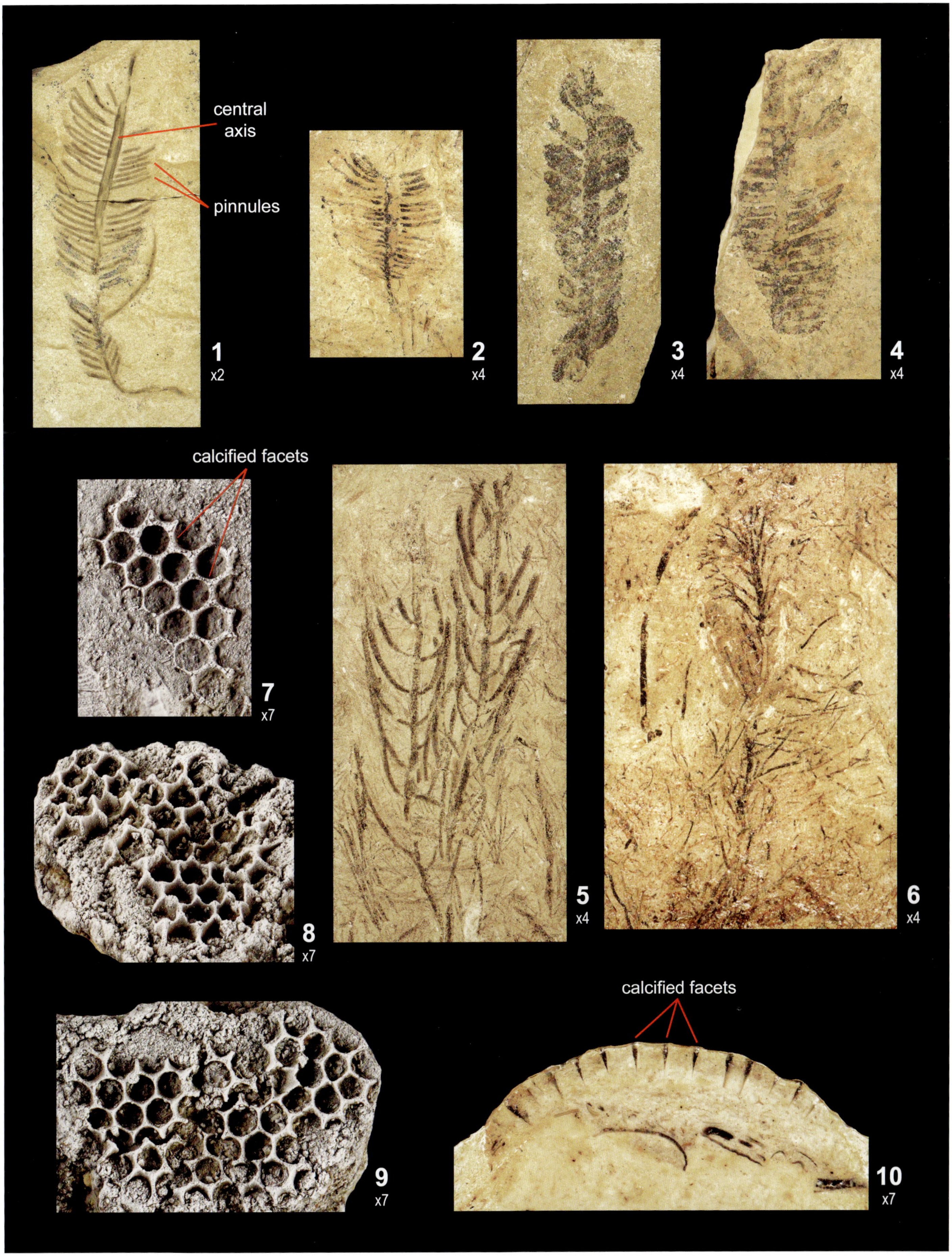
central
axis
pinnules
1
x2
2
x4
3
x4
4
x4
calcified facets
7
x7
8
x7
5
x4
6
x4
calcified facets
9
x7
10
x7

Buthograptus gundersoni LoDuca, 2019

Figures 15-3-3 and 15-3-4

Description. A species of *Buthograptus* with stout pinnules typically 2–4 mm long, 0.5–0.7 mm in maximum width, and with a length-to-width ratio of no more than 7 for the longer pinnules on a given thallus; interval between adjacent pinnules typically less than two times the proximal pinnule width (LoDuca 2019).

Occurrence. Common locally. The type specimens are from the Pecatonica Member in a quarry south of New Glarus, Wisconsin (LoDuca 2019). Specimens are also known to occur in the shaley dolostone of the upper part of the Forreston Beds, Grand Detour Member at several localities in north-central Illinois, including South Beloit (locality 20) and Durand (locality 11).

Buthograptus meyeri LoDuca, 2019

Figures 15-3-5

Description. A species of *Buthograptus* with slender, widely spaced pinnules, the interval between adjacent pinnules typically being more than three times the pinnule width; pinnules strongly upcurved, with distal portions nearly parallel to the central axis (LoDuca 2019).

Occurrence. Common locally. The type specimens are from the lower part of the Pecatonica Member in a reclaimed quarry near Columbus, Wisconsin.

Order DASYCLADALES

Family TRIPLOPORELLACEAE

Genus CALLITHAMNOPSIS

Callithamnopsis fruticosa (Hall, 1865)

Figure 15-3-6

Description. Thallus noncalcified, consisting of a main axis with monopodial branching (terminal bud grows as a central leader branch and lateral branches remain subordinate) arranged in whorls spaced at intervals of 1–2 mm; main axis uniaxial, cylindrical, without segmentation; laterals branched up to six times (LoDuca 2019).

Remarks. This species was first reported by Hall (1865) from the Platteville Formation of southwestern Wisconsin. He interpreted it to be a graptolite and assigned it to the genus *Oldhamia*. The type material was later reinterpreted by Whitfield (1894) to be algae and reassigned by him to the new genus *Callithamnopsis*.

Occurrence. Relatively rare. The type species is from the Platteville Formation near Platteville, Wisconsin. Additional material reported by LoDuca (2019) is from the Pecatonica Member of the Platteville Formation in a quarry near Sun Prairie, Wisconsin, and a small pit west of Columbus, Wisconsin. The shaley beds in the upper part of the Forreston Beds, Grand Detour Member, Platteville Formation, in northern Illinois and southern Wisconsin have good potential for future discoveries.

Genus CHAETOCLADUS

Chaetocladus plumula Whitfield, 1894

For figures, see LoDuca (1997)

Description. Thallus noncalcified, morphologically simple, bottlebrush shaped, composed of an unbranched, parallel-sided central axis surrounded by a multitude of laterals that are slender, unbranched, sharply terminated, without rounding or tapering, and disposed as discrete whorls around the axis; thallus 45 mm in length and 2 mm wide (LoDuca 1997).

Occurrence. Rare in the Platteville Formation. The type species is from the Platteville Formation near Platteville, Wisconsin (Whitfield 1894). The shaley dolostones in the upper part of the Forreston Beds, Grand Detour Member, Platteville Formation in northern Illinois and southern Wisconsin have good potential for future discoveries.

Family Not Determined

Genus CYCLOCRINITES

Description. Cyclocrinitids are a small group of problematic fossils that range in age from Middle Ordovician through Middle Silurian. They are known largely from North America and Europe. Researchers have assigned them to a variety of taxa, including protozoans, sponges, corals, bryozoans, mollusks, algae, and problematica (Nitecki 1970).

The cyclocrinitid **thallus** (plantlike vegetative body) consisted of a **central axis** or stem with a somewhat inflated apical end bearing numerous branches called **laterals** arranged in whorls that radiated outward in all directions. The laterals expanded distally, with each terminating in a lateral head. The lateral heads were tightly packed together, forming an outer cortex. The **facet** is a concave structure at the base of the lateral head commonly surrounded by six adjacent facets that form the characteristic hexagonal surface pattern. The calcified hexagonal facets are typically the only part of the thallus that is preserved. Being lightly calcified, the facets appear to have provided minimal structural support for cyclocrinitids. Like extant dasyclads, cyclocrinitids appear to have been light sensitive and probably were restricted to the marine photic zone (uppermost 80 m) in warm, low-energy waters (Beadle 1988).

Cyclocrinites sp.

Figures 15-3-7 through 15-3-10

Description. Thallus slightly flattened sphere, approximately 20 mm in diameter, lateral heads nonoverlapping; facets typically 1 mm wide at surface, only inner calcareous wall of facets preserved, walls taper from outer surface inward, pinching out a short distance from surface; wall replaced with sparite and lacking microstructure; prominent hexagonal surface pattern.

Remarks. Overall shape of the thallus and prominent hexagonal surface pattern are most like that of *C. darwini* (Miller, 1874).

Occurrence. Locally common in the Pecatonica, Mifflin, and Grand Detour Members in northern Illinois and southern Wisconsin.

CHAPTER 16

MISCELLANEOUS MICROFOSSILS (CONODONTS, FORAMINIFERA, SCOLECODONT-BEARING POLYCHAETES, CHITINOZOANS, ACRITARCHS)

This chapter includes a diverse assemblage of largely unrelated microfossils (Appendix 2). Except for the conodonts, most Platteville microfossils have received little research attention. Much remains to be learned about the diversity and distribution of the associated foraminifera, scolecodonts, chintinozoans, and acritarchs. The following is a brief overview of the Platteville Formation microfossils.

CONODONTS

Conodonts range in age from the Late Cambrian to the end of the Triassic and are abundant in the Upper Ordovician succession of the upper Midwest USA. Features of the soft-part anatomy have been revealed in nearly complete specimens from Carboniferous-age rocks of Scotland (Briggs et al. 1983) and partial specimens from the Late Ordovician of South Africa (Aldridge and Theron 1993). These eel-like animals possessed teeth-bearing feeding apparatuses, bilateral symmetry, a segmented body that is laterally compressed, finlike extensions, and bilobed structures at the anterior end. Overall, the morphological features suggest that conodonts are probably primitive chordates (Briggs 1992). As fossils, conodonts are generally preserved as microscopic tooth-, bar-, or cone-shaped elements ranging in length from 0.3 to 3 mm. The overall length of the conodont animal typically was less than 10 cm, based on the relative size of known feeding apparatuses.

The Platteville Formation contains a diverse and abundant conodont fauna known in large part from the work of Webers (1966), who compiled data on the bed-by-bed distribution and frequency of species in southeastern Minnesota. Webers' work is notable because he was one of the early pioneers who recognized that individual conodonts had multiple phosphatic toothlike elements of variable shape forming a distinctive feeding apparatus. Although rare, intact multielement conodont apparatuses (assemblages) have been found in the fossil record (Scott 1934; Rhodes 1952; Smith et al. 1987; Liu et al. 2017). Prior to the work of Webers (1966) and Bergström and Sweet (1966), most conodont taxonomy was based on formally naming individual conodont elements. After the 1960s, most conodont specialists distinguished between single-element species and multielement assemblages.

Platteville conodonts identified by Webers (1966) include the following multielement assemblages:

Belodina compressa (Branson and Mehl, 1933)
Drepanodus suberectus (Branson and Mehl, 1933)
Ozarkodina obliqua (Stauffer, 1930)
Phragmodus cognitus Stauffer, 1935
Phragmodus undatus Branson and Mehl, 1933
Polyplacognathus ramosa Stauffer, 1935
Pravognathus idonea (Stauffer, 1935)
Trichonodella recurva (Branson and Mehl, 1933)
Zygognathus illustris (Stauffer, 1935)

Sweet (1984, 1987) utilized Webers' data to produce a graphic correlation of Late Ordovician rocks in the upper Midwest showing that the Platteville Formation lies within the *Phragmodus undatus* conodont zone of the North American Midcontinent Province, thus establishing the relative age of the formation and its place in regional correlations. These time–stratigraphic relations have been confirmed in more recent work by Leslie (2000) and Leslie and Bergström (1995, 2005).

FORAMINIFERA

Foraminifera are single-celled organisms that have a shell or test composed of calcium carbonate or, as in early Paleozoic forms, very fine agglutinated sand grains. Platteville species typically range in length from 0.5 to 1.0 mm. They evolved during early Cambrian time and are still present today in nearly all marine environments. Little is known of the foraminifera in the Platteville Formation outside of the brief report by Gutschick (1986), who described and illustrated a variety of chambered and globular agglutinated siliceous foraminifera from the Mifflin Member at Dixon, Illinois, consisting of the following species:

Psammosphaera cava Moreman, 1930
Sorosphaerella cooperensis (Conkin et al., 1968)
Hemisphaerammina thola (Moreman, 1933)
***Tolypammina*? sp.**
Reophax blackriverensis Gutschick, 1986

The latter was regarded by Vachard et al. (2010) as "more likely a representative of the algospongia" (p. 217). Multichambered foraminifera-like specimens also have been found in association with putative fossil peat mosses in the Platteville Formation near Mt. Horeb, Wisconsin (Cardona-Correa et al. 2016).

SCOLECODONT-BEARING POLYCHAETES

Polychaetes are a class of free-moving, segmented, marine worms belonging to the phylum Annelida, ranging in age from earliest Cambrian to the present. Some members possess pharyngeal jaws composed of an assemblage of hardened, organic, microscopic elements (<1 mm in length) called scolecodonts that are diverse and common in the Upper Ordovician succession of North America, including the Platteville Formation. Despite the rich fossil record, progress in understanding taxonomic diversity has been relatively slow because, like the conodonts, most taxa described prior to the 1950s were based on single-element species rather than whole-animal multielement assemblages (Eriksson and Bergman 2003).

One of the most comprehensive taxonomic assessments of North American Ordovician scolecodonts to date is that of Stauffer (1933). His report includes descriptions and illustrations of single-element species from the Glenwood, Platteville, and Decorah Formations from Minneapolis, Minnesota. The reported Platteville species include the following:

Arabellites acutidentatus Stauffer, 1933
***Arabellites* sp. a**
Oenonites spiculatus Stauffer, 1933
***Oenonites* sp. a**
***Oenonites* sp. b**
Protarabellites grandis Stauffer, 1933
Protarabellites intermedius Stauffer, 1933
Lumbriconereites affinis Stauffer, 1933
Lumbriconereites cameratus Stauffer, 1933
Lumbriconereites plattevillensis Stauffer, 1933
Lumbriconereites superbus Stauffer, 1933
Dinoscolites mirabilis Stauffer, 1933
Ungulites aculeatus Stauffer, 1933
Ungulites tridentatus Stauffer, 1933

Staurocephalites dentatus Stauffer, 1933, was reported by Walker (1953) to occur in the Platteville Formation at Lancaster, Wisconsin.

CHITINOZOANS

Chitinozoans are an extinct group of cryptic microfossils whose systematic position is presently uncertain. Suggested affinities include protists, fungi, metazoans, or possibly eggs of a marine animal (Paris and Nõlvak 1999). They are widely distributed in Ordovician through Devonian marine sedimentary rocks and are useful in regional and global correlations (Jansonius and Jenkins 1978). Their distinguishing features include radial symmetry about a central axis and a hollow vase- or tube-shaped body with an outer wall consisting of organic (pseudochitinous) material. Most specimens range between 0.05 and 2.0 mm in length. Cutler and Sloan (1983) described five chitinozoan species from the Platteville Formation of Minnesota:

Conochitina elegans Eisenack, 1931
Conochitina* (*Rhabdochitina*?) *minnesotensis (Stauffer, 1933)
Cyathochitina campanulaeformis (Eisenack, 1931)
Cyathochitina kuckersiana (Eisenack, 1931)
Desmochitina minor Eisenack, 1931

ACRITARCHS

Acritarchs are organic microfossils (0.015–0.080 mm in size) composed of acid-insoluble carbon compounds. They are an informal polyphyletic group of microfossils without nomenclatural rank or status and are considered a palynological catchall (Servais et al. 2004b). They were the primary producers at the base of the food chain in Proterozoic and Paleozoic marine environments and are widespread in the fossil record (Servais et al. 2004a). They are believed to be the resting cysts of phytoplanktonic algal protists, with strong similarities to dinoflagellate cysts (Vecoli and Hérissé 2004).

Important contributions to North American Ordovician acritarch literature include Ruedemann and Wilson (1936), Jacobson (1978, 1979, 1987), and Jacobson and Achab (1985). Much remains to be done regarding the biostratigraphy, taxonomic diversity, and evolution of acritarchs in the Ordovician stratigraphic succession of North America.

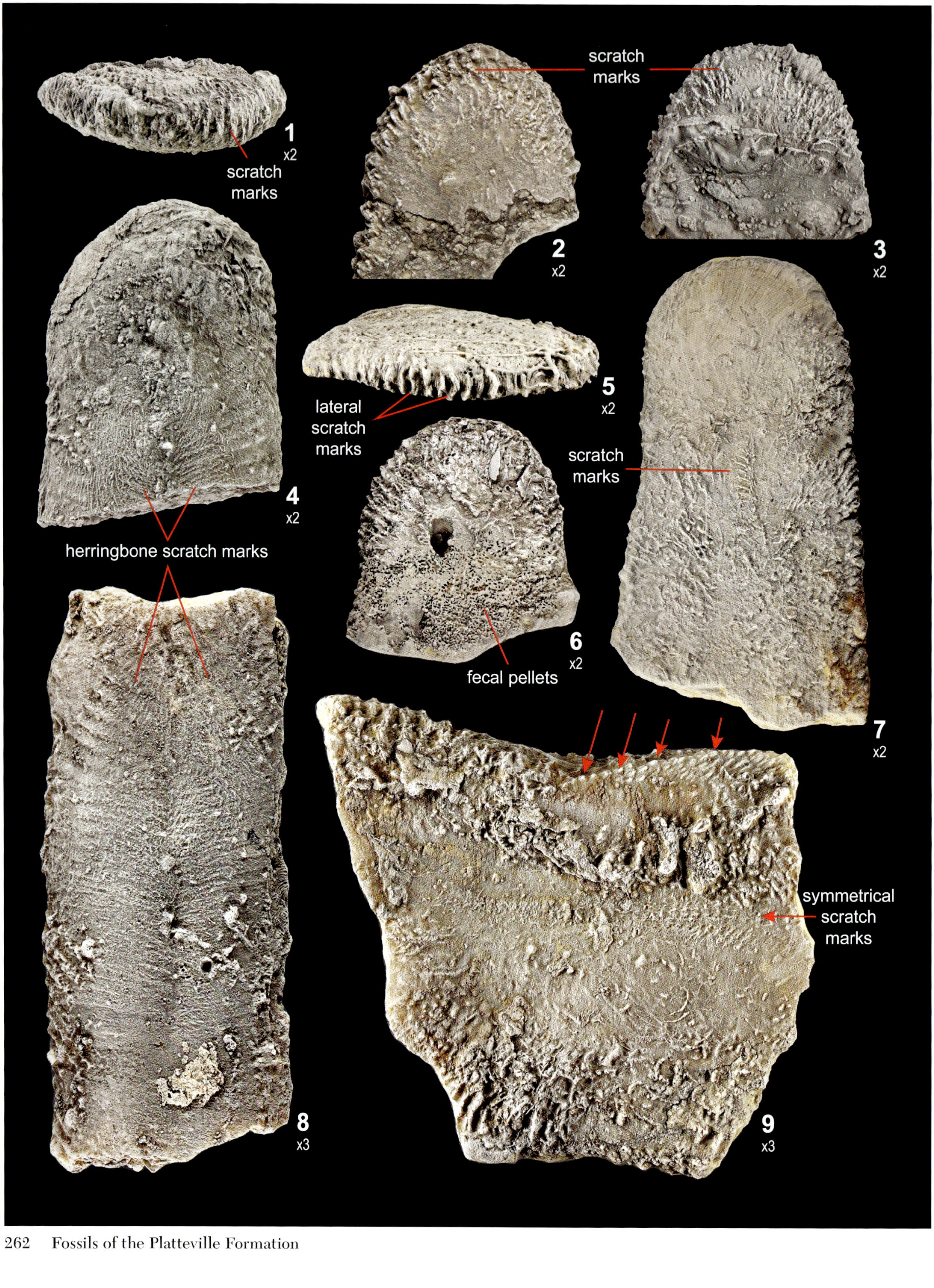
1
x2
scratch marks
scratch marks
2
x2
3
x2
4
x2
herringbone scratch marks
5
x2
lateral scratch marks
6
x2
fecal pellets
scratch marks
7
x2
8
x3
9
x3
symmetrical scratch marks

CHAPTER 17
TRACE FOSSILS

Trace (ichno) fossils are diverse, abundant, and locally very well preserved in the Platteville Formation (Appendix 2). They include trackways, burrows, borings, and fecal pellets, all of which record a variety of organism activities and behaviors. Observed Platteville ichnogenera include *Arachnostega*, *Balanoglossites*, *Cruziana*, *Chondrites*, *Gyrolithes*, *Petroxestes*, *Phycodes*, *Rauffella*, *Ropalonaria*, *Rosselia*, *Rusophycus*, *Sulcolithus*, *Taenidium*, *Thalassinoides*, *Trypanites*, and *Zoophycos* (Appendix 2). The overall assemblage is indicative of the "*Cruziana* ichnofaces," an environment characterized by poorly sorted and unconsolidated muddy substrates in a shallow, subtidal open-marine setting (Seilacher 2007). No attempt is made here to identify ichnospecies, and even some ichnogenera are questionable because of the considerable variations in individual trace fossils.

Trace fossils come in a bewildering array of morphological variations, making it a challenge to group and classify them systematically. Several classification schemes are currently in use, including (1) ethological (behavior) classification (Seilacher 1964), applied and discussed in numerous studies over the past 50 years; (2) morphological classification based on detailed features of burrows, bioerosional traces, trackways, imprints, and trails (Knaust 2012); and (3) classification based on the categories of architectural designs (CADs) of Buatois and Mángano (2011, 2013, 2016) and Buatois et al. (2017). The Platteville trace fossils are here organized first according to the morphologic classification of Knaust (2012), followed by the CADs of Buatois et al. (2017) and then ichnogenera.

Figure 17-1 (Opposite page)
1–7, 9. ***Rusophycus* isp.** (**1**) Terminal end of burrow showing prominent scratch marks and flattened profile of burrow; (**2**) terminal end of burrow showing prominent scratch marks; (**3**) terminal end of burrow showing prominent scratch marks; (**4**) terminal end of burrow showing undertrack with fine herringbone-shaped scratch marks that are oblique to the long dimension and meet along the axis, forming a subtle V-shaped pattern; (**5**) lateral edge of terminal end showing marginal scratch marks; (**6**) terminal end of burrow showing marginal scratch marks and parts of burrow filled with fecal pellets; (**7**) terminal end of burrow showing undertrack with fine scratch marks; (**9**) undertrack of burrow cast showing an array of fine scratch marks (red arrow) symmetrically aligned and possibly created by the exopodite brushings of an arthropod; red arrows point to a linear series of marks suggestive of indentations made by the edges of trilobite pleural segments. Mifflin Member, Dixon, Illinois (locality 2), **1**, **3**, ISGS 1454; **2**, ISGS 1458; **4**, ISGS 1298; **5**, ISGS 1466; **6**, ISGS 1464; **7**, ISGS 1516; **9**, ISGS 1501.

8. ***Cruziana* isp.** (**8**) Undertrack with fine herringbone-shaped striations that are oblique to the long dimension and meet along the axis, forming a subtle V-shaped pattern. Mifflin Member, Dixon, Illinois (locality 2), **8**, ISGS 1295.

SYSTEMATIC DESCRIPTIONS

Subhorizontal Burrows

Bilobate Trails and Paired Grooves

Ichnogenus CRUZIANA

***Cruziana* isp.**

Figure 17-1-8

Description. Bilobate traces preserved as flattened burrow casts weathered out of matrix in full relief, 15–30 mm in width, 5–8 mm thick; lower surface (undertrack) with fine herringbone-shaped striations that are oblique to the long dimension and meet along the axis, forming a subtle V-shaped pattern; upper surface of cast lacking herringbone-shaped striations. Internal filling composed of sparite, lime mudstone, or both, commonly in the form of fecal pellets.

Remarks. *Cruziana* traces are bilobed structures formed by the physical action of a bilaterally symmetrical invertebrate animal, including, but not necessarily restricted to, trilobites. The Platteville traces are preserved as cavities filled with lime mudstone or sparite and appear to be resting structures that were excavated in dewatered and compacted firmground composed of lime mud. The convex shapes of the traces are similar in size and profile to several trilobite species found in the Platteville Formation. They are commonly interpreted to be a trilobite resting trace (Seilacher 2007).

Occurrence. Common in the Mifflin Member, particularly in the limestone facies of northern Illinois.

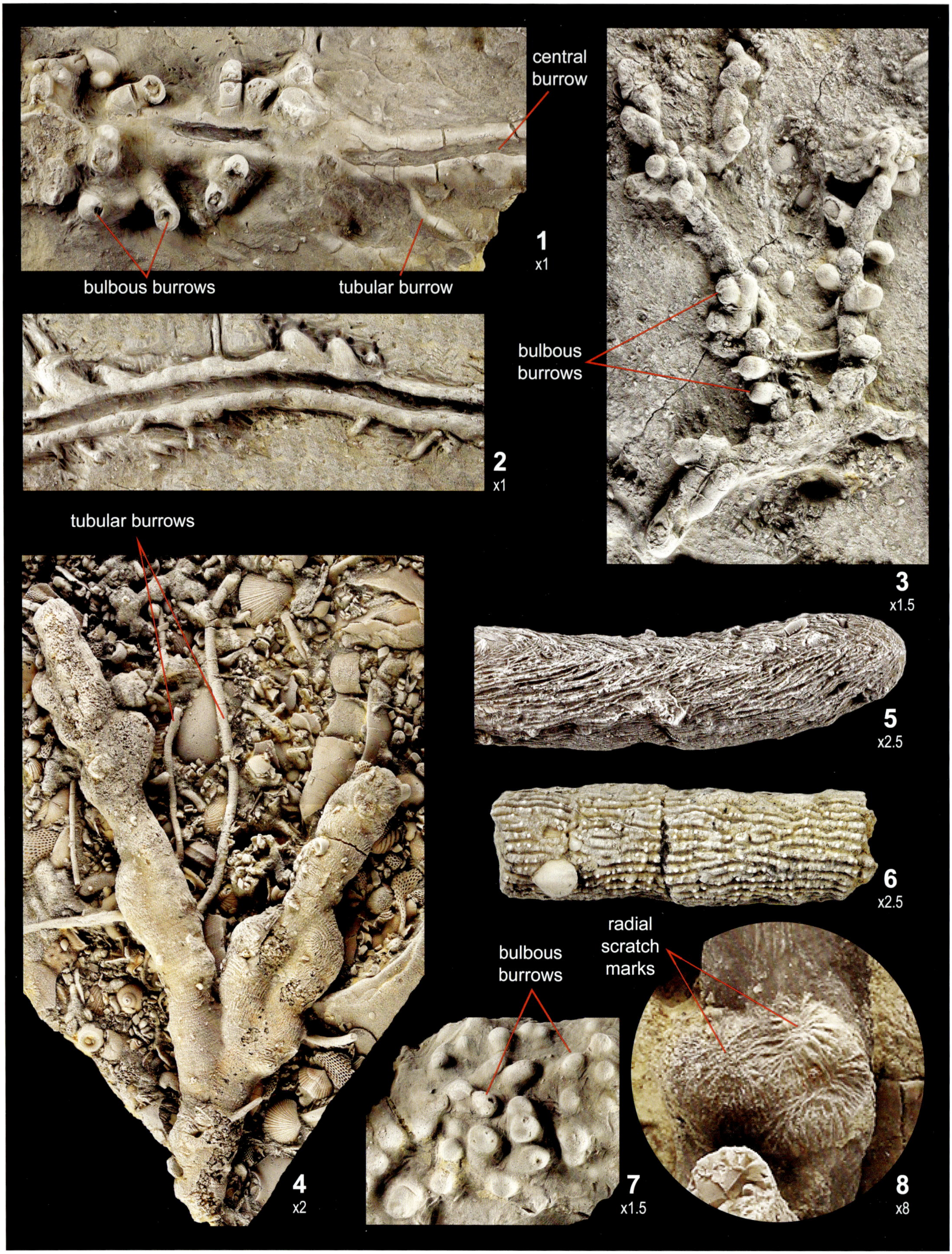

central burrow
bulbous burrows
tubular burrow
1
x1
2
x1
bulbous burrows
3
x1.5
tubular burrows
5
x2.5
6
x2.5
radial scratch marks
bulbous burrows
4
x2
7
x1.5
8
x8

Bilaterally Symmetrical Short, Scratched Impressions and Burrows

Ichnogenus RUSOPHYCUS

Rusophycus isp.

Figures 17-1-1 through 17-1-7 and 17-1-9

Description. Burrow casts, flattened, horizontal, bilaterally symmetrical, tongue shaped, weathered out of matrix in full relief, 15–30 mm wide, 5–8 mm thick; terminal end broadly rounded and commonly marked by prominent scratch marks and ridges along the lateral edge. Some specimens show markings that appear to have been created by sharp, laterally projecting spines, possibly by the edges of the pleural segments of a trilobite.

Remarks. Some specimens are a composite of *Cruziana* and *Rusophycus*.

Occurrence. Common in the Mifflin Member throughout the outcrop belt.

Passively Filled Horizontal Burrows

Ichnogenus RAUFFELLA

Rauffella isp. 1

Figures 17-2-1 through 17-2-8; and 17-3-1 through 17-3-4

Description. Burrows branched (not systematically) or unbranched, horizontal to inclined, cylindrical, thinly lined and passively filled with sediment like that of the surrounding matrix or preserved as sparite-filled cavities apparently formed in dewatered and well-compacted lime mud, thus preserving a cast of the original burrow exterior prior to lithification; surface distinctly sculptured with sharp, straight, or anastomosing threadlike scratch marks; burrow diameters range from 3 to 15 mm; some branches form distinctive bulb-like clusters; preserved in full relief or in convex semirelief.

Remarks. The Platteville species *Rauffella* isp. 1 is most like *R. filosa* (Ulrich, 1889b) from the Decorah Formation near Minneapolis, Minnesota. With some hesitation, Ulrich assigned the species to the sponges based on the prominent external longitudinal striations, which he thought were arranged like the spicules of the Dictyospongidae. Sardeson (1925) pointed out that *R. filosa* lacked an internal spicular sponge skeleton; therefore, he consigned its origin to that of a trace fossil created by a burrowing organism. He interpreted the longitudinal striations to be scratch marks.

Rauffella isp. 1 shows significant variations in burrow diameter, surface sculpture, and branching style, suggesting burrowing activity of adults and conspecific juveniles or possibly a composite burrow system formed by a variety of organisms. The horizontal to inclined, thinly lined, passively filled burrows of *Rauffella* isp. 1 are similar in architectural design to those of the Ordovician ichnogenus *Palaeophycus*. The frequent branching, however, is unlike that typically seen in *Palaeophycus*. Some specimens of *Rauffella* isp. 1 have straight surface striations like those in *Palaeophycus striatus* Hall, 1852, whereas others are anastomosing and interwoven like those in *P. sulcatus* (Miller and Dyer, 1878).

Traces of *Rauffella* isp. 1, like those of *Palaeophycus*, appear to have been formed by a predaceous or suspension-feeding wormlike animal mining along sedimentologic interfaces. The burrow galleries were subsequently filled by passive sedimentation or remained blocked and partially empty, later to be diagenetically filled in with sparite to preserve a cast of the burrow interior.

Occurrence. Particularly abundant in the Pecatonica, Mifflin, and Grand Detour Members. The marl beds in the Mifflin Member at Dixon, Illinois, have yielded large numbers of very well preserved full-relief specimens free of their matrix.

Rauffella isp. 2

Figures 17-3-5 through 17-3-9

Description. Thinly lined and passively filled horizontal burrow systems with regularly spaced bulbous chambers alternately branching in a distinctive zigzag manner on either side of a central burrow; bulb surfaces range from smooth to intricate brainlike pattern of lobes and scratch marks; chambers and central burrow passively filled with either lime mud matrix, sparite, or a combination of both; bulbs range

Figure 17-2 (Opposite page)
1–8. ***Rauffella* isp. 1.** (**1**) Full-relief burrow system with central burrow (partially collapsed) giving rise to short bulbous burrows and a single, slender, elongate tubule (lower right); (**2**) full-relief collapsed burrow giving rise to short bulbous burrows and slender, unbranched tubular burrows; (**3**) full-relief burrow system with numerous short bulbous branches; (**4**) an unusually well-preserved full-relief burrow system; matrix has weathered away, leaving a cast of interior surface showing minute scratch marks and subtle annulations; several slender unbranched tubular burrows project from sides of main burrows; (**5**) full-relief interior burrow cast showing threadlike scratch marks; (**6**) full-relief interior burrow cast showing prominent subparallel scratch marks with tiny tubercle-like marks; (**7**) full-relief interior burrow cast showing numerous short bulbous burrows; (**8**) full-relief interior burrow cast showing bulbous burrows with characteristic radiating scratch marks. Mifflin Member, Dixon, Illinois (locality 2), **1**, ISGS 1475; **2**, ISGS 547; **5**, ISGS 1461; **6**, ISGS 1497; **7**, ISGS 1478. Mifflin Member, Dixon, Illinois (locality 4), **3**, ISGS 1524; **4**, ISGS 1450; **8**, ISGS 466.

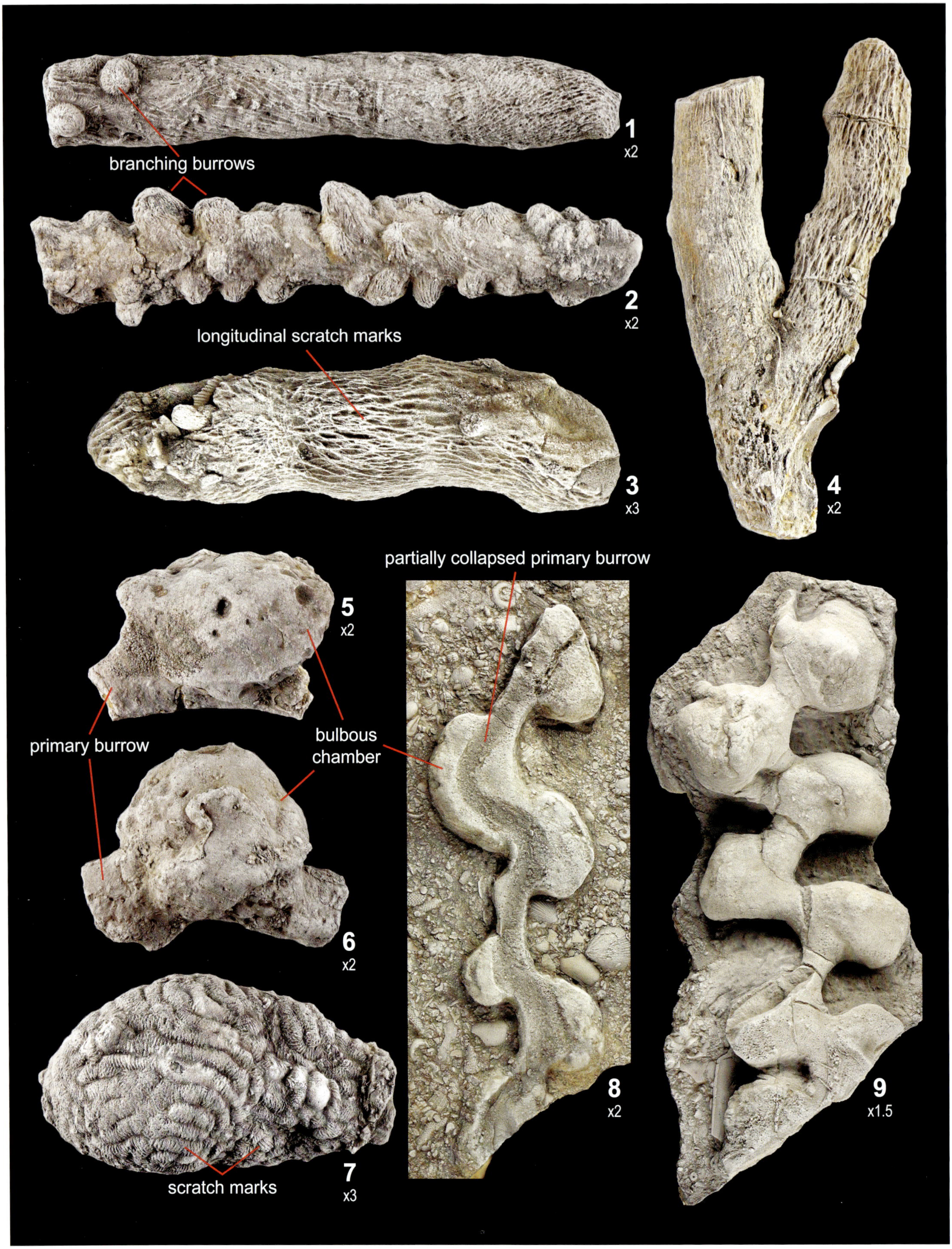
branching burrows
1
x2
2
x2
longitudinal scratch marks
3
x3
4
x2
partially collapsed primary burrow
5
x2
primary burrow
bulbous
chamber
6
x2
scratch marks
7
x3
8
x2
9
x1.5

from 10 to 15 mm in the longest dimension, central burrow approximately 8 mm in diameter.

Remarks. The overall size of the burrow and alternately branching bulbous chambers is very similar to that in the trace fossil *Rauffella ulrichana* Sardeson, 1925, from the overlying Decorah Formation of southeastern Minnesota.

These lined, horizontal to subhorizontal burrows have the characteristic features of passively filled burrows, but the alternating bulbous chambers are unique among ichnogenera assigned to this CAD. The presence of scratch marks on the external surface of the bulbs and the occasional presence of sparite in the bulbs and central burrows suggest that *Rauffella* isp. 2 traces were formed in dewatered, relatively compact firmground. The bulbous chambers likely were open cavities in a burrow system that was either passively filled in with sediment or sealed and diagenetically filled with sparite over time.

Occurrence. Common in the Mifflin and Grand Detour Members in northern Illinois and southern Wisconsin.

Horizontal Branching Burrow Systems

Ichnogenus ARACHNOSTEGA

***Arachnostega* issp.**

Figures 17-4-1 through 17-4-3

Description. Branched or unbranched burrows on interior of sediment-filled shells; preserved in full relief or concave epirelief on steinkerns, burrow filling consists of sparite, lime mudstone, or both; burrow diameters vary from 0.1 to 0.8 mm, depending on the burrowing biota involved.

Remarks. The Platteville specimens generally fit the original description of *Arachnostega* given by Bertling (1992) and closely resemble specimens of *Arachnostega* from the Middle and Upper Ordovician succession of Estonia described by Vinn et al. (2014). They likely were formed by a soft-substrate detritus feeder.

At least two types of *Arachnostega* burrows are present in the Platteville: (1) small serpentine burrows approximately 0.1 mm in diameter and (2) a straight or winding burrow that is 0.8 mm in diameter.

Occurrence. *Arachnostega* traces are abundant in brachiopods, gastropods, bivalves, and cephalopods in the Mifflin and Grand Detour Members in north-central Illinois and south-central Wisconsin and the McGregor Member of southwestern Wisconsin, northeastern Iowa, and southeastern Minnesota. Locally common in the Pecatonica Member in north-central Illinois.

Simple Actively Filled (Meniscate)

Horizontal to Oblique Structures

Ichnogenus TAENIDIUM

***Taenidium* issp.**

Figure 17-4-4 through 17-4-9

Description. Unlined, meniscate, backfilled burrows, straight or meandering, circular or elliptical in cross section, of variable dimensions and configurations, probably more than one species present; burrow infilling differs in color and texture from surrounding host rock. Branching, where present, exhibits no particular pattern, nor does overall burrow configuration, other than a tendency toward horizontal expansion.

Remarks. *Taenidium* is interpreted as a feeding structure produced by deposit-feeding vermiform organisms that actively filled their burrows (Buatois and Mángano 2011).

Knaust (2019b) assigned the specimen shown in Figure 17-4-9 to *Rhizocorallites articularis* Müller, 1955, arguing that this type of burrow is more conformable with an interpretation as the fecal string (i.e., coprolite) or intestinal content (i.e., cololite) left by an epibenthic organism. He further suggested holothurians as potential producers of *R. articularis*.

Occurrence. Abundant in the Mifflin and Grand Detour Members in north-central Illinois and south-central Wisconsin and the McGregor Member of southwestern Wisconsin, northeastern Iowa, and southeastern Minnesota.

Figure 17-3 (Opposite page)

1–4. ***Rauffella* isp. 1.** (**1**) Full-relief interior burrow cast showing threadlike striations and two bulbous branches; (**2**) full-relief interior burrow cast showing multiple budding bulbous branches; (**3**) full-relief interior burrow cast showing anastomosing threadlike scratches; (**4**) branched burrow cast showing anastomosing threadlike scratches. Mifflin Member, Dixon, Illinois (locality 2), **1**, ISGS 1463; **2**, ISGS 1460; **3**, ISGS 1511; **4**, ISGS 1515.

5–9. ***Rauffella* isp. 2.** (**5**) Isolated bulbous chamber protruding from primary burrow; (**6**) isolated bulbous chamber protruding from primary burrow; (**7**) isolated bulbous chamber showing prominent scratch marks; (**8**) lower surface of full-relief horizontal burrow system showing partially collapsed primary burrow; (**9**) upper surface of full-relief horizontal burrow system showing regularly spaced bulbous chambers alternately branching in a distinctive zigzag fashion on either side of a central burrow. Mifflin Member, Dixon, Illinois (locality 2), **5**, ISGS 1469; **6**, ISGS 1470; **7**, ISGS 1455; **8**, ISGS 549; **9**, ISGS 534.

Horizontal Burrows with Horizontal to Vertical Branches

Ichnogenus PHYCODES

***Phycodes* isp.**

Figure 17-5-1

Description. Unlined, horizontal to subhorizontal, arcuate burrows consisting of a large single branch giving rise to numerous divergent free branches filled with very fine shelly fragments; branches curve outward in fan shape, 2–4 mm wide, and up to 120 mm long.

Remarks. Platteville specimens of *Phycodes* are commonly preserved in convex hyporelief on the underside of a stratum. *Phycodes* is largely associated with shallow marine environments and is interpreted as a feeding structure produced by unknown organisms (Seilacher 2007).

Occurrence. Abundant in the Mifflin and Grand Detour Members in north-central Illinois and south-central Wisconsin and the McGregor Member of southwestern Wisconsin, northeastern Iowa, and southeastern Minnesota.

Subvertical Burrows

Burrows with Helicoidal Spreiten

Ichnogenus ZOOPHYCOS

***Zoophycos* isp.**

Figures 17-5-2 through 17-5-9

Description. Horizontal arcuate burrow system 10–25 mm in width and approximately 7 mm thick, some with tonguelike lobes; preserved in full relief; possessing coarse, backfilled spreiten (curved arcs of sediment) composed of bioclastic material. Some are marked by prominent scratch marks.

Remarks. *Zoophycos* is believed to be a feeding burrow formed in quiet-water paleoenvironments where the sediment was well mixed by bioturbation (Seilacher 2007).

Occurrence. Abundant in the Mifflin and Grand Detour Members in northern Illinois and southern Wisconsin.

Vertical Helicoidal Burrows

Ichnogenus GYROLITHES

***Gyrolithes* isp.**

Figures 17-6-1

Description. Lined spiraled burrow with coils that are not in contact; burrow filled with sparite or fine matrix lime mud; thin outer wall; burrow round in cross section and 4 mm in diameter, outer surface marked by fine scratch marks; whorl radius measured from middle of burrow structure is approximately 5 mm; the most nearly complete specimen has three whorls.

Remarks. The helical burrow of *Gyrolithes* is thought to be a feeding strategy to exploit organic-rich zones of the substrate in calm and protected environments (Netto et al. 2007).

Occurrence. A rare trace fossil in the Mifflin Member at Dixon, Illinois.

Vertical Single U- and Y-shaped Burrows

Ichnogenus BALANOGLOSSITES

***Balanoglossites* isp.**

Figures 17-6-2 and 17-6-3

Description. U- and Y-shaped vertical branches that form very irregular and winding galleries with acorn-, bulb-, or lance-shaped branches; margins unlined; branch size varies in the order of several magnitudes within a single gallery system.

Remarks. As noted by Knaust and Dronov (2013), *Balanoglossites* is commonly misinterpreted as the morphologically similar ichnogenus *Thalassinoides* but differs from the latter in having U- and Y-shaped vertical branches that are very irregular and winding and typically not anastomose.

Occurrence. *Balanoglossites* is a very abundant and widespread trace fossil occurring, most notably, in the strata immediately below the hardground surfaces at the top of the Pecatonica and Quimbys Mill Members in northeastern Iowa, southern Wisconsin, and northern Illinois, as well as throughout the Illinois Basin and into the outcrop belt of southeastern Missouri. Like the occurrences in the Middle

Figure 17-4 (Opposite page)
1–3. ***Arachnostega* issp.** (**1**) Full-relief, coiled burrow cast on interior surface of brachiopods *Hesperorthis*; (**2**) concave epirelief burrow on steinkern of gastropod *Sinuites*; (**3**) small, full-relief, serpentine burrows on interior surface of gastropod *Trochonema*. Mifflin Member, Dixon, Illinois (locality 2), **1**, ISGS 1445; **3**, ISGS 1504. Mifflin Member, Fennimore, Wisconsin (locality 26), **2**, ISGS 1537.

4–9. ***Taenidium* issp.** (**4**) Full-relief meniscate burrow system; (**5–7**) meniscate burrows showing clearly defined menisci; (**8**) limestone slab cut and polished parallel to bedding surface showing backfill of lime mudstone (light matrix) into fine-grained grainstone (tempestite); (**9**) meniscate burrows (dark brown) in lime mudstone (light matrix). Mifflin Member, Dixon, Illinois (locality 2), **4**, ISGS 1452; **5**, ISGS 1512; **6**, ISGS 1457; **7**, ISGS 1456; **8**, ISGS 1472. Quimbys Mill Member, Annaton, Wisconsin (locality 47), **9**, ISGS 1494.

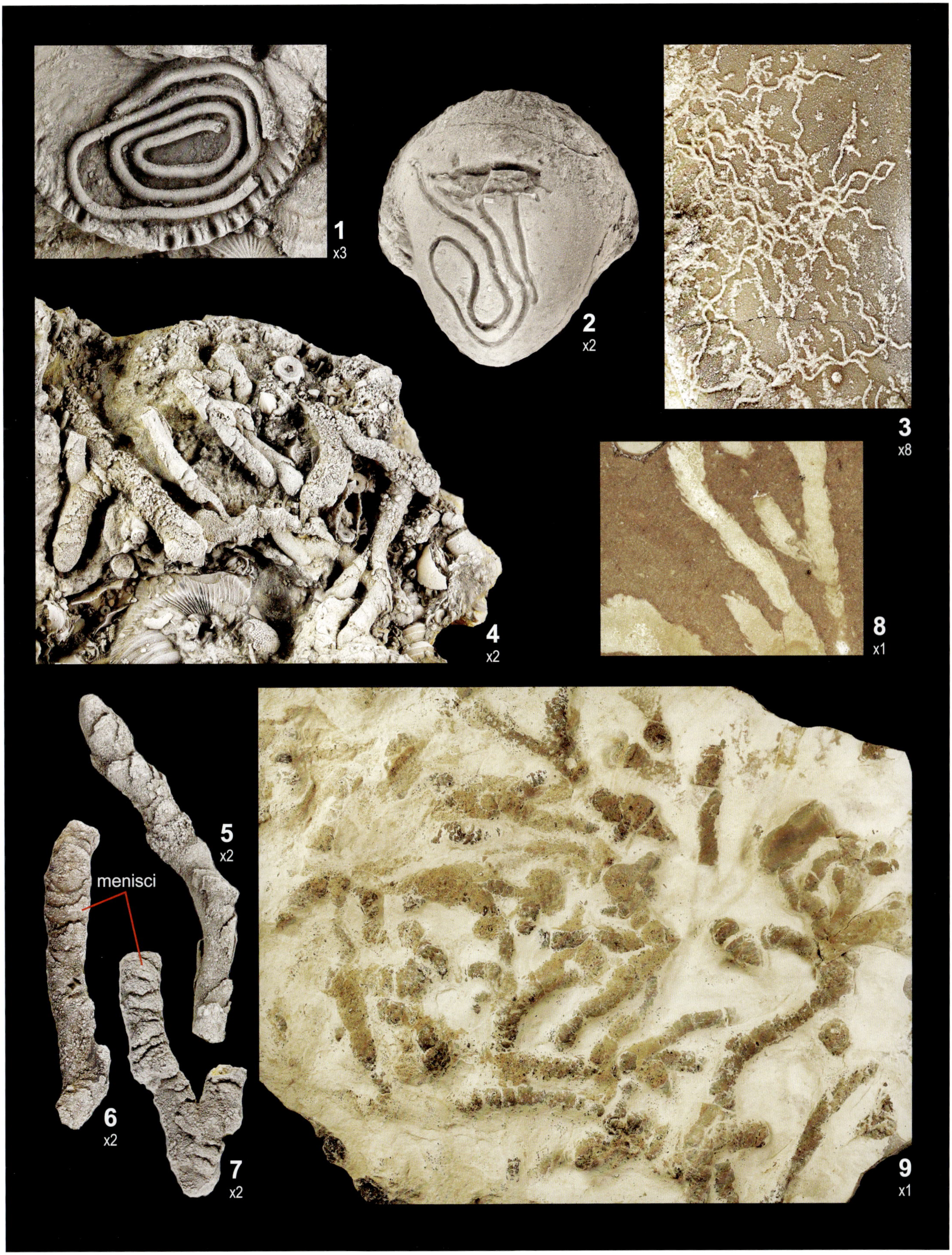
1
x3
2
x2
3
x8
4
x2
8
x1
5
x2
menisci
6
x2
7
x2
9
x1

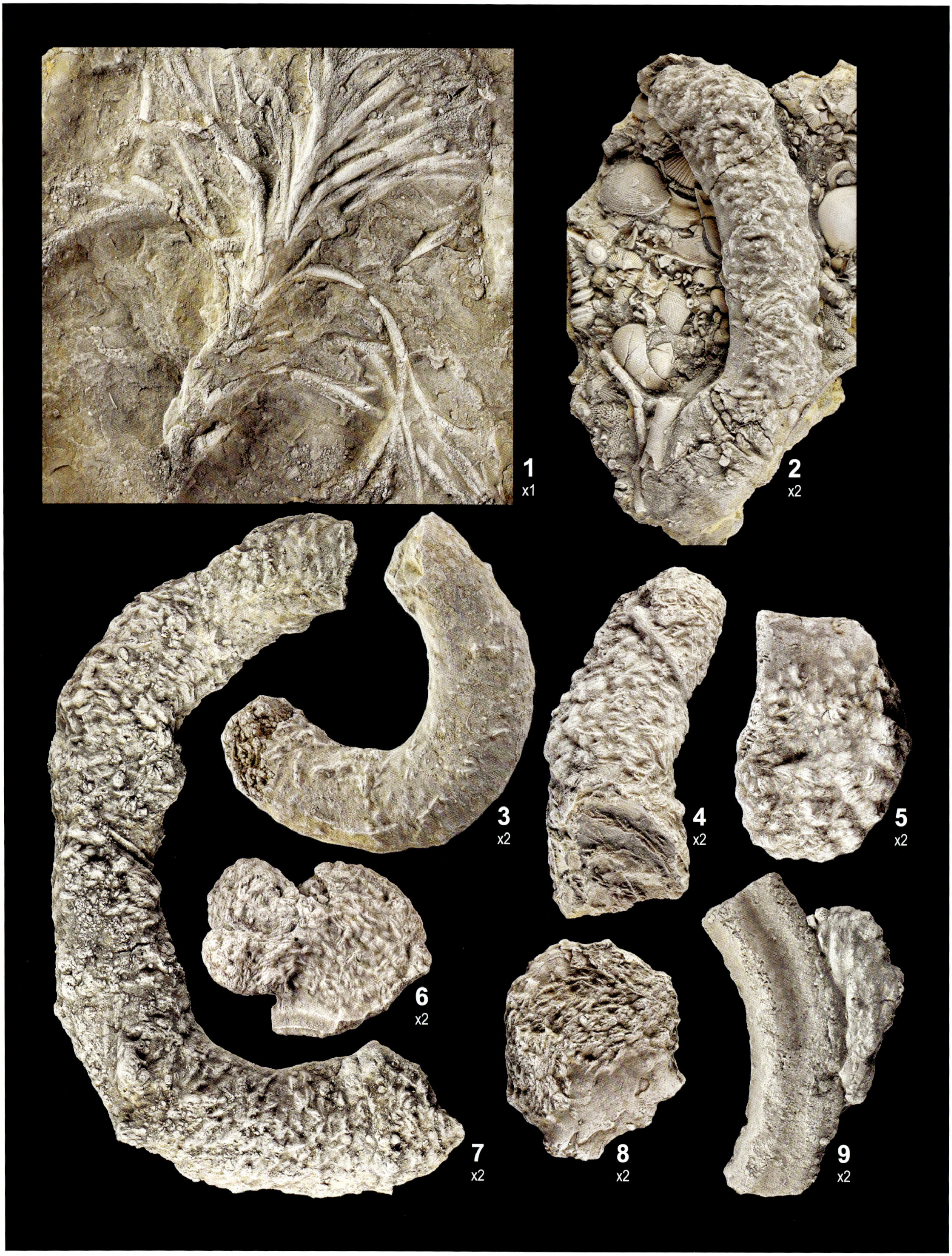
1
x1
2
x2
3
x2
4
x2
5
x2
6
x2
7
x2
8
x2
9
x2

Ordovician Volkhov Formation in the St. Petersburg (Russia) region (Knaust and Dronov 2013), Platteville *Balanoglossites* also is found in association with mineral-stained and *Trypanites*-perforated hardground surfaces that are bioeroded with elongate grooves assigned to the ichnogenus *Sulcichnus*. One of the more notable occurrences of *Balanoglossites* is in the limestone facies of the Pecatonica Member near Dixon, Illinois, where it commonly consists of selectively dolomitized burrows that weather out, leaving an outcrop surface of pronounced irregular cavities.

Burrows with Shaft and Downward-Radiating Probes

Ichnogenus CHONDRITES

***Chondrites* issp.**

Figures 17-7-1 and 17-7-2

Description. Unlined, systematically branched burrows and burrow networks, bifurcating at least three times; round to somewhat flattened horizontal networks preserved as convex full-relief burrows with diameters of 1–8 mm and regular symmetrical dichotomous branching at angles between 30° and 90°. Burrow systems are in close proximity to each other without interpenetration; fill variable in color and texture.

Remarks. The chondritids are believed to be feeding burrows formed by probing and backfilling processes (Seilacher 2007). The Platteville *Chondrites* traces vary in size, filling, and branching, thus suggesting that they were produced by a variety of invertebrate organisms. It is likely that several ichnospecies are present in the formation.

Occurrence. Abundant throughout the Platteville Formation in the Mifflin and Grand Detour Members in north-central Illinois and south-central Wisconsin and the McGregor Member of southwestern Wisconsin, northeastern Iowa, and southeastern Minnesota. Particularly abundant in the Nachusa Member in northern Illinois and southern Wisconsin.

Figure 17-5 (Opposite page)
1. ***Phycodes* isp.** (**1**) Burrow system on lower surface of limestone bed showing characteristic divergent burrows. Mifflin Member, Dixon, Illinois (locality 3), **1**, ISGS 1474.

2–9. ***Zoophycos* isp.** (**2**–**9**) Full-relief, backfilled arcs and tongue-like lobes of sediment composed of coarse bioclastic material. Mifflin Member, Dixon, Illinois (locality 2), **2**, ISGS 1505; **3**, ISGS 1523; **4**, ISGS 1507; **5**, ISGS 1468; **6**, ISGS 1508; **7**, ISGS 1506; **8**, ISGS 1514; **9**, ISGS 1521.

Maze and Boxwork Burrows

Ichnogenus THALASSINOIDES

***Thalassinoides* isp.**

Figure 17-7-3

Description. Cylindrical to elliptical burrows forming a complex three-dimensional, anastomosing to horizontal branching network with vertical shafts connected to the surface; burrows have diameters ranging from 3 to 50 mm and show regular branching with Y- or T-shaped bifurcations and intersections; burrow casts generally smooth; burrow systems preserved as both molds and casts.

Remarks. The *Thalassinoides* association is widespread in the Upper Ordovician carbonate succession through Laurentia from the Great Basin to North Greenland and was likely formed by a variety of deep-burrowing deposit feeders (Jin et al. 2012). The ichnofabric is commonly associated with hardgrounds and formed in firm sediment prior to cementation (Myrow 1995).

Occurrence. In northern Illinois and southern Wisconsin, the *Thalassinoides* ichnofacies is commonly developed below hardground surfaces at the tops of the Pecatonica and Quimbys Mill Members.

Vertical Concentrically Filled Burrows

Ichnogenus ROSSELIA

***Rosselia* isp.**

Figures 17-8-1 through 17-8-3

Description. Vertical conical burrow with cone-in-cone laminae on interior, 15–25 mm in diameter at top and about 10 mm at bottom, up to 50 mm long; outer surface of cone marked by longitudinal striations.

Remarks. Nara (1995) suggested that the burrow lining of *Rosselia* was produced by the outward expansion of mud produced by a detritus-feeding organism. Modern *Rosselia*-like examples have been attributed to the activity of polychaetes.

Occurrence. Very abundant in the Mifflin Member, particularly in the area of Fennimore, Grant County, Wisconsin.

Transitional Burrow/Bioerosion Traces

Ichnogenus SULCOLITHUS

Remarks. *Sulcolithus* is a transitional trace fossil occurring in both firmground and hardground substrates (Knaust

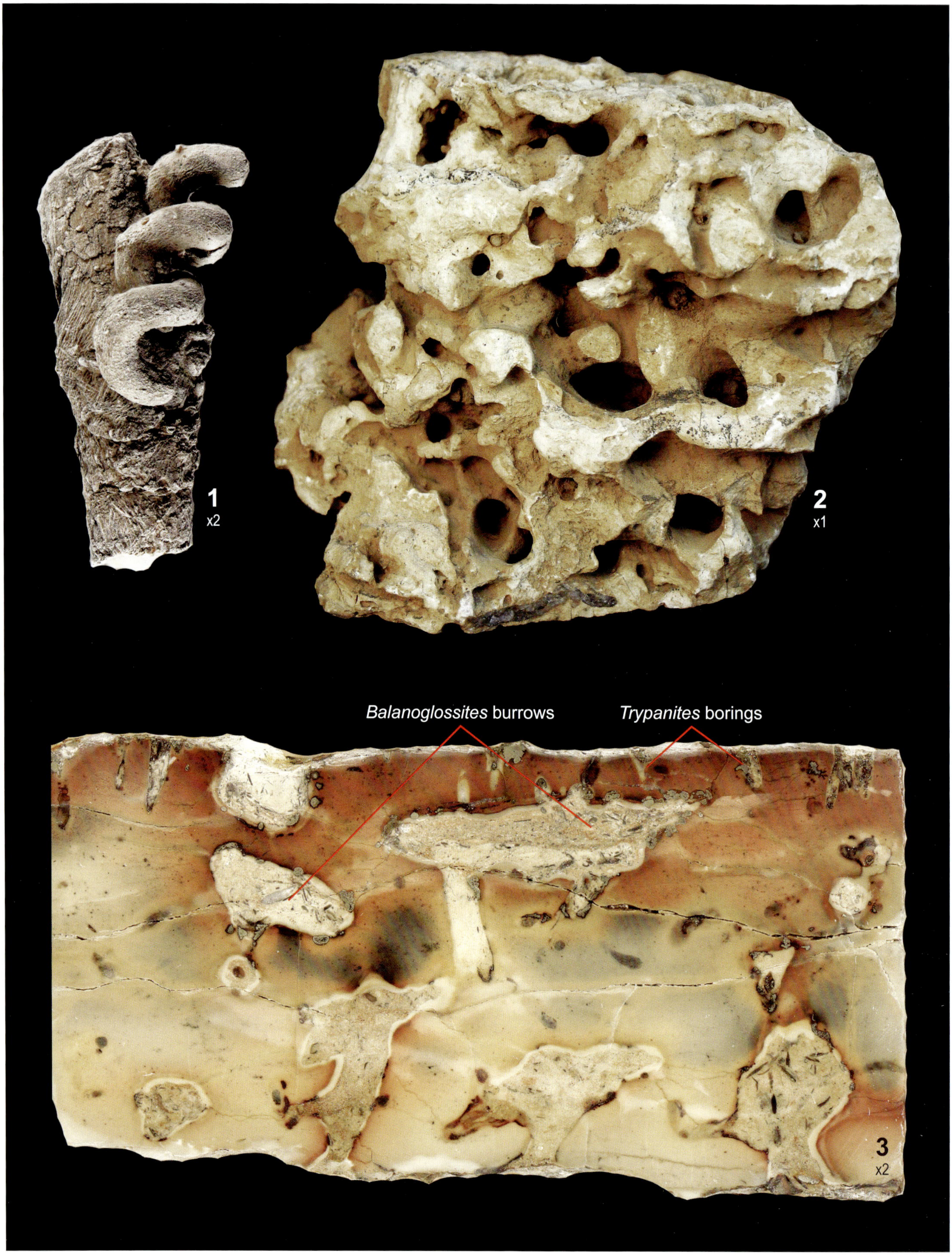
1
x2
2
x1
Balanoglossites burrows
Trypanites borings
3
x2

2019b). In terms of classification, it is transitional between a burrow and a bioerosion trace.

Sulcolithus isp.

Figure 17-8-4

Description. Burrow or boring elongate, superficial, relatively simple groove that is straight or slightly curved and has blunt ends; 10–15 mm wide and 5–10 mm deep, slightly wider at marginal bulges (incipient offshoots), some bifurcations present; often filled with comminuted shells and light-colored matrix.

Remarks. This *Sulcolithus* isp. is commonly found on hardground surfaces in the Platteville Formation in association with *Trypanites* borings and connected to subsurface burrow systems of *Balanoglossites*. It was formed on firm and hard lime mudstone substrates by burrowing or boring polychaete worms (Knaust 2019a).

Sulcolithus appears to be congeneric with *Scalaridomus* Palmer, 1978, from the Upper Ordovician Series, Chatfieldian Stage, Rivoli beds, Dunleith Member, Galena Formation near Decorah, Winneshiek County, Iowa. Knaust (2019a) concluded, "Judging from the figured (but not collected) holotype, *Scalaridomus* is likely a composite trace fossil containing components similar to *Sulcolithos variabilis* and *Balanoglossites triadicus*, and as such an invalid ichnotaxon" (p. 5).

Occurrence. Very abundant locally on the hardground surfaces at the top of Pecatonica and Quimbys Mill Members in northern Illinois and southern Wisconsin.

Figure 17-6 (Opposite page)

1. ***Gyrolithes* isp.** (**1**) Side view of spiral burrow in juxtaposition with *Rauffella* burrow. Mifflin Member, Dixon, Illinois (locality 2), **1**, ISGS 1467.

2, 3. ***Balanoglossites* isp.** (**2**) Vertical edge of a block of limestone from the upper 1 m of the Pecatonica Member showing prominent irregular cavities created by weathering of dolomitized burrows of *Balanoglossites*; (**3**) polished vertical section of hardground bed at top of Pecatonica Member showing *Trypanites* borings on upper surface and *Balanoglossites* burrows within the bed. Pecatonica Member, Dixon, Illinois (locality 2), **2**, ISGS 1526. Pecatonica Member, Dixon, Illinois (locality 4), **3**, ISGS 1536.

Bioerosional Trace Fossils

Cylindrical Vertical to Oblique Borings

Ichnogenus TRYPANITES

Trypanites isp.

Figures 17-8-4 and 17-8-5

Description. Borings circular in transverse section, 2 mm in diameter at surface, approximately 10 mm long, tapering downward to 1 mm in diameter; orientation of borings variable, ranging from vertical to subhorizontal; wall of boring unlined and smooth.

Remarks. Included here are borings preserved in hardground surfaces and mollusk shells.

Occurrence. Very abundant on hardground surfaces, particularly at the top of the Pecatonica and Quimbys Mill Members throughout the outcrop belt.

Pouch-Shaped Borings

Ichnogenus PETROXESTES

Petroxestes isp.

Figure 17-8-6

Description. Shallow to deep, unbranched, elongate borings in lithic substrates; longer than wide, 2 mm in width and up to approximately 10 mm in length; rounded ends and parallel, slightly convex, or slightly concave sides.

Remarks. Included here are borings in hardgrounds and molluscan shells.

Occurrence. Abundant locally in the Mifflin and Grand Detour Members in north-central Illinois and south-central Wisconsin and the McGregor Member of southwestern Wisconsin, northeastern Iowa, and southeastern Minnesota.

Camerate Network Borings

Ichnogenus ROPALONARIA

Ropalonaria isp.

Figures 17-8-7

Description. Linear array of small teardrop-shaped pits bored into mollusk shells; 0.5–1.0 mm long and 0.2 mm wide.

Remarks. *Ropalonaria* is believed to be a boring made by ctenostome bryozoans.

Occurrence. Common in the Mifflin and Grand Detour Members in north-central Illinois and south-central Wisconsin and the McGregor Member of southwestern Wisconsin, northeastern Iowa, and southeastern Minnesota.

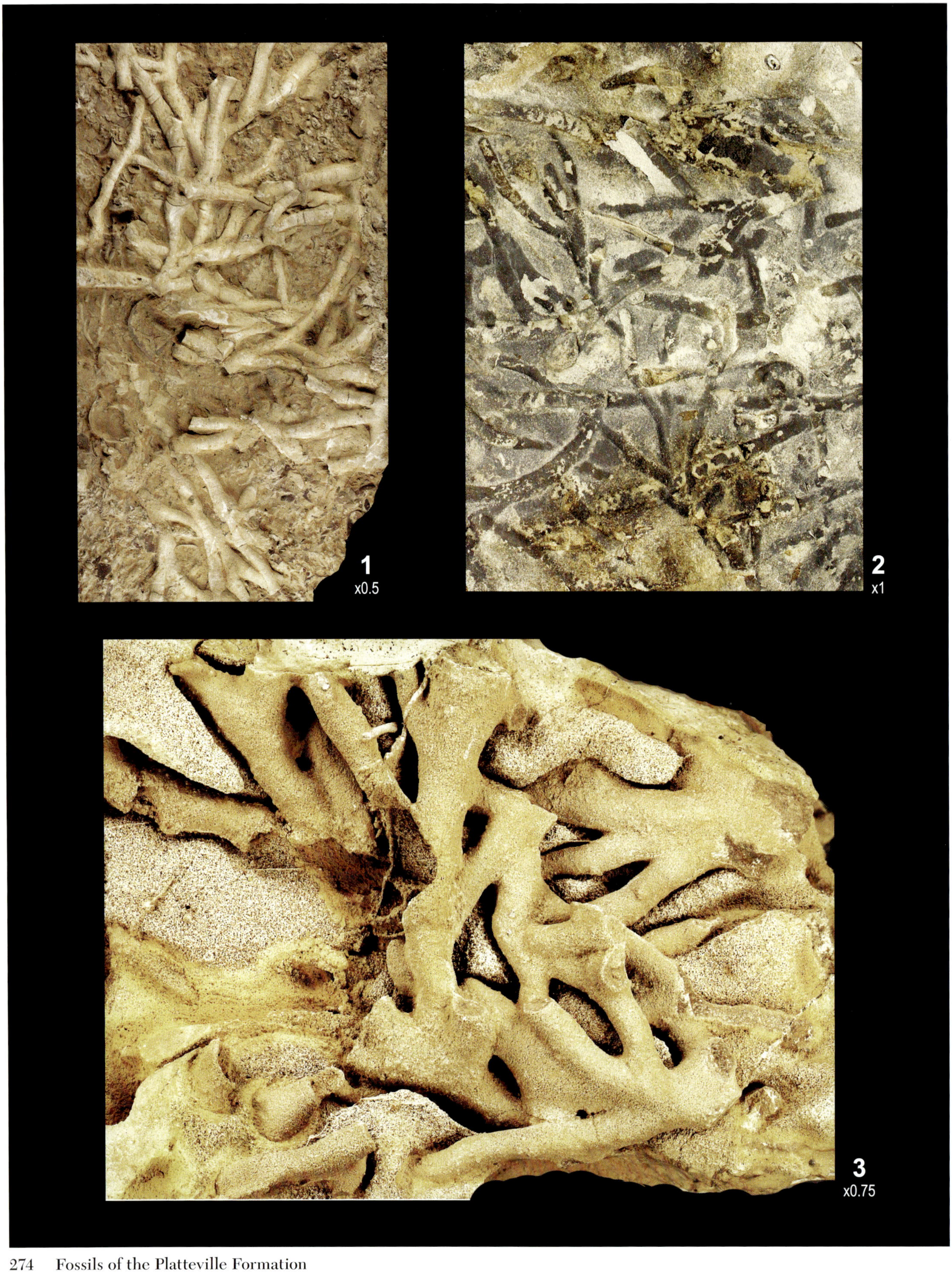
1
x0.5
2
x1
3
x0.75

Microcoprolites

Fecal Pellets

Figures 17-8-8 and 17-8-9

Description. Fecal pellets, homogeneous, conical to ovoid, 0.2–0.5 mm in diameter, composed of lime mudstone.

Remarks. Commonly preserved in mollusk shells, where they were protected from postdepositional compaction processes.

Occurrence. Abundant throughout the Platteville Formation, particularly well preserved in the limestone facies.

Figure 17-7 (Opposite page)
1, 2. ***Chondrites* issp.** (**1**) Bedding plane showing numerous branching horizontal burrows; (**2**) bedding plane showing numerous branching horizontal burrows. Mifflin Member, Dixon, Illinois (locality 2), **1**, ISGS 1472. Nachusa Member, Amboy, Illinois (locality 46), **2**, ISGS 1493.

3. ***Thalassinoides* isp.** (**3**) Full-relief burrow system from upper 1 m of Pecatonica Member showing typical Y- and T-shaped bifurcations and intersections. Pecatonica Member, Dixon, Illinois (locality 2), **3**, ISGS 1527.

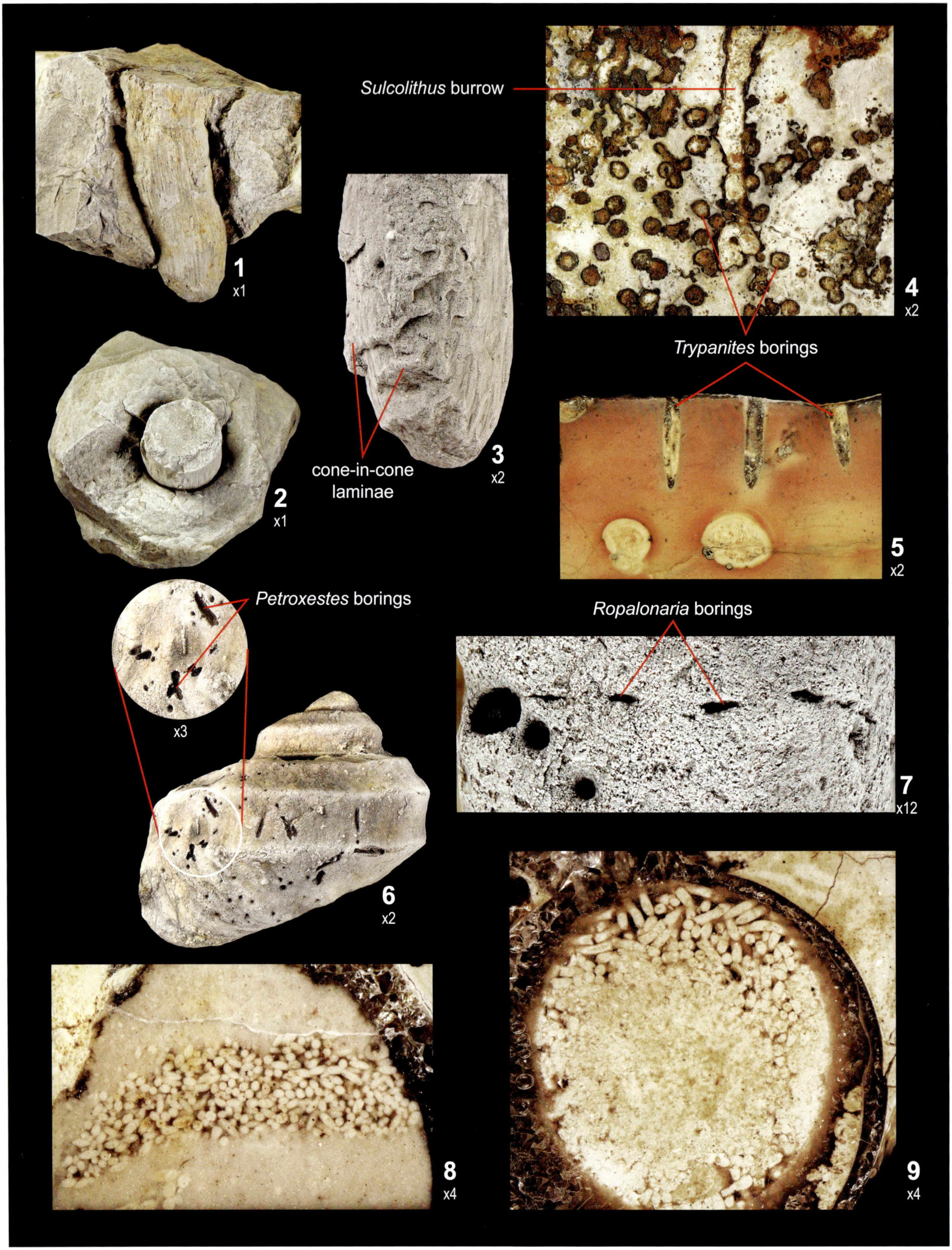
1
x1
2
x1
3
x2
cone-in-cone
laminae
Sulcolithus burrow
4
x2
Trypanites borings
5
x2
Petroxestes borings
x3
6
x2
Ropalonaria borings
7
x12
8
x4
9
x4

Figure 17-8 (Opposite page)
1–3. ***Rosselia* isp.** (**1**) Side view of an in situ burrow in near vertical orientation; (**2**) bedding plane view of in situ burrow; (**3**) side view of burrow showing cone-in-cone laminae. Mifflin Member, Fennimore, Wisconsin (locality 26), **1**, **2**, ISGS 1477; **3**, ISGS 1538.

4. ***Sulcolithus* isp.** (**4**) Horizontal burrow on hardground surface at top of Pecatonica Member showing associated *Trypanites* borings. Pecatonica Member, Dixon, Illinois (locality 4), **4**, ISGS 1541.

5. ***Trypanites* isp.** (**5**) Cut and polished vertical section of hardground at top of Pecatonica Member showing vertically oriented boring. Pecatonica Member, Dixon, Illinois (locality 4), **5**, ISGS 1540.

6. ***Petroxestes* isp.** (**6**) Outer shell of gastropod *Trochonema*; photographic enlargement showing several pouch-shaped borings. Mifflin Member, Dixon, Illinois (locality 2), **6**, ISGS 1510.

7. ***Ropalonaria* isp.** (**7**) Linear series of narrow bryozoan borings in surface of cephalopod siphuncle. Mifflin Member, Dixon, Illinois (locality 2), **7**, ISGS 361.

8, 9. **Fecal pellets.** (**8**) Pellets filling unlined burrow within gastropod shell; (**9**) pellets preserved within gastropod shells protected from postdepositional compactions. Mifflin Member, Dixon, Illinois (locality 2), **8**, ISGS 1528; **9**, ISGS 484.

REFERENCES

Aldridge, R.J., and Theron, J.N., 1993, Conodonts with preserved soft tissue from a new Ordovician Konservat-Lagerstätte: Journal of Micropalaeontology, v. 12, no. 1, p. 113–117.

Amati, L., 2014, Isoteline trilobites of the Viola Group (Ordovician: Oklahoma): Systematics and stratigraphic occurrence: Oklahoma Geological Survey Bulletin 151, 127 p.

Amati, L., and Westrop, S.R., 2004, A systematic revision of *Thaleops* (Trilobita: Illaenidae) with new species from the Middle and Late Ordovician of Oklahoma and New York: Journal of Systematic Palaeontology, v. 2, no. 3, p. 207–256.

Amati, L., and Westrop, S.R., 2006, Sedimentary facies and trilobite biofacies along an Ordovician shelf to basin gradient, Viola Group, south-central Oklahoma: Palaios, v. 21, no. 6, p. 516–529.

Badiozamani, K., 1973, The Dorag dolomitization model—Application to the Middle Ordovician of Wisconsin: Journal of Sedimentary Petrology, v. 43, p. 965–984.

Bain, H.F., 1905, Zinc and lead deposits of northwestern Illinois: U.S. Geological Survey Bulletin 246, 56 p.

Barrande, J., 1881, Système Silurian du centre de la Bohême, Vol. 6, Classe des Mollusques, ordre des Acéphalés: Prague, J. Barrande, 342 p.

Bassler, R.S., 1927, A new Early Ordovician sponge fauna: Journal of the Washington Academy of Sciences, v. 17, no. 15, p. 390–394.

Bassler, R.S., 1950, Faunal lists and descriptions of Paleozoic coral: Geological Society of America Memoir 44, 315 p.

Bassler, R.S., 1953, Part G: Bryozoa, *in* Moore, R.C., ed., Treatise on Invertebrate Paleontology: Boulder Colorado, Geological Society of America, and Lawrence, University of Kansas, 241 p.

Beadle, S.C., 1988, Dasyclads, cyclocrinitids and receptaculites: Comparative morphology and paleoecology: Lethaia v. 21, p. 1–12.

Bergenhayn, J.R.M., 1960, Cambrian and Ordovician loricates from North America: Journal of Paleontology, v. 34, no. 1, p. 168–178.

Bergström, S.M., Chen, X., Gutierrez-Marco, J.C., and Dronov, A., 2009, The new chronostratigraphic classification of the Ordovician System and its relations to major regional series and stages and to $\delta^{13}C$ chemostratigraphy: Lethaia, v. 42, no. 1, p. 97–107.

Bergström, S.M., and Sweet, W.C., 1966, Conodonts from the Lexington Limestone (Middle Ordovician) of Kentucky and its lateral equivalents in Ohio and Indiana: Bulletins of American Paleontology, v. 50, no. 299, p. 271–441, 8 pls.

Berner, R.A., and Kothavala, Z., 2001, Geocarb III: A revised model of atmospheric CO_2 over Phanerozoic time: American Journal of Science, v. 301, p. 182–204.

Bertling, M., 1992, *Arachnostega* n. ichnog.—Burrowing traces in internal moulds of boring bivalves (Late Jurassic, Northern Germany): Paläontologische Zeitschrift, v. 66 no. 1–2, p. 177–185.

Bethke, C.M., 1986, Hydrologic constraints on the genesis of the Upper Mississippi Valley mineral district from Illinois Basin brines: Economic Geology, v. 81, no. 2, p. 233–249.

Billings, E., 1854, On some new genera and species of Cystidea from the Trenton Limestone: Canadian Journal, v. 2, p. 215–219, 250–253, 268–274.

Billings, E., 1857, New species of fossils from Silurian rocks of Canada: Report for the year 1856, *in* Report of Progress for the Years 1853–54–55–56: Toronto, Geological Survey of Canada, p. 245–345.

Billings, E., 1858a, On the Asteriadae of the Lower Silurian rocks of Canada, *in* Figures and Descriptions of Canadian Organic Remains. Decade III: Montreal, Geological Survey of Canada, p. 75–85, 3 pls.

Billings, E., 1858b, On the Cystideae of the Lower Silurian rocks of Canada, *in* Figures and Descriptions of Canadian Organic Remains. Decade III: Montreal, Geological Survey of Canada, p. 9–74, 7 pls.

Billings, E., 1859a, Descriptions of some new species of trilobites from the Lower and Middle Silurian rocks of Canada: Canadian Naturalist and Geologist, v. 4, p. 367–383.

Billings, E., 1859b, On the Crinoideae of the Lower Silurian rocks of Canada, *in* Figures and Descriptions of Canadian Organic Remains. Decade IV: Montreal, Geological Survey of Canada, 72 p., 10 pls.

Billings, E., 1865, Palaeozoic Fossils, Volume I: Containing Descriptions and Figures of New or Little-Known Species of Organic Remains from the Silurian Rocks: Montreal, Geological Survey of Canada, 426 p.

Billings, W.R., 1887, A new genus and three new species of crinoids from the Trenton Formation with notes on a large specimen of *Dendrocrinus proboscidiatus*: The Ottawa Naturalist, v. 1, p. 49–54.

Blake, D.B., 1975, The order Cryptostomata resurrected, *in* S. Pouyet, ed., Bryozoa 1974: Proceedings of the 3rd Conference of the International Bryozoa Association: Lyon, France, Université Claude Bernard, Documents des Laboratoires de Géologie de la Faculté des Sciences de Lyon, Document HS 3, fasc. 2, p. 211–223.

Blake, D.B., 2018, Toward a history of the Paleozoic Asteroidea (Echinodermata): Bulletins of American Paleontology, v. 394, 96 p.

Blake, D.B., Gahn, F.J., Guensburg, T.E., 2020, Two new early Asteroidea (Echinodermata) and early asteroid evolution: Journal of Paleontology, no. 4, v. 5, p. 734–747.

Blake, D.B., Guensburg, T.E., and Lefebvre, B., 2016, New Early Paleozoic Asterozoa (Echinodermata) from the Armorican Massif, France, and the western United States: Annales de Paléontologie, v. 102, no. 3, p. 161–181.

Boardman, R.S., and Cheetham, A.H., 1983, Glossary of morphological terms, *in* Robison, R.A., ed., Treatise on Invertebrate Paleontology. Part G, Bryozoa, Volume 1: Boulder Colorado, Geological Society of America and Lawrence, University of Kansas Press, p. 304–320.

Boardman, R.S., Cheetham, A.H., Blake, D.B., Utgaard, J., Karklins, O.L., Cook, P.L., Sandberg, P.A., Lutaud, G., and Wood, T.S., 1983, Bryozoa (revised), *in* Robison, R.A., ed., Treatise on Invertebrate Paleontology, Part G, Bryozoa, Volume 1: Boulder Colorado, Geological Society of America, and Lawrence, University of Kansas Press, 625 p.

Bork, K.B., and Perry, T.G., 1967, Bryozoa (Ectoprocta) of Champlainian age (Middle Ordovician) from northwestern Illinois and adjacent parts of Iowa and Wisconsin: Journal of Paleontology, v. 41, no. 6, p. 1365–1392.

Bork, K.B., and Perry, T.G., 1968a, Bryozoa (Ectoprocta) of Champlainian age (Middle Ordovician) from northwestern Illinois and adjacent parts of Iowa and Wisconsin; Part 2, *Bythotrypa*, *Diplotrypa*, *Hemiphragma*, *Heterotrypa*, *Stigmatella*, *Eridotrypa*, and *Nicholsonella*: Journal of Paleontology, v. 42, no. 2, p. 337–355.

Bork, K.B., and Perry, T.G., 1968b, Bryozoa (Ectoprocta) of Champlainian age (Middle Ordovician) from northwestern Illinois and adjacent parts of Iowa and Wisconsin; Part 3, *Homotrypa*, *Orbignyella*, *Prasopora*, *Monticulipora*, and *Cyphotrypa*: Journal of Paleontology, v. 42, no. 4, p. 1042–1065.

Bouchet, P., Rocroi, J.P., Hausdorf, B., Kaim, A., Kano, Y., Nützel, A., Parkhaev, P., Schrödl, M., and Strong, E.E., 2017, Revised classification, nomenclator and typification of gastropod and monoplacophoran families: Malacologia, v. 61, p. 1–526.

Branson, B., 1909, The fauna of the Residuary Auburn Chert of Lincoln County, Missouri: Transactions of the Academy of Science of St. Louis, v. 18, p. 39–52.

Branson, E.B., and Mehl, M.G., 1933, Conodont studies: The University of Missouri Studies: A Quarterly of Research, v. 8, 349 p.

Bretsky, P.W., and Bretsky, S.S., 1975, Phenetic variation in some Middle Ordovician strophomenid brachiopods: Geological Society of America Memoir 142, p. 3–36.

Bretsky, P.W., Bretsky, S.S., and Schaefer, P.J., 1977, Molluscan and brachiopod dominated biofacies in the Platteville Formation (Middle Ordovician), upper Mississippi Valley: Bulletin of the Geological Society of Denmark, v. 26, p. 115–132.

Brett, C.C., Whiteley, T.E., Allison, P.A., and Yochelson, E.L., 1999, The Walcott-Rust Quarry: Middle Ordovician Trilobite Konservat-Lagerstätten: Journal of Paleontology, v. 73, no. 2, p. 288–305.

Briggs, D.E.G., 1992, Conodonts: A major extinct group added to the vertebrates: Science, v. 256, no. 5061, p. 1285–1287.

Briggs, D.E.G., Clarkson, E.N., and Aldridge, R.J., 1983, The conodont animal: Lethaia, v. 16, p. 1–14.

Brower, J.C., 1994, Camerate crinoids from the Middle Ordovician (Galena Group, Dunleith Formation) of northern Iowa and southern Minnesota: Journal of Paleontology, v. 68, p. 570–599.

Brower, J.C., 2007, Upper Ordovician crinoids from the Platteville Limestone of northeastern Iowa: Journal of Paleontology, v. 81, p. 103–115.

Brower, J.C., and Veinus, J., 1974, Middle Ordovician crinoids from southwestern Virginia and eastern Tennessee: Bulletins of American Paleontology, v. 66, no. 283, 1125 p.

Brower, J.C., and Veinus, J., 1978, Middle Ordovician crinoids from the Twin Cities area of Minnesota: Bulletins of American Paleontology, v. 74, no. 304, p. 371–506.

Brower, J.C., and Veinus, J., 1982, Long-armed cladid inadunates, *in* Sprinkle, J., ed., Echinoderm Faunas from the Bromide Formation (Middle Ordovician) of Oklahoma: The University of Kansas Paleontological Contributions Monograph 1, p. 129–144.

Brown, A., 1894a, I.—On the structure and affinities of the genus *Solenopora*, together with descriptions of new species: Geological Magazine, v. 1, no. 4, p. 145–151.

Brown, A., 1894b, II.—On the structure and affinities of the genus *Solenopora*, together with descriptions of new species: Geological Magazine, v. 1, no. 5, p. 195–203.

Brown, G.D., Jr., 1965, Trepostomatous Bryozoa from the Logan and Jessamine Limestones (Middle Ordovician) of the Kentucky Bluegrass region: Journal of Paleontology, v. 39, no. 4, p. 974–1006.

Browne, R.G., 1965, Some Upper Cincinnatian (Ordovician) colonial corals of north-central Kentucky: Journal of Paleontology, v. 39, no. 6, p. 1177–1191.

Buatois, L.A., and Mángano, M.G., 2011, Ichnology: Organism–Substrate Interactions in Space and Time: Cambridge, UK, Cambridge University Press, 358 p.

Buatois, L.A., and Mángano, M.G., 2013, Ichnodiversity and ichnodisparity: Significance and caveats: Lethaia, v. 46, p. 281–292.

Buatois, L.A., and Mángano, M.G., 2016, Recurrent patterns and processes: The significance of ichnology in evolutionary paleoecology, *in* Mángano, M.G., and Buatois, L.A., eds., The Trace-Fossil Record of Major Evolutionary Events, Volume 2: Mesozoic and Cenozoic: Berlin, Springer, p. 449–473.

Buatois, L.A., Wisshak, M., Wilson, M.A., and Mángano, M.G., 2017, Categories of architectural designs in trace fossils: A measure of ichnodisparity: Earth-Science Reviews, v. 164, p. 102–181.

Bulman, O.M.B., 1945, A monograph of the Caradoc (Balclatchie) graptolites from limestones in Laggan Burn, Ayrshire: Monograph of the Palaeontological Society 98, pt. 1, 78 p.

Bulman, O.M.B., 1962, On the genus *Amplexograptus* Lapworth, Elles & Wood: Geological Magazine, v. 99, no. 5 p. 459–467.

Bunker, B.J., Ludvigson, G.A., and Witzke, B.J., 1985, The Plum River fault zone and the structural and stratigraphic framework of eastern Iowa: Iowa Geological Survey Technical Information Series 13, 126 p.

Buschbach, T.C., 1964, Cambrian and Ordovician strata of northeastern Illinois: Illinois State Geological Survey Report of Investigation 218, 90 p.

Byers, C.W., and Stasko, L.E., 1978, Trace fossils and sedimentologic interpretation; McGregor Member of Platteville Formation (Ordovician) of Wisconsin: Journal of Sedimentary Research, v. 48, no. 4, p. 1303–1310.

Cardona-Correa, C., Piotrowski, M.J., Knack, J.J., Kodner, R.E., Geary, D.H., and Graham, L.E., 2016, Peat moss-like vegetative remains from Ordovician carbonates: International Journal of Plant Sciences, v. 177, no. 6, p. 523–538.

Carlucci, J.R., and Westrop, S.R., 2012, Trilobite biofacies along an Ordovician (Sandbian) carbonate buildup to basin gradient, southwestern Virginia: Palaios, v. 27, no. 1, p. 19–34.

Carlucci, J.R., Westrop, S.R., and Amati, L., 2010, Tetralichine trilobites from the Upper Ordovician of Oklahoma and Virginia and phylogenetic systematics of the Tetralichini: Journal of Paleontology v. 84, no. 6, p. 1099–1120.

Carlucci, J.R., Westrop, S.R., Amati, L., Adrain, J.M., and Swisher, R.E., 2012, A systematic revision of the Upper Ordovician trilobite genus *Bumastoides* (Illaenidae), with new species from Oklahoma, Virginia and Missouri: Journal of Systematic Palaeontology, v. 10, no. 4, p. 679–723.

Carrera, M.G., and Rigby, J.K., 1999, Biogeography of the Ordovician sponges: Journal of Paleontology, v. 73, p. 26–37.

Carrera, M.G., and Sumrall, C.C., 2020, Ordovician sponges from the Lenoir Limestone, Tennessee: New evidence for a differential sponge distribution along the margins of Laurentia: Journal of Paleontology, v. 94, no. 1, p. 1–11.

Carter, J.G., Altaba, C.R., Anderson, L.C., Araujo, R., Biakov, A.S., Bogan, A.E., Campbell, D.C., Campbell, M., Jin-hua, C., Cope, J.C.W., et al., 2011, A synoptical classification of the Bivalvia (Mollusca): Paleontological Contributions, v. 4, p. 1–47.

Carter, J.G., Harries, P.J., Malchus, N., Sartori, A.F., Anderson, L.C., Bieler, R., Bogan, A.E., Coan, E.V., Cope, J.C.W., Cragg, S.M., and García-March, J.R., 2012, Illustrated glossary of the Bivalvia, *in* Treatise on Invertebrate Paleontology, Part N, Treatise Online No. 48, Revised Volume 1, Chapter 31: Boulder, Colorado, Geological Society of America, and Lawrence, University of Kansas Press, p. 1–209.

Catalani, J.A., 1987, Biostratigraphy of the Middle and Late Ordovician cephalopods of the Upper Mississippi Valley area, *in* Sloan, R.E., ed., Middle and Late Ordovician Lithostratigraphy and Biostratigraphy of the Upper Mississippi Valley: Minnesota Geological Survey Report of Investigations 35, p. 187–189.

Catalani, J.A., 2010, The Platteville molluscan fauna: MAPS (Mid-America Paleontology Society) Digest, v. 33, no. 1, p. 64–74.

Chatterton, B.D.E., and Ludvigsen, R., 1976, Silicified middle Ordovician trilobites from the south Nahanni river area, District of Mackenzie, Canada: Palaeontographica Abteilung A, Band A154, Lieferung 1–3, p. 1–106.

Clarke, J.M., 1897a, *Nanno*, a new cephalopodan type: American Geologist, v. 14, no. 4, p. 205–272.

Clarke, J.M., 1897b, The Lower Silurian Cephalopoda of Minnesota: Minnesota Geological Survey Final Report 3, Part 2, p. 761–812.

Clarke, J.M., 1897c, The Lower Silurian trilobites of Minnesota: Minnesota Geological Survey Final Report 3, Part 2, p. 695–759.

Congreve, C.R., 2013, Evolutionary patterns of trilobites across the end Ordovician mass extinction [Ph.D. thesis]: Lawrence, University of Kansas, 255 p.

Congreve, C.R., Krug, A.Z., and Patzkowsky, M.E., 2015, Phylogenetic revision of the Strophomenida, a diverse and ecologically important Palaeozoic brachiopod order: Palaeontology, v. 58, no. 4, p. 743–758.

Congreve, C.R., Krug, A.Z., and Patzkowsky, M.E., 2019, Evolutionary and biogeographical shifts in response to the Late Ordovician mass extinction: Palaeontology, v. 62, no. 2, p. 267–285.

Conkin, J.E., Conkin, B.M., and Canis, W.F., 1968, Mississippian foraminifera of the United States; Part 3, The limestones of the Chouteau Group in Missouri and Illinois: Micropaleontology, v. 24, no. 2, p. 133–178.

Conrad, T.A., 1843, Observations on the lead-bearing limestone of Wisconsin and descriptions of a new genus of trilobites and fifteen new Silurian fossils: Proceedings of the Academy of Natural Sciences of Philadelphia, v. 1, no. 33, p. 329–335.

Cooper, G.A., 1930, The brachiopod genus *Pionodema* and its homeomorphs: Journal of Paleontology, v. 4, p. 369–382.

Cooper, G.A., 1956, Chazyan and related brachiopods: Smithsonian Miscellaneous Collections, v. 127, p. 1–1245, pl. 1–269.

Cope, J.C.W., 1997, The early phylogeny of the class Bivalvia: Palaeontology, v. 40, p. 713–746.

Coryell, H.N., 1921, Bryozoan faunas of the Stone River Group of central Tennessee [Ph.D. thesis]: Bowling Green, Western Kentucky University, reprinted from Proceedings of the Indiana Academy of Science for 1919, p. 261–340.

Cox, L.R., Newell, N.D., Boyd, D.W., Branson, C.C., Casey, R., Chavan, A., Coogan, A.H., Dechaseaux, C., Fleming, C.A., Haas, F., Hertlein, L.G., et al., 1969, Bivalvia, *in* Moore, R.C., ed., Treatise on Invertebrate Paleontology, Part N: Mollusca 6, Volume 1, Boulder, Colorado, Geological Society of America, and Lawrence, University of Kansas Press, p. 1–129.

Crowley, T.J., and Baum, S.K., 1995, Reconciling Late Ordovician (440 Ma) glaciation with very high (143) CO_2 levels: Journal of Geophysical Research, v. 100, p. 1093–1101.

Cuif, J.P., Feuillée, P., Fischer, J.C., and Pascal, A., 1973, Présence d'astrorhizes chez les Chaetetida mésozoïques: Comptes Rendus de l'Académie des Sciences, Paris (series D), v. 277, p. 2473–2476.

Cutler, B., and Sloan, R.E., 1983, Chitinozoa from the Glenwood and Platteville Formations (Mid Ordovician): Geological Society of America Abstracts and Programs, v. 15, p. 260.

Darby, D.G., and Stumm, E.C., 1965, A revision of the Ordovician trilobite *Asaphus platycephalus* Stokes: Contributions from the Museum of Paleontology (University of Michigan), v. 20, no. 3, p. 63–73.

Decker, C.E., 1935, The graptolites of the Simpson group of Oklahoma: Proceedings of the National Academy of Sciences of the United States of America, v. 21, no. 5, p. 239–243.

Decker, C.E., and Frederickson, E.A., Jr., 1941, A new graptolite horizon in Wisconsin: Journal of Paleontology, v. 15, no. 2 p. 157–159.

Delo, D.M., 1940, Phacopid trilobites of North America: Geological Society of America Special Paper 29, 135 p., 13 pls.

DeMott, L.L., 1963, Middle Ordovician trilobites of the upper Mississippi Valley [Ph.D. thesis]: Cambridge, Massachusetts, Harvard University, 236 p.

DeMott, L.L., Sloan, R.E., Shaw, F.C., and Tripp, R.P., 1987, Platteville and Decorah trilobites from Illinois and Wisconsin: Minnesota Geological Survey Report of Investigations 35, p. 63–98.

Donovan, S.K., and Paul, C.R.C., 1985, A new possible armoured worm from the Tremadoc of Sheinton, Shropshire: Proceedings of the Geologists' Association, v. 96, no. 1, p. 87–91.

Durham, J.W., 1966, Evolution among the Echinoidea: Biological Reviews, v. 41, no. 3, p. 368–391.

Dybowski, W., 1877, Die Chaetitiden der ostbaltischen Silur-Formation: Russisch-Kaiserliche Mineralogische Gesellschaft zu St. Petersburg Verhandlungen, Series 2, v. 14, p. 1–134.

Ebbestad, J.O.R., Kolata, D.R., and Mare, I., 2021, New onychochilid mollusks from the Middle and Late Ordovician of Baltica and Laurentia: Journal of Paleontology, v. 96, no. 1, p. 106–122.

Eisenack, A., 1931, Neue Mikrofossilien des baltischen Silurs. I: Palaeontologische Zeitschrift, v. 13, p. 74–118.

Elias, R.J., 1983, Middle and Late Ordovician solitary rugose corals of the Cincinnati Arch region: U.S. Geological Survey Professional Paper 1066-N, 13 p.

Elles, G.L., and Wood, E.M.R., 1907, A monograph of British graptolites: Monograph of the Palaeontographical Society, 539 p.

Eriksson, M., and Bergman, C.F., 2003, Late Ordovician jawed polychaete faunas of the type Cincinnatian region, U.S.A.: Journal of Paleontology, v. 77, no. 3, p. 509–523.

Esker, G.C., III, 1964, New species of trilobites from the Bromide Formation (Pooleville Member) of Oklahoma: Oklahoma Geological Survey Notes, v. 24, no. 9, p. 195–209.

Fenton, C.L., 1928, Forms of *Strophomena* from Black River and Richmond Formations: The American Midland Naturalist, v. 11, no. 3–4, p. 144–159.

Finks, R.M., and Rigby, J.K., 2004, Paleozoic demosponges, *in* Kaesler, R.L., ed., Treatise on Invertebrate Paleontology, Part E, Porifera (Revised), Volume 3: Boulder, Colorado, Geological Society of America, and Lawrence, University of Kansas Press, p. 9–274.

Flower, R.H., 1946, Ordovician cephalopods of the Cincinnati region, Part 1: Bulletins of American Paleontology, v. 29, no. 116, 738 p.

Flower, R.H., 1952, New Ordovician cephalopods from eastern North America: Journal of Paleontology, v. 26, no. 1, p. 24–59.

Flower, R.H., 1955, Status of endoceroid classification: Journal of Paleontology, v. 29, no. 3, p. 329–371.

Flower, R.H., 1957, Studies of the Actinoceratida: New Mexico Bureau of Mines and Mineral Resources Memoir 2, 59 p.

Flower, R.H., 1958, Some Chazyan and Mohawkian Endoceratida: Journal of Paleontology, v. 32, no. 3, p. 433–458.

Flower, R.H., 1963, New Ordovician Ascoceratida: Journal of Paleontology, v. 37, no. 1, p. 69–85.

Flower, R.H., 1964a, Nautiloid shell morphology: New Mexico Bureau of Mines and Mineral Resources Memoir 13, 62 p.

Flower, R.H., 1964b, The nautiloid order Ellesmerocerida: New Mexico Bureau of Mines and Mineral Resources Memoir 12, 234 p.

Flower, R.H., 1968, An endoceroid from the Mohawkian of Quebec: New Mexico Bureau of Mines and Mineral Resources Memoir 21, p. 8–17.

Foerste, A.F., 1912, *Strophomena* and other fossils from Cincinnatian and Mohawkian horizons, chiefly in Ohio, Indiana and Kentucky: Journal of the Scientific Laboratories of Denison University, v. 17, p. 17–174.

Foerste, A.F., 1917, Notes on Richmond and related fossils: Journal of the Cincinnati Society of Natural History, v. 22, no. 2, p. 42–55.

Foerste, A.F., 1928, A restudy of some of the Ordovician and Silurian cephalopods described by Hall: Journal of the Scientific Laboratories of Denison University, v. 23, p. 173–230.

Foerste, A.F., 1932, Black River and other cephalopods from Minnesota, Wisconsin, Michigan, and Ontario (Part 1): Journal of the Scientific Laboratories of Denison University, v. 27, p. 47–136.

Foerste, A.F., 1933, Black River and other cephalopods from Minnesota, Wisconsin, Michigan, and Ontario (Part 2): Journal of the Scientific Laboratories of Denison University, v. 28, p. 1–146.

Frey, R.C., 1995, Middle and Upper Ordovician nautiloid cephalopods of the Cincinnati Arch region of Kentucky, Indiana, and Ohio: U.S. Geological Survey Professional Paper 1066-P, 126 p.

Frisk, Å.M., and Ebbestad, J.O.R., 2007, Paragastropoda, Tergomya and Gastropoda (Mollusca) from the Upper Ordovician Dalby Limestone, Sweden: Geologiska Föreningen i Stockholm Förhandlingar, v. 129, p. 83–99.

Frizzell, D.L., and Exline, H.I., 1955, Monograph of fossil holothurian sclerites: Rolla, Missouri School of Mines, Technical Series No. 89, 204 p.

Frýda, J., and Rohr, D.M., 2004, Gastropoda, *in* Webby, B.D., Paris, F., Droser, M.L., and Percival, I.G., eds., The Great Ordovician Biodiversification Event: New York, Columbia University Press, p. 184–195.

Galle, A., and Parsley, R.L., 2005, Epibiont relationships on hyolithids demonstrated by Ordovician trepostomes (Bryozoa) and Devonian tabulates (Anthozoa): Bulletin of Geosciences, v. 80, no. 2, p. 125–138.

Goldman, D., Campbell, S.M., and Rahl, J.M., 2002, Three-dimensionally preserved specimens of *Amplexograptus* (Ordovician, Graptolithina) from the North American mid-continent: Taxonomic and biostratigraphic significance: Journal of Paleontology, v. 76, no. 5, p. 921–927.

Golonka, J., 2002, Plate-tectonic maps of the Phanerozoic, *in* Kiessling, W., Golonka, J., and Flugel, E., eds., Phanerozoic Reef Patterns: Society for Sedimentary Geology Special Paper 72, p. 21–75.

Grant, J., 1881, Description of a new species of *Porocrinus* from the Trenton Limestone: Ottawa Field Nature Club Transactions, no. 2, p. 42.

Gray, D.I., 1980, Spicule pseudomorphs in a new Palaeozoic chaetetid, and into sclerosponge affinities: Palaeontology, v. 23, p. 803–820.

Guensburg, T.E., 1984, Echinodermata of the Middle Ordovician Lebanon Limestone, central Tennessee: Bulletins of American Paleontology, v. 86, no. 319, 100 p.

Gutschick, R.C., 1954a, A new species of *Astraeospongia* from the Middle Ordovician of northern Illinois: Journal of Paleontology, v. 28, p. 430–433.

Gutschick, R.C., 1954b, Holothurian sclerites from the Middle Ordovician of northern Illinois: Journal of Paleontology, v. 28, p. 827–829.

Gutschick, R.C., 1986, Middle Ordovician agglutinated foraminifera including *Reophax* from the Mifflin Formation, Platteville Group of Illinois: Journal of Paleontology, v. 60, no 2, p. 233–248.

Hall, J., 1847, Palaeontology of New-York. Volume 1. Containing Descriptions of the Organic Remains of the Lower Division of the New-York System: Albany, New York, C. Van Benthuysen, 338 p. [equivalent of the Lower Silurian rocks of Europe]

Hall, J., 1851, Description of new, or rare species from the Paleozoic series, *in* Foster, J.W., and Whitney, J.D., eds., Report on the Geology of the Lake Superior Land District: United States Congress Special Session, Senate Executive Documents, Part 2, p. 203–231.

Hall, J., 1852, Paleontology of New-York. Volume 2. Containing Descriptions of the Organic Remains of the Lower Division of the New-York System: Albany, New York, C. Van Benthuysen, 368 p.

Hall, J., 1861, Report of the Superintendent of the Geological Survey [of Wisconsin] Exhibiting the Progress of the Work. January 1, 1861: Madison, Wisconsin Geological Survey, 52 p.

Hall, J., 1865, Figures and Descriptions of Canadian Organic Remains. Decade II. Graptolites of the Quebec Group: Montreal, Geological Survey of Canada, 151 p.

Hall, J., and Clarke, J.M., 1892, Paleontology: Volume 8. An Introduction to the Study of the Genera of Palaeozoic Brachiopoda. Part 1: Albany, New York, C. Van Benthuysen, 367 p.

Harazim, D., Callow, R.H.T., and McIlroy, D., 2013, Microbial mats implicated in the generation of intrastratal shrinkage ('synaeresis') cracks: Sedimentology, v. 60, no. 7, p. 1621–1638.

Hartman, W.D., and Goreau, T.F., 1975, A Pacific tabulate sponge, living representative of a new order of sclerosponges: Postilla (Peabody Museum of Natural History), v. 167, p. 1–21.

Hill, D., 1981, Coelenterata, *in* Teichert, C., ed., Treatise on Invertebrate Paleontology, Part F, Rugosa and Tabulata, Supplement 1, Volumes 1 and 2: Boulder, Colorado, Geological Society of America, and Lawrence, University of Kansas Press, 762 p.

Hoare, R.D., and Mapes, R.H., 1995, Relationships of the Devonian *Strobilepis* and related Pennsylvanian problematica: Acta Palaeontologica Polonica, v. 40, no. 2, p. 111–128.

Hoare, R.D., and Pojeta, J., Jr., 2006, Ordovician Polyplacophora (Mollusca) from North America: The Paleontological Society Memoir 80, 27 p.

Holmden, C., Creaser, R.A., Muehlenbachs, K., Leslie, S.A., and Bergstrom, S.M., 1998, Isotopic evidence for geochemical decoupling between ancient epeiric seas and bordering oceans: Implications for secular curves: Geology, v. 26, p. 567–570.

Horne, D.J., 2003, Key events in the ecological radiation of the Ostracoda: The Paleontological Society Papers, v. 9, p. 181–202.

Huff, W.D., Bergström, S.M., and Kolata, D.R., 1992, Gigantic Ordovician volcanic ash fall in North America and Europe: Biological, tectonomagmatic, and event-stratigraphic significance: Geology, v. 20, p. 875–878.

Huff, W.D., and Kolata, D.R., 1990, Correlation of the Ordovician Deicke and Millbrig K-bentonites between the Mississippi Valley and the southern Appalachians: American Association of Petroleum Geologists Bulletin, v. 74, no. 11, p. 1736–1747.

Jacobson, S.R., 1978, Acritarchs from Middle and Upper Ordovician rocks in New York state and the Cincinnati region in Ohio and Kentucky [Ph.D. thesis]: Columbus, The Ohio State University, 303 p.

Jacobson, S.R., 1979, Acritarchs as paleoenvironmental indicators in Middle and Upper Ordovician rocks from Kentucky, Ohio and New York: Journal of Paleontology, v. 53, no. 5, p. 1197–1212.

Jacobson, S.R., 1987, Middle Ordovician acritarchs are guide fossils for the Upper Ordovician: Lethaia, v. 20, no. 1, p. 91–92.

Jacobson, S.R., and Achab, A., 1985, Acritarch biostratigraphy of the *Dicellograptus complanatus* graptolite zone from the Vaureal formation (Ashgillian), Anticosti Island, Quebec, Canada: Palynology, v. 9, no. 1, p. 165–198.

James, U.P., 1878, Descriptions of newly discovered species of fossils from the Lower Silurian Formation, Cincinnati Group: The Paleontologist, v. 1, p. 1–8.

Jansonius, J., and Jenkins, W.A.M., 1978, Chitinozoa, *in* Haq, B.U., and Boersma, A., eds., Introduction to Marine Micropaleontology: New York, Elsevier, p. 341–357.

Jin, J., Harper, D.A., Rasmussen, J.A., and Sheehan, P.M., 2012, Late Ordovician massive-bedded *Thalassinoides* ichnofacies along the palaeoequator of Laurentia: Palaeogeography, Palaeoclimatology, Palaeoecology, v. 367, p. 73–88.

Karklins, O.L., 1969, The cryptostome bryozoa from the Middle Ordovician Decorah Shale, Minnesota: Minnesota Geological Survey Special Publication SP-6, 121 p.

Karklins, O.L., 1983, Ptilodictyoid Cryptostomata bryozoa from the Middle and Upper Ordovician Rocks of Central Kentucky: The Paleontological Society Memoir 14, 31 p.

Karklins, O.L., 1984, Trepostome and cystoporate bryozoans from the Lexington Limestone and the Clays Ferry Formation (Middle and Upper Ordovician) of Kentucky: U.S. Geological Survey Professional Paper 1066-I, 192 p.

Karklins, O.L., 1985, Bryozoans from the Murfreesboro and Pierce Limestones (Early Blackriveran, Middle Ordovician), Stones River Group, of Central Tennessee: The Paleontological Society Memoir 15, p. 1–42.

Kay, G.M., 1931, Stratigraphy of the Ordovician Hounsfield metabentonite: Journal of Geology, v. 39, no. 4, p. 361–376.

Kay, G.M., 1934, Mohawkian Ostracoda: Species common to Trenton faunules from the Hull and Decorah Formations: Journal of Paleontology, v. 8, no. 3, p. 328–343.

Kay, G.M., 1940, Ordovician Mohawkian Ostracoda: Lower Trenton Decorah fauna: Journal of Paleontology, v. 14, no. 3, p. 234–269.

Keller, S.J., and Abdulkareem, T.F., 1980, Post-Knox unconformity—Significance at Unionport Gas-Storage Project and relationship to petroleum exploration in Indiana: Indiana Geological Survey Occasional Paper 31, 19 p.

Kesling, R.V., 1960, Middle Ordovician Black River ostracods from Michigan: Part 2, *Levisulculus* and *Eurychilina*: Contributions from the Museum of Paleontology (University of Michigan), v. 15, no. 15, p. 349–363.

Kesling, R.V., 1967, Cystoids, *in* Moore, R.C., ed., Treatise on Invertebrate Paleontology, Part S, Echinodermata, Supplement 1, Volumes 1 and 2: Boulder, Colorado, Geological Society of America, and Lawrence, The University of Kansas Press, p. 85–267.

Kier, P.M., 1982, Echinoids, *in* Sprinkle, J., ed., Echinoderm Faunas from the Bromide Formation (Middle Ordovician) of Oklahoma: The University of Kansas Paleontological Contributions Monograph 1, p. 310–314.

Klappa, C.F., and James, N.P., 1980, Small lithistid sponge bioherms, early Middle Ordovician Table Head Group, Western Newfoundland: Bulletin of Canadian Petroleum Geology, v. 28, no. 3, p. 425–451.

Knaust, D., 2012, Trace-fossil systematics, *in* Knaust, D., and Bromley, R.G., eds., Trace Fossils as Indicators of Sedimentary Environments: Developments in Sedimentology, v. 64, p. 79–101.

Knaust, D., 2019a, *Rhizocorallites* Müller, 1955 from the Triassic and Jurassic of Germany: Burrow, coprolite, or cololite? Paläontologische Gesellschaft, p. 1–17.

Knaust, D., 2019b, *Sulcolithos variabilis* igen. et isp. nov.: Grooves on firm and hard bedding surfaces: Paläontologische Zeitschrift, p. 1–12.

Knaust, D., and Dronov, A., 2013, *Balanoglossites* ichnofabrics from the Middle Ordovician Volkhov formation (St. Petersburg Region, Russia): Stratigraphy and Geological Correlation, v. 21, no. 3, p. 265–279.

Knight, J.B., 1941, Paleozoic gastropod genotypes: Geological Society of America Special Paper 32, 510 p.

Knight, J.B., Cox, L.R., Keen, A.M., Batten, R.L., Yochelson, E., and Robertson, R., 1960, Systematic descriptions, *in* Moore, R.C., ed., Treatise on Invertebrate Paleontology Part I, Mollusca, Volume 1: Boulder, Colorado, Geological Society of America, and Lawrence, The University of Kansas Press, p. 169–324.

Koken, E.F.R.K., 1896, Die Leitfossilien. Ein Handbuch für den Unterricht und für das Bestimmen von Versteinerungen: Leipzig, Germany, Chr. Herm. Tauchnitz, 848 p.

Kolata, D.R., 1975, Middle Ordovician echinoderms from northern Illinois and southern Wisconsin: The Paleontological Society Memoir 7, 74 p.

Kolata, D.R., 1982, Camerates, *in* Sprinkle, J., ed., Echinoderm Faunas from the Bromide Formation (Middle Ordovician) of Oklahoma: The University of Kansas Paleontological Contributions Monograph 1, p. 179–205.

Kolata, D.R., 2010, Cambrian and Ordovician Systems (Sauk Sequence and Tippecanoe I Subsequence), *in* Kolata, D.R., and Nimz, C.K., eds., The Geology of Illinois: Champaign, Illinois State Geological Survey, p. 136–157.

Kolata, D.R., 2013, Bedrock geology of Dixon West Quadrangle, Lee County, Illinois: Illinois State Geological Survey IGQ Dixon West-BG, scale 1:24,000, 2 sheets; report, 5 p.

Kolata, D.R., Frost, J.K., and Huff, W.D., 1986, K-bentonites of the Ordovician Decorah Subgroup, upper Mississippi Valley: Correlation by chemical fingerprinting: Illinois State Geological Survey Circular 537, 30 p.

Kolata, D.R., and Graese, A.M., 1983, Lithostratigraphy and depositional environments of the Maquoketa Group (Ordovician) in northern Illinois: Illinois State Geological Survey Circular 528, 49 p.

Kolata, D.R., Huff, W.D., and Bergström, S.M., 1996, Ordovician K-bentonites of eastern North America: Geological Society of America Special Paper 313, 84 p.

Kolata, D.R., Huff, W.D., and Bergström, S.M., 1998, Nature and regional significance of unconformities associated with the Middle Ordovician Hagan K-bentonite complex in the North American midcontinent: Geological Society of America Bulletin, v. 110, no. 6, p. 723–739.

Kolata, D.R., Huff, W.D., and Bergström, S.M., 2001, The Ordovician Sebree trough: An oceanic passage to the Midcontinent United States: Geological Society of America Bulletin, v. 113, no. 8, p. 1067–1078.

Kolata, D.R., and Nelson, W.J., 1991, Tectonic history of the Illinois Basin, *in* Leighton, M.W., Kolata, D.R., Oltz, D.F., and Eidel, J.J., eds., Interior Cratonic Basins: American Association of Petroleum Geologists Memoir 51, p. 561–601.

Kunk, M.J., Sutter, J., Obradovich, J.D., and Lanphere, M.A., 1985, Age of biostratigraphic horizons within the Ordovician and Silurian Systems: The chronology of the geological record: Geological Society of London Memoir 10, p. 89–92.

Lasemi, Z., 1983, Recognition of original mineralogy in micrites and its genetic and diagenetic implications [M.S. thesis]: Champaign, University of Illinois, 131 p.

Leslie, S.A., 2000, Mohawkian (Upper Ordovician) conodonts of eastern North America and Baltoscandia: Journal of Paleontology, v. 74, no. 6, p. 1122–1147.

Leslie, S.A., and Bergström, S.M., 1995, Revision of the North American late Middle Ordovician standard stage classification and timing of the Trenton transgression based on K-bentonite bed correlation, *in* Cooper, J.D., Droser, M.L., and Finney, S.C., Ordovician Odyssey: Short Papers for the Seventh International Symposium on the Ordovician System, Fullerton, CA: San Diego, California, Society for Sedimentary Geology (SEPM)–Pacific Section, p. 49–54.

Leslie, S.A., and Bergström, S.M., 2005, Conodont biostratigraphy across the Turinian–Chatfieldian stage transition (Late Ordovician, Mohawkian) in the Upper Mississippi Valley, *in* Ludvigsen, G.A., and Witzke, B.J., eds., Facets of the Ordovician Geology of the Upper Mississippi Valley Region. Guidebook for the 35th Annual Field Conference of the Great Lakes Section, Society for Sedimentary Geology (SEPM), September 23–25, 2005: Iowa Geological Survey Guidebook Series No. 24, p. 30–34.

Lespérance, P.J., and Desbiens, S., 1995, Selected Ordovician trilobites from the Lake St. John District of Quebec and their bearing on systematics: The Paleontological Society Memoir 42, 19 p.

Lewis, R.D., 1982, Holdfasts, *in* Sprinkle, J., ed., Echinoderm faunas from the Bromide Formation (Middle Ordovician) of Oklahoma: The University of Kansas Paleontological Contributions Monograph 1, p. 57–64.

Liu, H.P., Bergström, S.M., Witzke, B.J., Briggs, D.E., McKay, R.M., and Ferretti, A., 2017, Exceptionally preserved conodont apparatuses with giant elements from the Middle Ordovician Winneshiek Konservat-Lagerstätte, Iowa, USA: Journal of Paleontology, v. 91, no. 3, p. 493–511.

LoDuca, S.T., 1997, The green alga *Chaetocladus* (Dasycladales): Journal of Paleontology, v. 71, no. 5, p. 940–949.

LoDuca, S.T., 2019, New Ordovician marine macroalgae from North America, with observations on *Buthograptus*, *Callithamnopsis*, and *Chaetocladus*: Journal of Paleontology, v. 93, no. 2, p. 197–214.

Lucas, S.G., 2012, The extinction of the conulariids: Geosciences, v. 2, no. 1, p. 1–10.

Luczaj, J.A., 2006, Evidence against the Dorag (mixing-zone) model for dolomitization along the Wisconsin arch—A case for hydrothermal diagenesis: American Association of Petroleum Geology Bulletin, v. 90, no. 11, p. 1719–1738.

Ludvigsen, R., and Chatterton, B.D.E., 1982, Ordovician Pterygometopidae (Trilobita) of North America: Canadian Journal of Earth Sciences, v. 19, p. 2179–2206.

Ludvigsen, R., and Westrop, S.R., 1983, Trilobite biofacies of the Cambrian–Ordovician boundary interval in northern North America: Alcheringa, v. 7, no. 4, p. 301–319.

Ludvigson, G.A., Jacobson, S.R., Witzke, B.J., and González, L.A., 1996, Carbonate component chemostratigraphy and depositional history of the Ordovician Decorah Formation, Upper Mississippi Valley, *in* Witzke, B.J., Ludvigson, G.A., and Day, J., eds., Paleozoic Sequence Stratigraphy: Views from the North American Craton: Geological Society of America Special Paper 306, p. 67–86.

Ludvigson, G.A., and Witzke, B.J., 2005, The emerging record of Late Ordovician global change, *in* Ludvigsen, G.A., and Witzke, B.J., eds., Facets of the Ordovician Geology of the Upper Mississippi Valley Region: Guidebook for the 35th Annual Field Conference of the Great Lakes Section, Society for Sedimentary Geology (SEPM), September 23–25, 2005: Iowa Geological Survey Guidebook Series No. 24, p. 67–71.

Ludvigson, G.A., Witzke, B.J., González, L.A., Carpenter, S.J., Schneider, C.L., and Hasiuk, F., 2004, Late Ordovician (Turinian–Chatfieldian) carbon isotope excursions and their stratigraphic and paleoceanographic significance: Palaeogeography, Palaeoclimatology, Palaeoecology, v. 210, no. 2–4, p. 187–214.

Maletz, J., 2014a, Graptolite reconstructions and interpretations: Paläontologische Zeitschrift, v. 89, no. 3, p. 271–286.

Maletz, J., 2014b, The classification of the Pterobranchia (Cephalodiscida and Graptolithina): Bulletin of Geosciences, v. 89, no. 3, p. 477–540.

Malinky, J.M., 1990, *Solenotheca*, new hyolitha (Mollusca) from the Ordovician of North America: Proceedings of the Biological Society of Washington, v. 103, no. 2, p. 265–278.

Marques, A.C., and Collins, A.G., 2004, Cladistic analysis of Medusozoa and cnidarian evolution: Invertebrate Biology, v. 123, no. 1, p. 23–42.

McAlester, A.L., 1968, Type species of Paleozoic nuculoid bivalve genera: Geological Society of America Memoir 105, 143 p.

McKerrow, W.S., Dewey, J.F., and Scotese, C.R., 1991, The Ordovician and Silurian development of the Iapetus ocean: Special Papers in Palaeontology 44, p. 165–178.

McKinney, F.K., 1971, Trepostomatous Ectoprocta (Bryozoa) from the lower Chickamauga Group (Middle Ordovician), Wills Valley, Alabama: Paleontological Research Institution, v. 60, no. 267, p. 193–337.

McKinney, F.K., and Jackson, J.B.C., 1989, Bryozoan Life Histories: Bryozoan Evolution: Chicago, University of Chicago Press, p. 97–119.

Meek, F.B., and Worthen, A.H., 1865, Descriptions of new Crinoidea, etc., from the Carboniferous rocks of Illinois and some of the adjoining states: Proceedings of the Academy of Natural Sciences of Philadelphia, p. 155–166.

Meek, F.B., and Worthen, A.H., 1868, Lower Silurian species: Fossils of the Trenton Group, *in* Part II. Palaeontology, *in* Volume III. Geology and Palaeontology: Geological Survey of Illinois, p. 291–565.

Merriam, D.F., and Atkinson, W.R., 1956, Simpson filled sinkholes in eastern Kansas: Kansas Geological Survey Bulletin 119, p. 61–80.

Miller, S.A., 1874, Genus *Pasceolus* (Billings): Cincinnati Quarterly Journal of Science, v. 1, no. 1, p. 4–7.

Miller, S. A., 1889, North America Geology and Paleontology for Use of Amateurs, Students, and Scientists: Cincinnati, Ohio, Western Methodist Concern, 664 p.

Miller, S.A., and Dyer, C.B., 1878, Contributions to paleontology, no. 2: Journal of the Cincinnati Society of Natural History, v. 1, p. 24–39.

Mitchell, C.E., Melchin, M.J., Cameron, C.B., and Maletz, J., 2013, Phylogenetic analysis reveals that *Rhabdopleura* is an extant graptolite: Lethaia, v. 46, no. 1, p. 34–56.

Moore, R., 1961, Crustacea Ostracoda, *in* Treatise on Invertebrate Paleontology Part Q, Arthropoda Volume 3: Boulder, Colorado, Geological Society of America, and Lawrence, The University of Kansas Press, p. 1–442.

Moreman, W.L., 1930, Arenaceous foraminifera from Ordovician and Silurian limestones of Oklahoma: Journal of Paleontology, v. 4, p. 42–59.

Moreman, W.L., 1933, Arenaceous foraminifera from Lower Paleozoic rocks of Oklahoma: Journal of Paleontology, v. 7, no. 4, p. 393–397.

Morgan, W.A., 2012, Sequence stratigraphy of the great American carbonate bank. The Great American carbonate bank, *in* Derby, J.R., Fritz, R.D., Longacre, S.A., Morgan, W.A., and Sternbach, C.A., eds., The Geology and Economic Resources of the Cambrian–Ordovician Sauk Megasequence of Laurentia: American Association of Petroleum Geologists Memoir 98, p. 37–79.

Moss, D.K., and Westrop, S.R., 2014, Systematics of some Late Ordovician encrinurine trilobites from Laurentian North America: Journal of Paleontology, v. 88, no. 6, p. 1095–1119.

Mossler, J.H., 1983, Paleozoic lithostratigraphy of southeastern Minnesota: Minnesota Geological Survey Miscellaneous Map M-51, scale 1:500,000, 8 pls.

Mossler, J.H., 1985, Sedimentology of the Middle Ordovician Platteville Formation, southeastern Minnesota: Minnesota Geological Survey Report of Investigations 33, 27 p.

Mossler, J.H., 1987, Paleozoic lithostratigraphic nomenclature for Minnesota: Minnesota Geological Survey Report of Investigations 36, 32 p., 1 pl.

Mossler, J.H., 2008, Paleozoic stratigraphic nomenclature for Minnesota: Minnesota Geological Survey Report of Investigations 65, 83 p.

Moysiuk, J., Smith, M.R., and Caron, J.B., 2017, Hyoliths are Palaeozoic lophophorates: Nature, v. 541, no. 7637, p. 394–397.

Müller, A.H., 1955, "Helminthoide" Lebensspuren aus der Trias von Thüringen: Geologie v. 4, p. 407–415.

Myrow, P.M., 1995, *Thalassinoides* and the enigma of Early Paleozoic open-framework burrow systems: Palaios, v. 10, no. 1, p. 58–74.

Nara, M., 1995, *Rosselia socialis*: A dwelling structure of a probable terebellid polychaete: Lethaia, v. 28, p. 171–178.

Nelson, W.J., 2010, Structural features, *in* Kolata, D.R., and Nimz, C.K., eds., The Geology of Illinois: Champaign, Illinois State Geological Survey, p. 90–104.

Netto, R.G., Buatois, L.A., Mángano, M.G., and Balistieri, P.R.M.N., 2007, *Gyrolithes* as a multipurpose burrow: An ethologic approach: Revista Brasileira de Paleontologia, v. 10, no. 3, p. 157–168.

Nitecki, M.H., 1970, North American cyclocrinitid algae: Fieldiana: Geology v. 21, 182 p.

North American Commission on Stratigraphic Nomenclature, 2005, North American Stratigraphic Code: American Association of Petroleum Geologists Bulletin, v. 89, no. 11, p. 1547–1591.

Nunn, J.R., 1986, The petrology and paleogeography of the Starved Rock Sandstone in southern, Iowa, western Illinois, and northeastern Missouri [M.S. thesis]: Iowa City, University of Iowa, 144 p.

Okulitch, V.J., 1935, Fauna of the Black River Group in the vicinity of Montreal: Canadian Field Naturalist, v. 49, p. 96–107.

Okulitch, V.J., 1938, Some Black River corals: Royal Society of Canada Transactions, 3rd ser., v. 32, sec. 4, p. 87–111, 2 pls.

Öpik, A.A., 1930, Beiträge zur Kenntnis der Kukruse- (C_2-C_3-) Stufe in Eesti, IV: Acta et Commentationes Universitatis Tartuensis (Dorpatensis) A 19, p. 1–34, 6 pls.

Ostrom, M.E., 1967, Paleozoic stratigraphic nomenclature for Wisconsin: Wisconsin Geological and Natural History Survey Information Circular 8, 4 p., 1 pl.

Ostrom, M.E., 1969, Champlainian Series (Middle Ordovician) in Wisconsin: American Association of Petroleum Geologists Bulletin, v. 53, p. 672–678.

Ostrom, M.E., 1970, Sedimentation cycles in the Lower Paleozoic rocks of western Wisconsin, *in* Field Trip Guidebook for Cambrian–Ordovician Geology of western Wisconsin: Wisconsin Geological and Natural History Survey Information Circular 11, p. 10–34.

Ostrom, M.E., 1987, Middle Ordovician Platteville Formation, Hoadley Hill, Wisconsin: Geological Society of America Centennial Field Guide, North-Central Section of the Geological Society of America, v. 3, p. 205–208.

Palmer, T.J., 1978, Burrows at certain omission surfaces in the Middle Ordovician of the Upper Mississippi Valley: Journal of Paleontology, v. 52, p. 109–117.

Palmer, T.J., and Wilson, M.A., 2004, Calcite precipitation and dissolution of biogenic aragonite in shallow Ordovician calcite seas: Lethaia, v. 37, p. 417–427.

Paris, F., and Nõlvak, J., 1999, Biological interpretation and paleobiodiversity of a cryptic fossil group: The "chitinozoan animal": Geobios, v. 32, no. 2, p. 315–324.

Parsley, R.L., 1982, Paracrinoids, *in* Sprinkle, J., ed., Echinoderm Faunas from the Bromide Formation (Middle Ordovician) of Oklahoma: The University of Kansas Paleontological Contributions Monograph 1, p. 212–223.

Parsley, R.L., and Mintz, L.W., 1975, North American Paracrinoidea (Ordovician, Paracrinozoa, New, Echinodermata): Bulletins of American Paleontology, v. 68, no. 288, p. 1–115.

Perry, T.G., 1962, Spechts Ferry (Middle Ordovician) bryozoan fauna from Illinois, Wisconsin, and Iowa: Illinois State Geological Survey Circular 326, p. 1–36.

Pojeta, J., Jr., 1971, Review of Ordovician pelecypods: U.S. Geological Survey Professional Paper 695, 46 p.

Pojeta, J., Jr., 1978, The origin and early taxonomic diversification of pelecypods: Philosophical Transactions of the Royal Society of London B: Biological Sciences, v. 284, no. 1001, p. 225–246.

Pojeta, J., Jr., and DuFoe, J., 2008, New information about *Echinochiton dufoei*, the Ordovician spiny chiton: American Malacological Bulletin, v. 25, p. 25–34.

Pojeta, J., Jr., Eernisse, D.J., Hoare, R.D., and Henderson, M.D., 2003, *Echinochiton dufoei*: A new Ordovician chiton: Journal of Paleontology, no. 4, v. 77, p. 646–654.

Pojeta, J., Jr., and Runnegar, B., 1976, The paleontology of rostroconch mollusks and the early history of the phylum Mollusca: U.S. Geological Survey Professional Paper 968, 88 p.

Pojeta, J., Jr., Runnegar, B., Morris, N.J., and Newell, N.D., 1972, Rostroconchia: A new class of bivalve mollusks: Science, v. 177, p. 264–267.

Pojeta, J., Jr., Vendrasco, M.J., and Darrough, G., 2010, Upper Cambrian chitons (Mollusca, Polyplacophora) from Missouri, USA: Bulletins of American Paleontology, v. 2010, no. 379, p. 1–88.

Polechová, M., 2013, Bivalves from the Middle Ordovician Šárka Formation (Prague Basin, Czech Republic): Bulletin of Geosciences (Czech Geological Survey), v. 88, no. 2, p. 427–461.

Portlock, J.E., 1843, Report on the Geology of the County of Londonderry, and Parts of Tyrone and Fermanagh: Dublin, Ordnance Survey of Ireland, p. 274.

Pratt, W.P., and Wandrey, C.J., 1996, Maps showing thickness and limestone–dolostone ratios of selected Paleozoic carbonate units in the northern Midcontinent, USA; folio of the northern Midcontinent area: U.S. Geological Survey Miscellaneous Field Studies Map 1835-I, report, 10 p.; 3 pls.

Raymond, P.E., 1920, Some new Ordovician trilobites: Bulletins of the Museum of Comparative Zoology at Harvard University, v. 64, no. 2, p. 273–296.

Raymond, P.E., 1931, Notes on invertebrate fossils, with descriptions of new species: Bulletin of the Museum of Comparative Zoology at Harvard University, v. 55, no. 6, p. 165–212.

Raymond, P.E., and Narraway, J.E., 1910, Notes on Ordovician trilobites, III, Asaphidae from the Lowville and Black River: Annals of the Carnegie Museum, v. 7, p. 46–59.

Raymond, P.E., and Okulitch, V.J., 1940, Some Chazyan sponges: Bulletin of the Museum of Comparative Zoology at Harvard University, v. 86, no. 5, p. 197–214.

Reso, A., and Wegner, K., 1964, Echinoderm (Holothurian?) sclerites from the Bromide Formation (Black Riverian) of southern Oklahoma: Journal of Paleontology, v. 38, no. 1, p. 89–94.

Rhodes, F.H.T., 1952, A classification of Pennsylvanian conodont assemblages: Journal of Paleontology, v. 26, no. 6, p. 886–901.

Riding, R., 2004, *Solenopora* is a chaetetid sponge, not an alga: Palaeontology, v. 47, no. 1, p. 117–122.

Rigby, J.K., 1967, Two new Early Paleozoic sponges and the sponge-like organism *Gaspespongia basalis* Parks, from the Gaspé Peninsula, Quebec: Journal of Paleontology, v. 41, p. 766–775.

Rigby, J.K., 1977, Two new Middle Ordovician sponges from Foxe Plain, Southeastern District of Franklin: Geological Survey of Canada Bulletin, v. 269, p. 121–129.

Rigby, J.K., and Desrochers, A., 1995, Lower and Middle Ordovician lithistid demosponges from the Mingan Islands, Gulf of St. Lawrence, Quebec, Canada: The Paleontological Society Memoir 41, 35 p.

Robson, S.P., and Young, G.A., 2013, Late Ordovician Conulariids from Manitoba, Canada: Journal of Paleontology, v. 87, no. 5, p. 775–785.

Roemer, F., 1861, Die fossile Fauna der silurischen Diluvial-Geschiebe von Sadewitz bei Oels in Nieder-Schlesien: Eine palaeontologische Monographie: Breslau, Poland, Nischkowsky, 79 p.

Rohr, D.M., 1993, Middle Ordovician carrier shell *Lytospira* (Mollusca, Gastropoda) from Alaska: Journal of Paleontology, v. 67, p. 959–962.

Rohr, D.M., Blodgett, R.B., and Furnish, W.M., 1992, *Maclurina manitobensis* (Whiteaves) (Ordovician Gastropoda): The largest known Paleozoic gastropod: Journal of Paleontology, v. 66, p. 880–884.

Rohr, D.M., and Yochelson, E.L., 1999, Life association of shell and operculum of Middle Ordovician gastropod *Maclurites*: Journal of Paleontology, v. 73, p. 1078–1080.

Rosen, F., 1867, Ueber die Natur der Stromatoporen und über die Erhaltung der Hornfaser der Spongien im fossilen Zustande [Ph.D. thesis]: Dorpat, Estonia, Universität Dorpat.

Rowan, E.L., Goldhaber, M.B., and Hatch, J.R., 2002, Regional fluid flow as a factor in the thermal history of the Illinois basin: Constraints from fluid inclusions and the maturity of Pennsylvanian coals: American Association of Petroleum Geologists Bulletin, v. 86, no. 2, p. 257–277.

Rudkin, D.M., Tripp, R.P., Ludvigsen, R., and Landing, E., 1994, The Ordovician trilobite *Hemiarges* (Lichidae: Trochurinae) from North America and Greenland: Studies in Stratigraphy and Paleontology in Honor of Donald W. Fisher: New York State Museum Bulletin, v. 481, p. 289–306.

Ruedemann, R., and Wilson, T.Y., 1936, Eastern New York Ordovician cherts: Geological Society of America Bulletin, v. 47, no. 10, p. 1535–1586.

Salmon, E.S., 1942, Mohawkian Rafinesquinae: Journal of Paleontology, v. 16, no. 5, p. 564–603.

Salter, J.W., 1859, Fossils from the base of the Trenton Group, *in* Figures and Descriptions of Canadian Organic Remains: Geological Survey of Canada, Decade 1, p. 1–47.

Salter, J.W., 1858, On *Cyclocystoides*, a new genus of Echinodermata from the Lower and Middle Silurian rocks, *in* Figures and Descriptions of Canadian Organic Remains. Decade III: Montreal, Geological Survey of Canada, p. 86–90.

Sardeson, F.W., 1892, The range and distribution of the lower Silurian faunas of Minnesota with descriptions of some new species: Minnesota Academy of Natural Sciences Bulletin, v. 3, p. 326–343.

Sardeson, F.W., 1899, *Lichenaria typa* W. & S.: American Journal of Science (1880–1910), v. 8, no. 44, p. 101.

Sardeson, F.W., 1908, Discoid crinoidal roots and *Camarocrinus*: The Journal of Geology, v. 16, p. 239–254.

Sardeson, F.W., 1925, Ordovicic kelp, sponges, and sea worms in Minnesota: Pan-American Geologist, v. 43, no. 4, p. 271–286.

Sardeson, F.W., 1928, Derivation of the Calceocrinidae: Pan-American Geologist, v. 49, p. 35–46.

Scotese, C.R., 2002, Paleomap: C.R. Scotese, http://www.scotese.com.

Scott, H.W., 1934, The zoological relationships of the conodonts: Journal of Paleontology, v. 8, p. 448–455.

Scrutton, C.T., 1997, The Palaeozoic corals, I: Origins and relationships: Proceedings of the Yorkshire Geological Society, v. 51, pt. 3, p. 177–208.

Scrutton, C.T., 1998, The Palaeozoic corals, II: Structure, variation and palaeoecology: Proceedings of the Yorkshire Geological Society, v. 52, pt. 1, p. 1–57.

Seilacher, A., 1964, Sedimentological classification and nomenclature of trace fossils: Sedimentology, v. 3, no. 3, p. 253–256.

Seilacher, A., 2007, Trace Fossil Analysis: Berlin, Springer, 226 p.

Semeniuk, V., 1971, Subaerial leaching in the limestones of the Bowan Park Group (Ordovician) of central western New South Wales: Journal of Sedimentary Petrology, v. 41, p. 939–950.

Servais, T., Li, J., Stricanne, L., Vecoli, M., and Wicander, R., 2004a, Acritarchs, *in* Webby, B.D., Paris, F., Droser, M.L., and Percival, I.G. eds., The Great Ordovician Biodiversification Event: New York, Columbia University Press, p. 348–360.

Servais, T., Stricanne, L., Montenari, M., and Pross, J., 2004b, Population dynamics of galeate acritarchs at the Cambrian–Ordovician transition in the Algerian Sahara: Palaeontology, v. 47, no. 2, p. 395–414.

Shaw, F.C., 1968, Early Middle Ordovician Chazy trilobites of New York state: New York State Museum Memoir, v. 17, p. 1–162.

Shrock, R.R., and Raasch, G.O., 1937, Paleontology of the disturbed Ordovician rocks near Kentland, Indiana: American Midland Naturalist, v. 18, no. 4, p. 532–607.

Sirenko, B., 2006, Report on the present state of our knowledge with regard to the chitons (Mollusca: Polyplacophora) of the Magellan Strait and Falkland Islands: 2nd International Chiton Symposium, Journal of the Malacological Society of Japan, v. 65, no. 1–2, p. 81–89.

Siveter, D.J., 2008, Ostracods in the Palaeozoic?: Senckenbergiana Lethaea, v. 88, p. 1–9.

Sloan, R.E., 1987, History of study of the Middle and Late Ordovician rocks of the upper Mississippi Valley, *in* Sloan, R.E., ed., Middle and Late Ordovician Lithostratigraphy of the Upper Mississippi Valley: Minnesota Geological Survey Report of Investigations 35, p. 3–6.

Sloan, R.E., Kolata, D.R., Witzke, B.J., and Ludvigson, G.A., 1987, Description of major outcrops in Minnesota and Iowa, *in* Sloan, R.E., ed., Middle and Late Ordovician Lithostratigraphy and Biostratigraphy of the Upper Mississippi Valley: Minnesota Geological Survey Report of Investigations 35, p. 197–223.

Smith, A.G., 1960, Amphineura, *in* Moore, R.C., ed., Treatise on Invertebrate Paleontology, Part I, Mollusca, Volume 1: Boulder, Colorado, Geological Society of America, and Lawrence, University of Kansas Press, p. 41–76.

Smith, M.P., Briggs, D.E.G., and Aldridge, R.J., 1987, A conodont animal from lower Silurian of Wisconsin, USA, and the apparatus architecture of panderodontid conodonts, *in* Aldridge, R.J., ed., Palaeobiology of Conodonts: Chichester, UK, Ellis Horwood Ltd., p. 91–104.

Spencer, W., and Wright, C., 1966, Asterozoans, *in* Moore, R.C., ed., Treatise on Invertebrate Paleontology, Part U, Echinodermata, Asterozoa–Echinozoa, Volumes 1 and 2: Boulder, Colorado, Geological Society of America, and Lawrence, University of Kansas Press, p. 78–107.

Sprinkle, J., ed., 1982a, Echinoderm Faunas from the Bromide Formation (Middle Ordovician) of Oklahoma: The University of Kansas Paleontological Contributions Monograph 1, 369 p.

Sprinkle, J., 1982b, Cylindrical and globular rhombiferans, *in* Sprinkle, J. ed., Echinoderm Faunas from the Bromide Formation (Middle Ordovician) of Oklahoma: The University of Kansas Paleontological Contributions Monograph 1, p. 231–273.

Stauffer, C.R., 1930, Conodonts from the Decorah shale: Journal of Paleontology, v. 4, no. 2, p. 121–128.

Stauffer, C.R., 1933, Middle Ordovician Polychaeta from Minnesota: Geological Society of America Bulletin, v. 44, p. 1173–1218.

Stauffer, C.R., 1935, The conodont fauna of the Decorah Shale (Ordovician): Journal of Paleontology, v. 9, no. 7, p. 596–620.

Stauffer, C.R., and Thiel, G.A., 1941, The Paleozoic and related rocks of southeastern Minnesota: Minnesota Geological Survey Bulletin 29, 261 p.

Steele-Petrovich, H.M., 2009, Biological affinity, phenotypic variation and palaeoecology of *Tetradium* Dana, 1846: Lethaia, v. 42, no. 4, p. 383–392.

Steele-Petrovich, H.M., 2011, Replacement name for *Tetradium* Dana, 1846: Journal of Paleontology, v. 85, no. 4, p. 802–803.

Steinová, M., 2012, Probable ancestral type of actinodont hinge in the Ordovician bivalve *Pseudocyrtodonta*: Bulletin of Geosciences (Czech Geological Survey), v. 87, no. 2, p. 333–346.

Stokes, C., 1824, On a trilobite from Lake Huron, *in* Bigsby, J.J., ed., Notes on the geography and geology of Lake Huron: Transactions of the Geological Society of London, ser. 2, v. 1, pt. 2, p. 175–209.

Stolarski, J., 2000, Origin and phylogeny of Guyniidae (Scleractinia) in the light of microstructural data: Lethaia, v. 33, p. 13–38.

Swain, F.M., and Cornell, J.R., 1987, Ostracoda of the superfamilies Drepanellacea, Hollinacea, Leperditellacea, and Healdiacia from the Decorah Shale of Minnesota, *in* Sloan, R.E., ed., Middle and Late Ordovician Lithostratigraphy and Biostratigraphy of the Upper Mississippi Valley: Minnesota Geological Survey Report of Investigations 35, p. 102–130.

Swain, F.M., Cornell, J.R., and Hansen, D.L., 1961, Ostracoda of the families Aparchitidae, Aechminidae, Leperditellidae, Drepanellidae, Eurychilinidae and Punctaparchitidae from the Decorah Shale of Minnesota: Journal of Paleontology, v. 35, no. 2, p. 345–372.

Sweet, W.C., 1964, Nautiloidea–Orthocerida, *in* Moore, R.C., ed., Treatise on Invertebrate Paleontology, Part K, Mollusca, Volume 3, Cephalopoda: Boulder, Colorado, Geological Society of America, and Lawrence, University of Kansas Press, p. 78–107.

Sweet, W.C., 1984, Graphic correlation of upper Middle and Upper Ordovician rocks, North American Midcontinent Province, U.S.A., *in* Bruton, D.L., Aspects of the Ordovician System: Totnes, UK, Scandinavian University Press, p. 23–35.

Sweet, W.C., 1987, Distribution and significance of conodonts in Middle and Upper Ordovician strata of the Upper Mississippi Valley region, *in* Sloan, R.E., ed., Middle and Late Ordovician Lithostratigraphy and Biostratigraphy of the Upper Mississippi Valley: Minnesota Geological Survey Report of Investigations 35, p. 167–172.

Swisher, R.E., Westrop, S.R., and Amati, L., 2015, The Upper Ordovician trilobite *Raymondites* Sinclair 1944 in North America: Journal of Paleontology, v. 89, no. 1, p. 110–134.

Teichert, C., 1964, Cephalopoda: General features, Endoceratoidea, Actinoceratoidea, Nautiloidea, Bactritoidea, *in* Moore, R.C., ed., Treatise on Invertebrate Paleontology, Part K, Mollusca 3: Boulder, Colorado, Geological Society of America, and Lawrence, University of Kansas Press, p. 4–505.

Templeton, J.S., and Willman, H.B., 1963, Champlainian Series (Middle Ordovician) in Illinois: Illinois State Geological Survey Bulletin 89, 260 p.

Tofel, J.E., and Bretsky, P.W., 1987, Middle Ordovician *Lophospira* (Archaeogastropoda) from the upper Mississippi Valley: Journal of Paleontology, v. 61, p. 700–723.

Tucker, R.D., and McKerrow, W.S., 1995, Early Paleozoic chronology: A review in light of new U–Pb zircon ages from Newfoundland and Britain: Canadian Journal of Earth Sciences, v. 32, no. 4, p. 368–379.

Ubaghs, G., Moore, R.C., and Teichert, C., 1978, Camerata, *in* Teichert, C., ed., Treatise on Invertebrate Paleontology Camerata, Part T, Echinodermata 2, Volumes 1–3: Boulder, Colorado, Geological Society of America, and Lawrence, University of Kansas Press, p. 408–519.

Ulrich, E.O., 1879, Descriptions of new genera and species of fossils from the Lower Silurian about Cincinnati: Journal of the Cincinnati Society of Natural History, v. 2, p. 8–30.

Ulrich, E.O., 1886, Report on the Lower Silurian Bryozoa with preliminary descriptions of some of the new species: Minnesota Geological and Natural History Survey Annual Report 14, p. 57–103.

Ulrich, E.O., 1889a, Contributions to the Micro-palaeontology of the Cambro-Silurian Rocks of Canada, Part II: Ottawa, Geological and Natural History Survey of Canada, p. 22–57.

Ulrich, E.O., 1889b, Preliminary description of new Lower Silurian sponges: American Geologist, v. 3, no. 4, p. 233–248.

Ulrich, E.O., 1890a, New Lamellilbranchiata, containing descriptions of new species of *Modiolopsis*: American Geologist, v. 5, no. 5, p. 270–284.

Ulrich, E.O., 1890b, New Lamellibranchiata, on two new genera and six new species: American Geologist, v. 6, no. 3, p. 173–181.

Ulrich, E.O., 1890c, New Lamellilbranchiata, descriptions of new species, with remarks on others: American Geologist, v. 6, no. 3, p. 382–389.

Ulrich, E.O., 1890d, Section VI. Palaeozoic Bryozoa, *in* Volume VIII. Geology and Palaeontology: Geological Survey of Illinois, p. 285–688.

Ulrich, E.O., 1892a, New Lamellibranchiata, descriptions of one new genus and eight new species: American Geologist, v. 10, no. 2, p. 96–104.

Ulrich, E.O., 1892b, New Lower Silurian Lamellibranchiata, chiefly from Minnesota rocks: Minnesota Geology and Natural History Survey Annual Report 19 [for 1890], p. 211–248.

Ulrich, E.O., 1892c, New Lower Silurian Ostracoda: American Geologist, v. 10, no. 5, p. 263–270.

Ulrich, E.O., 1892d, Two new Lower Silurian species of *Lichas* (subgenus *Hoplolichas*): American Geologist, v. 10, no. 5, p. 271–272.

Ulrich, E.O., 1893, On Lower Silurian Bryozoa of Minnesota: Minnesota Geology and Natural History Survey Final Report 3, p. 96–332.

Ulrich, E.O., 1894a, The Lower Silurian Lamellibranchiata of Minnesota, *in* The Geology of Minnesota: Minneapolis, Minnesota Geological and Natural History Survey Final Report 3, pt. 2, p. 475–628 [distributed separately in 1894].

Ulrich, E.O., 1894b, The Lower Silurian Ostracoda of Minnesota, *in* The Geology of Minnesota: Minnesota Geological and Natural History Survey Final Report 3, pt. 2, p. 629–693.

Ulrich, E.O., 1925, New classification of the "Heterocrinidae," *in* Foerste, A.F., ed., Upper Ordovician Faunas of Ontario and Quebec: Canada Geological Survey Memoir 138, p. 82–106.

Ulrich, E.O., and Bassler, R.S., 1904, A revision of the Paleozoic Bryozoa, pt. 2, on genera and species of Trepostomata: Smithsonian Miscellaneous Collections, v. 47, no. 1470, p. 15–55.

Ulrich, E.O., and Everett, O., 1890, Section V. Descriptions of Lower Silurian sponges, *in* Volume VIII. Geology and Palaeontology: Geological Survey of Illinois, p. 255–282.

Ulrich, E.O., and Scofield, W.H., 1897, The Lower Silurian Gastropoda of Minnesota: The Paleontology of Minnesota, v. 39, no. 2, p. 813–1081.

Utgaard, J., 1968a, A revision of North American genera of ceramoporoid bryozoans (Ectoprocta), pt. 1, Anolotichiidae: Journal of Paleontology, v. 42, no. 4, p. 1033–1041.

Utgaard, J., 1968b, A revision of North American genera of ceramoporoid bryozoans (Ectoprocta), pt. 2, *Crepipora*, *Ceramoporella*, *Acanthoceramoporella*, and *Ceramophylla*: Journal of Paleontology, v. 42, no. 6, p. 1444–1455.

Utgaard, J., 1969, A revision of North American genera of ceramoporoid bryozoans (Ectoprocta), pt. 3, The ceramoporoid genera *Ceramopora*, *Papillalunaria*, *Favositella*, and *Haplotrypa*: Journal of Paleontology: v. 43, no. 2, p. 289–297.

Utgaard, J., 1983, Paleobiology and taxonomy of the order Cystoporata: *in* Robison, R.A., ed., Treatise on Invertebrate Paleontology, Part G, Bryozoa (Revised): Introduction, Order Cystoporata, Order Cryptostomata: Boulder, Colorado, Geological Society of America, and Lawrence, University of Kansas Press, p. 327–357.

Vachard, D., Pille, L., and Gaillot, J., 2010, Palaeozoic foraminifera: Systematics, palaeoecology and responses to global changes: Revue de Micropaléontologie, v. 53, no. 4, p. 209–254.

Van Iten, H., de Moraes Leme, J., Simoes, M.G., Marques, A.C., and Collins, A.G., 2006, Reassessment of the phylogenetic position of conulariids (?Ediacaran-Triassic) within the subphylum Medusozoa (phylum Cnidaria): Journal of Systematic Palaeontology, v. 4, no. 2, p. 109–118.

Vecoli, M., and Le Hérissé, A., 2004, Biostratigraphy, taxonomic diversity and patterns of morphological evolution of Ordovician acritarchs (organic-walled microphytoplankton) from the northern Gondwana margin in relation to palaeoclimatic and palaeogeographic changes: Earth-Science Reviews, v. 67, no. 3–4, p. 267–311.

Vendrasco, M.J., and Runnegar, B., 2004, Late Cambrian and early Ordovician stem group chitons (Mollusca: Polyplacophora) from Utah and Missouri: Journal of Paleontology, v. 78, no. 4, p. 675–689.

Vendrasco, M.J., Wood, T.E., and Runnegar, B.N., 2004, Articulated Palaeozoic fossil with 17 plates greatly expands disparity of early chitons: Nature, v. 429, p. 288–291.

Vinn, O., 2005, A new cornulitid genus from the Silurian of Gotland, Sweden: Geologiska föreningen, v. 127, no. 3, p. 205–210.

Vinn, O., 2010, Adaptive strategies in the evolution of encrusting tentaculitoid tubeworms: Palaeogeography, Palaeoclimatology, Palaeoecology, v. 292, no. 1–2, p. 211–221.

Vinn, O., and Mutvei, H., 2005, Observations on the morphology and affinities of cornulitids from the Ordovician of Anticosti Island and the Silurian of Gotland: Journal of Paleontology, v. 79, no. 4, p. 726–737.

Vinn, O., Wilson, M.A., Zatoń, M., and Toom, U., 2014, The trace fossil *Arachnostega* in the Ordovician of Estonia (Baltica): Palaeontologia Electronica, v. 17, p. 1–9.

Vinn, O., and Zatoń, M., 2012, Phenetic phylogenetics of tentaculitoids—Extinct, problematic calcareous tube-forming organisms: Geologiska föreningen, v. 134, no. 2, p. 145–156.

Wagner, P.J., 1999, The utility of fossil data in phylogenetic analyses: A likelihood example using Ordovician–Silurian species of the Lophospiridae (Gastropoda: Murchisoniina): American Malacological Bulletin, v. 15, p. 1–31.

Wagner, P.J., 2002, Phylogenetic relationships of the earliest anisotrophically coiled gastropods: Smithsonian Contributions to Paleobiology, v. 88, p. 1–152.

Wagner, P.J., 2006, Taxonomic opinions recorded in the paleobiological database at Fossilworks: PaleoDB-Fossilworks, http://fossilworks.org/?page=paleodb.

Wagner, P.J., 2017, Taxonomic opinions recorded in the paleobiological database at Fossilworks: PaleoDB-Fossilworks, http://fossilworks.org/?page=paleodb.

Wahlman, G.P., 1992, Middle and Upper Ordovician symmetrical univalved mollusks (Monoplacophora and Bellerophontina) of the Cincinnati Arch region: U.S. Geological Survey Professional Paper 1066-O, 213 p., 45 pl.

Walcott, C.D., 1875, New species of trilobite from the Trenton Limestone at Trenton Falls, New York: Cincinnati Quarterly Journal of Science, v. 2, p. 347–349.

Walcott, C.D., 1876, Descriptions of new species of fossils from the Trenton Limestone: 28th Annual Report of the New York State Museum of Natural History, p. 93–97.

Walcott, C.D., 1877, Descriptions of new species of fossils from the Chazy and Trenton Limestones: 31st Annual Report of the New York State Museum of Natural History, p. 68–71.

Walcott, C.D., 1884, Fossils from the Groups of New York: 35th Annual Report of the New York State Museum of Natural History, p. 207–214.

Walker, M., 1953, The development of a diplograptid from the Platteville Limestone: Geological Magazine, v. 90, no. 1, p. 1–16.

Warn, J., and Strimple, H.L., 1977, The Disparid Inadunate Superfamilies Homocrinacea and Cincinnaticrinacea (Echinodermata, Crinoidea), Ordovician–Silurian, North America: New York, Paleontological Research Institution.

Webby, B.D., 2015, Part E, Revised, v. 4, Chapter 16B: Lebechiida: Treatise Online, v. 41, p. 1–51.

Webby, B.D., Elias, R.J., Young, G.A., Neuman, B.E.E., and Kaljo, D., 2004a, Corals, *in* Webby, B.D., Paris, F., Droser, M.L. and Percival, I.G., eds., The Great Ordovician Biodiversification Event: New York, Columbia University Press, 484 p.

Webby, B.D., Paris, F., Droser, M.L. and Percival, I.G., eds., 2004b, The Great Ordovician Biodiversification Event: New York, Columbia University Press, p. 484.

Webers, G.F., 1966, The Middle and Upper Ordovician conodont faunas of Minnesota: Minnesota Geological Survey Special Publication 4, 123 p.

Weiss, M.P., 1955, Some Ordovician brachiopods from Minnesota and their stratigraphic relations: Journal of Paleontology, v. 29, no. 5, p. 759–774.

West, R.R., and Clark, G.R., 1983, Chaetetids, *in* Studies in Geology, Notes for a Short Course, v. 7, p. 130–140.

Westphal, K.W., 1974, New fossils from the Middle Ordovician Platteville Formation of southwest Wisconsin: Journal of Paleontology, v. 48, p. 79–83.

Weyer, D., 1973, Über den Ursprung der Calostylidae Zittel 1879 (Anthozoa Rugosa, Ordoviz–Silur) [The origin of the Calostylidae; Anthozoa, Rugosa, Ordovician–Silurian]: Freiberger Forschungsheft, pt. 1, v. C282, p. 23–87.

Whatley, R.C., Siveter, D.J., and Boomer, I.D., 1993, Arthropoda (Crustacea: Ostracoda), *in* Benton, M.J., ed., The Fossil Record 2: London, Chapman & Hall, p. 343–356.

Whitfield, R.P., 1878, Preliminary descriptions of new species of fossils from the lower geological formations of Wisconsin: Annual Report of the Wisconsin Geological Survey for 1877, p. 50–89.

Whitfield, R.P., 1882, Palaeontology, Geology of Wisconsin: Wisconsin Geological Survey of 1873–1879, pt. 4, p. 163–363.

Whitfield, R.P., 1886, Notice of geological investigations along the eastern shore of Lake Champlain, with descriptions of the new fossils discovered: American Museum of Natural History Bulletin, v. 1, no. 8, p. 293–345.

Whitfield, R.P., 1894, On new forms of marine algae from the Trenton Limestone, with observations on *Buthograptus laxus* Hall: American Museum of Natural History Bulletin, v. 6, p. 351–358.

Whitfield, R.P., 1895, Republication of descriptions of fossils from the Hall Collection in the American Museum of Natural History, from the Report of Progress for 1861 of the Geological Survey of Wisconsin, by James Hall, with illustrations from the original type specimens not heretofore figured: American Museum of Natural History Memoirs, v. 1, pt. 2, p. 39–74.

Whittington, H.B., 1949, *Dolichoharpes* and the origin of the harpid fringe: American Journal of Science, v. 247, p. 276–285.

Whittington, H.B., 1953, North American Bathyuridae and Leiostegiidae (Trilobita): Journal of Paleontology, v. 27, p. 647–678.

Whittington, H.B., 1997, Morphology of the exoskeleton, *in* Kaesler, R.L., ed., Treatise on Invertebrate Paleontology, Part O: Arthropoda 1, Trilobita (Revised), Volume 1: Boulder, Colorado, Geological Society of America, and Lawrence, University of Kansas Press, p. 1–85.

Whittington, H.B., and Evitt, W.R., 1954, Silicified Middle Ordovician trilobites: Geological Society of America Memoir 59, p. 1–137.

Williams, A., Brunton, C.H.C., Carlson, S.J., Alvarez, F., Ansell, D., Baker, P.G., Basset, M.G., Blodgett, R.B., Boucot, A.J., Carter, J.L., et al., 2000, Part H: Brachiopoda (Revised), Volumes 2 and 3, *in* Kaesler, R.L., ed., Treatise on Invertebrate Paleontology: Boulder, Colorado, Geological Society of America, and Lawrence, University of Kansas Press, p. 1–423.

Williams, A., Carlson, S.J., Brunton, C.H.C., Holmer, L.E., and Popov, L., 1996, A supra-ordinal classification of the Brachiopoda: Philosophical Transactions of the Royal Society of London B: Biological Sciences, v. 351, no. 1344, p. 1171–1193.

Williams, A., James, M.A., Emig, C.C., MacKay, S., and Rhodes, M.C., 1997, Brachiopod anatomy, *in* Kaesler, R.L., ed., Treatise on Invertebrate Paleontology, Part H, Brachiopoda (Revised), Volume 1: Boulder, Colorado, Geological Society of America, and Lawrence, University of Kansas Press, p. 7–188.

Williams, M., and Vannier, J.M.C., 1995, Middle Ordovician Aparchitidae and Schmidtellidae: Significance of 'featureless' ostracods: Journal of Micropalaeontology, v. 14. p. 7–24.

Williams, S.H., 1994, Revision and definition of the *C. wilsoni* graptolite zones (Middle Ordovician) of southern Scotland: Transactions of the Royal Society of Edinborough, Earth Sciences, v. 85, p. 143–157.

Willman, H.B., Atherton, E., Buschbach, T.C., Collinson, C.W., Frye, J.C., Hopkins, M.E., Lineback, J.A. and Simon, J.A., 1975, Handbook of Illinois Stratigraphy: Illinois State Geological Survey Bulletin 95, 261 p.

Willman, H.B., and Kolata, D.R., 1978, The Platteville and Galena Groups in Northern Illinois: Illinois State Geological Survey Circular 502, 75 p.

Wilson, A.E., 1945, *Strophomena* and its homeomorphs *Trigrammaria* and *Microtrypa* from the Ottawa limestone of the Ottawa–St. Lawrence lowlands: Transactions of the Royal Society of Canada, v. 39, no. 4, p. 121–150.

Winchell, N.H., 1881, New Brachiopoda from the Trenton and Hudson River Formations in Minnesota: Minnesota Geological and Natural History Survey Annual Report 9, p. 115–122.

Winchell, N.H., and Schuchert, C., 1895a, Sponges, graptolites and corals from the Lower Silurian of Minnesota, *in* The Geology of Minnesota, Volume 3, Part 1, of the Final Report: Paleontology: Minneapolis, Minnesota Geological and Natural History Survey, p. 55–95.

Winchell, N.H., and Schuchert, C., 1895b, The Lower Silurian Brachiopoda of Minnesota, *in* The Geology of Minnesota, Volume 3, Part 1, of the Final Report: Paleontology: Minneapolis, Minnesota Geological and Natural History Survey, p. 333–437.

Witzke, B.J., Ludvigson, G.A., and Young, J.N., 2005, Stop 4—The Locust Road roadcut section along the south wall of Canoe Creek Valley, *in* Ludvigsen, G.A., and Witzke, B.J., eds., Facets of the Ordovician Geology of the Upper Mississippi Valley Region: Guidebook for the 35th Annual Field Conference of the Great Lakes Section, Society for Sedimentary Geology (SEPM), September 23–25, 2005: Iowa Geological Survey Guidebook Series No. 24, p. 110–113.

Worthen, A.H., and Meek, F.B., 1875, Palaeontology of Illinois, Section II. Descriptions of invertebrates, *in* Volume VI. Geology and Paleontology: Geological Survey of Illinois, p. 489–532.

Yochelson, E.L., 1994, *Macroscenella* (Mollusca) from the Middle Ordovician of Wisconsin—A reinterpretation and reassignment: Journal of Paleontology, v. 68, p. 1252–1256.

APPENDIX 1

LOCALITY REGISTER

The following is a list of localities referenced in this overview. Stratigraphic descriptions for many of these localities can be found in Templeton and Willman (1963) and Willman and Kolata (1978). Additional outcrop localities of the Platteville Formation can be found in Kolata et al. (1986) and Kolata et al. (1996).

It is always necessary to obtain permission on private property, even when the property is not posted as private. This applies to all the localities listed below. Many quarries are now off limits because of laws governing safe maintenance of mining sites. Collecting is also illegal in outcrops along interstate highways.

1. **Dixon Crusher East Quarry:** 4 km north of Dixon, Lee County, Illinois; 41°51′54.31″N, 89°26′29.76″W. Pecatonica, Mifflin, and Grand Detour Members, Platteville Formation.

2. **Dixon Crusher Northwest Quarry:** 4 km north of Dixon, Lee County, Illinois; 41°52′50.30″N, 89°27′10.34″W. Pecatonica, Mifflin, and Grand Detour Members, Platteville Formation.

3. **Dixon Northeast Quarry:** 5 km north of Dixon, Lee County, Illinois; 41°53′40.41″N, 89°26′56.07″W. Pecatonica, Mifflin, and Grand Detour Members, Platteville Formation.

4. **Dixon Northwest Quarry:** 5 km north of Dixon, Lee County, Illinois; 41°53′32.34″N, 89°27′58.97″W. Pecatonica, Mifflin, and Grand Detour Members, Platteville Formation.

5. **Gold Mine Road Quarry:** 4.5 km southeast of Shannon, Carroll County, Illinois; 42°6′54.81″N, 89°42′27.49″W. Pecatonica, Mifflin, and Grand Detour Members, Platteville Formation.

6. **Forreston Southwest Quarry:** Northeast corner of intersection of Freeport and West Grove Roads, 4 km southwest of Forreston, Ogle County, Illinois; 42°6′1.71″N, 89°37′30.10″W. Mifflin and Grand Detour Members, Platteville Formation.

7. **Lightsville Quarry:** Quarry on north side of road approximately 1.5 km west-northwest of Lightsville, Ogle County, Illinois; 42°9′6.77″N, 89°25′22.12″W. Pecatonica, Mifflin, and Grand Detour Members, Platteville Formation.

8. **Byron Quarry:** Quarry on North Barker Road on north side of Byron, Ogle County, Illinois; 42°8′15.54″N, 89°15′42.12″W. Pecatonica, Mifflin, and Grand Detour Members, Platteville Formation.

9. **Kishwaukee Road Quarry:** Small quarry on northwest side of Kishwaukee Road 3.5 km east-northeast of intersection of State Routes 72 and 2 in Byron, Ogle County, Illinois; 42°7′59.28″N, 89°12′55.93″W. Mifflin and lower Grand Detour Members, Platteville Formation.

10. **Davis Northeast Quarry:** Quarry at southeast intersection of Pieper Road and Goeke Road, 1.6 km northeast of Davis, Stephenson County (westernmost Winnebago County), Illinois; 42°25′49.33″N, 89°23′52.86″W. Grand Detour, Nachusa, and Quimbys Mill Members, Platteville Formation.

11. **Durand Southeast Quarry—Trask Bridge Road:** Quarry on north side of State Route 70, 5.8 km southeast of intersection of State Routes 70 and 75 near Durand, Winnebago County, Illinois; 42°22′35.72″N, 89°17′11.05″W. Uppermost Mifflin, Grand Detour, Nachusa, and Quimbys Mill Members, Platteville Formation; Guttenberg Member, Decorah Formation; and Dunleith Member, Galena Formation.

12. **Durand Northeast Quarry:** Quarry on Crowley Road 0.4 km northeast of Durand, Winnebago County, Illinois; 42°26′32.85″N, 89°19′8.33″W. Grand Detour Member, Platteville Formation.

13. **Durand North Quarry:** Quarry on Center Road 0.8 km north of Durand, Winnebago County, Illinois; 42°26′38.24″N, 89°19′41.26″W. Pecatonica and Mifflin Members, Platteville Formation.

14. **Best Road Quarry:** Quarry 2.4 km northeast of Davis, Winnebago County, Illinois; 42°25′40.69″N, 89°22′46.14″W. Mifflin Member, Platteville Formation.

15. **Rock City East Quarry:** Quarry on south side of State Route 75, 1.4 km east southeast of Rock City, Stephenson County, Illinois; 42°24′37.14″N, 89°27′5.56″W. Quimbys Mill Member, Platteville Formation through basal Dunleith Member, Galena Formation.

16. **Shirland East–Bymaster Quarry:** Old Bymaster Farm Quarry south of Shirland Road, 1.5 km east of Shirland, Winnebago County, Illinois; 42°26′38.75″N; 89°10′43.98″W. Mifflin Member, Platteville Formation.

17. **Shirland West Quarry:** Quarry on north side of West Winslow Road 0.9 km west of Shirland, Winnebago County, Illinois; 42°26′35.89″N; 89°12′38.11″W. Mifflin Member, Platteville Formation.

18. **Shirland North Quarry:** Quarry on east side of Boswell Road 2.7 km north of Shirland, Winnebago County, Illinois; 42°28′9.16″N; 89°11′35.66″W. Nachusa, Quimbys Mill Members, Platteville Formation; Guttenberg Member, Decorah Formation; and Dunleith Member, Galena Formation.

19. **Winslow Road Quarry:** Quarry on west side of North Winslow Road 6.7 km southwest of Shirland, Winnebago County, Illinois; 42°25′26.74″N; 89°14′36.54″W. Pecatonica and Mifflin Members, Platteville Formation.

20. **South Beloit Quarry:** Quarry on southwest side of South Beloit, Winnebago County, Illinois; 42°29′43.92″N; 89°4′35.57″W. Mifflin and Grand Detour Members, Platteville Formation.

21. **Beloit West Quarry:** Carl Bauer quarry on north side of State Route 81 (West Liberty Avenue) 6.7 km west of intersection with McKinley Avenue in Beloit, Wisconsin; 42°30′58.29″N; 89°8′14.98″W. Mifflin, Grand Detour, and Nachusa Members of Platteville Formation.

22. **Beloit–Newark Road/Route 81 Quarry:** Beloit–Newark Road/State Route 81 Quarry Section; quarry 0.6 km southeast of intersection of Beloit–Newark Road and State Route 81, Rock County, Wisconsin; 42°32′23.04″N; 89°15′2.23″W. Mifflin, Grand Detour, and Nachusa Members of Platteville Formation.

23. **Yager Quarry–Hilldale Road:** Quarry 4 km south of Mt. Ida, Grant County, Wisconsin; 41°56′07.97″N, 90°45′21.47″W. Mifflin and Grand Detour Members, Platteville Formation.

24. **Fennimore Southwest Mt. Ridge Road Quarry #1:** Small quarry in middle of field 2.6 km southwest of Fennimore, Grant County, Wisconsin; 42°58′13.32″N; 90°40′50.49″W. Mifflin Member, Platteville Formation.

25. **Fennimore Southwest Mt. Ridge Road Quarry #2:** Small roadside quarry on west side of Mt. Ridge Road 3.1 km southwest of Fennimore, Grant County, Wisconsin; 42°58′2.60″N; 90°41′7.90″W. Mifflin Member, Platteville Formation.

26. **Fennimore Southwest Mt. Ridge Road Quarry #3:** Small roadside quarry on east side of Mt. Ridge Road 3.9 km southwest of Fennimore, Grant County, Wisconsin; 42°57′34.62″N; 90°41′26.25″W. Mifflin Member, Platteville Formation.

27. **Lancaster North–Roadcut Route 61:** Roadcut on U.S. Highway 61, 7.6 km north of Lancaster, Grant County, Wisconsin; 42°54′31.86″N; 90°40′1.61″W. Mifflin Member, Platteville Formation.

28. **Bee Town Northeast Quarry–Route 35:** Quarry on northeast side of State Route 35, 4.2 km northeast of Bee Town, Wisconsin; 42°49′30.46″N; 90°51′14.94″W. Mifflin and Grand Detour Members (McGregor Member), Platteville Formation and Guttenberg Member, Decorah Formation.

29. **Lancaster West Roadcut:** Roadcut on County Road A, 6 km west of intersection of A with U.S. Highway 61 in Lancaster, Grant County, Wisconsin; 42°51′57.22″N, 90°46′39.71″W. Mifflin and Grand Detour Members (McGregor Member), Platteville Formation.

30. **Platteville Southwest Roadcut:** Roadcut on U.S. Highway 151, 9.5 km southwest of intersection of State Routes 80 and 151 near Platteville, Grant County, Wisconsin; 42°39′44.24″N, 90°33′6.92″W. St. Peter Sandstone, Platteville Formation and Decorah Formation. Suggested alternate type section of the Platteville Formation (Templeton and Willman 1963). Section described by Ostrom (1987).

31. **Butler Quarry–Lee Center:** Quarry on west side of Inlet Road 3 km east-southeast of Lee Center, Lee County, Illinois; 41°44′15.97″N, 89°14′43.62″W. Pecatonica, Mifflin, and Grand Detour Members, Platteville Formation.

32. **Avon East Quarry:** Quarry on north side of West Beloit–Newark Road 3 km east of Avon, Rock County, Wisconsin; 42°32′36.46″N, 89°17′38.68″W. Grand Detour Member, Platteville Formation.

33. **Chana Northeast Quarry:** Quarry on East Cottonwood Road 2.4 km northeast of Chana, Ogle County, Illinois; 41°59′21.06″N, 89°11′39.78″W. Mifflin Member, Platteville Formation.

34. **Magnolia Quarry:** Quarry on north side of Rock County Highway A, 1.6 km west of State Route 213 and the town of Magnolia, Rock County, Wisconsin; 42°43′6.44″N, 89°18′26.99″W. Mifflin and Grand Detour Members, Platteville Formation.

35. **Rochelle Macklin Quarry:** Quarry at southwest corner of intersection of State Route 38 and Interstate I-39, 3.3 km east-northeast of Rochelle, Ogle County, Illinois; 41°55′47.59″N, 89°1′47.73″W. Pecatonica, Mifflin, Grand Detour, Nachusa, and Quimbys Mill Members, Platteville Formation.

36. **Rochelle Skare Road Quarry:** Quarry on east side of Skare Road 8.34 km northwest of intersection of State Routes 38 and 251 in Rochelle, Ogle County, Illinois; 41°58′28.47″N, 89°9′5.46″W. Grand Detour Member, Platteville Formation.

37. **Stitzer Southeast–Ridge Road Quarry:** Quarry on north side of County Road E, 2.3 km southeast of intersection of County Roads E and F in Stitzer, Grant County, Wisconsin; 42°55′4.50″N, 90°35′51.70″W. Mifflin Member, Platteville Formation.

38. **Stitzer Southeast–Rock School Road Quarry:** Quarry on north side of County Road E, 4 km southeast of intersection of County Roads E and F in Stitzer, Grant County, Wisconsin; 42°54′36.10″N, 90°34′46.46″W. Mifflin Member, Platteville Formation.

39. **Fennimore North–Tower Road Quarry:** Quarry on east side of Tower Road, 9.12 km north of intersection of State Routes 18 and 61 in Fennimore, Grant County, Wisconsin; 43°3′51.15″N, 90°41′52.88″W. Pecatonica, Mifflin, and Grand Detour Members, Platteville Formation.

40. **South Wayne Quarry:** Quarry on east side of Township Route D, 0.9 km south of the intersection with State Route 11 in South Wayne, Lafayette County, Wisconsin; 42°33′38.64″N, 89°52′51.37″W. Pecatonica, Mifflin, and Grand Detour Members, Platteville Formation.

41. **Davis South Quarry:** Buss Quarry on east side of county road 4.2 km south-southwest of Davis, Stephenson County, Illinois; 42°23′5.31″N, 89°25′33.66″W. Mifflin, Grand Detour, and Nachusa Members, Platteville Formation.

42. **Hogan School Quarry:** Abandoned quarry at intersection of Conger and Lightsville Roads 5.7 km northwest of Byron, Ogle County, Illinois; 42°8′56.99″N, 89°19′1.35″W. Grand Detour Member, Platteville Formation.

43. **Mifflin Roadcut:** Old roadcut on north side of road just east of Mifflin, Iowa County, Wisconsin; 42°52′14.16″N, 90°20′52.93″W. Upper part of Mifflin Member, Platteville Formation.

44. **Shirland Northwest Quarry:** Abandoned quarry on north side of Yale Bridge Road 4.7 km northwest of Shirland, Winnebago County, Illinois; 42°28′41.57″N, 89°13′46.72″W. Mifflin Member, Platteville Formation.

45. **State Line Road Quarry:** Abandoned quarry on north side of State Line Road 5.3 km south-southeast of Avon, Green County, Wisconsin; 42°29′41.56″N, 89°18′12.30″W. Lower part of Mifflin Member and all the Pecatonica Member as well as upper part of Glenwood Formation.

46. **Amboy West Quarry:** Renner Quarry on east side of Robbins Road 2.4 km west of Amboy, Lee County, Illinois; 41°43′0.26″N, 89°21′35.07″W. Lower Galena Formation as well as Quimbys Mill and Nachusa Members, Platteville Formation.

47. **Annaton North Quarry:** Quarry on north side of Ebenezer Road 5 km northeast of Annaton, Grant County, Wisconsin; 42°57′11.74″N, 90°30′55.56″W. Lower Galena Formation and Quimbys Mill, Grand Detour, and Mifflin Members, Platteville Formation.

48. **Dixon South Renner Quarry:** Quarry on west side of State Highway 26, 4 km south of intersection of State Route 26 and Interstate 88 in Dixon, Lee County, Illinois; 41°47′0.71″N, 89°27′39.91″W. Lower Galena Formation and Quimbys Mill, Nachusa, Grand Detour, and Mifflin Members, Platteville Formation.

49. **Dixon Southwest Quarry:** Quarry on west side of Rock Island Road on southwest side of Dixon, Lee County, Illinois; 41°49′30.06″N, 89°31′4.15″W. Mifflin, Grand Detour, Nachusa, and Quimbys Mill Members, Platteville Formation.

APPENDIX 2

FAUNAL LIST WITH RELATIVE ABUNDANCE ESTIMATES

Listed here are the species known to be present in the Pecatonica, Mifflin, Grand Detour, Nachusa, and Quimbys Mill Members of the Platteville Formation. The relative abundance of each species is indicated as **rare** (X) when <10 specimens are known to occur in the formation, **common** (XX) from ≥10 to <100, and **abundant** (XXX) when ≥100. Abundance is based on the author's 60+ years of collecting and observations of private and institutional collections. Because fossil accumulations invariably represent a biased subset of the original community, this abundance scale should be viewed only as an expression of the range from least likely to most likely to be found in any given outcrop. It is a scale that will need to be revised on a continuing basis as new specimens are discovered.

Abundance information for the foraminifera, scolecodonts, and chitinozoans is not available, and species are indicated only as present (P) within a given stratigraphic member. The sponge fauna described by Ulrich and Everett (1890) as occurring 7.6 m above the base of the Platteville Formation at Dixon, Illinois, would place the horizon in the Pecatonica Member at 3.1 m below the top. A bed of diverse anthaspidellid sponges has been documented in the lower part of the Grand Detour Member at Dixon, Illinois (Willman and Kolata 1978); however, no such sponge bed has been observed in the Pecatonica Member in the Dixon area. Consequently, sponge occurrences in the Pecatonica Member are indicated with a question mark (?).

Species	Quimbys Mill Member	Nachusa Member	Grand Detour Member	Mifflin Member	Pecatonica Member
Poriferans					
Anthaspidella firma			X	X	?
Anthaspidella obliqua			X	X	?
Anthaspidella mammulata			X	X	?
Anthaspidella florifera			X	X	?
Anthaspidella parvistellata			X	X	
Anthaspidella scutula					?
Anthaspidella grandis					?
Anthaspidella fenestrata			X	X	?
Anthaspidella magnifica					?
Calathium? (Zittelella) *infelix*					?
Zittelella typicalis (*Cnemidium*? *trentonensis*)			X	X	?
Zittelella lobata			X	X	?
Zittelella inosculata			X	X	?
Strotospongia maculosa					?
Patellispongia cf. *oculata*			X		
Streptosolen obconicus			X	X	?
Edriospongia basalis					?
Hindia inaequalis			XXX	XXX	X
Dystactospongia minor			X	X	X
Dystactospongia rudis					?
Astraeospongia dixonensis				X	
Astraeospongia cf. *dixonensis*	X				
Thamnobeatricea sp.					X
Stromatocerium rugosum			X		
Pachystylostroma cf. *ungerni*			X		
Solenopora cf. *spongioides*			X	XXX	
Cnidarians					
Lambeophyllum profundum			XXX	XX	
Cyathophylloides cf. *stellata*			X		
Foerstephyllum magnificum			X		
Lichenaria cf. *typa*				XX	
Saffordophyllum undulatum			X		
Rhabdotetradium sp.				X	
Paratetradium sp.		X	X	X	
Conularia cf. *trentonensis*			X	X	

Table continues

Species (Continued)	Quimbys Mill Member	Nachusa Member	Grand Detour Member	Mifflin Member	Pecatonica Member
Bryozoans					
Monticulipora sp.				X	
Atactoporella sp.				X	
Orbignyella sublamellosa			XXX	XXX	XX
Peronopora sp. 1				XX	
Peronopora sp. 2			X		
Homotrypa exilis	XX		XX	XX	
Heterotrypa sp.			X	X	
Leptotrypa hexagonalis			XX	XXX	
Hemiphragma spp.			XX	XX	XX
Diplotrypa catenulata			XXX	XXX	
Eridotrypa subtilis			XXX	XXX	
Amplexopora winchelli spinulosum				XXX	
Monotrypella aequalis grandis			X	X	XXX
Tarphophragma incontroversa			XX	XX	
Tarphophragma cf. *multitabulata*			XXX	XX	
Parvohallopora cf. *spissata*			XX	XX	X
Nicholsonella cf. *ponderosa*			XXX		XXX
Ceramopora sp.			XXX	XXX	
Ceramoporella sp.			XXX	XXX	
Ceramoporid spp.				XXX	
Bythotrypa cf. *laxata*			X		
Escharopora angularis			XXX	XXX	
Pachydictya sp.			XXX	XXX	
Athrophragma cf. *foliata*				XX	
Stictopora spp.			XXX	XXX	
Stictoporella spp.			XX	XX	
Phylloporina sp.			XX	XX	XX
Arthroclema cf. *armatum*			XX	XX	
Corynotrypa cf. *delicatula*			XX	XX	
Brachiopods					
Pseudolingula sp.			X	X	
Plectoglossa cf. *oklahomensis*				X	
Acanthocrania granulosa			XX	XX	
Petrocrania dixonensis			XX	XX	

Table continues

Species (Continued)	Quimbys Mill Member	Nachusa Member	Grand Detour Member	Mifflin Member	Pecatonica Member
Philhedra cf. *minor*				X	
Strophomena plattinensis	X	X	XXX	XXX	XXX
Trigrammaria winchelli			XXX	XXX	
Oepikina minnesotensis	X	X	XXX	XXX	XXX
Bellimurina subquadrata				XX	XX
Sowerbyella sp.				X	
Triplesia cf. *carinata*			X		
Triplesia cf. *subcarinata*			X		
Onychoplecia sp.			X		
Skenidioides anthonensis				XXX	
Hesperorthis concava			XXX	XXX	XXX
Campylorthis deflecta			XXX	XXX	XXX
Eridorthis costellata			X	X	
Rostricellula minnesotensis			XXX	XXX	XXX
Camerella gregeri			X		
Protozyga nicolleti			XXX	XXX	
Doleroides pervetus			XXX	XXX	
Polyplacophorans					
Echinochiton dufoei			XX	XX	
Litochiton cf. *marginatus*				XX	
Multiplacophoran sp. 1				X	
Multiplacophoran sp. 2				X	
Gastropods					
Scalites peracuta			XX	XX	
Raphistomina sp.			XX	XX	
Asgardaspira undulata				XX	
Helicotoma umbilicata			XX	XX	
Ophiletina angularis			XX	XX	
Ophiletina sublaxa			XX	XX	
Lesueurilla beloitensis			XX	XX	
Straparollina sp.				XXX	
Maclurina bigsbyi			XX	XX	
Maclurites nitidus				XX	
Trochonema umbilicata			XXX	XXX	

Table continues

Species (Continued)	Quimbys Mill Member	Nachusa Member	Grand Detour Member	Mifflin Member	Pecatonica Member
Trochonema beloitense			XXX	XXX	
Globonema niota			X		
Lophospira milleri			XXX	XXX	
Lophospira perangulata			XX	XX	
Lophospira serrulata			XXX	XXX	
Eunema helicteres			XX	XX	
Eunema centralis				XX	
Gyronema semicarinatum				XX	
Catalanispira plattevillensis				XX	
Holopea ampla			XX	XX	
Holopea cf. *insignis*			XX	XX	
Paraliospira supracingulata			XX	XX	
Eotomaria vicina			XX	XX	
Clathrospira subconica			XX	XX	
Cataschisma convexum				XXX	
Liospira obtusa			XXX	XXX	XX
Hormotoma gracilis			XXX	XXX	
Ectomaria prisca			XX	XX	
Pseudophorid sp.				XX	
Omospira cf. *laticincta*				X	
Cyclonema textilis				XX	
Subulites dixonensis			XX	XX	
Subulites beloitensis			XX	XX	
Cyrtospira tortilis				XX	
Fusispira sp.			X	X	
Bucania cf. *emmonsi*			XX	XX	
Salpingostoma buelli			XX	XX	
Tetranota sexcarinata			XX	XX	
Tetranota wisconsinensis			XX	XX	
Pterotheca attenuata			XX	XX	
Sinuites rectangularis				XXX	
Phragmolites fimbriatus			XXX	XXX	
Phragmolites triangularis				X	
Monoplacophorans					
Micropileus cf. *variabilis*			XX	XX	

Table continues

Species (Continued)	Quimbys Mill Member	Nachusa Member	Grand Detour Member	Mifflin Member	Pecatonica Member
Macroscenella superba			X	X	
Scenella compressa				XX	
Vallatotheca cf. *unguiformis*				X	
Archinacellids					
Archinacella cf. *rugatina*			X	X	
Archinacella sp.			X	X	
Bivalves					
Homilodonta cf. *scofieldi*				XXX	
Ctenodonta nasuta			XX	XX	XX
Tancrediopsis gibberula			XX		
Similodonta cf. *similis*				XX	
Cyrtodonta janesvillensis			XX	XX	
Cyrtodonta billingsi			XX	XX	
Cyrtodonta cf. *grandis*				XX	
Cyrtodontula subcarinata			XX	XX	XX
Vanuxemia dixonensis			XXX	XXX	XXX
Vanuxemia obtusifrons			XX	XX	
Vanuxemia rectirostris			XX	XX	
Vanuxemia rotundata			XX	XX	
Vanuxemia terminalis			XX	XX	
Cleionychia lamellosa				XX	
Pterinea sp.				X	
Colpomya sp.				XXX	
Pseudocyrtodonta sp.				XXX	
Aristerella cf. *nitidula*				XX	
Goniophora sp.				XX	
Modiolopsis spp.				XX	
Cymatonota orthonotum			XX	XX	
Rostroconchs					
Eopteria cf. *conocardiformis*				XX	
Hippocardia cf. *cooperi*				XX	
Cephalopods					
Cartersoceras shideleri			XXX	XXX	XX

Table continues

Species (Continued)	Quimbys Mill Member	Nachusa Member	Grand Detour Member	Mifflin Member	Pecatonica Member
Cyrtocerina crenulata			X	X	
Anaspyroceras cylindricum			XX	XX	XX
Anaspyroceras cf. *lesueuri*			XX	XX	XX
Whitfieldoceras mumiaforme			XXX	XXX	
Gorbyoceras sp.			XX	XX	XX
Monomuchites annularis			XX	XX	XX
Monomuchites? sp.			XX	XX	
Proteoceras cf. *tyronensis*			XX	XX	XX
Proteoceras sp. 1			X	X	
Proteoceras sp. 2			XX	XX	
Redpathoceras cf. *clarkei*			XX	XX	
Cameroceras cf. *alternatum*			XX	XX	XX
Endoceras annulatum			XX	XX	XX
Endoceras sp.			XX	XX	XX
Nanno cf. *aulema*				XX	
Endocerid apical siphuncle sp. 1			XX	XX	
Actinoceras beloitense			XX	XX	XX
Gonioceras occidentale			XXX	XXX	
Centrocyrtoceras cf. *annulatum*				XX	
Centrocyrtoceras sp.				XX	XX
Plectoceras occidentale			XX	XX	XX
Plectoceras robertsoni			XX		XX
Oncoceras spp.			XX		
Beloitoceras janesvillense			XX		
Beloitoceras lycum			XX		
Beloitoceras pandion			XX		
Beloitoceras plebeium			XX		
Cyrtorizoceras cf. *minneapolis*				XX	
Loganoceras cf. *regulare*			XX	XX	
Richardsonoceras simplex			XXX		
Richardsonoceras beloitense			XX		
Zittelloceras beloitense			XX		
Zittelloceras clarkeanum			XX	XX	
Tripteroceras planoconvexum			XX	XX	
Allumettoceras planodorsatum			XX	XX	

Table continues

Species (Continued)	Quimbys Mill Member	Nachusa Member	Grand Detour Member	Mifflin Member	Pecatonica Member
Manitoulinoceras cf. *neleum*			XX		
Diestoceras? sp.			XX		XX
Ulrichoceras beloitense			XX	XX	
Trilobites					
Thaleops ovata			XXX	XXX	XX
Thaleops cf. *longispina*			X		
Thaleops (*Illaenus*) sp. 1			X		
Thaleops (*Illaenus*) sp. 2			XX	XX	
Thaleops (*Illaenus*) sp. 3			XX	XX	
Bumastoides milleri			XXX	XXX	
Bumastoides cf. *graffhami*			XX	XX	
Failleana indeterminatus			XX	XX	
Amphilichas cf. *subpunctatus*			XX	XX	
Probolichas cf. *robbinsi*			X	X	
Hemiarges cf. *paulianus*			XX	XX	
Hemiarges sp.					X
Ceraurinella scofieldi			XX	XX	
Ceraurinella templetoni			XX	XX	
Ceraurus mifflinensis			XXX	XXX	XX
Ceraurus jenlorum			XXX	XXX	XX
Eubufoceraurus scheerorum				XX	
Acanthoparypha cf. *evitti*				X	
Cybeloides cimelia			XX	XX	
Anapliomera shirlandensis			XX		
Walencrinuroides rarus			XXX	XXX	
Sceptaspis lincolnensis			XX	XX	
Calyptaulax plattevillensis			XX	XX	
Bathyurus (*Raymondites*) *longispinus*			XX	XX	
Basiliella barrandi	XX		XX	XX	
Isotelus aff. *homalonotoides*			XX	XX	
Vogdesia simplex			XX	XX	
Isotelus platycephalus			XX	XX	
Nahannia cf. *mediocrisulcata*				X	
Hypodicranotus cf. *striatulus*			X	X	
Dolichoharpes cf. *reticulata*			XX	XX	

Table continues

Species (Continued)	Quimbys Mill Member	Nachusa Member	Grand Detour Member	Mifflin Member	Pecatonica Member
Ostracodes					
Eoleperditia fabulites	XX		XXX	XXX	XX
Eurychilina subradiata				XXX	
Bullatella granilabiata			XX	XX	
Leperditella cf. *tumida*			XX	XX	
Leperditella canalis				XXX	
Schmidtella crassimarginata			XXX	XXX	
Krausella inaequalis			XX	XXX	
Echinoderms					
Cleiocrinus cf. *bromidensis*			X		
Reteocrinus spinosus			X		
Reteocrinus mahlburgi			X		
Rhaphanocrinus buckleyi			X		
Abludoglyptocrinus charltoni			XX	XX	
Abludoglyptocrinus pustulosis			X		
Cremacrinus arctus			XX	XX	
Isotomocrinus minutus			X		
Hybocrinus cf. *bilateralis*			X		
Carabocrinus cf. *micropunctatus*			X		
Porocrinus pentagonius			XX	XX	
Porocrinus cf. *smithi*			XX	XX	
Quinquecaudex springeri			X	X	
Merocrinus britonensis			X		
Cupulocrinus gracilis			XXX	XXX	
Cupulocrinus plattevillensis			XX	XX	
Oklahomacystis cf. *bibrachiatus*			X		
Coronocystis durandensis			XX		
Praepleurocystis cf. *ranaformis*			X		
Schuchertia sp.				X	
Kolataster perplexus				X	
Palasterinid sp.				X	
Genus and species undetermined			X		
Hudsonaster cf. *rugosa*			X	X	
Cyclocystoides cf. *halli*			XX	XX	

Table continues

Species (Continued)	Quimbys Mill Member	Nachusa Member	Grand Detour Member	Mifflin Member	Pecatonica Member
Bothriocidaris solemi			XX	XX	
Neobothriocidaris sp.			XX	XX	
Unibothriocidaris cf. *bromidensis*				X	
Thuroholia croneisi				XX	
Thuroholia cribriformis				XX	
Graptolites					
Amplexograptus maxwelli			XX	XX	
Hyolithids					
Solenotheca (=*Hyolithes*) *baconi*			XX	XX	
Cornulitids					
Cornulites sp.			XX	XX	
Macroalgae					
Buthograptus laxus	XX				XX
Buthograptus gundersoni			XX		XX
Buthograptus meyeri					XX
Callithamnopsis fruticosa					XX
Chaetocladus plumula	?				
Cyclocrinites sp.			XX	XX	XX
Conodonts					
Belodina compressa			XX	XX	XX
Drepanodus suberectus			XX	XX	XX
Ozarkodina obliqua			XX	XX	XX
Phragmodus cognitus			XX	XX	XX
Phragmodus undatus			XX	XX	XX
Polyplacognathus ramosa			XX	XX	XX
Pravognathus idonea			XX	XX	XX
Trichonodella recurva			XX	XX	XX
Zygognathus illustris			XX	XX	XX
Foraminifera					
Psammosphaera cava				P	

Table continues

Species (Continued)	Quimbys Mill Member	Nachusa Member	Grand Detour Member	Mifflin Member	Pecatonica Member
Sorosphaerella cooperensis				P	
Hemisphaerammina thola				P	
Tolypammina? sp.				P	
Reophax blackriverensis				P	
Scolecodonts					
Arabellites acutidentatus			P	P	P
Arabellites sp. a			P	P	P
Oenonites spiculatus			P	P	P
Oenonites sp. a			P	P	P
Oenonites sp. b			P	P	P
Protarabellites grandis			P	P	P
Protarabellites intermedius			P	P	P
Lumbriconereites affinis			P	P	P
Lumbriconereites cameratus			P	P	P
Lumbriconereites plattevillensis			P	P	P
Lumbriconereites superbus			P	P	P
Dinoscolites mirabilis			P	P	P
Ungulites aculeatus			P	P	P
Ungulites tridentatus			P	P	P
Staurocephalites dentatus			P	P	P
Chitinozoans					
Conochitina elegans			P	P	
Conochitina (*Rhabdochitina*?) *minnesotensis*			P	P	
Cyathochitina campanulaeformis			P	P	
Cyathochitina kuckersiana			P	P	
Desmochitina minor			P	P	
Trace Fossils					
Cruziana isp.				XX	
Rusophycus isp.				XX	
Rauffella isp. 1			XXX	XXX	XXX
Rauffella isp. 2			XX	XX	
Taenidium issp.			XXX	XXX	
Arachnostega issp.			XXX	XXX	XX

Species (Continued)	Quimbys Mill Member	Nachusa Member	Grand Detour Member	Mifflin Member	Pecatonica Member
Phycodes isp.			XXX	XXX	XXX
Zoophycos isp.			XXX	XXX	
Balanoglossites isp.	XXX				XXX
Gyrolithes isp.				X	
Chondrites issp.	XXX	XXX	XXX	XXX	XXX
Thalassinoides isp.	XXX				XXX
Rosselia isp.				XXX	
Trypanites isp.	XXX				XXX
Petroxestes isp.			XXX	XXX	
Ropalonaria isp.			XX	XX	
Sulcolithus isp.	XXX				XXX
Fecal pellets			XXX	XXX	XXX

INDEX

Q

R